SCIENCE AND POLITY IN FRANCE AT THE END OF THE OLD REGIME

SCIENCE AND POLITY IN FRANCE AT THE END OF THE OLD REGIME

◇◇◇◇◇◇◇◇◇◇◇◇◇◇◇◇◇◇◇◇◇◇◇◇◇◇◇◇◇◇◇◇◇◇◇◇◇

Charles Coulston Gillispie

◇◇◇◇◇◇◇◇◇◇◇◇◇◇◇◇◇◇◇◇◇◇◇◇◇◇◇◇◇◇◇◇◇◇◇◇◇

PRINCETON UNIVERSITY PRESS
PRINCETON, NEW JERSEY

Published by Princeton University Press, Princeton, New Jersey
In the United Kingdom:
Princeton University Press, Guildford, Surrey

Library of Congress Cataloging in Publication Data will be found on the last printed page of this book

This book has been composed in V-I-P Garamond

Clothbound editions of Princeton University Press books are printed on acid-free paper, and binding materials are chosen for strength and durability

Printed in the United States of America by
Princeton University Press,
Princeton, New Jersey

À mon amie
ERCG

CONTENTS

◇◇

PART THREE. Applications

PREFACE

◇◇◇

The word *polity* in the title of this work has excited some apprehension among colleagues, friends, and even family. I mean it in either or both of the first two senses given in the *Oxford English Dictionary*: (1) "Civil organization (as a condition); civil order," and (2) "Administration of a state, civil government (as a process or course of action)." Perhaps one exemplification is pertinent, from J. Brown, *Poetry and Mus.* (1763) iv, 40: "In the Course of Time, and the Progress of Polity and Arts, a Separation of the several Parts . . . would naturally arise." My book is written in awareness that much of science has in general little or nothing to do with government, and that much of government has little or nothing to do with science, but that there are intersections. This is a history of the intersections, when they began to assume a form characteristic of the modern state and of modern science. It is not meant to be either an internal or an external history of science. It is a civil history of work-a-day French science late in the enlightenment, and is meant to be complete. If I have omitted important instances pertaining to the theme, the reason is inadvertence rather than selection.

In part I have thought to explain the vitality of all French science at the end of the old regime through exhibiting the extent of its involvement in affairs. Defining the relationship between science and polity, between men of knowledge and men in power, had better be left to the conclusion, after the evidence has been set forth. I shall there argue that the pattern has become a general one, that its origin lies farther back in history than the period of this book, and that it inheres in the nature of science and of politics. What was particular to France two centuries ago is that the interactions became regular and frequent enough to be called systematic rather than episodic.

Variant approaches to the public history of science would no doubt be appropriate for other times and places. The themes would overlap to some extent with this one and with each other. If the milieu were Britain in the industrial revolution, the relations of science with private enterprise might be the most rewarding aspect. If the locale were Russia or Japan in the last hundred years, then modernization or westernization might be the process to be studied. If it were Germany in the nineteenth century, culture could well be the context. If attention were centered on the United States in the twentieth century, social mobility, economic growth, and power politics would probably be the topics. Such has been the importance of the state in modern history, however, that there is perhaps some justification for taking special interest in France during the time of French scientific preeminence, which carried over from the last decades of the old regime into the revolutionary changes that transformed polity in France and throughout all Europe.

The book has been long in the writing and would not have been written at all without the hospitality of the Center for Advanced Study in Behavioral Sciences, Stanford, California; and generous support from the National Science Foundation, the John Simon Guggenheim Foundation, and the Dayton-Stockton Fund in Princeton University. Among the many people who have helped, I should like specially to acknowledge the enthusiastic and accurate services of three research assistants, William R. Reiter and Wayne Walker in the early stages, and JoAnn Morse in the verification of quotations and compilation of the index. Gail Filion is a fine editor, who knows when to assist, when to insist, and when to desist. I am grateful on all counts. Words of praise for the most important libraries on which I have relied would be initials on the Parthenon: the Bibliothèque nationale and the collections of the American Philosophical Society in Philadelphia and of the Firestone Library in Princeton. It is an agreeable feature of working in French history that one often is concerned with the origins or earlier phases of agencies that preserve their own records and archives, and the persons currently responsible generally respond to serious inquiries with a courtesy that bespeaks the French respect for history at its most sympathetic. I shall not anticipate here all the repositories mentioned in the footnotes, but I should like to say that their curators greatly added to the joy of the research, and to the pleasure of studying in France.

My wife, who joins me in these and other matters, makes a joy of everything she touches.

Princeton, New Jersey
January 1980

A NOTE ON THE CITATIONS

◇◇

Secondary sources are cited by the name of the author and the date of publication, which will identify the work in the bibliography. Contemporary sources and archival materials are specified in the footnotes, which are also covered by the index. Rather than weight the bibliography with detail about all the collections and primary writings consulted, I have thought it would be more convenient to include bibliographical footnotes discussing the important materials, primary and secondary, that exist for the study of each topic. These notes generally appear early in the appropriate sections, and their location is given in the index by there italicizing the page references.

A word is needed about the organization and dating of the memoirs of the Academy of Science. An annual volume was printed under the title *Histoire et mémoires de l'Académie Royale des Sciences de Paris*. The *Histoire* consists of announcements and abstracts. The *Mémoires* consist of the scientific papers themselves. The two sections are paginated and (for this reason) cited separately. Confusion in dating often arises because publication was always two to four years in arrears. Thus the volume for 1780 appeared in 1784 and contained memoirs submitted at any time between the nominal and actual dates. I indicate that volume as MARS (1780/84), with the latter date that of publication. The same convention is observed with the *Mémoires de la Société Royale de Médecine* and other collections.

The following abbreviations are employed in the footnotes:

AN	Archives Nationales
BANM	Bibliothèque de l'Académie Nationale de Médecine
BFP	Bibliothèque de la Faculté de Pharmacie, Université de Paris
BMHN	Bibliothèque du Musée National d'Histoire Naturelle
BMS	Bibliothèque de la Manufacture Nationale de Porcelaine de Sèvres
BN	Bibliothèque Nationale
BPC	Bibliothèque de l'École Nationale des Ponts et Chaussées
DSB	*Dictionary of Scientific Biography*, ed. C. C. Gillispie, 16 Volumes, New York: 1970-1980.
HARS	*Histoire de l'Académie Royale des Sciences de Paris*
HSRM	*Histoire de la Société Royale de Médecine*
MARS	*Mémoires de l'Académie Royale des Sciences de Paris*
MSRM	*Mémoires de la Société Royale de Médecine*

SE *Mémoires de mathématique et de physique présentés . . . par divers sçavans.* (This collection is usually referred to as the *Savants étrangers*).

In rendering the names of institutions and organizations, I have followed my ear rather than consistency. In each case, the first mention gives the French designation, in order to avoid any possible ambiguity. Thereafter, it seems natural to say Academy of Science rather than Académie des sciences, Ministry of Finance rather than Contrôle-Générale des Finances, and similarly with many agencies and organizations. With others, however, the French seems so intrinsic to one's sense of the thing, and the English so artificial, that I have left it in the original—Ponts et chaussées, for example, rather than Roads and Bridges, which is absurd, or Civil Engineers, which is ambiguous if not anachronistic.

PART ONE

INSTITUTIONS

CHAPTER I

The State and Science

A great man, whose teaching, whose example, and above all whose friendship I shall always miss, was convinced that the truths of political and moral science are capable of the same certainty as those that form the system of physical science, even in those branches like astronomy that seem to approximate mathematical certainty.

He cherished this belief, for it led to the consoling hope that humanity would inevitably make progress toward a state of happiness and improved character even as it has already done in its knowledge of the truth.

The marquis de Condorcet
alluding to Turgot in *Essai sur l'application de l'analyse à la probabilité des décisions rendues à la pluralité des voix* (1785, p. i)

1. TURGOT AND HIS CIRCLE

This history begins naturally in 1774 with the ministry of Turgot, who drew upon science and systematic knowledge in formulating policies intended to rehabilitate the French monarchy on the accession of Louis XVI. Largely frustrated as reforms, main elements of his program reemerged fifteen years later in the design of revolutionary institutions. When Anne-Robert-Jacques Turgot, baron de Laune, entered high office, he was a novelty among statesmen. Men of science and men of letters acclaimed him one of theirs. "I hope for protection from him for men of thought," wrote Voltaire from Les Délices, "because he is an excellent man of thought himself."[1] Just under two years later the king, in an access of the weakness that may have already doomed the reign, left him fall victim to the enemies of change.

The juncture that brought Turgot to office was more crucial than a

[1] Voltaire to d'Argental, 5 September 1774, *Voltaire's Correspondence*, ed. Theodore Besterman, 107 vols. (Geneva, 1953-1965), 89, 5-6, No. 18002.

change of ministry or even reign. Either it was the beginning of the revolutionary movement or the last chance to avert it. Historians have adopted both views. At all events, it was one of the profound crises of regime that mark stages in the history of France. At those moments of paralysis, interests come to oppose one another so categorically that they bring the normal processes of politics to a stop, and the working of institutions no longer answers to the necessities of the country. On the resolvability of this impasse turned the question whether the monarchy might retrieve its grasp over matters of state: whether it might win through to solvency by drawing upon the whole wealth of the nation, and whether ministers might make the public interest prevail over the constitutional prerogatives of the privileged orders and corporate bodies, ending or abating their exemption from direct taxation and from what would later be thought the common obligations of citizenship in the modern state.[2]

This much was evident at the time. A deeper problem, at once more emotional and more practical, has been less noticed by historians. Was it ever possible for the expert knowledge and direction animating the reforms of a Turgot to be congruent with the popular will? "Some evil to abate, some piece of good to do, that is all he saw before him,"[3] observed a childhood friend and confidant, the abbé de Véri, who mediated between Turgot and the comte de Maurepas, the old politician whom a young king had chosen for personal mentor.

When Maurepas proposed Turgot's appointment to the king, first as minister of maritime affairs (the navy, the merchant marine, and the colonies) and, after a month, as controller-general of finance, he did so with a view to bringing into government a magistrate known for probity and talent. Court circles were less conscious of the complementary—the intellectual— aspect of Turgot's reputation. Or else, if conscious of it, they came to like it less through actual exposure. "Monsieur," Maurepas is said to have told him some months later, "Concentrate on our current finances. Try to make provision for the present without wracking your brains over changing things fundamentally."[4] Maurepas's disenchant-

[2] For a brief discussion of the significance of the Turgot ministry, see Palmer (1959-1964), *1*, 448-458. The literature on Turgot is very large. The most important items are Schelle's introductory narratives prefaced to each volume of his edition of the *Oeuvres de Turgot* (1913-1923) (hereafter cited as Turgot, *Oeuvres*); Dakin (1939); Faure (1961); A.-N.-C. marquis de Condorcet, *The Life of M. Turgot* (London, 1787; 1st publ. 1786); Pierre-Samuel Dupont de Nemours, *Mémoires sur la vie et les ouvrages de M. Turgot* (Philadelphia, 1782). Meek (1973) has translated a selection from his writings.

[3] Joseph-Alphonse de Véri, *Journal*, 2 vols. (1928-1930), *1*, 336; on the early career of Maurepas, and its closeness to a previous scientific milieu and generation, see below, Chapter V, Section 2.

[4] Faure (1961), 325, quoting François Métra, *Correspondance Secrète* . . . , 18 vols. (London, 1787-1790), *1*, 125, for 7 December 1774.

ment was natural. The two traditions of administration and intelligentsia had been quite distinct and often antipathetic in previous times—as, indeed, they since have largely been. For that reason it is the retrospect of the *philosophe* in office that has intrigued posterity. True, Turgot soon proved to be insufficiently in power, but it was in his generation of the Enlightenment, in the context of its civic and governmental problems, and most consciously in his own entourage that the coupling of these incompatibles turned out to be fertile, leaving as offspring the tribe that has ever increased, that of the expert in public affairs.

Both branches of the lineage are germane. On the one side Turgot was a nobleman whose ancestors for centuries had served the crown high in the responsible magistracy. Among such officials the tone of civil service was that of a knowledgeable discipline dignified by the formality appropriate to a governing class, or better an administering class, constituting itself arbiter of the public interest. His father, Michel-Etienne Turgot, had been prévot des marchands—municipal administrator, in effect—of the city of Paris. He has been compared to Haussmann for his improvements and embellishments in the capital. To this day prints of the pictorial map he commissioned display for casual lovers of the city the street plan of the 1730s. Such service had its code, animating the best of its representatives with a Roman sense of order and duty. "Too honest a man," complained the duc d'Orléans of a predecessor of the elder Turgot, Charles Trudaine (the father of Turgot's own patron),[5] whom the regent removed for defending the interests of the city all too stiffly against the speculative ventures of John Law.[6] Just after Turgot himself, thirty-four years old in 1761, was posted intendant to Limoges, the generality of Rouen fell vacant. He might have had it. It carried more responsibility and honor. There he would have been administering his ancestral province and would have been close to the ambiance of Paris. He refused to retract his commitment, however, and remained in the remote and relatively barbarous Limousin to make the reputation that won him governmental office. For he made his generality a laboratory of the enlightenment, in which measures formed from progressive civic and economic analysis were tried on the affairs of a backward region.

Before going to Limoges, Turgot, a young lawyer in Paris, had consorted with leading Encyclopedists. He frequented the salon of Madame Geoffrin and later that of Mademoiselle de Lespinasse. D'Alembert singled him out, a grave and coming young man, and wrote a letter introducing him to Voltaire.[7] On d'Alembert's persuasion he contributed five articles to the *Encyclopédie*, writing on *Etymologie*—"Locke and after him the abbé de Condillac have shown that language is really a kind of calcu-

[5] Below, this section. [6] Delorme (1950), 101. [7] Turgot, *Oeuvres* 2, 89.

lus, of which the grammar and even in large part the logic are only rules, but this calculus is far more complicated than that of numbers."[8] On *Existence*—"We know of one class of proofs in which we habitually place confidence; indeed, we have no others which can assure us of the existence of objects not actually present in our senses, and about which we nevertheless make no doubt: that mode is *induction*, which proceeds from effects and rises to the cause."[9] On *Expansibilité*—"Property of certain fluids by which they tend incessantly to occupy a greater space,"[10] a phenomenon to be attributed to the repulsive effect of heat, which Turgot took to be an imponderable fluid. On *Foire*—"We conclude that *great fairs* are never so useful as the constraints they presuppose are inhibiting, and that far from being the proof of a flourishing commerce, they can on the contrary exist only in states where trade is burdened, overweighted with impositions, and consequently mediocre."[11] And, finally, on *Fondations*—"A founder is a man who wishes to eternalize the effect of his will. . . . We conclude that no work of man is made for immortality; and since the *Fondations*, ever multiplied by vanity, would in the long run absorb all funds and all private property, it is incumbent that after a time they be abolished."[12]

Such was the tenor of Turgot's youthful opinions. They were commonplace enough among enlightened spirits, and it was rather the actualities of his education and career that set him apart from his elders among the philosophes and encyclopedists. Turgot drew away from the latter when the ban of the censor came down on the *Encyclopédie*. He had in view a career in government, rather than against it, and was averse to the sectarian spirit, whether manifested in religion or counter to it. A third son, and a lad of intellectual temper, Turgot had originally looked to an ecclesiastical career and taken two years at the Sorbonne. In contrast to the scorn often expressed for the university in the 18th century, Turgot always considered his formal education to have been valuable. "My dear Abbé," he would sometimes say to Morellet, a comrade at school and a lifelong friend, "it is only we, who have studied for our *licence*, who know how to reason accurately."[13] His was a trained intellect and in that respect

[8] *Ibid.*, 473-516, 506-507.

[9] *Ibid.*, 517-538, 536-537.

[10] *Ibid.*, 538-576, 538. In 1783 the abbé Rochon, a member of the Academy of Science and an experimentalist, credited Turgot with having been the first to distinguish between the processes of evaporation and vaporization, pointing out that air is an impediment to the latter and essential to the former, and that evaporation resembles solution. It was thus, in Rochon's view, that "M. Turgot a jeté les premiers fondemens des vrais principes de cette partie importante de la Physique et de la Chimie." The ideas were "si neuves et si profondes," however, that the editors themselves failed to understand the distinction. Alexis-Marie de Rochon, *Recueil de mémoires sur la mécanique et la physique* (1783), x-xiv.

[11] Turgot, *Oeuvres 1*, 577-583, 583.

[12] *Ibid.*, 584-593, 585, 593.

[13] Dakin (1939), 12.

unlike the looser minds characteristic of the philosophes, whose thinking, however active, had been formed rather by literature and by personal experiences impressed emotionally upon sensibility. It may be that, like literary people in other milieux, the philosophes tended to be critical of authority mainly because of its existence. Not so Turgot, whose view of authority was critical of it rather in its exercise.

Under the early discipline of higher education, Turgot made an intellectual commitment to history and to science in a reciprocity that distinguished his career. While a student at the Sorbonne, he composed a stated oration in Latin sketching the scheme of progress that the reflections of his mature years developed into a philosophy of history.[14] The vein is neither moralistic like that of Voltaire nor naturalistic like that of Montesquieu. Instead, successive states of scientific knowledge are what mark the stages of historical development. Turgot's philosophy of history forms one of the sources of modern positivism, the other having been the phenomenalist account of knowledge itself in the mode of d'Alembert and Condillac. A full philosophical synthesis of historicism and phenomenalism awaited August Comte, however.

When attention turns from Turgot's intellectual legacy to his conduct of administrative work, the interesting feature is its scientific aspect rather than the idea of progress that inspired it. Not that he ever took himself for a scientist, although among his student memoirs are essays on cosmology and mechanics exhibiting a degree of comprehension high enough that he clearly could have done science had he wished.[15] Instead, his prospects in the civil service led him to develop his own knowledge along lines of political economy and social science.

Developing his bent brought Turgot into a position of intellectual leadership fortified by practice in a manner that distinguished the work of his contemporaries from the writings of their predecessors, Voltaire, Rousseau, Diderot, and their fellow philosophes. In his generation, the second of the Enlightenment, the movement of culture we associate with the Age of Reason passed over from criticism into action. That was its characteristic aspect. The influential figures dealt in knowledge and techniques rather than ideas and affirmations. In their mentality, science no longer figured as an enigmatic summons to naturalistic criticism of society. Their knowledge of it was substantial, and it set the example for the kind of social science that would draw upon political economy and technology, ready now in their turn to become rational bodies of information and instruments of analysis. The contrast pertains to more than intellectual

[14] Translated in *Oeuvres 1*, 214-235, "Tableau philosophique des progrès successifs de l'esprit humain."

[15] "Lettre à Buffon sur son sytème de formation de la terre," *ibid.*, 109-113.

styles. The period was also that of early industrialization, after all, even if mainly in Britain, and of enlightened despotism, even if mainly elsewhere on the continent. In these matters, too, where craftsmanship was becoming engineering and political effectiveness was presupposing knowledge, men of administration and men of practice counted for more than did the men of books. The same was true of the less familiar interactions of science and politics with which we will be concerned in the context of French institutions.

The *éloge* of an eminent predecessor has often offered the occasion for an ambitious young Frenchman to come before the public. Turgot's first published writing of moment was an essay on the life and works of his mentor in the 1750s, Vincent de Gournay.[16] Gournay's actual post was that of intendant in the Bureau du commerce.[17] The situation was equivalent to an undersecretaryship in a modern ministry. From it Gournay exerted a profound influence upon economic theory and governmental practice. It is disputed whether the very phrase "Laissez faire, laissez passer" was of his coining. There is no dispute about his having founded the school that made this the cardinal precept in administration of a national economy.[18] The heads of the teaching sound very familiar in Turgot's exposition, an exposition amounting to injunction: let knowledge of commerce be a proper science of facts and relations; let its object be comparison of the productions of nature and artifice in value and price; let it model itself upon statics and state the laws by which commercial prices reach a natural balance comparable to the equilibrium of physical bodies arranged in the order of specific gravities; let it be axiomatic that every industrious person has a right to make, buy, or sell what he pleases, knowing "his own interest better than another to whom it is indifferent"; let government, therefore, appreciate that regulation could only be self-defeating; let the state remove obstacles to trade, the most serious in France being the high rate of interest and the thicket of taxes; let the encouragement of industry be restricted to rewarding inventors, domesticating foreign manufactures, and improving the skill and knowledge of artisans. Gournay, in short, "thought that the Bureau of Commerce was much less useful for conducting trade, which ought to go its own course, than for defending it against the schemes of the financiers."[19]

In comparing these positions to the views adopted simultaneously in the more articulate and famous school of economic analysis, that of Ques-

[16] *Ibid.*, 595-623.

[17] On the organization, functioning, and importance of the Bureau du commerce in the eighteenth century, see the introduction to the published inventory of *Procès-verbaux, Bureau du commerce, 1700-1791*, ed. P. Bonassieux and E. Lelong (1900); Biollay (1885), pt. 3; Parker (1965); and below, Chapter VI, Section 1.

[18] On Gournay, see Schelle (1897).

[19] Turgot, *Oeuvres 1*, 609.

nay and the physiocrats, who came to very similar conclusions, historians ascribe the differences between the two groups to their disagreement over the source of wealth. The physiocrats placed it entirely in the earth and rested fiscal prescriptions upon the single tax on land. The followers of Gournay, on the other hand, rejected Quesnay's opinion that manufacturing and commerce are sterile pursuits that merely transform and exchange objects of value. They adopted instead a pluralistic view of wealth emphasizing industry. The difference in orientation goes back to a difference in persons. Quesnay and his entourage were writers and men of doctrine: Mirabeau, Abeille, Fourqueux, Chreptowicz, Mercier de la Rivière, the abbés Baudeau, Roubeaud, Le Trosne, de Saint-Peravy and de Vauvilliers, together with Pierre-Samuel Dupont de Nemours, whose listing this is.[20] Latter-day economic platonists, they tended to put their confidence in persuading princes—the grand duke of Baden, the Archduke Leopold of Tuscany, the Emperor Joseph II. Not so the Gournay connection, most of whom filled responsible positions.

Gournay himself, an adventurous trader in his youth, had been brought into the Bureau of Commerce in 1751 at the instance of Maurepas, the same who a quarter of a century later recommended Turgot to the king. There he transformed the spirit of administration. He appears to have had, and perhaps it was he who imparted to Turgot, the power of attracting disciples, inculcating in them less a doctrine than a mode of discharging responsibility. He would deal with particular matters as they came before the staff of the Bureau, where the intendants met in weekly conference with the controller-general. There his fellow officials found themselves progressively impressed by the coherence, consistency and good sense of his reasoning, and above all by his knowledge of how merchants actually traded. He enjoyed discussion. He poured out memoranda and memoirs on particular matters. He never became heated or imperious.[21] He thrived in committee.

It may be illuminating to think of this small circle—Gournay, Daniel Trudaine, his son Trudaine de Montigny, Morellet, Malesherbes, Turgot himself, together with several of the physiocrats who tended into their company, notably Dupont de Nemours, Fourqueux, and Mercier de la Rivière—as Gallic Fabians of Free Trade, comparable to their socialist counterparts a century and a half later in Britain in that they had their effect through expertness in detail, flexibility in tactics, administrative tenacity in the layers of civil service below the surface of politics, and membership of agencies and committees where they prevailed in a

[20] Pierre-Samuel Dupont de Nemours, "Sur les économistes," in *Oeuvres de M. Turgot . . .* , (Dupont edition) *3*, 309-320.

[21] Turgot, "Eloge de Gournay," *Oeuvres 1*, 611.

hundred or a thousand small decisions taken in offices on specific problems. In the end it thus appeared as if generalizations in the science of political economy had shaped events, whereas in fact it was knowledge of political economy that permitted those who troubled to be expert and informed to anticipate events, to guide policy in relation to reality, and to condition opinion in the school of use.

Of them all, only Morellet never held administrative office prior to Turgot's ministry (in which, however, he and Dupont did serve in the capacity of secretaries). His reputation was that of an economic analyst specializing somewhat mordantly in problems of commerce.[22]

Of the rest, Daniel Trudaine was the most considerable. "Le grand Trudaine," he was called by contemporaries familiar with the workings of the state. To Turgot he was an elder patron in a family closely associated with his own and in the same tradition of service. Intendant of finance from 1734, Trudaine was Gournay's official chief since he had the Bureau of Commerce in his "detail" (it is curious how old French bureaucratic usage persists in modern military terminology). Earlier in Trudaine's career he had been intendant of Auvergne. There he concentrated his effort upon improving communications in order to bring that ill-traveled province into the rhythm of commercial and cultural progress.

Roads were no humdrum subject in the eighteenth century, the greatest age of road-building since the fall of Rome. When he was intendant of finance in Paris, Trudaine expanded the interest he had started locally in Auvergne, and in addition to directing the Bureau of Commerce put in hand a program of modernization of the highways nationally. Under his administration the haphazard old trade of constructing roads and bridges developed into a branch, or (not to be anachronistic) a nourishing root, of the future profession of civil engineering. More will be said of the famous Corps des ponts et chaussées,[23] founded at Trudaine's behest in 1747. Given his temperament and interests, he felt a lively affinity for the scientific community, encouraged members of the Académie royale des sciences (for a "grand commis d'état" was in a position to do so), and was accorded the recognition that the scientific community extended to open-minded and open-handed patrons, election to honorary membership.

His son, Charles-Philibert, called Trudaine de Montigny, was Turgot's lifelong friend. He married the daughter of Fourqueux, an official under Turgot who served a brief and disastrous turn as controller-general in 1787. In succession to his father after 1769, Trudaine de Montigny continued to direct both the Bureau of Commerce and the Ponts et chaussées,

[22] For Morellet, see Proteau (1910). For Morellet's observations on his friendship with Turgot, see his *Mémoires* (1821-1823), esp. *1*, 11-17, 28, 31-32, 36-37, 230-231, 233-238.

[23] Below, Chapter VII, Section 1.

though not (it may be sensed) with the same power to command respect. He mingled dilettantism with patronage in his relation to the scientific community; presenting a burning glass to the Academy, of which he too held honorary membership, accommodating early experiments by Lavoisier on combustion in the laboratory installed at Montigny, establishing prizes for optical devices.[24]

Of the other friends of Turgot's youth, Lamoignon de Malesherbes remains the most famous in his own right. He was the only one Turgot brought into government at the ministerial level, and came to high place out of enlightened *noblesse oblige*. Malesherbes preferred judicial to executive authority and spent his official life administering institutions so as to mitigate their inherent abuses. He sought to combine liberty with responsibility of the press in the office of director of publication (*la Librairie*). President of the Cour des aides, he pressed for disallowal of the religious laws and limitation of arbitrary confinement by letter of cachet. Almost alone among progressive statesmen, Malesherbes, to the parliamentary manner born, remained loyal to the sovereign courts during their exile in the last years of Louis XV, as much later he did to his king when he rose before the Convention to the defense of Louis XVI, and in consequence shared his condemnation to death. A brave, a selfless, a sympathetic though not an energetic man, he wished nothing better when young than to second the career of Turgot. "It is in speaking of Turgot," remarked a contemporary, "that he abandoned himself to his natural eloquence. In all that they had done, conceived, and meditated together, he forgot himself in order to enlarge the part of his friend."[25] Turgot's was an ambition that his friends admired without resenting, recognizing his right to it. "Between M. Turgot and myself," Malesherbes observed after both had fallen from power, "there is one great difference. M. Turgot always intended to pursue a great career. It is not in the least acceptable to him to have been discharged."[26]

By the time Turgot went to Limoges as intendant in 1761, he was ready for that career. Gournay had died two years previously. Turgot in his turn became the inspiration of a continuing, eventually a far-reaching connection of experts and reformers. Condorcet, one of the closest of the younger friends who began to form ranks around him, evoked in a passage worthy of Tocqueville the responsibilities borne by an intendant in the eighteenth century:

> Government sees but with his eyes, and acts but by his hands. It is upon the information he collects, upon the memorials he dispatches,

[24] On the Trudaines, see Delorme (1950), Petot (1958), Serbos (1964).

[25] Quoted from Charles de Lacretelle, *Testament politique et littéraire*, in Grosclaude (1961), 15.

[26] *Ibid.*, 12-13.

and upon the accounts he renders in, that ministers decide upon everything, and that in a country where every political power centers in administration, and where a legislation, imperfect in all its parts, compels it to unintermitted activity, and to reflection upon every subject.[27]

2. THE LIMOUSIN

The former generality of Limoges now comprises the departments of Haute-Vienne and the Corrèze, most of the Charente, and part of the Creuse. In Turgot's time it contained almost a thousand parishes and a population of over half a million, declining in number and condition and dependent upon a backward and decaying agriculture. Turgot set himself to reverse the deterioration. The most immediate problem was financial. On the one hand he pressed the controller-general to reduce the quota for the generality in the annual subdivision of the *taille*. On the other hand he moved to put collections on a rational footing in order that those subject to taille might pay impositions, if not according to the income of their properties (a counsel of perfection), then at least according to their real value instead of in the grossly unfair allocations prevailing through a tangle of tradition, corruption, and faulty records. For the purpose of establishing a tax base, Turgot wished to commission an accurate *cadastre* or land-register founded on a survey of property lines and an impartial assessment. His very limited success depended on creating a body of tax officials, a provincial bureaucracy, to carry on the functions that local persons had formerly been coopted to discharge, usually by a process of mutual delation in each village.[28]

Displacing obligation by officialdom and payment was also the crux of his most unqualified success, abolition of the *corvée* within his generality. Through that survival of servitude, the peasant owed the crown a certain number of days each year of labor on the roads, an obligation generally discharged in a grudging and malingering spirit. Turgot took counsel with Trudaine and proposed to furnish engineers from the Ponts et chaussées with hired labor financed by the whole generality instead of hands impressed from communes bordering the highway. Cannily he paid the peasants enough in advance to allay their suspicions and then withheld sufficient funds in the form of payment or tax rebate until the job was done. The entire reform he brought off at a saving. In its combination of

[27] Condorcet, *Life of Turgot*, 44-45. See Ardashev (1909) for the intendants under Louis XVI, and Bordes (1960) for the intendants under Louis XV. Gruder (1968) is an up-to-date social and administrative study of the entire subject.

[28] Dakin (1939), 33-62, and for detail, see Turgot, *Oeuvres* 2, 1-19, 81-115, 407-416, and *passim*.

engineering, reasonableness, and assault upon a scandal, Turgot's local suppression of the corvée impressed liberal opinion nationally and more than any other measure made his reputation.[29]

It was ever his way to combine administrative attentiveness with provision of official services and introduction of modern technique in large things and small. He furnished the generality with a veterinary school.[30] He had potatoes planted and dispelled mistrust by having them served at his own table.[31] He brought from Paris a better rat trap. He favored formation of a Society of Agriculture, one among the many that were proliferating to indoctrinate the provinces with the principles of physiocracy under the guidance of a central body in Paris. It is true that Turgot did not take this movement altogether seriously, considering it an affair of landowners and theorists rather than farmers and cultivators, though one unlikely to do harm. It might even do good.[32]

Turgot never rusticated there in the manifold detail of provincial administration. He regularly spent two months of the year in Paris and maintained a correspondence with Hume, with Rousseau, with Voltaire, with Diderot, and more regularly with Condorcet and Dupont de Nemours, which appears to have been the recreation, not of leisure, for he took none, but of intellect. In all the time of his intendancy, his main analytic preoccupation lay with a matter of political economy concerning the regimen of the entire nation. The movement to free the trade in grain proved, indeed, to be the deepest of Turgot's commitments, and acting upon it was certainly the most daring of his policies.

Nothing in his eyes more shamefully distinguished the economy as it was from the economy as it should be than the practice of consigning the grain trade to the control of the police as a function of civil order rather than leaving it subject to natural laws of supply and demand. By a vast and detailed structure of regulations, the city of Paris had to be kept provisioned with a minimum supply of flour at a fixed price no matter what the circumstances of the surrounding countryside. All other municipalities stock-piled to forestall shortages, a practice often indistinguishable in spirit or fact from hoarding against their neighbors. Controls were forever being invoked by one authority or another to forbid exportation and prevent internal traffic in grain. Everywhere speculators bided ready to profit from a dearth and perhaps to cause it. The consequence, according to en-

[29] Dakin (1939), 63-74; Turgot, *Oeuvres* 2, 118-121, 183-224, 319-321, 333-354, 421, 477, 626.

[30] Turgot, *Oeuvres* 2, 44, 435-436.

[31] Dakin (1939), 87.

[32] On the Society of Agriculture, see Lavergne (1870), app. 439-474; Dakin (1939), 79-91; Turgot, *Oeuvres* 2, 225-228, 430-434; and for a fuller discussion, below, Chapter V, Section 5.

lightened economic analysis, was that in time of glut the local producer was discouraged by slack demand and reduced his production, while in time of famine the peasantry and laboring population were at the mercy of how effectively and honestly the regulations could protect their access to supplies of grain and bread thus artificially constrained.[33]

On the imperative of freeing the grain trade, Turgot and the "economists" were at one with the physiocrats. The campaign kept him closely in touch with opinion in the capital. Among other associations, it brought him the friendship of Pierre-Samuel Dupont de Nemours, the still very young author of a strong physiocratic tract, *De L'Exportation et de l'importation des grains* (1764). Impressed thereby, Turgot sought him out during a visit to Paris.[34] Turgot, Dupont de Nemours, and the two Trudaines, father and son, collaborated in drafting the provisions of an edict to propose to the controller-general, Laverdy, extending a qualified liberalization that had been introduced in 1763 by his predecessor, Bertin. In Laverdy's judgment they went too far, and the decree he issued in 1764[35] was a much moderated version of their proposal. Even that never went fully into effect. Local authorities resisted, demanding continuation of their ancient right to buy up grain at set prices before sale was opened to merchants and bakers. The police of Paris objected, ever alert to the temper of the city. The populace became alarmed, there and elsewhere. Demonstrations cropped out against bakers and grain merchants. Mobs attacked granaries. Paranoia spread as rumors ran to the effect that the king himself had entered into a plot to profiteer from rising prices, an imaginary episode later magnified by revolutionary polemics into a *pacte de famine*. The intensity of the reaction might have given Turgot a warning of what lay in store for the application of principles that cut too deep and touched the population on the quick.[36] In the country generally, the government retreated from this policy favored only by liberals and speculators. Not so Turgot, who persisted in his own generality—and made it work!

It was a major achievement. Sustained above popular suspicion by his local reputation, Turgot did as he had done with the corvée, exceeded his authority under the edict of 1764 and abolished the regional regulation of commerce in grain and bread. He brought profits to the farmers by allowing exportation in good years. Most difficult of all, he persuaded merchants to form a consortium for purchasing grain abroad in years of shortage and selling it locally on the free market. By these measures, Turgot

[33] On the grain trade, see Schelle's introduction, Turgot, *Oeuvres* 2, 45-63; Dakin (1939), 92-103; Biollay (1885); and especially Kaplan (1976), the definitive work.

[34] Saricks (1965), 39, 56. On Turgot and Dupont, see also Jolly (1956), esp. 24-33.

[35] Turgot, *Oeuvres* 2, 405-406.

[36] Dakin (1939), 94-103; Biollay (1885), 105-125.

carried the Limousin safely, and for himself triumphantly, through the famine that befell in 1769 and lasted until 1772. The merchants did incur a small deficit, but the subsidy paid to make it up was far less costly than the relief normally and ineffectively undertaken by government under comparable conditions.[37]

Such were the problems with which Turgot dealt throughout thirteen years of assiduous attention to the details of tax-gathering, farming, road-building, education, subsistence, and public health, constantly informing himself, drafting circulars, framing reports, and supervising and inspecting the actions of the provincial subordinates he had to find and train.[38] There in the Limousin amid a suspicious, ignorant peasantry, and a defensive, scarcely less ignorant nobility, Turgot instituted a policy that may have been nine parts good, traditional public service and one part novel principle. But that one part, deriving from knowledge of technology and economic theory, lifted his administration above honest routine and turned bureaucracy toward expertise.

"You are born . . . to be the savior of France," an admiring friend wrote to Turgot when the king appointed him controller-general: "Your enlightened views, your formal knowledge, profound meditation, twenty years of experience, recognition by everyone of your patriotic zeal and above all integrity—true citizens are all persuaded of the possibility that a second Sully exists for the welfare of France, a second Colbert."[39]

3. THE MINISTRY OF TURGOT

Practically, though not formally, in the late eighteenth century the controller-general was head of the Ministry. In that office he combined functions now pertaining to the presidency of the Council with those of the modern Ministries of Finance and the Interior. This book is not the place to dwell upon the merits of Turgot's program nor to recount its failure in detail. The former have been extolled to the point of idolization beginning with his immediate disciples.[40] The latter may be followed in an admirable study, *La Disgrâce de Turgot*, by Edgar Faure, who held similar

[37] Dakin (1939), 104-117; Turgot, *Oeuvres 3*, 111-154, 256-357, 425-459.

[38] It is with a certain piquancy that the historian of science recognizes among the names of his clerks that of J.-E. Montucla, later the first historian of mathematics of whom it may be said that he wrote the subject in a modern fashion. Turgot, *Oeuvres* 2, 440, 515. This association is more than a mere curiosity. It shows that the linking of science and history into a historiography of science goes right back to the origin of the intellectual movement that became positivism in Comte's philosophy of science. On Montucla, see Sarton (1936).

[39] Quoted in Faure (1961), 228.

[40] See, for example, Pierre François Boncerf, *Les Inconvénients des droits feudaux* (1789); Foncin (1877) is still useful.

office amid comparable rigidities and inadequacies of state in the Fourth Republic, and who was a member of the Mendès-France government among others. He writes from inside the constraints put upon an intelligent and informed, indeed an expert, statesman by the configuration of French political life. Monsieur Faure sees the life of such governments transpiring in a double movement. The curve rises in the minister's ability to extort measures, not from widespread support, which in French politics is rarely obtainable, but by imposing his will and reputation in the extremity of the regime amid fears of major groups lest worse befall. Then as each successive measure offends or alienates some set of interests or opinions, resistance and resentment accumulate, the curve breaks over into the descendant, and government steers a doomed course past one reef after another until the mutually incompatible elements of opposition momentarily agree that the minister is less tolerable than the danger that brought him to office, now in any case receded.

Faure finds in Turgot less the fiscal than the economic reformer, and argues persuasively that the financial crisis has been exaggerated by historians moralizing at the expense of Turgot's predecessor, the devious and ingenious abbé Terray, and over-dramatized by their tendency to represent all affairs as backdrop to the events of 1789. To be sure, Turgot warned the king of the dangers inherent in extravagance of taste and weakness of will. The central thrust of policy was directed not toward the situation of the treasury, however, but toward liberalizing and stimulating the whole economy, wherein agriculture was the largest sector and grain the most important commodity.

With his credit still strong, Turgot addressed himself to freeing the grain trade, his favorite measure and meant to be the first in a policy of economic growth through liberalism. The scope was moderate, for he did not initially touch the provisioning of Paris or the controls over exportation. The preparation was cautious and circumspect, for he had not yet lost his following and with it his judgment. Nevertheless, his decision to press ahead even in limited degree bespeaks a certain dogmatism. The decision was taken in the face of Turgot's knowledge that the prospects for the harvest was unsure. It disregarded the evidence of widespread popular feeling, whether right or wrong, that the old regulations were an essential protection against famine and speculation. In deference to that feeling and to the realities involved in regulation of the bread supply, previous administrations had largely let lapse the edicts of 1763 and 1764. Turgot overrode the adherents of controls. He ignored those well disposed to his principles who counseled greater caution in applying them. He carried the day with his colleagues, the crown, and the newly restored parlement. The decree was issued in September 1774. He had prevailed by drawing upon adherence to himself, however, rather than to grounds of policy.

When policy miscarried, among the assets lost, therefore, was the magic of his reputation.

For miscarry it did in the short run, and there was no long run to redeem it. The harvest was worse than expected. Turgot had gambled that prices would rise high enough to bring food to market but not to create panic. He lost. A cattle plague, a virulent murrain, broke out to compound the trouble. Famine threatened in the areas worst affected. Confronted with shortage and no longer protected by regulation, people in certain regions took matters of subsistence into their own hands and touched off a series of grain and bread riots afterward called the *guerre des farines*. The object of the leaders was the classic one of commandeering provisions before they got into the hands of speculators.

Given the conjunction of reforming intent on the part of government with material deprivation among the populace, it is possible to see in these manifestations, not mere disorders in the time of dearth, but revolution yet unborn, its program fixed in embryo. They spread in a set of social chain-reactions prophetic of the Great Fear of the summer of 1789. One of the new school of social historians has traced out a physiology of the riots on a map exhibiting the movement from one locality to the next, down the valley of the Oise, for example, from Beaumont through L'Isle-Adam, Pontoise and Poissy and across the Seine to the nerve center of Versailles itself.[41]

Simultaneously with the miscarriage of high-minded policy, the excuse was put forward that it had been frustrated by a plot, one fomented to discredit the government by its reactionary enemies, and notably by the prince de Conti, by a plot or the fear of brigands or both. If this theory is correct, and Faure follows Rudé and Lublinski in thinking it may well be, then Turgot may have been, not the last statesman who could have prevented revolution, but the first to fall victim in his prescriptions to the incommunication between official policy and popular psychology that was a feature of the Revolution itself. For it already appears at the very dawn of liberalism that reformers failed to comprehend the concrete sense for the conditions of survival felt among the people themselves, to whom liberalism was quite irrelevant in the lives they actually led.

From the point of view of this history, that interpretation is a tempting one. It presages the unpopularity in actual revolution of the reforming scientific and administrative impulse animating Turgot and later his disciples. Even in the person of its founder, this tradition never had the common touch. He put down the riots firmly, not to say severely. His victory in the guerre des farines destroyed his reputation for being an incarnation of the popular will in a regime lost but for the probity he would bring to

[41] Rudé (1956); see also Lublinski (1959).

bear. He might yet, and no doubt he did, stand for the public interest. But to do so against what might now be taken for the people's will in regard to the most fundamental of their concerns, their very subsistence, dissipated the moral credit he had brought to office. At least it did so in circles beyond the intelligentsia. Henceforth, he was fair game like any of his predecessors or successors cast in the common mold of minister, busy keeping order.

To follow the further application of Turgot's program, each measure attended by its increment of opposition, is to observe the decline of his ministry. The circumstances that had brought him to office receded—the prospect of a new reign and an inexperienced king; the stalemate between administration and magistracy; the confusion and insufficiency of the finances; the obsolescence of economic policy; the fiscal untouchability of wealth and privilege; himself the administrator unspotted by corruption, distinguished by success, and seized of knowledge. All that diminished in importance before his tendency to bore the king by expecting of him too much attention and character; before the resentment of the Catholic party, the *dévots*, offended by the reconciliation of the crown with the parlements, to which they rightly attributed the expulsion of the Jesuits ten years previously; before the worldliness of the following of Choiseul, reproaching the government for an inglorious foreign policy and an abject military posture and putting it about that Turgot was, after all, a man of systems rather than political realities; before the increasing friction with his colleagues in the ministry, Maurepas and Saint-Germain, the latter at the War Office and together with Marie-Antoinette and others of the royal family always at the king's ear; before the automatic, instinctive resistance of financiers, speculators, and monopolists reinforced by the great majority of the propertied and noble classes, ever keener in their sense of privilege than in their public spirit; before the very parlements he had recalled, whose members had not returned to abdicate their claim to represent the nation in the face of a willful government. They were unappeased by the addition of Malesherbes to the ministry, and their leading spirit, the prince de Conti, fought Turgot from his deathbed, and won.

Turgot did hold the king firmly enough on course to push his central legislation into law by means of a *lit de justice*. The six edicts, the measures were called. The most significant abolished the corvée nationally, ended the regulation of grain in Paris, too, and suppressed the guilds in all municipalities. These three items were repealed upon Turgot's fall from power. For fall he did, his reforms calling into being no positive political resources to overbalance a combination of the weight now running against him. All that remained of the major provisions was reorganization of certain of the farms into direct *régies*, administrations licensed by the state and responsible to it. The first and greatest of the farms Turgot was not ready to touch, the famous tax farm to which the Treasury consigned collection of indirect taxation with a profit to itself. A second, that of stagecoaches and posts—Messageries—was transformed into a national enterprise that left France with a type of diligence called a "turgotine." The third, concerned with supplying gunpowder, is of central importance to this history, for Turgot's reform placed Lavoisier at the head of the Régie des poudres. More will be said of that in Section 6.

It is not surprising that the euphoria in which Turgot came to office was so soon dissipated and the hopes of the enlightened disappointed. His fall must have been as inevitable as anything in politics, for which in the partisan sense Turgot turned out to be ill suited. Only rarely can the intelligence to understand the economic and technical factors required to redeem a polity have been coupled with the political address and dexterity to carry them into effect. The Limousin might be administered, but France had to be governed, and Turgot's genius was of an administrative and scientific rather than a governmental order. We know very little about him inwardly, but a great deal about the effect he produced on others, which like a science teacher in a backward class was that of a man unforgivably in the right who put them in the wrong. He had the misfortune for a politician that his expression always announced his sentiments, which were frequently unflattering to associates.[42] He never married. His respect for women, according to Dupont de Nemours, whose own conduct was free, partook of "straightforwardness (*honnêteté*), which differs a bit from gallantry."[43] Nor had he any feeling for the character or sentiments of the uneducated. When a group of parishes in the Limousin failed to send a woman to attend the course in midwifery that he had brought a certain Madame du Coudray to Limoges to conduct for them, he scolded the parish priests in a circular and, though disappointed, announced that he would give their parishioners one more chance "to repair their negligence."[44]

Those he did touch deeply were his own kind. They consoled themselves over his fall—it is not recorded that he complained himself—by recognizing that the times were not ready, the society not worthy. Amid such reflections do intellectuals intervening in politics ever rescue self-esteem from frustration, and in such moods do disciples carry on—the like of Condorcet, Dupont de Nemours, Morellet, Malesherbes, Trudaine de Montigny, and others now to be identified. In public affairs, they represented the positions and aspirations that in the politics of the Revolution

[42] Condorcet, *Life of Turgot*, 374; and Véri says of his friend: "Incapable de tournures adoucissantes, sa physionomie porte, sans qu'il le sache, un air de dédain envers ceux qui lui paraissent avoir tort. Ce n'est pas un moyen d'obtenir le concours de ses collègues" (*1*, 392-393). See also Morellet (1898), 115.

[43] Turgot, *Oeuvres 1*, 41.

[44] *Ibid.* 2, 323.

became at their most advanced those of the Girondists. Pierre Vergniaud, for one, orator and epitome of the Girondists, arrived in Paris from Limoges in 1771, a protégé of Turgot.[45] The comte d'Angiviller, for another, had received the post of directeur-général des Batiments du roi when Turgot was appointed controller-general, and turned the office from a near sinecure into the equivalent of a ministry of fine arts, culture, and science in the last years of the old regime. It was he who made the Louvre a public art gallery and who after 1780 exercised supervision over the manufactories of porcelain at Sèvres, of tapestries at the Gobelins, and of carpets at the Savonnerie in Chaillot. Under his jurisdiction came the Jardin du roi, the Observatory, and the academies, including the Academy of Science.[46]

Among the others were would-be writers who in the previous generation would have been lesser philosophes. Several among them, although they never developed much literary resonance, became ministerial timber when the Revolution opened governmental careers to educated people of no specific vocation. Dominique-Joseph Garat succeeded Danton in the Ministry of Justice in 1792 and Roland in the Ministry of the Interior in early 1793.[47] François de Neufchateau acquired competence in agronomy and became procurator of the governing council in Santo Domingo from 1783 to 1787. He was minister of the interior for a few months in 1797 and again in the last years of the Directory, and in the interval was himself a member of the Executive Directory.[48] Fellow writers since forgotten were personally prominent and influential in dispensing literary patronage in the last years of the Old Regime: J. B. Suard, the leader of literary Anglophilia whom Turgot had appointed official historiographer,[49] and Saint-Lambert, the poet of rational pastoralism, of whom Diderot is somewhere reported to have said that he had every attribute of a poet except verve and vision.[50]

Such were the men who foregathered with the fallen minister in the foyer of his reputation, the salon of Madame Helvétius in Auteuil, with whom Turgot's name had been linked before her marriage to the philosophe and again after her husband's death in 1771.[51] They loved him, said Garat, somewhat hyperbolically perhaps, "as Socrates was loved by his disciples and Cato by Brutus."[52] Turgot died in 1781, his life just overlapping the youth of those who at the end of the century developed

[45] Bowers (1950), esp. 26-30; Lintilhac (1920), 2; Verdière (1866).
[46] Silvestre de Sacy (1953), esp. 53-57, 144-154, 180-182.
[47] See Garat (1862), esp. 353, and (1821) *2*, 326.
[48] François de Neufchateau, *Le Conservateur* (1799) *1*, x-xiii.
[49] Hunter (1925), 133. See also Darnton (1971).
[50] Saint-Lambert (1801) *5*, 350-352.
[51] Guillois (1897), 96.
[52] Garat (1821) *1*, 296.

his principles and inspiration into the political and philosophical school of thought called *idéologie*. Cabanis, Daunou, Destutt de Tracy, Maine de Biran, Volney, all took their start with his example. His entourage was not an organized group, much less a party, but a connection in the eighteenth-century Whig sense of the word, a ramification of friends, former colleagues, associates, and followers joined in a common outlook upon public affairs, letters, philosophy, and science, and with entrée into all those worlds. Benjamin Franklin became a familiar of Madame Helvétius and the circle in Auteuil, combining in his appeal to literate Paris similar elements of politics, patriotism, letters, and science.[53] So after him, though in a less intimate way, did Thomas Jefferson, who observed in a letter to Dupont de Nemours acknowledging the gift of Turgot's writings, "The sound principles which he established in his particular as well as general work are a valuable legacy to ill-governed men, and will spread from their provincial limit to the great circle of mankind."[54]

For although Turgot's ministry did fail of its ends politically and constitutionally, it did not in the end fail administratively or scientifically. Indeed, in those latter respects it exhibited the emergence in public affairs of developments more profound than the problems of the political moment and more dispersed through the whole texture of social, institutional, and cultural reality, developments of which Turgot's role in government was to be the prophetic agent rather than the cause. The days of the corvée, for example, were certainly numbered in any case by the economics of public works. Although reinstated in immediate reaction to his fall from power, it was again abolished in 1787 in conformity with the will of the Assembly of Notables, a body representing many of the elements that had overpowered Turgot eleven years earlier. Brief though his ministry had been, it did have form. To bring forward the expert and give him authority there where interest and routine misgoverned, that was the thrust of his administrative purpose, and it formed the school of followers just reviewed. From their kind derived the intellectuals of the Gironde and the Directory. Theirs became at its best the France of the July Monarchy and the Third Republic, in its excellent *lycées*, in its meticulous accountancy, in the rigidity of its spirit of patriotic rectitude.

4. ADMINISTRATIVE REFORM

Condorcet and Dupont de Nemours were Turgot's men of confidence in science and in economics, vectors in the impregnation of government with knowledge. Turgot stood godfather to Dupont's son and suggested the names Eleuthère Irénée after the Greek for liberty and peace. A bust of

[53] See: Guillois (1897), 96-98; Lopez (1966), 243-301.

[54] Jefferson to Dupont de Nemours, 29 November 1813, Chinard (1931), 206.

him still stands in the embrasure of Eleuthère's bedroom in his onetime house above the Brandywine in Delaware. Eleuthère had learned powder-making at Essonnes, having been recommended to the powder-mill there through the influence of Lavoisier. Turgot's administration drew upon members of the scientific community, most notably Lavoisier for munitions and Vicq d'Azyr for public health, and inserted a guiding hand into its concerns. He installed Condorcet at the Mint. The previous March (1773) the Academy had chosen that young, professionally unestablished, and highly ambitious mathematician to be acting permanent secretary in effect, although in form he was adjunct to the ailing Grandjean de Fouchy.[55]

The historian has become used to seeing a movement from aristocracy toward liberalism and democracy in all these developments, whereas what needs to be perceived is a movement from bureaucracy toward technocracy. It would be a mistake, however, to suppose that the latter process was simply the technical instrumentality for the former. Before exploring the systematic aspects of the program, it may prove persuasive to exhibit its reality in certain episodes and measures.

At the Mint Turgot created the post of supernumerary inspector for Condorcet and charged him to concert efforts with the veteran functionary Tillet to bring in a project acceptable to the Academy of Science for standardization of weights and measures.[56] Turgot took direct interest in the problem. The idea of basing the linear unit upon something in nature appealed to his taste. Some say that La Condamine first suggested that the pendulum with a frequency of one oscillation per second, being about three feet long, would make a convenient standard, though others attribute the idea to Christopher Wren. In any case, Turgot commissioned an astronomer, Messier, to determine its precise length at the 45th parallel.[57] His preference opened a disagreement with traditionalists, among them Tillet, who, when reform of weights and measures revived like other reforms early in the Revolution, thought it more practical to standardize conventional units than to define new ones.[58]

In high ministerial office Turgot found time to continue the practice he had begun in Limoges on a provincial scale of issuing directives and dispatching agents on technical matters and missions of many sorts: having

[55] Cahen (1904), 8. On Condorcet's service as adjunct to Fouchy, see Baker (1967b) and (1975), 35-47.

[56] Turgot, *Oeuvres 4*, 132-133; see also exchange of letters between Condorcet and Turgot in the summer of 1775, Henry (1883), nos. clxxxii-clxxxv, 232-236. For Turgot's relations with members of the Academy, see also Foncin (1877).

[57] Turgot, *Oeuvres 5*, 31-33; on La Condamine's proposal, see below, Chapter II, Section 3, n. 127 and on Wren, J. F. Scott, DSB *14*, 509-511.

[58] On the metric system, see Gillispie, "Laplace," DSB *15*, 333-336.

sea water desalinated aboard men-of-war in port at Lorient; underwriting botanical explorations in Peru; arranging for an ice breaker to be tested in winter navigation of the Seine and the Marne; developing processes for conversion of iron into steel; abating pollution of air and water caused by the retting of hemp;[59] initiating tea plantations and an agricultural school in Corsica. In 1774 A.-L. de Jussieu set out a Corsican pine grown from a cone imported by Turgot. It still flourishes in the Ecole de botanique of the Jardin des plantes. During Turgot's brief tenure of the maritime ministry, he acted on Condorcet's suggestion that a translation of Euler's treatises on naval architecture and artillery practice be put in hand in order to improve instruction at the service schools. It was characteristic that he should have further proposed that, since translation was being undertaken without the author's consent, it would be gracious to recompense Euler with what he would have earned had the original work been commissioned in France. He would charge the sum on the discretionary fund of his ministry.[60]

Both civil and military needs required that the technical grasp of engineering be strengthened and its reach extended, and what with hostility to the corvée and intimacy with the Trudaines, Turgot kept up something like a family connection with the Corps des ponts et chaussées even before becoming its director *ex officio* in his capacity of controller-general. During his ministry he enhanced the standing of its school by an edict according it the designation "Royal" like the technically more prestigious School of Military Engineering at Mézières, the famous Ecole royale du génie. The Ponts et chaussées neither achieved nor deserved a reputation for scientific eminence in the eighteenth century. It lacked both the discipline and the financial resources that the military regime entailed for Mézières. Nevertheless, the attainments and employments of its graduates mark a definite stage in the evolution of a trade, that of construction, into a profession, that of civil engineering. Its practitioners learned by formal schooling now instead of by apprenticeship and had their knowledge attested by an institution instead of their craft admitted by a master. Of that, more later.[61]

Canals appealed yet more urgently to Turgot's imagination, alert to the

[59] Turgot, *Oeuvres* 4, 88-89, 236-237, 240-241.

[60] *Ibid.*, 92-93. The work on naval architecture was the *Scientia navalis, sive tractatus de construendis ac dirigendis navibus*, 2 vols. (St. Petersburg, 1749) translated as *Théorie complète de la construction et de la maneouvre des vaisseaux* (1783), and that on artillery was a commentary by Euler on the work of the English expert, Benjamin Robins, *Neue Grundsätz der Artillerie, aus dem Englischen des Herrn Benjamin Robins übersetzt* (Berlin, 1745), which occupies vol. 14, ser. 2, of Euler's *Opera Omnia* (Leipzig, 1922). The original was entitled *New Principles of Gunnery*, 3 vols. (London, 1742), and the French translation, *Nouveaux principes d'artillerie* (Dijon, 1783).

[61] Below, Chapter VII, Section 1.

advantages that an adequate network of waterways complementing the highway system would bring to internal commerce. Hydrodynamics had become a highly developed branch of the science of rational mechanics following out of Daniel Bernoulli's treatise of the subject in 1738.[62] To minds that were overly impatient at the difference between problems of theory and execution and without experience of either, it seemed scandalous that this extensive literature should have found no application in actual waterworks. In 1775 Turgot appointed a commission to advise the government on the improvement of internal navigation. Its membership consisted of Condorcet, d'Alembert, and the abbé Charles Bossut, author of textbooks of mechanics and examiner of engineering cadets for Mézières. As usual the recourse was to education, and on the commission's recommendation a professorship of hydrodynamics was instituted and Bossut named to the chair. What with the play of faction and personality, the teaching had to be instituted in the Académie royale d' architecture, and though the course met with an uneven reception, its purpose was to impart the theoretical knowledge that would transform ditch-digging into hydraulic engineering.[63]

In the emergency of the cattle catastrophe, Turgot appealed directly to the Academy to address itself to the problem of these fearful epidemics, or epizootics (to use the contemporary and proper term) that periodically devastated livestock in France and throughout Europe in the eighteenth century. Whether the maladies themselves were more virulent than in previous centuries or the perception of them merely more acute probably cannot be known since much uncertainty attends the history of all disease, human and animal. Observation and record-keeping became systematic only in the twentieth century. Those who treated illness connected symptoms differently. Diseases themselves change over time, and so does the very conception of disease.[64]

None of these problems is to the present purpose, however, for even were it possible, it would be unnecessary to translate the *maladie sur les bestiaux* that swept the country in 1714 and 1754 as rinderpest, to distinguish the *fièvre charbonneuse* of 1757 from the *charbon symptomatique* of 1762 and decide which might have been anthrax, or to diagnose the *mal sous la langue* in the Limousin and Lyonnais in 1763 as hoof-and-mouth disease. It is enough to know a cattle plague for the disaster it surely was, the

[62] *Hydrodynamica, sive de viribus et motibus fluidorum commentarii* (Strasbourg, 1738). For the role of hydrodynamics in eighteenth-century mechanics, see Gillispie (1971a); Clifford A. Truesdell, *Rational Fluid Mechanics*, the Introduction to Leonhard Euler, *Opera Omnia*, 2nd ser., *12*, *13* (Zurich, 1954-1955).

[63] Hahn (1962) and (1964); see also the draft and text of Turgot's instruction to Bossut, Henry (1883), 237-240.

[64] An informative monograph is Dronne (1965).

severity depending on morbidity, duration, and extent, and to imagine the anxieties of proprietors and peasants. Even when no epizootic raged, these diseases were endemic. The dreaded listlessness, staggers, sores, fever, black carbuncle, diarrhea, bloody urine, or clotted stool might appear at any time. The village was then fortunate where there was a cool head and skilled hand in the person of the farrier or surgeon. The art of both depended on manual dexterity, though frequently country doctors in their closeness to the life of the localities would take an interest and a part. Often, however, recourse was to the magician, sorcerer or quack, or to the priest. Country people were ever quick to believe in spells cast by enemies.[65]

The historian might be tempted to attribute to Turgot's unlucky star, or to the incantations of the dévots, the coincidence of the most persistent of these recurrent plagues with the grain crisis during the nearly two years of his ministry. The virus (for that word was used) invaded the Southwest in May 1774. It may have come through the Basque country from Spain. It may have entered the port of Bayonne carried by a shipload of milch cows from Protestant Holland destined for Huguenot localities. In a short time it was raging through all of Aquitaine south of the Garonne. The following year it broke out in the North, in Normandy in the spring (where it was contained within a few villages) and then in the region of Calais in October, whence it spread throughout Flanders and Artois and down into Picardy.[66]

Faced with these threats, the government was not wholly without experience or personnel. The Académie royale de chirurgie owed official support for its chartering in 1748 to Trudaine's desire in the wake of the cattle plague of 1745 to enlarge the competence of surgeons in caring for farm animals.[67] More directly to the point, the state had patronized creation of the new vocation of veterinary medicine. In 1762 Claude Bourgelat opened the first veterinary school at Lyons. A horse lover and friend of d'Alembert, he had run an Académie royale d'equitation in that city. In *Eléments d'Hippiatrique*[68] he elaborated a scheme of veterinary medicine, and the school at Lyons received the designation "royale" when in its first year a delegation of seven students, sent to cope with an outbreak of

[65] For contemporary accounts, see the memoir on epizootics by Paul Bosc d'Antic printed in his *Oeuvres*, 2 vols. (1780) 2, 192-235; Jean-Jacques Paulet, *Recherches historiques et physiques sur les maladies épizootiques*, 2 vols. (1775); Félix Vicq d'Azyr, *La Médecine des bêtes à cornes*, 2 vols. (1781); and *Exposé des moyens curatifs et préservatifs qui peuvent être employés contre les maladies pestilentielles des bêtes à cornes* (1776), esp. pt. I, a discursive but vivid history and running account compiled under the pressure of events.

[66] Dronne (1965), 146-154, 198-203.

[67] Hours (1957), 64; Weulersse (1910) 2, 196; and below, Chapter III, Section 3. Gelfand (1973a) is an excellent thesis on the education of surgeons.

[68] Bourgelat, 3 vols. (Lyons, 1750-1753).

plague in Dauphiny, returned triumphant, having saved fifty-three out of the sixty-two head of cattle confided to their care while all the beasts stricken before their arrival died. Over five hundred veterinarians were trained there before the school disappeared amid the disasters that struck Lyons in the Revolution. In 1765 a second school opened in Paris, moving the next year to the chateau of Alfort on the eastern outskirts. What with larger funds and its proximity to the capital, it quickly became the more famous and drew more students. Bourgelat himself moved there, leaving the direction at Lyons to the abbé Rozier, later famous for his journal.[69]

Henri Bertin, the minister originally responsible for encouraging these developments, deserves more notice than he has been accorded in the histories.[70] He was one of those excellent administrators whose intelligence, probity and devotion to the public interest make a striking contrast with the fecklessness of privilege, politics, and sovereignty in the old regime. He tended to be overlooked even at the time, however. Bertin was a tiny man physically, and the department of state he headed was called familiarly "le petit Ministère." It had a jurisdiction over agriculture, mining, transportation, communications, manufacturing, and lotteries, except for what came under the Bureau of Commerce in the controller-general's department.[71] Somehow Louis XV and Madame de Pompadour, who were not always bad judges, had come to appreciate Bertin's qualities of reliability and good sense. As lieutenant-general of police in Paris, he sat alongside Vincent de Gournay and Trudaine in the Conseil de commerce from 1757 to 1759, and he followed Silhouette in the post of controller-general from 1759 to 1763. There at the top of the government the political winds blew too strongly for him in those last years of the Seven Years War, and on his resignation, Louis XV, wanting to continue him in the Conseil d'etat, instituted his little ministry and allocated to it the affairs that specially interested him. Bertin and it survived the change of reign, somewhat to the surprise of gossips. Perhaps Louis XVI was also sensible of the advantages of unassuming dependability. When the king dismissed Turgot, Bertin was the one who carried word to the fallen minister and acted as controller-general again pending the appointment of Bernard de Clugny. In July 1774 he had also filled in at Foreign Affairs during the absence of Vergennes. He resigned from government in 1780,

[69] E. Leclainche, "La Médecine vétérinaire," in Laignel-Lavastine (1949), *3*, 661-668; on Bourgelat, see especially Dronne (1965), 63-73; on the school at Lyons, Hours (1957); and on that at Alfort, Railliet and Moulé (1908). For Rozier's journal, see below, Chapter III, Section 1.

[70] On Bertin's career and ministry, see Dronne (1965), esp. 1-28; Bourde (1967), 1079-1289; Bloch (1930); Michel Antoine, "Le Secrétariat d'état de Bertin, 1763-1780" (1948), an unpublished thesis that I have not been able to consult, is said to be at the Ecole des Chartes.

[71] Dronne (1965), 21-22.

under pressure from Necker, and his ministry was abolished, its functions reverting to the controller-general's charge.[72]

Bertin had been intendant of Lyons from 1750 to 1759 and there had known Bourgelat well. In providing for the training of veterinarians, their intention was to constitute a cadre of qualified persons. The students at Alfort and Lyons were normally sent to study there by administrators in their regions, who had, however, no notion of then setting them up in practice, one vet to a village. Even had the scale been so ambitious, jealousy on the part of the farriers would have precluded such a program. Defining precisely what their status would be, and where careers would lie, was an unsolved problem, one bound to bedevil the creation of any new vocation.[73] As for the curriculum, Bourgelat was often criticized for its being too horsey in subject matter and too elementary in expectations. The latter stricture was probably unjust. Following Bourgelat's death in 1779 and Bertin's retirement in 1780, the school at Alfort came under the intendant of Paris, Bertier de Sauvigny. A friend of Condorcet, he wished the education to be fortified scientifically and created chairs of anatomy, chemistry, and natural history for three leading scientists, Vicq d'Azyr, Fourcroy, and Daubenton. Unfortunately, the new teaching was largely ineffective, since the typical student was less likely to be some village Newton or Descartes than he was to be the village blacksmith's boy, still unequal to a higher education.[74]

Not that Bertin was indifferent to science and progressive thought: on the contrary, he was an honorary member of the Academy of Science and a central figure in the network of societies of agriculture. He respected science without ideology, however, and promoted agriculture without physiocracy. He had little sympathy for the doctrine of laissez-faire, whether emanating from the school of Quesnay or from the economists. His was the outlook of the administrator concerned with actualities of subsistence and civic order, a condition to be watched over and maintained, even apprehensively. Science in such a view was nothing of a cause or force, but a body of techniques and information on which government might draw when, for example, it needed accurate surveys for a new cadastre or land registry, as indeed it did.[75]

He called on members of the Academy to advise his ministry, notably Duhamel du Monceau, the leading agronomist. In consultation with him, with progressive cultivators, with certain intendants in the provinces, with Bourgelat and the personnel of Alfort, Bertin's ministry developed measures to recommend to local authorities when cattle plague appeared.

[72] *Ibid.*, 18-19, 24.

[73] Hours (1957), 62-63.

[74] Dronne (1965), 73-79; Hours (1957), 68; Bourde (1967), 1220-1225.

[75] Dreux (1933).

In normal times the policy had three aspects. The first was circumspection. Intendants were to have cases dealt with quietly lest the fear of drastic measures prompt proprietors to hide sick cattle. The second was sanitation and hygiene. Tainted beasts were to be isolated, utensils and implements cleaned, and cadavers buried eight feet deep. The third was therapeutic. Purges, oral and nasal irrigations, bleeding, cauterization of blisters, herbal decoctions of many kinds, mustard plasters laced with cantharides—such was the sequence of remedies visited upon those suffering cows, all the while gasping for death amid fumes of camphor, assa foetida, nitre, sulfur, and spirit of ammonia.[76]

No one supposed such measures to be more than palliative. When the intendant of Bayonne, d'Aisne, reported plague in May 1774, Bertin sent a student from Alfort, one Guyot, to investigate. D'Aisne had been a little tardy, and once the full account came back, Bertin concluded immediately that the most rigorous precautions must be invoked. It was agreed within the ministry that slaughter was the only way to arrest such a fully epizootic outbreak as this one now appeared to be. Not only infected cattle but every herd that had been exposed to a single case must be sacrificed, and a military *cordon sanitaire* thrown around the affected region to prevent all passage of cattle, hides, or salt meat. His ministry lacked authority to impose such draconic regulations or to get out the troops. Turgot had already learned of the threat, however, first through a request from d'Aisne for funds for Guyot's mission, and second through the draft of a decree directing that the slaughter commence. It had been drawn by Esmangart, intendant of Bordeaux, and submitted for his approval on 5 November 1774.[77]

Any chief minister newly arrived in office and taken for a reformer would have been reluctant to impose the wholesale slaughtering of cattle on distant provinces by martial law. It was characteristic of Turgot, however, that he should have responded by taking the whole matter out of Bertin's hands and into his own; that he should have rebuked the several parlements that registered the regulations of provincial estates for trespassing on administration; that he should have assumed that existing practice was routine and uninformed and have turned, therefore, to the Academy of Science to discover the basis for a rational policy; that having started in the liberal mood of countermanding arbitrary measures, he ended by ordering slaughter far more widely than Bertin had imagined doing; and that all this should have issued in the assumption of responsibility for public health on the part of the medical profession under authority from the state, long after he was gone from office.

Turgot consulted Condorcet before requesting the Academy to name a

[76] Dronne (1965), 109-118.

[77] *Ibid.*, 149-150, 158-162, 168.

commission.[78] From his own entourage, he proposed that Malesherbes, Trudaine, and Condorcet himself be of the number, and from the scientific ranks of the Academy he requested Duhamel, Lenoir, and Tenon, together with any others qualified. The Academy added Daubenton, its senior naturalist, himself a sheep-breeder, and reached outside its ranks to enlist Philibert Chabert, director of the Ecole d'Alfort, a veterinarian not to be confused with the astronomer J.-B. Chabert de Cogolin. It had certainly been decided ahead of time between Condorcet and Turgot that the chairman would be Félix Vicq d'Azyr, physician and anatomist and so much the most brilliant of the younger people in the medical sciences that he bade fair to become to them what Lavoisier soon was to chemistry and Laplace to the exact sciences. Meanwhile, the cattle plague commission gave him his chance to serve his country and make a name outside the closed and sometimes vicious circle of the Paris medical community. Turgot had stipulated that at least two commissioners, a doctor and a man of science, must repair to the localities where plague was raging to conduct investigations on the spot. The Academy, wrote Vicq d'Azyr, understood all along that the assignment was to be his, and "did me the honor of giving me both jobs." In combining them, he went on to observe, "I have two objects to fulfill. The first concerns public welfare; the second is a matter of pure curiosity."[79]

Arresting the plague was more urgent if not more important than understanding it better, and application preceded science. Armed with plenary powers, Vicq d'Azyr left Paris for Bordeaux on 2 December 1774. Already by the 8th he had ready the main line of policy—slaughter, indemnification, and disinfection—and issued his instructions in a booklet.[80] All regions were to be classified in one of three categories. First, in those that were free of plague, a regime of preventive sanitation very like Bourgelat's was enjoined. Second, in those where scattered cases had begun to appear, no time was to be wasted in trying for cures. Sick animals were to be dispatched instantly and the herd quarantined. Third, those where plague was widespread were to be isolated by *cordon sanitaire*. Inside, where contagion was already abroad, it was permissible to try to cure sick animals, though Vicq d'Azyr preferred that they be killed. Outside, every new case must at once be slaughtered. In all instances, owners of the animals thus sacrificed for the general good were to be indemnified by the state at a fair proportion of their value. Cadavers were to be incinerated or

[78] Turgot to Condorcet, 18 November 1774, Henry (1883), 208-209; Dronne (1965), 168.

[79] Quoted in Dronne (1965), 168, 170.

[80] *Observations sur les moyens que l'on peut employer pour préserver les animaux sains de la contagion et en arrêter les progrès* (Bordeaux, 1774); for his credentials, see Turgot to Esmangart, intendant of Bordeaux, 29 November 1774, Turgot, *Oeuvres 4*, 250.

buried in quicklime ten feet under the ground. Stables and barnyards were then to be disinfected drastically with fire, lime, and sulfuric acid fumes, and the premises left empty for months. The touch of this new generation of experts appears in Vicq d'Azyr's recognition that the country people were not to be trusted to carry out a thorough decontamination. The authorities must establish an inspection.[81]

After seeing these measures instituted in the generality of Bordeaux, Vicq d'Azyr turned to research. His instructions had been "to determine by all possible and chemical methods whether it is possible to purify and restore the putrid air that carries the contagion from place to place."[82] Turgot was if anything more eager for a solution than Vicq d'Azyr himself, or perhaps more optimistic about the possibility of one. That was, Turgot reminded Vicq d'Azyr on 28 December, "one of the principal objects of his mission and the most important," and urged him to multiply his experiments.[83] None of them affected policy at the time, however, and we will save an account of them and of his findings for the discussion of Vicq d'Azyr's further medical and scientific career.[84]

Policy became increasingly severe. Turgot's initial and liberal thought, based on Austrian practice, had been to slaughter the first eight or ten beasts that fell ill in a given parish, and if that failed to stop the disease, to leave the proprietors the hope that some proportion would nevertheless survive.[85] Those half-measures were never published, and instead Turgot by a decree of the Council of State of 30 January adopted Vicq d'Azyr's thorough-going strategy, which was to be executed by the army "because of its effectiveness and disinterestedness."[86] First an entire region would be cordonned off. Next detachments accompanied by a veterinarian, farrier, or surgeon would inspect every parish. Where they found plague, the commander would order slaughter or isolation according to its extent and enforce the sanitary regulations.

[81] Dronne (1965), 171.

[82] Turgot to Esmangart, 29 November 1774, *Oeuvres 4*, 251.

[83] Turgot to Vicq d'Azyr, 28 December 1774, quoted in Dronne (1965), 171, from correspondence AN, F[12].151.

[84] Below, Chapter III, Section 2; Vicq d'Azyr published his matured conclusions six years later, *La Médecine des bêtes à cornes*, 2 vols. (1781).

[85] Draft of a letter to the intendants of Languedoc and Montauban, December 1774, Turgot, *Oeuvres 4*, 251-252; Turgot to archbishop of Narbonne, 1 January 1775, 5, 34-35.

[86] Vicq d'Azyr, *Instruction sur la manière de désinfecter une paroisse* (1775), 3, dated 8 January; Dronne (1965), 178. For detail of the regulations, see Vicq d'Azyr's "Mémoire instructif sur l'exécution du plan adopté par le Roi pour parvenir à détruire entièrement la maladie . . ." printed in Turgot, *Oeuvres 5*, 50-58; and also in the contemporary compilation in which Vicq d'Azyr collected the recommendations, reports and regulations issued under the pressure of events, *Exposé des moyens curatifs et préservatifs qui peuvent être employés contre les maladies pestilentielles des bêtes à cornes* (1776), 630-640.

These measures succeeded only in Navarre and the environs of Bayonne. As new outbreaks continued elsewhere throughout the summer and autumn, Turgot and Vicq d'Azyr agreed on tightening the vise and sealing off all the country south of the Garonne, where the epizootic had evidently to run its course. Outside that zone, all sick animals were to be slaughtered and every herd that had been exposed to even a single case was to be driven across the river to take its chances in the zone already contaminated. When the epizootic relentlessly spread further into Languedoc and appeared also in the North, extreme preventives were ordered everywhere. As soon as one beast fell ill, the entire herd was to be slaughtered and buried in one of Vicq d'Azyr's deep ditches. Proprietors were to receive one third of the normal value. Lest they think the indemnity niggardly, they were told that it was based on the traditional expectation of saving one animal in three, which the commission had proved to be false. In fact, not one in twenty or perhaps fifty would survive, and hence owners should take the compensation to be an act of royal grace and favor.[87]

This final and ultra-caustic phase of the Turgot-Vicq d'Azyr epizootic regulation became the basis of the code of rural sanitation applied in France from Napoleonic times into the twentieth century.[88] It has worked, and it would certainly have stopped the epizootic of 1774-1776 much sooner except that the governmental machinery was incapable of enforcing it. The army disliked its part and cooperated half-heartedly. Local populations disliked having troops sent among them for this or any reason, and now had to fear for their wives and daughters as well as their livestock. The greatest obstacle was the resentment of every proprietor, unwilling to abandon what he desperately hoped to save of his herd in the larger interest of eradicating plague. This resistance was fortified by the widely recognized development of a measure of immunity among surviving cattle in the regions worst infested.[89] The owner would drive away the snooping veterinary student if he could, and hide his blistered cows in a distant wood if he could not. Cupidity, of course, but not merely that—the arrival of agents of the royal government for whatever purpose never augured well.

It is an almost invariable rule of any program of reform in the eighteenth century that the more enlightened its agents, the greater the irritation they developed with what they would soon begin calling the prejudices and indiscipline of the population. Turgot thought that such

[87] Vicq d'Azyr, "Mémoire instructif," in Turgot, *Oeuvres* 5, 54-55; for the further development of policy, see "Second mémoire instructif sur l'exécution du plan adopté par le roi" (28 November 1775), in *Exposé des moyens curatifs* . . . , 640-666. Cf. Dronne (1965), 178-188.

[88] Dronne (1965), 204-205, 210-211.

[89] *Ibid.*, 183-186; Vicq d'Azyr, "Mémoire instructif," in Turgot, *Oeuvres* 5, 51-52.

resistance, distrust, and devious obstinacy would prove less serious in the North, where "your peasants are not so barbarous and difficult to manage as those of Guyenne and Gascony."[90] Maybe so. In Artois the authentic voice of the sturdy French farmer did speak out openly in the substantial person of one Baclu, *echevin* or alderman in the village of Audrwicq. He delivered an ultimatum to Turgot's agent, a surgeon called Le Breton who arrived with instructions to apply Vicq d'Azyr's final solution. Let him just try it—thus Baclu—and the church bell would be rung, the village called to arms, Le Breton knocked over the head instead of the cows, and a hole dug big enough for him and all the cattle he wanted to fell.[91]

In these tense affairs, Vicq d'Azyr was more than an expert consultant to the government. He was its agent, and only formally the Academy's, a Commissaire du roi from early 1775, actually directing operations in the field, receiving instructions directly from Turgot, and reporting back to Paris.[92] Turgot's orders, he wrote during one tour of inspection,

> are fully executed in Flanders, but not at all in Artois. . . . The greatest anarchy reigns everywhere. The inhabitants don't know whether they are to take orders on the epizootic from the Estates or from the Intendants. . . . After the slaughtering, the Estates closed their financial exchange. The countryman believes himself deprived of support and abandoned to the terrors of a massacre that is unjust because the Estates never ordered it, and that he refuses to allow as long as they do not order it. . . . At the least word they arm themselves with pitchforks. . . . There is no inspection of stables, the dogs are not shut up, the ditches are dug in the middle of the arable, there are no soldiers, no militia and not even a single commissioner from the Estates. . . . In a word no one knows who is in charge, and nobody obeys.[93]

Though there is reason to think this report exaggerated, the frustrations were real. Simultaneously with such experiences, Vicq d'Azyr came to feel the need of a more regular basis for professional authority and readier channels of communication between that authority in Paris and the officials and medical men in their local situations throughout the provinces. It was at his suggestion that Turgot, in what proved the last days of his ministry, created by a decree of the Council of State (29 April 1776) a new commission to be headed by François de Lassone, first physician-elect to the king and hence the most highly placed of medical courtiers.[94]

[90] Turgot to Le Peletier, intendant of Soissons, 17 December 1775, *Oeuvres* 5, 80.
[91] Dronne (1965), 200-201.
[92] See their correspondence in vols. 4 and 5 of Turgot, *Oeuvres*.
[93] Dronne (1965), 201, citing a letter, AN, H[1]. 46.
[94] Enemies on the Faculty of Medicine accused him of having planted the idea out of

Six of the leading doctors of Paris were to constitute the commission, which was to be a serious working body assembling information, not just on epizootics now, but also and even mainly on epidemic diseases afflicting man. The commissioners were to meet weekly in company with an inspector detailed from the controller-general's office to attend to correspondence with the provinces on public health, animal and human.

It was almost surely Vicq d'Azyr's intention from the beginning, or very near it, to prolong this body into the permanence of an academy of medicine. Fortunately for that design, Turgot's immediate successor as controller-general was none other than the very intendant of Bordeaux, Bernard de Clugny, under whose eyes Vicq d'Azyr had battled the outbreaks of plague in the previous year.[95] In the summer of 1776, the new commission began calling itself the "Société et correspondance royale de médecine." A the outset Vicq d'Azyr designated himself "Premier correspondant."[96] His post was then converted to the academic one of permanent secretary in August 1778, when letters patent formally incorporated the Société royale de médecine. It thereupon assumed with the blessings of the state a responsibility for public health on behalf of a medical profession that had had no say in the matter. Its normal spokesmen, the Faculty of Medicine of the University of Paris, had never been consulted, and had indeed been deceived about these purposes. That, however, is also a subject best reserved for later, one to be discussed along with Vicq d'Azyr's further career, which took a turn different from what might have been predicted of his scientific talent.[97]

5. CONSTITUTIONAL REFORM

Exploring now certain interests more programmatic in nature than the foregoing episodes will lead more deeply into a structure of relations between science and politics, one nonetheless revealing for having remained largely latent until the Revolution. Thereupon an interplay between science and educational innovation appears and reappears, for that became one of the two main sets of issues in which those relations developed. The other, warfare and weaponry, we reserve for later discussion. The pattern

---

personal ambition, and, whatever the motives, the probability of his initiative is strengthened to virtual certainty by his pupil and admirer, J.-L. Moreau de la Sarthe, who collected and edited his *Oeuvres*, 6 vols. (1805) and prefaced them with a "Discours sur la vie et les ouvrages de Vicq d'Azyr" *1*, 1-88; see esp. 18-19; and specifically the Réponse of Saint-Lambert (*1*, 45) to Vicq d'Azyr's inaugural at the Académie française in 1788, an "Eloge de Buffon."

[95] Vicq d'Azyr, *Exposé des moyens curatifs* . . . , xiv.

[96] *Ibid.*, xi.

[97] Below, Chapter III, Section 2.

may already be discerned in Turgot's scheme for a thorough constitutional reformation of the monarchy. That scheme had no place among the actual political intentions of his ministry. He unfolded it before neither his colleagues nor the king. Even had he remained longer in office, he would not soon have advanced it, since its adoption would have attenuated the royal authority. Were he a private citizen, he might welcome such an eventuality, but a minister of the crown could scarcely propose it with propriety, and certainly not until the king was older and more experienced.[98] Our knowledge of it must remain slightly tendentious, therefore. It derives largely from a memorandum by the hand of Dupont de Nemours, about which Turgot was displeased.[99] The draft, he complained, contained altogether too much detail for a mere sketch, which was all that his young associate could properly have attempted since he could have no notion of the host of ideas that Turgot had been meditating for at least fifteen years. The whole thing would have to be redone, he said irritably.[100]

It never was, but we may take the memorandum as an approximation, however imperfect, to what Turgot had in mind, reflecting that there was nowhere else that Dupont could have got it and that it fits the pattern of his thought. Essentially, Dupont's version of Turgot's model for a constitution called for reordering the French state on a representative basis. In each village the proprietors would elect an assembly. Village assemblies would then be grouped together for the election of cantonal or district assemblies; cantonal assemblies would elect provincial assemblies; and finally a national assembly would be elected by those of the provinces. Property qualifications for eligibility increased substantially at each level. At no stage, were the estates to be the basis of organization. Turgot intended his "municipal" scheme to supplant ordering by estates, which would be eliminated for the obstructive and regressive bodies they were.

The Anglo-Saxon reader of French history must beware of concluding that a plan for representation would necessarily have entailed representative government. For legislation would have formed no part of the new functions. Turgot disbelieved in the capacity of society to give itself laws.

[98] Véri's account of a conversation with Turgot in July or August 1778, *Journal 2*, 147.

[99] "Mémoire sur les municipalités," Turgot, *Oeuvres 4*, 568-621. On the tortuous route by which the piece became known, see Dakin (1939), 343, n. 4; Turgot, *Oeuvres 4*, 568-574. The chicanery of Mirabeau led to the publication of a garbled version as Turgot's own in *Oeuvres posthumes de Turgot* (1787). Mirabeau had obtained a copy of the text, passed it to Calonne as his own composition and, when that failed to win him credit, conveyed it for a price to the publisher of the compilation just mentioned with certain additions of his own fabrication. This piece is discussed at length by Cavanaugh (1969), though I do not find his discussion of Turgot's motives or political theories convincing. Baker (1978) also treats the work in an excellent essay published when the present book was already in press. See also Condorcet, *Life of Turgot* (London, 1787), 193-237.

[100] Turgot to Dupont de Nemours, 23 September 1775, *Oeuvres 4*, 676.

That responsibility would remain in the crown. Ever the reformer, Turgot was thinking of modifying its power rather than its prerogative. The function of his new assemblies would be administrative and fiscal. They would operate much as the controller-general already did in relation to the several intendants and as they did in relation to their subordinates. Each elective body would apportion to its constituent assemblies at the next lower level their share of the revenues to be raised annually within its region until down in the village an equitable liability for every taxpayer would be assured by the knowledge each representative would have of the relative worth of his neighbors. On economic grounds it would, of course, always be to a proprietor's interest to conceal or undervalue his wealth. That tendency was to be counteracted by weighting representation proportionally to property. Thus would an equilibrium between the opposing forces of economy and influence in all localities generate a land registry fitting the truth more accurately than the most extensive survey or inquisitive assessment could ever do.

Suffice it to observe of the constitutional importance of this plan that the influence of the thinking behind it appears in Revolutionary provisions for the vote (even while any implication that a people must look to experts rather than itself for laws was passionately rejected just one stage later than was the traditional notion that it should look to prescription and the past). Its relevance to the history of the franchise gives it an interest similar to that held by Turgot's actual measures in their various contexts. In this instance, too, the last years of the old regime show an underlying continuity instead of merely a hiatus pending revolution. In 1778 Necker instituted an experimental local assembly for fiscal purposes in the single province of Berry. In 1787 Calonne proposed to the Assembly of Notables a project similar in many respects to Turgot's plan.[101]

More to the present point, any constitution that would represent the nation in ordering its affairs, whether governmentally or administratively, would be feasible only on the presupposition of an educated population. Such did not exist, and the Dupont de Nemours memoir went on to design, in effect, a Ministry of Education. Like other budding departments of state, the projected Conseil de l'instruction nationale would have emanated from the Council. It would set a national educational policy. Under its charge would be gathered all educational, scientific, and literary institutions—universities, colleges, schools, academies, whatever. It would have the authority to bend the *corps littéraires* to useful and civic purposes. By that phrase Dupont de Nemours had in mind mainly the academies, both provincial and Parisian, and notably the Académie française, the Académie royale des sciences, and the Académie royale des inscriptions et

[101] Goodwin (1946); Dakin (1939), 278, 341 n. 16, 343 n. 9.

belles-lettres. Of them all, he expressed the criticism agreed upon by virtually the entire intelligentsia, whether of their membership or not: "Their efforts at present tend only to the education of scientists, poets, and men of intellect and taste; those who cannot aspire to such a goal remain neglected and amount to nothing."[102] Under the new administration, by contrast, access to scientific and literary distinction would develop out of a national educational system whose purpose would be civic and moral, the tone austere and virtuous, and the incidence uniform and universal. Those suited to advanced studies would then be personally prepared to exhibit in their work a more masculine and consequential character than in the existing play of chance and favoritism. The conception is similar to that of the plan that in 1792 Condorcet put before the National Legislative Assembly and that must probably have evolved out of the discussions amid Turgot's circle.[103] Indeed, the Dupont de Nemours memoir may be read as the earliest recorded draft working these ideas up into a system.

Condorcet for his part was then addressing himself to a set of interests, the promotion of the calculus of probabilities, that he related to the ordering of electoral assemblies in a manner even more revealing of the deeper persuasions informing this entire system of reformist thought. He did not publish his main work upon the subject until 1785: *Essai sur l'application de l'analyse à la probabilité des décisions rendues à la pluralité des voix*. Turgot had died four years before. The epigraph at the beginning of the present chapter is taken from the preface, which tells how Condorcet had undertaken the work for him, and recalls the great man's faith in the prospect for investing political and social sciences with a certainty approaching that of mathematics.[104]

Such was, indeed, Turgot's belief. Although lacking in mathematical facility, himself, he kept informed in science, in large part through his correspondence with Condorcet, especially during the years at Limoges.[105] From the post of intendant, Turgot, administrator and man of affairs, would write of natural phenomena, perhaps by way of recreation: meteorology, the aurora borealis, crystallography. In one astonishing letter of 16 August 1771 Turgot suggested in a ruminative way that the increase in weight exhibited by metals upon calcination (oxidation) might most probably be the consequence of their combination with something

[102] Turgot, *Oeuvres* 4, 579.

[103] *Procès-verbaux du Comité d'instruction publique de l'Assemblée législative*, ed. James Guillaume (1889), 188-246.

[104] Above, 3.

[105] The Condorcet papers are conserved in the Bibliothèque de l'Institut de France, MSS 855. From among them, Charles Henry selected and published the letters he exchanged with Turgot between 1770 and 1779, largely on scientific subjects (1883).

atmospheric. The remark is nothing short of tantalizing to the historian of chemistry in its offhand anticipation of the famous hypothesis with which a year later Lavoisier, in all the urgency of a sealed note confided to the Academy establishing priority, ambitiously inaugurated his research into combustion and with it the Chemical Revolution. But Turgot's chemical letter is interesting in its own right. It exhibits his familiarity with recent experimental work and with affinity theory and also with the classic writings of Boerhaave and Hales.[106] All the while, Condorcet, rising young savant about town, wrote (much more frequently) from Paris the news of politics and letters: the health of Mlle de l'Espinasse, the exile and maneuvers of the parlement, cabals and rumors of cabals among writers or ministers or both. It would appear, each writing of the other's domain, that Condorcet saw in Turgot a means of access to public affairs, and Turgot saw in Condorcet a man Friday for his own interests in the world of science. Turgot's was the dominant influence, bordering on patronage. Throughout, Condorcet's attitude was deferential and his vein a continuing appeal for counsel and direction, well before Turgot became minister.

A problem central to the present history is to know where to look for the inwardness of the career of Marie-Jean-Antoine-Nicolas Caritat, marquis de Condorcet, and what to make of his reputation. The lineaments are well known. Condorcet was an aristocrat by birth, of a swashbuckling family without much fortune, originally of Dauphiny. A certain mathematical aptitude revealed itself in his education. By virtue of d'Alembert's patronage Condorcet was elected to the Academy at the age of twenty-seven. He met Turgot in the salons. He was literate. He was well born. He was possessed of mathematical knowledge. It was a victory for d'Alembert and the progressive elements over Borda and Buffon when they were able to choose him in 1773 to fulfill the functions of Grandjean de Fouchy, the permanent secretary of the Academy. In that capacity he made himself spokesman for the post-Encyclopedic reformers. Normally treated by historians as the Girondist *savant* in the Revolution and last of the philosophes, he might better be taken for an intermediary figure between the encyclopedists and positivism, between the psychology of Condillac and that of the idéologues. He completed the only piece of his writ-

[106] Henry (1883), 59-63. It is not beyond the bounds of possibility that Lavoisier might have learned of this letter and had his views suggested to him by it. The personal circumstances would have admitted of that. It would not have been out of character. There is in the sealed note and in the earliest form of Lavoisier's theory of combustion the same confusion between the "fixation" of air as carbon dioxide and its "combination" in oxidation, though no one could make that distinction prior to Priestley's isolation of the gas he called dephlogisticated air—i.e., oxygen—in 1774. On the beginning of Lavoisier's researches, see Guerlac (1961b), and on their continuation, Daumas (1955).

ing still read, the *Esquisse d'un tableau historique des progrès de l'esprit humain*,[107] an optimistic treatise on the imminent perfectibility of man and society, in hiding from the guillotine. It is said that he committed suicide to escape his execution. Except for the irony of the contrast between the message and circumstances of his book, it is doubtful whether his intellectual stature or his political talents would have won him the historical standing of a leading revolutionary personage. His conduct at the end was moving and his devotion to his ideas no less than heroic. Nevertheless, a certain ambiguity remains concerning his personal character and quality. He was married at the age of forty-two to a notable charmer, Sophie de Grouchy, a girl of twenty-three. The same doubt persists about the vigor of his masculinity that attended the reputation of d'Alembert in his association with Mlle de l'Espinasse and, for that matter, Turgot in his friendship with Madame Helvétius. All three seem to have been given to more powerful emotions in their relations with intellectual programs and intellectuals than with women.

In these days of psychological inquisitiveness, the reader will wish to know that Condorcet's mother, having vowed him to the Virgin from infancy on the death of his father, dressed him in girl's clothes until his ninth year. She was a pious bourgeoise who had already been once widowed when Condorcet's father, a cadet of the family and a soldier of fortune, married her, probably for her dowry. Condorcet's uncle was bishop successively of Auxerre and Lisieux, where he took the stern stand on church discipline befitting a blooded ancestry. He had charge of the education of his gauche and disappointing nephew and put him into a school kept by the Jesuits at Rheims and later into the collège de Navarre in Paris. There mathematics was his refuge from the things he was supposed to be learning, the scion on whom rested the hopes of a noble family, and who never acquired a modicum of grace. As a young man biting his nails about Paris, he felt at ease only in the company of the elder patrons he sought out—d'Alembert, Helvétius, Quesnay, Turgot.[108]

[107] (Paris, 1795).

[108] The biography by François Arago that introduces volume 1 of the twelve-volume *Oeuvres de Condorcet* (1847-1849) was prepared for delivery before the Academy in 1841. Its author, who had not known Condorcet, was then permanent secretary. The edition was undertaken jointly by Arago and Condorcet's son-in-law, General A. Condorcet-O'Connor. Motivated by family piety rather than scholarship, it is by no means complete, omitting all of Condorcet's mathematical writings and garbling such of the correspondence as the editors chose to print. On the state in which the Condorcet papers were left, see Cahen (1904), the first scholarly biography. For more recent studies, see Schapiro (1934) and Bouissounouse (1962). Baker (1967a) has written on the interrelations of Condorcet's scientific and political interests, and has now completed a full-scale study of his thought and its place in intellectual history (Baker, 1975). It is only fair to Condorcet to acknowledge that Baker, who has studied his work more exhaustively than anyone else, places a higher value on its intrinsic merit than I have ever been able to do. One of Baker's notes,

An interesting monograph by Gaston Granger discusses Condorcet's "social mathematics"—mainly the probabilistic analysis of voting procedures—and takes it to have been an early, and most likely the earliest recognizable, instance of the use that modern economic and sociological analysis makes of quantifiable models.[109] The argument could be accepted fully only by those who agree that analogy in the absence of filiation is a convincing mode of demonstration in intellectual history. Still, it would be too pat a solution simply to write off Condorcet along with Fontenelle and Bailly, making him into an exponent of science rather than a participant in it. The temptation to do so is undeniable. He was the only permanent secretary whose éloges can be compared to Fontenelle's in the historical literature of science,[110] whereas the later literature of science itself and of mathematics is virtually empty of references to his contributions. Yet, an equitable judgment of Condorcet must do justice to his reputation in his own time, which was not simply that of philosophe but also of savant like his mentor d'Alembert, at once intellectual and mathematician. Perhaps those two are the only ones of whom that may be said, although the importance of Condorcet's contribution to mathematics cannot be compared to that of d'Alembert in mechanics. Moreover, Granger is accurate in seeing him to have been an encyclopedist in temperament, a man whose deepest and most natural interests bore upon the social process and the role of science in that rather than upon problems of science itself.[111]

Nevertheless, Condorcet began with mathematics. He first made himself known in his capacity of *géomètre*. Two sorts of problems then attracted his interest: techniques of integrating differential equations and methods for finding approximate solutions by the expansion of functions into infinite series.[112] The analytic formulation of problems of rational mechanics

---

expressing puzzlement at my skepticism, suddenly made me realize the locus of the ambiguity one feels. It is that Condorcet made his career in the two worlds of science and politics, claiming importance in each on the basis of his promise or importance in the other, while exhibiting real effectiveness in neither.

[109] Granger (1956).

[110] Among his credentials in seeking election was a booklet in the style of Fontenelle, *Eloges des académiciens de l'Académie Royale des Sciences morts depuis 1666, jusqu'en 1699* (Paris, 1773), that he composed on Roberval, Huyghens, Mariotte, Roemer, and other, lesser known worthies of the early Academy, thus exhibiting that he could gracefully discharge this responsibility of a permanent secretary, which the current secretary, Grandjean de Fouchy, was finding uncongenial.

[111] Granger (1956); see the appendix.

[112] Readers interested in judging Condorcet's mathematics for themselves may consult his maiden work, *Traité du calcul intégral* (1765); a series of three memoirs on analysis published in MARS (1770/73), "Sur les équations aux différences finies," 108-136; "Sur les équations aux différences partielles," 151-178; and "Sur les équations différentielles," 191-231; and a memoir on approximation, "Réflexions sur les méthodes d'approximation connues jusqu'ici pour les équations différentielles," MARS (1771/74), 281-306.

had been the signal achievement of the generation led by d'Alembert, Euler, and Daniel Bernoulli in the mid-century, and it was left to their successors to devise means for manipulating and solving the equations. Mathematicians who have studied Condorcet's memoirs on analysis have criticized them for obscurity, abstractness, and prolixity. He employed a notation both unconventional and inconstant. He yearned for generality, wishing in the spirit of the times to classify and to systematize the problems of integration and approximation. Since he had no depth, his taste for the abstract obscured rather than illuminated his contributions. He seldom illustrated them in examples, and d'Alembert deplored his style. "I could wish," he wrote to Lagrange, still in Berlin, "that our friend Condorcet, who is ingenious and clever, worked in a different way. I have told him that several times. But evidently it's in the nature of his mind to work in that style: he must be let be."[113]

For whatever he was in the round, it cannot be denied that inside Condorcet there was a mathematician. Diffuse and misuse his talent he may have done, but it was known and recognized in his own generation. Laplace, never one for indulgence in these or any matters other than political, cited Condorcet several times on purely mathematical points. Laplace complimented him, for example, on his work on the integration of differential equations with finite differences.[114] Although the sincerity of the remark might be suspect, since it was written in 1773 when Laplace was still seeking election to the Academy, he again observed in 1782, by which time he was fully established, that his own mode of employing discontinuous functions in the resolution of partial differential equations confirmed what Condorcet had found by another method.[115]

Here is the place, therefore, and this the context in which to introduce the individual whose active career spanned the half-century or more in which French science was pre-eminent, and who in the fullness of that time became its law-giver, for of his competence in exact subjects no one ever dared doubt—Pierre-Simon de Laplace.[116] Laplace was born on 23

[113] Quoted in Granger (1956), 58, from *Oeuvres de Lagrange* *13*, 232.

[114] "Recherches: 1° sur l'intégration des équations différentielles aux différences finies, & sur leur usage dans la théorie des hasards; 2° sur le principe de la gravitation universelle, & sur les inégalités séculaires des planètes qui en dépendent," SE 7 (1776), 37-232. The reference to the treatise and analytic memoirs of Condorcet cited in note 112 above occurs on pp. 38-39, and later in the discussion of problems of universal gravitation, he describes (p. 201) as a "fort beau mémoire" Condorcet's paper on methods of approximation (MARS [1771/74], 281-306).

[115] "Mémoire sur les suites," MARS (1779/82), 207-309, the reference (210-211) being to Condorcet, "Sur la détermination des fonctions arbitraires qui entrent dans les intégrales des équations aux différences partielles," MARS (1771/74), 49-74.

[116] For a detailed and technical biography, see C. C. Gillispie, Robert Fox, and Ivor Grattan-Guinness, "Laplace," DSB *15*, 273-403.

March 1749 in Beaumont-en-Auge, nowadays in the department of Calvados. His father, Pierre de Laplace, was a prosperous landowner, a dealer in cider, and a syndic of the town. The family of his mother, born Marie-Anne Sochon, was comparably well off. They intended the boy for the Church and enrolled him in the excellent *collège* that had been opened by the Benedectines in Beaumont early in the eighteenth century. It prepared lads for the army, the magistracy, or the cloth. Laplace had his education in their school from his seventh through his sixteenth years, leaving with his teachers there and later at the University of Caen the recollection of a penetrating intelligence and an extraordinary memory. His higher education at Caen in the Collège des arts, like that of Turgot at the Sorbonne, is an instance that the eighteenth-century university did not invariably extinguish mind and talent. Two of the dons, Christophe Gadbled and Pierre Le Canu, professed natural philosophy and mathematics. With their encouragement, Laplace recognized his ability and taste for mathematical investigation and abandoned all thought of orders. Le Canu furnished him with a letter to d'Alembert, and in 1768 a nineteen-year-old Laplace arrived in Paris to make his scientific fortune.[117]

Too many hopeful country boys were thus addressed to d'Alembert. His first reception of Laplace was brusque. Accounts differ about precisely what the task was, whether solution of a problem or mastery of a text, that Laplace handily discharged to win the great man's esteem. "Monsieur," d'Alembert is said to have written a few days after the initial audience, "you see that I attach little importance to recommendations. You didn't need one. You have made yourself known even better, and for me that suffices. My support is your due."[118] Good as his word, d'Alembert thereupon arranged that Laplace be appointed professor at the Ecole royale militaire. There in that triumph of architectural proportion overlooking the Champ de Mars, Laplace began his career teaching elementary mathematics to indifferent cadets and setting them examinations.

Approach to the Academy followed, in which an urgency beyond what was usual appeared to possess Laplace. Unfortunately, we shall never see very fully into his private thoughts, for all surviving correspondence was burned in 1925 in a fire that destroyed the chateau of Mailloc belonging to his great-great grandson, and indications can only be pieced together from the writings of contemporaries and later persons who had seen the documents.[119] Certainly, however, he was pouring out mathematics. Never, observed Condorcet in a prefatory note to the first two papers that the Academy published in the *Savants étrangers*, had that body "yet seen so

[117] Simon (1929) and (1936).

[118] Fourier, *Eloge de Laplace* (1829), lxxxiii.

[119] The comte de Colbert-Laplace recounted what he could recall from those papers and from family tradition in a letter dated 16 February 1929 to Karl Pearson; see Pearson (1929), 203-204.

young a person present in so brief a time so many important memoirs, and on such diverse and difficult matters."[120] Laplace had then just been elected (on 31 March 1773) in the section of mechanics. In May 1771 the Academy had preferred Vandermonde's candidacy and in March 1772 that of Cousin, professor at the Collège royale, now the Collège de France. Both were men of parts scientifically and his elders by many years, and to have to wait his turn was normal for a young savant, no matter how able and rightfully ambitious.

Evidently Laplace chafed. Following the second rebuff, d'Alembert wrote Lagrange, then director of the mathematical section of the Academy of Berlin, to bespeak his interest in finding Laplace a place in the Prussian capital. Lagrange was characteristically disinclined to risk his credit with Frederick II by asking so much as a favor. Nothing came of that démarche, and Laplace was not lost even temporarily to Paris.[121] Hints survive that the domineering quality evident in his maturity was already resented in these early thrusts of his ambition. Apparently Condorcet had complained in a letter to Lagrange, who answered "I am a little surprised by what you tell me of M. de la Place. It is often a fault in young men, or so it seems to me, to puff themselves up with their first success, but later their presumptuousness diminishes at the rate their science grows."[122]

Laplace's strength and agility in mathematics may be compared in their virtuosity to the physical coordination and power of a young athlete enabling him to excel in any sport he pleases on any field where it may be played. He early saw that the fields where winning fame would count were celestial mechanics and the calculus of probability. To know all that guided him in the choice would be to understand the whole environment of intellectual opportunity wherein he operated, appreciating both the appeal of Newton to its scientific and the appeal of quantification to its civic aspect. It is the victorious Laplace of astronomy whose exploits have lived in the general historical awareness, vindicating the Newtonian picture of the world by calculations that showed how even the apparent irregularities of planetary motion are explained by the mutual attractions of the heavenly bodies. His early investigations into probability were largely simultaneous, however, and always involved equal ingenuity and perhaps a greater conceptual originality. Whoever reads all his memoirs in sequence will be astonished to observe how rapidly, once he saw the possibilities, he enlarged his grasp and made two entire sciences his own, distinct branches of analysis joined only by the approach Laplace developed to them. In both he came upon the subject matter through solving mathematical problems. He first discussed the inequalities of planetary motion

[120] SE 6 (1774), preface, 19.

[121] Bigourdan (1931a), 382-384.

[122] Lagrange to Condorcet, 18 July 1774, quoted in Andoyer (1922), 22, who had had access to the Laplace papers before their destruction.

in the complement to a memoir on integrating equations in finite differences in a manner applicable to the theory of chance, "*hasards*" as it was usually called before he transformed it into probability.[123] There he made epistemology the transition between his two subject matters. Concluding a sequence of problems of excruciating technicality in the theory of games, he broke off and quite unexpectedly introduced that contrast between a deterministic view of nature and a probabilistic view of knowledge that became his philosophical hallmark in the great works of his maturity, the *Mécanique céleste* and the *Théorie analytique des probabilités* with their accompanying popularizations.[124] In his case, the other interests he developed enhanced his mathematical stature instead of, like Condorcet's, detracting from it.

Already in 1773 the shape of Laplace's future concerns was evident in the complementarity between analysis in celestial mechanics, the mathematics that embraces the operations of nature in a fashion closest to the divine, and analysis in probability, the mathematics that pertains to the operations of a faulty human intelligence and mitigates the incompleteness of knowledge. "We owe to the weakness of the human mind one of the most delicate and ingenious of mathematical theories, the science of chance or probability,"[125] he then wrote, and in a later memoir first employed the famous language frequently quoted from the *Essai philosophique sur les probabilités* of 1814: "The word 'chance' then expresses only our ignorance of the causes of the phenomena that we observe occurring and succeeding one another in no apparent order. Probability is relative in part to that ignorance and in part to our knowledge."[126]

In the perspective of this book, moreover, probability is the more central of Laplace's interests, for the external opportunity that drew him to the subject was precisely its pertinence to politics.[127] Politics having been a domain to which Laplace was ever quite indifferent, except for its bearing on his own career, it needed the intervention of Condorcet, newly the acting permanent secretary, to show him the prospect. In the pair of pa-

[123] See note 114, above.

[124] In *Exposition du système du monde* (1796) (republished in *Oeuvres complètes* 6), Laplace gave a verbal summary of the subject of *Traité de mécanique céleste*, 5 vols. (1798-1825). Similarly the 2nd edition (1814) of *Théorie analytique des probabilités* (1812) is prefaced by an essay frequently republished separately under the title *Essai philosophique sur les probabilités*. It appears in *Oeuvres complètes* 7.

[125] SE 7 (1776), 114; for a detailed discussion of Laplace's statement in this memoir, see "Laplace," DSB *15*, 279-286.

[126] "Sur les approximations des formules qui sont fonctions de très-grands nombres," MARS (1782/85), 1-88, cont. (1783/86), 423-467; the quotation is from the latter volume, 424.

[127] Gillispie (1972) contains a detailed account of the development of Laplace's work on probability and of its interactions with Condorcet's academic influence and interest in civic applications.

pers first selected by the Academy from among the many with which Laplace had been besieging it, the earlier that was published defined a type of series he had discovered for the integration of differential equations in two independent variables, of which he noted in passing that the form might potentially be a useful one in the theory of chance.[128] Condorcet's imagination responded more enthusiastically to the companion paper, a "Memoir on the probability of causes taken from events."[129] He observed in the preface already quoted, which brought Laplace before the scientific public, that it

> treats a branch of the analysis of chances, much more important and less known than that which forms the subject of the former Memoir; here the probability is unknown, that is to say the number of chances for or against a possible event is undetermined. It is known only that in a given number of experiments, this event occurred a certain number of times, and it is required to know how from that information alone the probability of what is going to happen in the future can be stated. It is obvious that this question comprises all the applications that can be made of the doctrine of chances to the uses of ordinary life, and of that whole science, it is the only useful part, the only one worthy of the serious attention of Philosophers. The ordinary calculus is good for nothing except computing probabilities in games of chance and lotteries, and is no use even for undoing the popularity of these amusements, equally harmful to industry and morality. The men who know how to make the calculations are not the ones who ruin their fortunes in gambling and lotteries.[130]

We need not here be concerned with technicalities. In that paper Laplace laid the analytic foundations for statistical inference. What must be noticed historically is that it was Condorcet who seized upon the prospect for making of probability a social mathematics. Laplace never mentioned civic matters in his earliest papers, having started the problems out of mathematical ambition, not public spirit. Condorcet could not see very far into the problems until Laplace showed the way. But he could see the subject. His correspondence with Turgot began alluding to it, and for his part he showed Laplace the way toward winning appreciation for mathematical talent from men charged with the affairs of government by finding problems in its concerns. Condorcet headed his prefatory note with precisely the phrase "*calcul des probabilités*" that Laplace henceforth preferred to the traditional "*théorie des hasards*," employed from the time of

[128] "Mémoire sur les suites récurro-récurrentes et sur leurs usages dans la théorie des hasards," SE 6 (1774), 353-371.

[129] "Mémoire sur la probabilité des causes par les événemens," *ibid.*, 621-656.

[130] Preface, *ibid.*, 18.

Girolamo Cardano, the Renaissance gambler who began it. In all the intervening analytic development that its problems had been given by Pascal, Jakob Bernoulli, Daniel Bernoulli, and De Moivre, it had been a mathematics lacking in worthy subject matter and restricted largely to games of chance and conjecture. To be sure, civic applicability was sometimes invoked, but in fact there was none, and only Laplace had the ingenuity and resources to bring it to bear upon really interesting objects: upon philosophy, with respect to theory of knowledge; upon scientific method, with respect to theory of observational error; and upon political and social science, with respect both to analytic demography and to electoral and judicial procedures.

Laplace took up demography in a "Mémoire sur les probabilités" that he submitted to the Academy in July 1780,[131] some six years after his paper on estimating the probability of causes from events. He had in the meantime been working in celestial mechanics and also calculating the probabilities that the apparently random motions and distributions of the comets have a cause different from that regulating the planetary system.[132] Demography as a science derives largely from the fiscal preoccupations of eighteenth-century public administration. In principle, records of births, marriages, and deaths were maintained in parish registers. In 1771 the abbé Terray, then controller-general, required all intendants to compile the figures for their generalities and to return reports annually to Paris. Turgot brought in the Academy of Science in the first year of his ministry, and it published a summary of the figures for the city of Paris and the *faubourgs* reaching back almost to the beginning of the century.[133] Recapitulation showed that from 1745 to 1770, 251,527 boys and 241,945 girls were born. The proportion of 105 to 101 remained almost constant year by year. Figures also existed for London, where, too, more boys were born than girls though in a slightly greater ratio, i.e. 19 to 18.

Laplace's memoir applied a method he had developed in the earlier, largely technical papers for determining the probability on the basis of past experience that the occurrence of future events of an either/or nature lies within given limits. He seized the opportunity to try the formulas on the birth of boys and girls, a real numerical example, rather than the hackneyed old fiction of black and white balls picked out of an imaginary urn. Among the problems that interested him was one calling for the probability of complex events compounded from simple ones of which the

[131] MARS (1778/81), 227-332; *Oeuvres complètes* 9, 383-485.

[132] "Mémoire sur l'inclinaison moyenne des orbites des comètes . . . ," SE 7 (1776), 503-540.

[133] Jean Morand, "Récapitulation des baptêmes, mariages, mortuaires et enfans trouvés de la ville & faubourgs de Paris, depuis l'année 1709, jusques & compris l'année 1770," MARS (1771/74), 830-848.

respective probabilities are unknown, and reciprocally for estimating how many observations have to be made for the predicted result to have a specified probability of being correct. In furnishing data for such calculations the population figures, with their slight but known disproportion between male and female babies, provided just that full and actual statistical basis lacking in hypothetical runs of heads or tails in coins of which the slight assymetries were unmeasured.[134] It was out of such a lineage of mathematical problems that Laplace finally in 1786 came to demography as a subject in its own right and not merely a convenient repository of examples and data. For he then published a memoir on the vital statistics of the city of Paris from 1771 to 1784, together with an estimate of the population of all France over a two-year period.[135]

The actual counting of heads in a periodic census began in France only in 1801, an early measure of Napoleonic administration. In the late eighteenth century, the administration proposed to arrive at estimates of the population through determining the factor by which the average number of annual births was to be multiplied in order to approximate the total.[136] Laplace's memoir applied to this problem his technique for predicting future events from the observation of those past. The question was precisely the sort that could be managed by a prediction of probable error coupled with a computation of how many observations would need to be made in order to restrict its range within given limits. Samplings showed the number 26 to be the appropriate multiplier. Applied to the average annual birth figure for the years 1781 and 1782, it gave the product of 25,299,417 for the population of the French kingdom. In order to reduce to a thousand to one the odds against making an error no greater than half a million in the estimate, the sampling that established the factor of 26 would have had to consist of 771,469 inhabitants. If by comparison the multiplier had been taken to be 26½, then the figure for the population would have been 25,785,944, and the sampling would have had to be 817,219 in order to maintain the same odds against incurring an error larger than half a million. These figures were disconcerting, and Laplace recommended that a count be carried to 1,000,000 or even 1,200,000 in order to assure a degree of accuracy appropriate to the importance of the information.

For the information was important. Population, Laplace observed in a quantitative statement of social felicity, is an index to the prosperity of

[134] "Sur les approximations des formules qui sont fonctions de très-grands nombres," MARS (1782/85), 1-88; (1783/86), 423-467; *Oeuvres complètes* 10, 209-291.

[135] "Sur les naissances, les mariages & les morts à Paris, depuis 1771 jusqu'en 1784; & dans toute l'étendue de la France, pendant les années 1781 & 1782," MARS (1783/86), 693-702; *Oeuvres complètes* 11, 35-46.

[136] Reinhard (1965).

the nation. Observing its variations in the light of events could serve to measure the effect of physical or moral agencies upon human welfare. Impressed by the guidance that such information might provide to those responsible for public policy, the Academy (Laplace noted) had decided to insert in its annual volume of memoirs the summary of births, marriages, and deaths throughout the kingdom.[137] Thus did the Academy's last volumes of memoirs in the old regime come to contain successive installments of an "Essai pour connaître la population du royaume."[138] In its columns, the populations of municipalities and regions marked out on the Cassini map of France[139] were estimated through multiplying by 26 the average number of births in each locality. The author, a "magistrate to be commended for his public spirit,"[140] in the words of the Academy's commission (Laplace, Dionis du Séjour, and Condorcet), was La Michodière, successively intendant in Auvergne, the Lyonnais, and Rouen, and thereby a former colleague of Turgot, with whom he had in fact corresponded about population problems as early as 1760.[141]

Clearly, Laplace's interest in the social application of mathematics was of a different order from that of Condorcet. In the case of Laplace, the motivation was mathematical and professional, and the phenomena only happened to arise in political and civic realms, grist to the mill. In the case of Condorcet, the emphasis was reversed. His motivation was sociopolitical, and mathematics was an instrumentality.But the interplay between them was more complex from the outset than it would have been in the merely obvious roles of Laplace cast as performer and Condorcet as publicist. The evidence is that again at a later juncture, this one occurring after Laplace had brought the subject to a high level of mathematical development, Condorcet managed to identify the locus of interesting problems that Laplace then acted upon. It is well known to historians of mathematics that among the many topics in Laplace's master treatise on the subject, the *Théorie analytique des probabilités* of 1812, is the analysis Laplace gave to the organization of electoral procedures, the credibility of

[137] MARS (1783/86), 693.

[138] "Essai pour connaître la population du royaume," MARS (1783/86), 703-718; (1784/87), 577-592; (1785/88), 661-689; (1786/88), 703-717; (1787/89), 601-610; (1788/91), 755-767.

[139] See below, Chapter II, Section 3.

[140] MARS (1783/86), 703.

[141] Turgot, *Oeuvres* 2, 82-83; 4, 35, 330. The identity of La Michodière is not mentioned in print but may be known from the "plumitif" or rough minutes kept by Condorcet as permanent secretary. These may be consulted in the Archives of the Academy of Science, Institut de France. The earliest entry mentioning this project is of 2 July 1785. Under the *nom de plume* of Messance, La Michodière published *Recherches sur la population des généralités d'Auvergne, de Lyon, de Rouen . . . depuis 1674 jusqu'en 1764* (Paris, 1766), and *Nouvelles recherches sur la population de la France . . .* (Lyons, 1788).

witnesses, and the process of decision by juries and judicial panels.[142] Those were subjects that he never touched prior to the Revolution. They had been started by Condorcet, at the behest of Turgot, so he said. It is quite evident, therefore, that Laplace himself in the years of his maturity came to take the problems seriously enough to give them the clarification and precision that lay in his power to achieve, but not in Condorcet's.

Although Condorcet in his *Essai sur l'application de l'analyse à la probabilité des décisions rendues à la pluralité des voix*[143] brought no technical improvement to this, his favorite branch of mathematics, he did mark out the new ground delineated by his title, albeit blurrily. His interest had originally grown out of the discussions reported by Dupont de Nemours in the Turgot circle about constitution making and the scheme for substituting elective assemblies at the local level for corporative estates in the discharge of administrative and fiscal functions.[144] Given a community, how are the facts to be determined on which to base rational administration? That was the problem to which Turgot's disciples addressed their several versions of his own plan of representative assemblies. Given a society in which many views are held, how are those that are true to be determined? That was the problem posed to political theory by the unacceptability of the traditional recourse to custom, law, and usage. Their accretions constituted the basis on which all reform had been resisted. The latter was, of course, the problem to which Rousseau intuitively responded with the concept of the General Will. But even in an analytic view of politics, a mere democratic counting of heads was no answer. The possession of truth could not well be attributed to an unenlightened and uninstructed majority—witness the popular resistance to freedom for the grain trade. The model of the British House of Commons held no appeal, for it was no better than an arena where localities, classes, and corporate groups voiced their contending views and secured their often abusive interests. There were no whiggish illusions in Turgot's mind, or in Condorcet's, about that body, which they saw to be another survival of precisely that corporative structure in society that a truly representative assembly should supplant.

The question, then, was one of constructing for the multiple elements of society a rational way of determining its collective will and interest, or of distinguishing between the true interests of the community and special or false interests. An example is the operation of a judicial tribunal, for though not the most general or primary problem, it was one that readily brings out the point of view from which Condorcet thought to bring the process of collective decision making within the purview of mathematical

[142] *Oeuvres complètes* 7, xc-xciv, 277-279, 453-470, 520-530; see also Gillispie (1963), 432-436.

[143] (Paris, 1785).

[144] See above, Section 4.

analysis. What confidence do the findings of a panel of judges or jurors warrant when they condemn a criminal? In dealing with this question, Condorcet conceived what was certainly his most original idea in mathematics. One might consider the verdict an event and calculate the probability that it had been caused by the guilt of the accused. He saw the problem, in other words, to be a potential application of Laplace's technique for determining the probability of causes from effects.

More generally, Condorcet sought to subject to probabilistic analysis the problem of election to assemblies of all sorts, and also to analyze the voting procedures of collective bodies. It was possible to consider any group, whether of electors or of representatives, to be a voting mechanism defined by three parameters—(a) the number of voters, (b) the majority required for a decision, (c) the probability that each vote expresses a correct judgment—and constrained within a system of five variable probabilities—(1) that the decision will not be contrary to truth, (2) that some decision will be reached, (3) that the decision will be conformable to truth, (4) that a majority of unknown magnitude will be correct, (5) that a given majority will be correct.[145] It will hardly be worthwhile to follow the construction that Condorcet concocted with these factors. One small and practical result did follow. He recognized, as indeed Borda had before him, that although majority decision may be reasonable in a choice of two mutually exclusive alternatives, in a selection among three or more options, the decision by simple plurality may often traduce rather than express the wish of the largest number of voters: the use of preferential ballots and the practice of runoff elections wherever instituted since that time derive ultimately from Condorcet's analysis.[146]

The yield was small, and its interest historically lies in the insight that the effort gives into the political mentality, not of Condorcet alone, but of Turgot and the entire tradition of enlightened, expert reform he inspired and led. For in their thinking, it was not the purpose of assemblies to represent, nor was it the purpose of voting to express, the various conflicting groups and interests within a society.[147] The analysis could never be relevant to a judicial panel like an English jury or a representative body like the American Congress. Jury trial, after all, is supposed to get at the facts by an adversary process, and not at some social truth behind the facts. Turgot's belief in the possibility of an exact science of politics makes his hostility to the basis and operation of the provincial estates not only politically understandable but theoretically consistent. For in the new assemblies that he imagined, he conceived the process of voting to be a collective device for determining the truth. If we ask what was meant prob-

[145] For a detailed discussion, see Granger (1956), 104-106.

[146] See Black (1958), 159-180.

[147] See Guilbaud (1952).

abilistically by the truth, a definition that Laplace later gave in a popular lecture may serve. It was drawn out of just this ambiance. "Truth, justice, humanity," he told an audience of students in 1795, "there are the eternal laws of the social order which ought to rest uniquely on the true relations of man with his own kind and with nature. They are as necessary to his maintenance as universal gravitation is to the existence of the physical order."[148]

The assumptions make it easier to understand the irritation that Turgot provoked among less theoretical persons, for clearly the new assemblies, had they ever come into existence, were to be extensions of his capacity to embody the truth in politics as over against error, which is to say the reformer against the selfish interests and corrupt factions that would exploit ignorance and superstition for the purposes of power. We will also better understand a certain dogmatism in a Condorcet once introduced in revolutionary circumstances into actual politics, where he saw issues as those of truth against error and corruption. Indeed, it is not too much to say that if we take Turgot and Condorcet to have been examples of an analytic and rationalist approach to political science, and Rousseau to have been the epitome of a sentimental and intuitive approach, the opposites touched in their instinct for considering the will of society to be an embodiment of truth. The general will and the probabilistic will were alike in that, and both were quite irrelevant to any actual practice of majority government in representative assemblies. Turgot's opponents were not mistaken in smelling the technocrat animating his projects. Like many apocryphal sayings, his "Give me five years of despotism and France will be free," has the ring of the à propos.[149] What is astonishing is how deeply into a technical question may be traced the divisions of temperament that distinguish a Turgot from a Rousseau in their political theory, a quantifiable will from a general will; and yet how similar were the habits, drawn from immersion in the French corporate state, about the way in which individual persons and subsidiary interests are to participate in the life of society.

6. MUNITIONS

In a practical sense, however, it was for producing gunpowder and munitions of war that Turgot enlisted science most successfully in these last fifteen years of the old regime. Here the agent of a rationalized administration was Lavoisier, who, in the post of gunpowder administrator (Ré-

[148] *Oeuvres complètes* 14, 173. The lecture was given before the École normale of the year III.

[149] Garat, *Mémoires sur Suard* 2, 330.

gisseur des poudres) and simultaneously at work in the chemical laboratory that he installed in the Arsenal of Paris, made himself the protagonist of the chemical revolution and the most influential person in the scientific world at large. At that time Laplace, some six years younger and still largely involved in mathematics, was little known in public. The story of Lavoisier's reform of the munitions industry presents an interest at once administrative and technological, exhibiting in both aspects what modernization entailed. To bring that out will require detail, both of the old arrangements and of those that replaced them.

In eighteenth-century France, domestic provision of saltpetre, the major item of raw material in making gunpowder, remained the monopoly of a hoary trade, a guild of gothic harpies, the Salpêtriers du roi, to whose forerunners the crown had confided exploitation of its right of eminent domain when the new weapons began to be employed in the fifteenth century. Everywhere in the old Europe the actual fabrication of explosives derived legally from the sovereign's prerogative, "un droict souverain," went an edict of Charles IX, "à nous seul appartenant pour la tuition et deffense de nostre royaume."[150] In France the crown exercised this right in its own arsenals until the reign of Louis XIV. It happened then, as it also did in the historically more famous matter of indirect taxation, that the requirements of the state increased beyond the capacity of a rudimentary bureaucracy to meet them, and Colbert adopted the expedient of "farming" or leasing the facilities into the capable hands of private entrepreneurs who, armed thereby with sanctions by the state, operated them for their own profit in return for furnishing the services. The relations between the Gunpowder Farm and the Saltpetremen that evolved out of this arrangement in the reign of Louis XV were the crux of the problem that

[150] Payan (1934), 17-21; for the administrative history of the gunpowder industry, see in general Payan, and for its technical history, Multhauf (1971), and Bottée and Riffault (1811), the latter being a contemporary work by two officials of the service. Riffault occasionally collaborated in Lavoisier's laboratory. The Academy of Science devoted volume 11 (1786) of its *Savants étrangers* series to memoirs submitted in its competition, discussed below. Lavoisier edited the entire volume, which has the special title *Recueil de mémoires et de pièces sur la formation et la fabrication du salpêtre*, and the introductory "Histoire" is from his hand. He also took upon himself primary responsibility for a publication by the Régie des poudres, easy to confuse with the above because of the similarity of title: *Recueil de mémoires et d'observations sur la formation et sur la fabrication du salpêtre* (Paris, 1776). Lavoisier's memoirs and treatises relating to the Régie des poudres are published in volume 5 of his *Oeuvres*. At the Bibliothèque nationale, brochures and pamphlets relating to saltpetre and gunpowder are catalogued in series L f[65], and at the Archives nationales documents concerning the Régie des poudres are classified in series AD VI, 16 and AD VI, 17 for the pre-revolutionary period, and in AD VI, 79 for the early revolutionary measures. The registers of the Régie des poudres are conserved in the administrative offices of the Laboratoire central des poudres, Boulevard Morland. Microfilm copies of certain portions are deposited in the Firestone Library, Princeton University.

needed to be resolved if these once sovereign functions were to be retrieved from private hands and exercised in the public interest.

Gunpowder is a mixture of "true" saltpetre or nitre (potassium nitrate), charcoal, and sulfur in proportions varying about a classic mean of "six, ace, ace"—75% saltpetre, 12½% charcoal, and 12½% sulfur. It explodes on ignition because the latter two ingredients are completely combustible, and potassium nitrate is an active oxidizing agent. The trade that extracted saltpetre was not a large one. In all France there were between seven and eight hundred saltpetremen employing twelve to fifteen hundred laborers. They were most closely knit in the capital, where in the time of Turgot's ministry and afterward during Lavoisier's administration, some twenty masters, the Salpêtriers de Paris, formed with their dependents a restrictive circle. Their facilities occupied sheds and warehouses situated at intervals outside the circumference of the old walls at the radius of Porte Saint-Denis and the Bastille.

The word itself—*sal-petrae*—was rather of genetic than chemical significance, and the method of harvesting the salt differed in Paris and the provinces according to the mode of its occurrence. At its most accessible it formed on limestone in the course of the nitrogen cycle, an efflorescence mainly of calcium nitrate. The leprous scale penetrated into porous walls of cellars and the lower masonry of buildings where surfaces were exposed to a damp circulation laden with exhalations of animals, men, and organic refuse. In the morning clumsy wagons belonging to the saltpetremen of Paris would rumble empty into the city for the day's prospecting. Each crew consisted of a driver, three laborers, and a foreman, the Homme de ville, who wore a dirty bandolier bearing the lilies of France to signify his right to enter and search in the name of the king in cellars and courts, damp passages and privies, debris and demolitions. He might commandeer rubble without payment wherever he found it and transport it back to his master's workyard. In practice the search had become less spontaneous and less enterprising than the statutes contemplated. The Homme de ville relied on private information instead of systematic intrusion. His master furnished him with pocket money from which to dispense tips to the concierges along his beat and to others who might tell of the whereabouts of buildings about to come down or other fruitful deposits.

When the prospecting party brought a load into the master's yard, he would set two husky laborers with sledge hammers onto pulverizing the charge of stone and plaster. In Lavoisier's opinion, no feature more stubbornly enmeshed the industry in toils of surly routine. By 1785 a stamping mill had been devised; the laborers successfully obstructed it as they did every innovation.[151] Their work done, they would shovel the gritty

[151] Early in the Revolution the Salpêtriers prepared a memorial that tells of their way

mass into water barrels for leaching out the salt. Potash (potassium carbonate) was thrown in, sometimes in the form of wood ashes, in order to convert the "earthy based" to "true" saltpetre. Now this mixture went into great copper cauldrons for cooking out the grosser impurities, among them common salt. As the concentration increased, a dose of Flanders paste was thrown in to clarify the liquor. On additional evaporation, the brew precipitated muddy yellow crystals of crude saltpetre (*de première cuite*). No farther did the law allow the saltpetremen to go in purifying their product, lest they be in a position to sell it privately. Their privilege required them to deliver it crude into the Arsenal of Paris or in the provinces to other royal magazines, all under lease to the Gunpowder Farm. There the personnel refined it by means of two further recrystallizations, after which it came out of solution, white as flour, to be dried in loaves and shipped to powder mills for corning, together with sulfur and charcoal, into gunpowder.[152] The saltpetremen of Paris formed the most important single sector of the trade. From 1783 to 1790, years for which full figures exist, they extracted from the powdered stones of the capital an annual average of 750,000 pounds. All the rest of the nation yielded only four to five times that amount to their cohorts of the provinces.[153]

The provincial saltpetreman led a moist and fumy life in contrast to the dry and gritty lot of his Parisian counterpart. Saltpetre "is born and grows," it was said in the country, in the earth of compost and manure

---

of doing business, *Mémoire à l'assemblée nationale, pour les vingt salpêtriers du roi, établis dans les Ville, Faubourgs, et Banlieu de Paris . . . par Me Lavaux* (BN, L f[65].11. It is one of those rare documents that admit the historian right into the workshop. For descriptions of the trade from the point of view of the authorities, consult Bottée and Riffault (1811), 1-77, esp. 23-25; Lavoisier, "Etablissement de nitrières" (1777), *Oeuvres* 5, 414-419; and the account of an officer of the service, Chevraud, "Observations sur les moyens d'augmenter la récolte du salpêtre en France," SE *11*, (1783/86), 323-370. Chevraud was inspector of powder and saltpetre at Besançon and afterwards commissioner at Essonnes, where he worked in touch with Lavoisier. He makes this remark (p. 327): "L'aveugle routine dans laquelle les Ouvriers ont vieilli, & dont on ne peut, pour ainsi dire, les détacher, y met encore un autre obstacle.

"Les Salpêtriers de Paris dépendent en quelque façon de leurs Ouvriers; & si on parvient à persuader le Maître, les Ouvriers ne l'étant point, ils conservent obstinément leur procédé défectueux. J'ai vu des Maîtres écouter avec plaisir les principes qu'on leur donnoit, mais qui n'osoient les pratiquer, dans la crainte d'être abandonnés de leurs Ouvriers.

"Quelles sont les sources de cette crainte?

"Le battage des plâtras à bras d'hommes, opération pénible qui s'exécute dans les ateliers. Oui, c'est cette opération qui asservit le Maître à l'Ouvrier, parce que ce premier ne peut que très difficilement remplacer des Ouvriers accoutumés à ce dur exercice. . . ."

[152] Bottée and Riffault (1811), 83.

[153] Tables in Lavoisier, "Titre et Qualité de Salpêtre Brut" (1792), *Oeuvres* 5, 648-649, and Bottée and Riffault (1811), unpaginated following the "Exposé historique." By 1811, however, the dependence on exploitations in Paris had increased until the capital accounted for 35 percent of the national supply (*ibid.*, 16).

heaps and in the humid under-reaches of farm buildings, and the trade was plied amid the rotting refuse of barnyard, stable, dovecote, and household. We are best informed about its practice in the Franche-Comté because as early as 1766 the Academy of Besançon sought to modify it in the knowledge of more orderly and rational methods employed in Switzerland and beyond the Rhine.[154] A band of one hundred and thirty saltpetremen ranged the province. They went armed with the right of entry and search, the *droit de fouille*, which extended right into the living quarters of the farm. They might ransack the barns, disturb the stock, make off with manure, and scrape walls and foundations. They had authority to commandeer lodgings, tools, provisions, fuel, and even meals at prices fixed by custom in times long gone by. Villages were liable for providing them with transport. Persons who refused them assistance were subject to heavy fines. In principle the saltpetremen would descend upon a locality every three years and scavenge in all its farms. In practice, they would overlook particular properties for a price, and many a farmer preferred giving a bribe to putting up with intrusion. Often whole villages would compound together to buy off the obligation to find lodgings, furnishings, and wagons. The saltpetremen themselves came to prefer this form of income to the stinking work of extraction. If treated parsimoniously, or if some outraged proprietor stood on the letter of the regulations, they might increase their nuisance value by installing vats and barrows right in the living quarters of the farmhouse. When they did do their proper work, their technique was no more enlightened than that of their fellows in the cities.

Nothing was simple about the supply of saltpetre, and before 1775 nothing was of good economy. On delivering their crude saltpetre *de première cuite* to an arsenal, the saltpetremen received a fixed price from the agents of the Farm regardless of its quality. A tradition that impurities must amount to no more than thirty percent by weight was supposed to protect the interest of the administration. According to lore a crafty eye could judge the strength of the salt from the flaring of a sample fused in a flame.[155] In fact there were no reliable means for estimating the percentage of saltpetre in the yellow lumps shoveled out onto the scales of the refinery. The point is one to be borne in mind, for this practical matter of economy, and not pure patriotic zeal, later led to the technical developments permitting the revolutionary extraction of saltpetre by all civic hands in the military extremity of the year II of the Republic.

The arrangement was unsatisfactory in the eyes of buyer and seller, for

[154] SE, *11* (1786) 13-19; Lavoisier, "Observations . . . sur la récolte du salpêtre" (1776), *Oeuvres* 5, 682-683.

[155] Bottée and Riffault (1811), 100.

if the Farm was thus deprived of any control over the quality of raw material, the set price of seven sous per pound was far below the saltpetreman's cost of production, which by 1775 reached near to twelve. It was not in any profit on his product that his reward consisted, but in a manifold set of privileges and minor exemptions from which he managed to overcompensate his losses and eke out a livelihood. He bought potash from the Farm at the uneconomic price of thirty livres the hundredweight. The Farm also sold back to him from the refinery mother-liquor that yielded one pound and ten ounces of good saltpetre in every tun. Although his byproduct of salt was in fact inedible, the General Tax Farm was obliged to buy it from him on the pretense of protecting its own monopoly of the salt trade. Despite all this, the saltpetremen were sinking beneath the rising price level by midcentury, and the crown, alarmed for the supply of munitions, came to their rescue with a further subsidy of two sous per pound. This subvention was to be paid by the Treasury, not the Farm, regularly in Paris and during wartime in the provinces. Minor privileges had accrued in earlier times: exemption from dues and tariffs while transporting materials; exemption from the taille in certain provinces; its payment at reduced rates in others; and provision in the latter that the amount be set by the intendant directly and not by local officials who might share the grudge of the populace against the corps. Estimating the actual expense of all this in 1776, Lavoisier calculated that saltpetre that had cost the Farm seven sous per pound in the crude state and eleven to twelve sous refined, had in fact cost society eleven to twelve and fifteen to sixteen sous, respectively, and he observed that this difference was the source of the profit, the very handsome profit, of the powder farmers.[156] In Dupont de Nemour's estimate they made thirty percent on their money.[157]

The profits had risen while the industry decayed. The first of the Powder Farmers, François Berthelot, took his first lease in 1665. "Berthelot *des poudres*," he was called, known for enterprise, instant wealth, and vulgarity. He had begun as an actual powder-maker. He pounded his product in his own mills, and he always delivered. His direction of the industry actually did supply the armies and fleets of Louis XIV, for he took four successive leases.[158] A different sort came to succeed him, financiers who had never seen a powder mill. At the expiration of a contract, nine years in duration ordinarily, the candles would be lit in the Salle d'adjudication

[156] "Observations impartiales sur la récolte du salpêtre," Lavoisier, *Oeuvres* 5, 680-692; Lavoisier drew largely on the information for the Franche-Comté assembled by the Academy of Besançon (above, n. 154). For further economic detail assembled by the salpêtriers themselves, see BN Lf^{65}.11.

[157] *Mémoires sur . . . Turgot* (Philadelphia, 1782), 82.

[158] Payan (1934), 75-81.

in Versailles, and the Council of State would invite bids. The Farmer always agreed to furnish a fixed quantity of powder at a set price in return for a lease on the royal monopoly over the fabrication of gunpowder and the supply of saltpetre. The apparent saving to the crown derived from the Farmer's accepting a price for finished powder below that of commerce and even lower than the cost of production. He made his profit in part from monopoly over sales of hunting and blasting powder. The slave trade provided an increasingly important vent. Always, however, the primary asset was the economy in outlay owing to the privileges of the saltpetre corps, to whose more ancient rights and organization the Farmer's interests were thus bound in a symbiosis obstructive of all rational technology.

The Powder Farmers of the reign of Louis XV, occupying themselves in finance and speculation, took no interest in technology. They preferred to contract out to local entrepreneurs the fabrication of much of the powder they were required to supply, and allowed the powder magazines and refineries to fall into neglect. They were under no obligation to furnish more than the powder stipulated in their contract. If war came, the crown might find additional supplies where it could. In the extremity of the Seven Years War, the French had to turn to the Netherlands and pay twice the domestic price for saltpetre, which, however, had cost the Dutch less to ship from India than the Farm had to pay in France. Given this differential, the Farm itself began importing saltpetre from Holland in normal times, thus further sapping the vigor of the industry they leased. The embarrassment to French arms in the Seven Years War did finally bring the Farm under severe criticism, which military leaders exaggerated to the extent of attributing defeat primarily to the failure of logistics, the dependence on saltpetre from overseas, the burden thus forced upon the Treasury, and the inferior quality of powder issued to the troops.[159] It was true that no other major power depended on any such obsolete arrangements for assuring the supply of munitions, nor did any other government subject property owners to the depredations of a saltpetre corps. The maritime powers drew their saltpetre from India and the land powers from rational manufacture. In Sweden and Prussia the state had commissioned construction of nitrification plants that composted wastes to the order of the armed forces.

So closed was the trade in France, however, closed both to scientific and to foreign influences, that scientists in Paris learned of these methods only after Turgot had referred the state of the art of gunpowder to the Academy of Science with a request for counsel. On 17 August 1775 he addressed a letter to Grandjean de Fouchy charging the Academy to institute what, in the parlance of a much later chapter in the history of military explo-

[159] *Ibid.*, 130-132.

sives, would be called a crash program. His urgency, however, was less a matter of power politics than of relieving private persons as soon as possible from the irritation occasioned by the *fouille* and other intrusions of the saltpetre corps. The Academy was to appoint commissioners forthwith in order that terms and instructions might be drafted before the holidays and the prize announced at the next public meeting. Its amount was to be the extraordinary figure of 4,000 livres, and there were to be two honorable mentions of 1,000 livres each. The Academy responded in the spirit of the hour and dispensed with its normal procedure of laying important subjects on the table for a week.[160]

Turgot had not awaited enlistment of the Academy in the technological problem to reorganize the industry administratively and to make it an exemplar of the policy of retrieving public functions from private hands and exercising them for the benefit of the Treasury. A decree of the Council of 28 May 1775 revoked the *Bail* Demont, the last contract of the Powder Farmers, and replaced it by a new administration, the Régie des poudres. Not yet nationalization, the measure created an intermediary form, a régie interessée, or privately financed commission, chartered to serve the public interest and responsible to the controller-general. Three definite principles were to guide its conduct of business. First, what profit could be made from exploitation of saltpetre and fabrication of gunpowder must remain with the crown. The régisseurs were legally bondsmen who furnished capital and found their return in royalties based upon production. Second, the declining slope of production must be reversed and France restored to self-sufficiency in munitions of war. Third, property owners must no longer be harrassed by the abuse, or even by the exercise, of the *droit de fouille* by the saltpetre corps. The régie was to devise some more acceptable means for providing the industry with crude saltpetre.[161]

The enterprise over which the régie presided was a considerable one. When it was fully reorganized, the personnel numbered more than eleven hundred. They worked at one or another of some forty installations, each in the charge of a commissioner and subject to appropriate inspectors. There were powder mills combined with refineries in the vicinity of Colmar, Nancy, Besançon, Dijon, Toulouse, Bordeaux, Tours and Rouen; powder mills alone near Saint-Omer, Mézières, Metz, Marseilles, Montpellier, Perpignan, Saint-Jean d'Angély, Brest and Essonnes; and refineries in Lyons, Saumur, Orléans, and Châlons-sur-Marne, and also in the Arsenal of Paris. There was a great naval warehouse at Nantes, lesser depots elsewhere, and outlets for sales in principal cities. Under the four

[160] "Histoire du prix proposé sur la formation du salpêtre," SE, *11* (1786), 1-3; for the identity of this volume, see note 150 above.

[161] The text of the decree is printed in Turgot, *Oeuvres 4*, 364-378, and in the Dupont edition, 7, 297-300.

régisseurs served a general manager, one J.-B. Bergault, who had executive authority over installations and operations. Two of their number, Le Faucheux and Clouet, were experienced in the industry. Le Faucheux had been director-general of manufacturing in Paris for the Farm and, before that, commissioner at Verdun, a post in which Clouet, younger and an enterprising chemist from the Ardennes, had succeeded him. A third colleague, Barbault de Glatigny, also carried over from the old company, in which he had had an interest under the *Bail* Demont. He was a man of finance and administration and held a post also under Turgot in the ministry.[162] But from the outset the leading part was Lavoisier's. Indeed, it is probable that the legislation was drawn to his specification.[163] Henceforth he made the munitions industry his main occupation and dominated it by virtue of competence, of attention, of intelligence, and of quality.

Less than most men, and less even than most scientists, could Antoine-Laurent Lavoisier bear to have things wrong or to be other than the one who sets them right. The Lavoisier who took in hand the affairs of the Gunpowder Administration was already well into the main theoretical and experimental work of his career, though still ten years and more away from the fully generalized reform that he and the school he formed brought to the science of chemistry: his famous paper "Reflections on Phlogiston" advancing the oxygen theory of combustion was read to the Academy in 1785;[164] the modern system of chemical nomenclature was issued in 1787;[165] and his *Traité élémentaire de chimie* was published in January 1789, on the very eve of the Revolution.[166] Alone among the leading

[162] Payan (1934), 150-157; Bottée and Riffault (1811), who print tables of organization (for the Napoleonic period, to be sure) in an unpaginated section following the historical introduction; Lavoisier, "Mémoire de la Régie des Poudres," *Oeuvres* 5, 714-728.

[163] In his defense of the régie in 1791, Lavoisier writes, "Un des régisseurs actuels lui [i.e. Turgot] présenta un plan de régie infiniment économique et propre à procurer aux peuples un soulagement qu'on ne pouvait espérer sous le régime d'une entreprise quelle qu'elle fût." "Mémoire de la Régie des Poudres," Lavoisier, *Oeuvres* 5, 715. Guillaume says, though without citing a source, that the notion of an extraordinary prize contest was also proposed to Turgot by Lavoisier, *ibid.*, 461.

[164] "Réflexions sur le phlogistique, pour servir de suite à la théorie de la combustion et de la calcination, publiée en 1777," MARS (1783/86) and Lavoisier, *Oeuvres* 2, 623-655 (read to the Academy, 28 June 1785).

[165] *Méthode de nomenclature chimique proposée par MM. de Morveau, Lavoisier, Berthollet & de Fourcroy* (Paris, 1787). On this subject, see Crosland (1962).

[166] Though outdated in many respects, the point of departure for biographical study of Lavoisier is still Grimaux (1888). See also McKie (1935) and Guerlac (1954). For the development of his scientific work, see particularly Daumas (1955) and Gillispie (1960), ch. 6 (based largely on a paper contributed to the "Colloque international sur l'histoire de la chimie au XVIIIe siècle" held in Paris, 11-13 September 1959). For the beginning of Lavoisier's chemical work, see Guerlac (1961b), the most important single contribution to Lavoisier scholarship in recent times. The volume of that scholarship is large: Smeaton (1963) summarized its content in the preceding decade.

scientists of his generation, Lavoisier was a Parisian born and educated. He lived most of his life and did most of his work within half an hour's brisk walk from the cul-de-sac Pecquet, his birthplace in the Marais, and the rue du Four-St. Eustache, near what was until recently Les Halles. There his father moved after his mother's early death, to make their home with his grandmother Punctis. Like many a well-to-do bourgeois ménage, the Lavoisier family kept up its association with the nearby country region from which they had come several generations back. Their ancestral town was Villers-Cotteret, and there to the north of Paris they would repair in the summers and recharge their energies for the city. Later, when Lavoisier had succeeded in achieving an affluence considerably greater than his father's solid prosperity, it was in the pattern of his class that he should buy a manor in the village of Fréchines near Blois, and in the pattern of his science that he should make it into an experimental farm.[167]

His was a highly protected childhood in a rigid family. They centered on him, the only surviving child (for his sister also died), all the hopes of a widowed father and all the worries about the effects of fatigue, study, and exposure to weather of a maiden aunt who had given up her own chance of marriage to nurse her sister's children. Like his father, he had his secondary education at the Collège des Quatre-Nations, often called the Collège Mazarin because it occupied the palace that had been left it by that statesman and that since Napoleonic times has housed the Institut de France. Like his father, too, he seemed to be headed for the law, and received his licence to practice from the Parlement of Paris in 1764. It might be thought that the career he made instead in science had the configuration of the magistracy at its best and most progressive. The decisive scientific influence of his formative years was certainly that of his teacher in chemistry, G.-F. Rouelle, famous for the charm and contagious enthusiasm of his lectures at the Jardin du roi, although it appears that Lavoisier learned the science in the master's own laboratory.[168] For his taste did draw him to science. The earliest recognition he received was a gold medal from the Academy of Science for a memoir that interested Sartine, the lieutenant-general (or minister) of police, on various means for illuminating the streets of a large city.[169] It is characteristic of Lavoisier that among the first items in his published *Correspondance* should be the draft (unsigned) of a letter to Grandjean de Fouchy, permanent secretary of the Academy, deploring the lack of a section for *physique expérimentale* and transmitting a plan by which that body might be reorganized, the exist-

[167] See Lenglen (1936); McKie (1952), 169-173; Smeaton (1956b); and below, Chapter V, Section 5.

[168] Guerlac (1956) and (1961b), 29, 32-34.

[169] "Mémoire sur les différents moyens qu'on peut employer pour éclairer une grande ville," Lavoisier, *Oeuvres 3*, 1-70.

ing membership distributed among the various ranks, and a new one created for *aspirans* or candidates. His own name is the first of these suggested. The letter was dated 12 April 1766, two years before Lavoisier did win election.[170] His chance came after the veteran mineralogist, J.-E. Guettard, a crusty and alarmingly candid old scientist, picked him for assistant on a field trip in the Jura and the Vosges. The journey may not have been to Lavoisier quite what the voyage of the *Beagle* was to Darwin, but it made the break between home and manhood and brought him into view. In May 1768 there was an academic election to a vacancy in the chemistry section of the Academy. Lavoisier won the vote by a very small margin over the nearest competitor, Gabriel Jars. It was regular procedure for the Academy to submit the names of the two leading postulants to the crown. In this instance, the minister preferred the claims of Jars, eleven years Lavoisier's senior in age and a mining expert who had just completed important missions for the government. By way of compromise the minister authorized creation of a supernumerary membership until a further vacancy should occur. One soon did. Jars died later the same year.[171]

Even then the direction that Lavoisier's ambition would finally take was not fully evident. Just before his election, he entered upon another set of duties of a sort not usually compatible with science. In March 1768 Lavoisier became a financier, subscribing the sum of 520,000 livres, 340,000 in cash and the rest in bills at 5 percent endorsed by his father, in order to take a one-third share of the investment of François Baudon, one of the sixty Farmers-General. That company, enormously more important than the Powder Farm, was then negotiating the normal six-year renewal of its *Bail* or contract with the Treasury. From their capital the Treasury was to be advanced the sum of 90 million livres, and various perquisites too complicated to enumerate were to be paid to certain pensioners and favorites of the court. The General Farm would then continue its administration and collection of a congeries of excise taxes, internal and external customs duties, and the state monopoly over sale of salt and tobacco.[172]

The historian is bound to ask himself why a brilliant young man, already well-to-do and with a strong bent for science, should have entered on that association? Not that there was anything discreditable about it at the time: on the contrary, but still, its object was profit, not knowledge or reputation; and the inference is hard to resist that money was greatly important to Lavoisier. It is an explanation confirmed by many features of the way he led his life, a remark that must at once be redeemed by noting that he handled his wealth with scrupulous probity, often with generosity, and always with a high sense of responsibility for persons depending

[170] Lavoisier, *Correspondance* 1, 9-12.

[171] Grimaux (1888), 29-30. On the career of Jars, see Chapter VI, Section 3.

[172] *Ibid.*, 62-82; a standard work on the General Farm is Matthews (1958).

on him. Prominence was also important to him, for Lavoisier's participation in the General Farm was more than a mere investment. It was an activity, and he liked to take authority. It is clear from his correspondence that in 1768, 1770, and 1771 the major portion of his time and energy went into this work. Like many institutions of the old regime, the General Farm operated in chambers and committees. The personnel were numerous—about 25,000 clerks, collectors, and officials—divided among the various administrations: frontiers, colonial importations, salt, tobacco, excise duties on goods shipped into Paris. Lavoisier made trips to the borders. He inspected the tobacco outlets. He sent back reports on the conduct and honesty of the personnel. Of the tobacconists in St.-Dizier, he noted Claude Chicart adulterated his snuff heavily with ashes; Claude Lucot and Jean Lucot lightly; the widow Gilbert and another four dealers not at all; among various "captains" or inspectors of customs, one Cadier at Chalons ought to be moved, he completely lacked "subordination"; Boyer at St.-Dizier was weak; Bevier at Dampierre was "full of zeal and good will."[173]

That was the sort of thing that occupied Lavoisier in the early days of his service. After 1774 when he took a larger share, he specialized in the excise on goods entering Paris. His was the suggestion in consequence of which Paris was girdled in 1786 with a smuggler-tight wall to eliminate the contraband that easily slipped around the old control point. A saying attributed to a marshal of France and much repeated by *tout Paris* held that the author of that innovation "ought to have been hanged; happily for M. Lavoisier this advice has not been followed."[174] By then Lavoisier was a Farmer-General in his own right, and even more deeply involved by association with his father-in-law. For in 1771 he had married Marie Paulze, the daughter of a financier and senior colleague in the Farm. Lavoisier was then twenty-eight. His bride was thirteen, an intelligent child being besieged by an elderly wooer whose sister was close to the abbé Terray, then controller-general. She brought Lavoisier a handsome dowry, learned English, and helped in the laboratory, drawing diagrams and making entries in the registers. They had no children.

In 1772, Lavoisier entered upon the train of chemical research that led him from the problem of combustion through the entire structure of the science to its reordering in modern form. On 1 November he registered a sealed envelope with Grandjean de Fouchy. By this practice the Academy safeguarded a scientist's most valuable property, his priority in a discovery. Lavoisier's note states that about a week previously he had found that sulfur, and also phosphorus, far from losing weight when burned, as commonly supposed, on the contrary gain considerably, and that the in-

[173] Lavoisier, *Correspondance 1*, 150, 159. [174] Grimaux (1888), 81.

crement comes from a prodigious quantity of air "fixed" during the combustion. He was led to think that the same might well be true of all bodies that gain weight on combustion, and equally so of metals in calcination (oxidation in the terminology that evolved out of these discoveries). "This discovery," he went on, "seems to me one of the most interesting that has been made since Stahl, and since it is difficult to keep from letting slip out (*de ne pas laisser entrevoir*) in conversations with friends something that might put them on the way to the truth, I have thought it right to place the present statement in the hands of the Secretary of the Academy until such time as I publish my experiments."[175] The species of avarice that enters into the mixed motivation of discovery is more understandable to the sociologist of science than it is acceptable to the self-knowledge of the scientist, and Lavoisier later altered the copy that remained among papers he intended for publication to read of the reason for depositing the note only, "I felt that I ought to secure my right" to the finding.[176]

Early in the next year, on 20 February, Lavoisier wrote out a memorandum to guide and orient his thoughts on the "long series of experiments that I intend to make" with the purpose of clarifying the role of "elastic fluids" in chemical combination. It was not yet evident to him, and became clear only in the development of his knowledge, that the differing properties of the fluids a chemist encounters represent anything more categorical than modifications of atmospheric air. All that was clear was the importance of the problem, which was such that he must review and do over all the work that had ever been done on it. It was a task "that seemed to me made to bring about a revolution in physics and in chemistry," and certainly the prescience with which he foresaw that prospect and the steadiness with which he ever held it in mind were the critical factors permitting him to realize that part of his ambition that lay in the modernization of chemistry.[177]

It was not until 1772 that Lavoisier channeled into chemistry the scientific energies that until then he had also devoted to mineralogy and geology. Guerlac's careful and judicious monograph on the work of that "crucial year" concludes that Lavoisier then came into chemistry formed by the pharmaceutical, mineralogical, and analytical concerns characteristic of a French approach to the science, and that what put him onto the role of air in chemical reactions was the phenomenon of effervescence when a sample of metal is plunged into an acid bath.[178] That the resulting "calx" (i.e. oxide) outweighs the parent metal was a puzzling appearance long suspected to be a general fact, but definitely established only

[175] Lavoisier, *Correspondance* 2, 389-390; for this episode, see Guerlac (1961b), 75.

[176] Lavoisier, *Oeuvres* 2, 103.

[177] Guerlac (1961b) prints the main portion of this memorandum, 228-230.

[178] *Ibid.*, 192-196; see also Gough (1968).

through the experiments of Louis-Bernard Guyton de Morveau. Guyton de Morveau was a provincial lawyer, a man of minor letters, and a scientific enthusiast still little known beyond the ambit of Dijon, one of a number of Burgundians whose entry into the scientific and literary world was facilitated by the Academy of that provincial capital. The most notable of the others was Lazare Carnot. Later on, when Guyton had fully understood and rallied to the oxygen theory of combustion, he became, although older than Lavoisier, one of his most effective lieutenants in campaigning for the new chemistry. Its system of nomenclature is largely of his designing, and he will be frequently encountered in detailed accounts of the revolutionary period, just behind the front rank both politically and scientifically, an attractive figure, cheerful and interested in the work that came his way in both domains.[179]

As for gas chemistry itself, it already existed in England and Scotland in the work of men who envisioned no general consequences for the fundamental theory or structure of the science. Joseph Black had identified "fixed air" (carbon dioxide) as a distinct chemical agent in 1756. Henry Cavendish isolated "inflammable air" (hydrogen) in 1766. Joseph Priestley, far more prolific than Lavoisier or any other in actual discovery, found nitrous oxide in 1772.[180] Scientific communications across a language barrier were still poor in the eighteenth century, when the continental world relied largely on French, and apparently Lavoisier learned of the work of the "pneumatic" chemists only after he had already outlined his own program of research into combustion. He did then learn. The results figure in his first book, the *Opuscules physiques et chimiques* of January 1774, where he could make the first distinction he needed in executing the program, that between fixed air (carbon dioxide), which is the product of effervescence and respiration, and respirable air, which supports combustion.[181]

The most important single thing he learned followed that book, however, and it also came from Priestley. In October 1774 Priestley was in Paris, visited Lavoisier's laboratory, and told him of discovering an air still more interesting for the vigor with which it supported combustion. He had obtained this "dephlogisticated" air the previous August by roasting the red calx of mercury (mercuric oxide), and had determined many of its properties. Lavoisier understood those properties far better than their discoverer. They were precisely what he needed to carry his program one stage further, the critical stage, for they permitted him to exhibit that calcination is indeed combination with something atmospheric. He re-

[179] Bouchard (1938).

[180] On Priestley's career, see Schofield (1966); it is still instructive to consult Conant (1957); see also Guerlac (1957).

[181] Guerlac (1961b), ch. 2. Lavoisier's *Opuscules* are republished in his *Oeuvres 1*, 437-666 (Table des Matières, 702-728; Plates XIV-XVI).

peated and extended experiments like those of Priestley, who had all innocently let word "slip out." They form the subject of the memoir Lavoisier read on Easter 1775 at the annual *séance publique* when the Academy tried to schedule something with wide appeal.[182] Unhappily, he never mentioned Priestley, who published his own experiments later in the year.[183] Unfortunately, too, Lavoisier had mistaken an essential point. He took the air that calcines metals to be the whole of the atmosphere, which error Priestley corrected, pointing out that in fact it comprises 20 percent.

Such was the point at which Lavoisier had arrived in his program of research when to his other responsibilities he added administration of the new Régie des poudres. A year later he and his wife moved into an apartment in the Arsenal and installed the laboratory there. Long afterward, his widow told in a memoir how he spent his days there, where they now worked and lived, and where he went on to order his science while also ordering the service he had taken into charge:

> Each day Lavoisier sacrificed some time to the new affairs for which he was responsible. Science always had a large part of his day. He arose at six o'clock in the morning and worked at science until eight, and in the evening from seven until ten. One whole day a week was devoted to experiments. It was, Lavoisier used to say, his day of happiness. Certain enlightened friends, certain young men proud to be admitted to the honor of cooperating in his experiments, foregathered in the laboratory in the morning. There they breakfasted, there they discoursed, there they worked, there they performed the experiments that gave birth to the beautiful theory that has immortalized its author. Ah, it was there that a person needed to be to see and hear that man endowed with so fine a mind, so just a judgment, so pure a talent, so lofty a genius. It was by his conversation that it was possible to judge of the beauty of his character, the elevation of his thought, the severity of his moral principles. If ever any of the persons whom he admitted to intimacy can read these lines, I think the memory will not cross their consciousness without their being moved!
>
> It was into these sessions that the best workmen were admitted to make the machines that Lavoisier invented.[184]

Compartmentalize his life though he might, one comes to recognize Lavoisier's distinctive touch in all the enterprises he put in hand. To fi-

[182] "Mémoire sur la nature du principe qui se combine avec les métaux," MARS (1775/78), 520-526; Lavoisier, *Oeuvres* 2, 122-128.

[183] On this much and acrimoniously discussed question of priority, see the fair and judicious summary in Daumas (1955), 67-90.

[184] Gillispie (1956b), 57.

nance, munitions, and science alike he brought the luminous accuracy of mind that was his signet, the spirit of accountancy raised to genius. His were the accounts, assembled under sentence of the guillotine, by which historians may now exonerate the Farmers-General from any imputation of dishonest speculation or personal corruption in the collection of taxes. Neither is there reason to question the returns he prepared when even earlier in the Revolution, his stewardship of the Régie des poudres came under scrutiny.[185]

The régie took responsibility on 1 July 1775. In the previous year production of saltpetre had fallen to 1,600,000 pounds in the face of requirements of 3,600,000, the difference met by purchase abroad. The new administration moved rapidly, stimulated by the American war. Already in 1776 and 1777, it found 1,700,000 pounds of gunpowder for the armies of the United States, and supplied smaller amounts to the Dutch and Spanish fleets after 1780. It steadily increased its sales to private persons for hunting pieces, and secured for its gunpowder an increasing share of the market provided by the slave trade. By 31 December 1788, the régie had retrieved French self-sufficiency. In the year then ending, its refineries produced 3,770,000 pounds of pure saltpetre. The gunpowder fabricated in its mills was reputed the best in Europe. It proved out at a carry of 115 to 130 *toises* instead of the 70 to 80 which had been normal in the Seven Years War (the *toise* or fathom measured just over six English feet). All this, moreover, was at a saving. Lavoisier valued the assets on that date at 4,474,848 livres/6s./9d. In its thirteen and a half years, the régie had paid into the Treasury 6,112,180 livres. It had funded the sums due the former Powder Farmers under their contract. It had continued paying the special subsidy due from the crown to the saltpetremen of Paris. It had found the prize money for the competition on the technology of saltpetre instigated by Turgot and sponsored by the Academy. Finally, it had relieved the nation of the indirect expense of the materials and furnishings formerly exacted by the saltpetre corps, while furnishing powder to the armed forces at a cost of 13 sous per pound when private customers were paying 20. All in all, Lavoisier estimated that, on the eve of the Revolution, the Régie des poudres had restored the French munitions industry at a combined profit and saving to the nation of some 20,000,000 livres. The compensation to the régisseurs themselves worked out at 16,000 to 17,000 livres annually for each, together with their quarters.[186]

The record was that of a success, not to say a triumph, a triumph less

[185] Lavoisier's summary of the accounts may be verified by reference to the original ledgers containing the registers of the Régie des poudres, conserved in the Laboratoire central des poudres, Boulevard Morland.

[186] Lavoisier, "Mémoire de la régie des poudres," *Oeuvres* 5, 714-728; "Mémoire sur la régie des poudres," *ibid.*, 732-734.

of science in its theory, however, than of scientific administration. The whole affair, indeed, displays the signal contribution that science, taken to be the vehicle of enlightenment and rationality, brought to technology in the eighteenth century.[187] The connection between renovation of the French munitions industry and the reformation of chemistry was personal in that both were the work of Lavoisier and both were in his style. It was also circumstantial, for deep theoretical chemistry, Lavoisier's or any other, is unthinkable except as sounding a great body of chemical practice.[188] It was institutional, finally, in the collaboration of the Academy of Science with the Régie des poudres, Lavoisier having gathered into his hand threads that worked them both. Thus it was that science permeated the industry: through persons, through circumstance, and through expert management. Its concepts did not always help, at least not much. In 1777 Lavoisier, intent upon persuading entrepreneurs to venture capital in the construction of artificial nitre plants, instructed them that saltpetre is a neutral salt formed of the combination of acid of nitre with vegetable alkali. Chemists knew a good deal about the behavior of those components, he observed, but nothing about their nature.[189] In 1789 they were not notably more advanced, reformed though the science then was in the fundamental theory of combustion and the industry in its management and organization. Of the same substance, Lavoisier could still say only that "acid of nitre seems like so many other productions to be the work of nature and of time."[190]

Insofar as Lavoisier's theoretical study in the interval bore on the work of the Régie des poudres at all, it did so either in regard to the role of air in chemical combination, a relation so broad as to be almost meaningless, or else in regard to its erroneous aspect, the role he also attributed to oxygen in acidification. In directions that Lavoisier drafted for the guidance of entrants in the Academy's saltpetre contest, he urged them especially to investigate the contribution of air to the formation of nitre.[191] Apparently, however, Lavoisier meant to be pressing theory deeper into practice. Throughout his long campaign as a theorist, his attention shifted back and forth between the two main fronts of combustion and acidification, and it was the crowning error of his strategy that he always believed that what forms acids is combination with oxygen in a sort of superoxidation.[192]

It is important to notice that his appointment to the Régie des poudres followed by a few months Priestley's correction of his mistake in taking

[187] See below, Chapters 5 and 6.

[188] Cf. Guerlac (1959b).

[189] "Instruction sur l'établissement des nitrières," *Oeuvres* 5, 391-392.

[190] "Mémoire sur . . . la Régie des poudres et salpêtres" (1789), *ibid.*, 694.

[191] "Histoire," SE, *11* (1786), 10-11.

[192] Gillispie (1960), 241-250.

the whole of the atmosphere for the active agent in combustion when presenting his theory at the open Easter meeting of the Academy in 1775. From this setback in theory of combustion proper, he turned to his other problem of acids, as if acting himself upon the injunction he was simultaneously writing into the terms of the saltpetre prize. It was the king's intention (so stated Turgot's charge directing the Academy to institute the prize) that only entries joining experiment to theory should be eligible for an award. The occasion being extraordinary, the commission appointed by the Academy was to verify all experiments communicated by the contestants, and to make its own trials in the meantime. Indeed, so severe was the crisis in the industry that the régisseurs des poudres were to be present with the academic commissioners during all experiments and to sign the register daily.[193] It is even more surprising, therefore, that Lavoisier saw fit to serve in both capacities, of which circumstance more in a moment.

Behaving more like a research team under Lavoisier's direction than a panel of judges, the commission rented a house and garden in Saint-Denis and fitted up a laboratory where they went to work. Lavoisier moved his apartment and his own laboratory to the Arsenal only in 1776, and it must have been there at Saint-Denis that he mounted the experiments on acids to which he turned with the chemistry of air never far from mind. The instructions to contestants go on to develop the suggestion that air may be more than merely a medium for the formation of saltpetre.[194] It may enter into the actual composition of acid of nitre. As usual Lavoisier acted on his own hints. Such was indeed the finding, just under a year later, of the memoir on nitric acid that he read to the Academy in April 1776,[195] almost on the anniversary of his false step on combustion in the Easter memoir of 1775.

The paper ranks among his most appealing. In it first appeared in full double movement his method of analysis verified by synthesis. Nitric acid does contain oxygen, and only the construction placed upon the fact, making oxygen the source of acidity, is wrong. That mistake was entailed by his whole course of thinking and was not yet correctible by the experiments. Shortly thereafter he published the memoir in a *Recueil* issued by the new Régie des poudres for the benefit of the industry in its first gesture of service to the public.[196] He admitted that his findings were more theoretical than practical, exhibiting only the most distant relation to any means of nitrification actually in use. He had included the memoir in the

[193] "Histoire," SE, *11* (1786), 2, 12-13.

[194] *Ibid.*, 10.

[195] Lavoisier, "Mémoire sur l'existence de l'air dans l'acide nitreux . . . ," *Oeuvres 2*, 129-138.

[196] See above, note 150.

Recueil in hopes that it might have "opened the eyes" of contestants on the operations of nature in forming nitric acid, for it did tend to prove that the materials are aeriform.[197]

Initial measures of policy adopted by the régisseurs were themselves too theoretical to make an impression on traditional procedures. Their first resort, enlisting the Academy of Science in the job of reforming the industry, was tactlessly done. Lavoisier seems never to have reflected that those who lived by the *fouille* would scarcely welcome an invitation to find ways to abolish it. All concerned were disappointed in the conduct of the saltpetre competition. Bad feeling among unsuccessful contestants echoes in Lavoisier's remark that "the commissioners will express themselves frankly and simply on the merits of each of the memoirs submitted to the competition; it has not been in their intention to humiliate anyone; but they cannot help being truthful.[198] At the very outset several members of the Academy had objected that Lavoisier ought to be disqualified as régisseur from doubling in function as one of its saltpetre prize commissioners. Confident in his own rectitude, ever superior to common carping, he let it be observed that since the Régie des poudres had ceased to seek a profit, "those who are in charge of it can have no other interest and no other purpose but the greatest advantage for the state and for the king's service."[199]

He alone was both régisseur and commissaire for the Academy. In the latter capacity his colleagues were Macquer, Sage, Baumé, and the chevalier d'Arcy (who died soon after). Macquer was the most capable of the chemists of the old school. Sage and Baumé were far less so, and both were soon on bad terms with Lavoisier, never one to suffer mediocrities gladly. Thirty-eight memoirs having been received without a worthy entry among them, the Academy had to extend the closing of the contest from the initial date of 1 April 1777 and to announce a second round in a further term of five years for a doubled prize of 8,000 livres. The result was anticlimax. It was 1786, and Turgot was ten years gone from office and five years dead, before the memoirs were finally in print accompanied by the comments and criticisms of the commissioners.[200] By then the Régie

[197] "Histoire," SE, *11* (1786), 30. [198] *Ibid.*, 35.

[199] *Ibid.*, 4.

[200] In the event, the prize went to the brothers Thouvenel of Nancy, the one a doctor of medicine and the other a commissioner of powder and saltpetre in charge of the magazine there. Theirs was a reasoned survey of the industry as a whole, as much economic as technical (SE, *11* [1786], 55-166). A name famous in the Revolution, that of Romme, appears on one memoir (421-478), subjecting a theory of nitrification to mathematical calculation and bearing for motto the sentiment, "Utile aux Gouvernements, funeste à l'humanité"—the author however, was not Gilbert, the future Jacobin and sponsor of the republican calendar, but his brother Charles, teacher of mathematics at a school for midshipmen.

des poudres had long since put into effect the less sweeping improvements in detail that, rather than some flight of ingenuity by an inspired inventor, were what actually reformed the industry.

Indeed, the most useful contribution came from one who for reasons of propriety could scarcely submit his discoveries in competition. The duc de la Rochefoucauld-d'Enville took seriously his honorary membership in the Academy of Science.[201] He had noticed abundant deposits of saltpetre occurring naturally along the chalk cliffs of the Seine hard by his chateau and village of La Roche-Guyon. Traces of an old enterprise indicated that they had been worked in olden times. La Rochefoucauld, the son of Turgot's great friend, the duchesse d'Enville, and the epitome of an enlightened nobleman, blamed the onset of feudality for its demise. He put in hand experiments and found that, given proper access of air, the mineral rapidly regenerated itself in these and smaller formations down the valley and along the *falaise* north of the estuary above Le Havre.[202] Lavoisier seized the chance for lessening the dependence of the régie on the fouille. With Clouet's help, he extended the experiments and sought out other natural occurrences.[203] They found the porous chalk (*tuffeau*) of the Loire Valley in Touraine and Poitou to be still more hospitable to saltpetrifaction. Already, in fact, the saltpetremen of those provinces had learned to exploit it and to spare the populace their unwelcome visits. The yield was small at first but the régie succeeded in developing the industry there until in the time of Napoleon those resources were supplying 10 percent of the needs of his armies.[204]

Turgot's legislation had given the Régie des poudres three years of grace in which to develop methods for procuring saltpetre to substitute for the fouille generally, which must thereafter become voluntary. From a technical point of view, the creation of plants for artificial saltpetre on the Swedish model held the greatest promise among the several possibilities. Lavoisier informed himself, again with the collaboration of La Rochefoucauld, who wrote to military acquaintances in that northern realm. The régie published an *Introduction* early in 1777.[205] The design was of a hangar rather like a tobacco barn in which carefully mixed accumulations of earth and organic matter would be arranged in layers so that saltpetre could form in a regular sequence. The style is that of laboratory

[201] Rousse (1892).

[202] La Rochefoucauld, "Mémoires sur la génération du Salpêtre dans la craie," SE, *11* (1786), 610-624.

[203] Lavoisier and Clouet, "Mémoire sur des terres naturellement salpêtrées en France," *ibid.*, 503-570; and "Mémoire sur des terres et pierres naturellement salpêtrées dans la Touraine et dans la Saintonge," *ibid.*, 571-609.

[204] Bottée and Riffault (1811), 10-16.

[205] "Instruction sur l'établissement des nitrières et sur la fabrication du salpêtre," Lavoisier, *Oeuvres* 5, 391-460.

directions, dry and precise. The procedures presuppose sterner sensibilities, and since they came very little into use, they do not demand description.

The failure of the program was one of incentives, not of chemistry. In Lavoisier's estimate, a plant of ten hangars employing six men would involve a capital outlay of 32,000 livres. Two thousand hangars would have to be operating continuously before the national requirement for saltpetre could be met. The risk was too great for a single entrepreneur, the scale too large for the crown. The policy, therefore, must be to tempt private entrepreneurs and local communities into the trade. A decree of the Council of 24 January exempted parishes from the fouille on condition that they construct plants with a capacity equal to their usual harvest. Religious houses and other communities were also invited to commute their obligations. The régie announced a program of premiums and rebates to be paid for saltpetre acquired from proprietors of plants.[206] All in vain. The little capital that ventured was lost. It would take stronger incentives and firmer measures than any known to the Régie des poudres to reach beyond the arsenals right down to the bottom of this old trade. Thus thrown back into dependence on the fouille, the régisseurs purged it of its worst abuses sufficiently to enrage the saltpetremen though not to reconcile property owners, and it is ironic to read the defense that, in the early necessities of the Revolution, Lavoisier was driven to advance of the practice that he and his collaborators had instituted the régie to obviate: a "slight nuisance" he called it in April 1789 and a little later, "a small sacrifice that every good Frenchman easily makes in the public interest, above all when he considers that it costs him nothing in his fortune, his person, or his liberty, which is never transgressed when it is the nation itself that lays down the regulation and necessity that dictates it."[207]

A miscarriage of chemistry was more shattering. In 1785 Claude-Louis Berthollet, a resourceful chemist never quite of Lavoisier's circle, prepared a new salt, a "muriate" containing a high proportion of oxygen and for that reason to be named potassium chlorate in the new nomenclature. He was a skilled, even an elegant operator in the laboratory. Like the English chemists, whom he admired, he enjoyed investigating individual substances: ammonia, tartaric acid, sulfur compounds, fulminating gold, and "dephlogisticated marine acid air" (i.e., chlorine). Widespread fame came to him from the discovery in 1785 of the bleaching properties of chlorine. Born in the village of Talloire in 1748, Berthollet derived from a French

[206] "Observations sur la récolte du salpêtre," *ibid.*, 690-691; Bottée and Riffault (1811), cviii-cxij; Payan (1934), 150-157; *Mémoire . . . pour les vingt salpêtriers du roi* (Paris, n.d.), BN, L f[65].11.

[207] "Mémoire sur . . . la Régie des poudres," Lavoisier, *Oeuvres* 5, 694; "Mémoire sur le service des poudres," *ibid.*, 707.

family settled in Savoy. He followed a medical course at the University of Turin and took his M.D. in 1768. Feeling that it was in him to be more than a mountain doctor, he came to Paris and found early patronage for his medical practice and scientific interests from the duc d'Orléans, whose scientific entourage in the Palais royal was given to freemasonry in a slightly *louche* way. He submitted many ingenious and original chemical investigations to the Academy, where they won esteem from Macquer and from Lavoisier, and on 21 April 1780 Berthollet was elected to membership. The chance that shaped his career toward the problems of industry occurred four years later on the death of Macquer, Nestor of the old chemistry, whose *Dictionnaire de chymie*[208] still gives the readiest explanation of its terminology and concepts, and who combined the function of professor at the Jardin du roi with those of director of dyes at the Gobelins and consulting chemist in the Manufactory of Porcelain at Sèvres. On 24 February 1784 Berthollet was appointed to succeed him at the Gobelins, which post came under the Bureau of commerce. Provided as Commissioner for Dyes with a laboratory and an assistant, Berthollet devoted most of his energy and all of his interest to chemistry and the scientific improvement of the products of chemical art. His treatise on the dye industry is a model of the rationalization that descriptive science could bring to the labyrinth of lore and thicket of recipes that constituted knowledge in an ancient trade.[209]

In the course of exploring the chemistry of chlorine compounds in his research on bleaching Berthollet discovered the properties of potassium chlorate. Because of its high oxygen content, it is a treacherous chemical, violently unstable in the presence of reducing materials, particularly carbon and sulfur. Might it not, thought Berthollet, after its extreme touchiness became evident in laboratory samples, be employed in place of saltpetre in order to fabricate a gunpowder more powerful by far than any yet known in warfare? Evidently he raised the question with Lavoisier, who wrote of the tests that ensued, "It is difficult to decide whether discoveries of this type are advantageous or not to humanity, but what is not doubtful is that great advantage can result from them for the nation that first makes use of new means of attack."[210]

The régisseurs made no doubt of their own duty and went directly to the controller-general with a program of experiment and development.

[208] *Dictionnaire de chymie, contenant la théorie & la pratique de cette science, son application à la physique, à l' histoire naturelle, à la médecine & à l'économie animale*, 2 vols. (Paris, 1766), first published anonymously. A second, revised edition appeared in 1778.

[209] *Eléments de l'art de la teinture*, 2 vols. (1791); on this work and Berthollet's early career in industrial chemistry, see below, Chapter VI, Section 2.

[210] "Sur l'accident d'Essonnes," Lavoisier, *Oeuvres* 5, 742-745; see also *Journal de Paris*, 31 October 1788.

Berthollet moved his quarters to the Arsenal for a time to join their efforts. In late September and throughout October 1788, the staff labored to produce a quantity sufficient for trial in the field. Before this time, Barbault de Glatigny had been replaced as régisseur by Le Tort, a veteran of the industry. On 26 October Lavoisier, Mme Lavoisier, Le Tort, and Berthollet took a carriage for Essonnes, the village near Corbeil where the powder-mill was situated that served the Arsenal of Paris. Le Tort, in charge of the tests, had had a special mill constructed in the open. It employed a single mortar-and-pestle powered gingerly by a hand-crank of which the shaft passed through a protective barrier of thick planks.

Composition of the new powder began at six o'clock the next morning. Although moderately moistened, the mixture of charcoal, sulfur, and chlorate persisted in sticking to the pestle on the upstroke. The batch amounted to sixteen pounds. To give the apparatus more to work on, Le Tort increased it to twenty. Still it stuck. Impatient, he took a piece of wood and scraped the material back off the pestle as it rose. To remonstrances, he answered that the powder had scarcely begun to work and that there was no danger. The party now consisted also of Chevraud, commissioner in the gunpowder service (and winner of the second prize in the Academy's recent competition); his young, unmarried sister; one Aledin, a powder-master of the establishment; and Mallet, an apprentice. All stood about Le Tort, peering at the mortar and its recalcitrant contents, the while he joked "*même avec gaieté*" on what an explosion would do.

Lavoisier was not amused. Everyone, he objected, must take shelter behind the barrier while the mill was working. The powder must not be stirred except when the machinery was at rest. Stamping must be interrupted for that purpose after every six or eight strokes of the pestle. He enjoined, he insisted, but the morning's mood was light. It continued so throughout mealtime, when Mallet and Aledin were left in charge of the milling, not without further admonition by Lavoisier. Breakfast was brisk, a quarter of an hour. Le Tort led the way back to the work together with Mlle Chevraud. Her brother with Lavoisier and his wife stopped behind a moment to explain one of the regular powder mills to Berthollet. The time was a quarter before nine. As they turned away, still some distance from the chlorate test, the new powder exploded amid a dense cloud of smoke. They ran to help. They found the mortar in shards, the pestle hurled afar, and Le Tort and Mlle Chevraud thrown thirty feet, crushed against a wall of millstones, and dying.

The Régie des poudres then abandoned its trials of potassium chlorate. For a time. Until the war. Not so Berthollet, however, whose initiation this was to the manufacture of munitions, and who met with further encouragement from the military engineers. But that is another story.

In the end it was by no single stroke either of science or of policy that the Régie des poudres won an ailing industry back to health. Rather, the régisseurs prevailed by dint of sound administration, by attention to personnel, by vigilance over costs, and by proper bookkeeping; by daily, weekly, and monthly measures of which the results, though not the history, may be read in the annually increasing production of saltpetre, fabrication of gunpowder, and payments into the treasury. Regular training courses were instituted to instruct the staff. A settled system of promotion through definite grades came into effect throughout the service. In 1787 uniforms were authorized. The apprentice personnel took examinations in geometry, elementary mechanics, and experimental physics and chemistry. Allied trades were regulated or encouraged. Potash, for example, was superior to wood ashes in converting the "earthy" nitrates into crude saltpetre, and the Régie des poudres approached the ministry with a plea to prohibit its exportation.[211] The supply of potash in turn depended on that of alkali. To liberate the former for munitions created incentives to develop an "artificial" manufacture of soda. The Leblanc soda process, which became one of the foundations of heavy chemical industry in the nineteenth century, developed out of the continuation of that effort during the Revolution.[212] The profession of industrial chemistry may itself be traced in large part to the staff and suppliers of the Régie des poudres. Chemical industrialists grew closer to the scientists in Paris—one of them Chaptal of Montpellier, won recognition in both realms. Others—Leblanc, Carny, Champy—formed the skills in virtue of which in the flux of the Revolution they aspired to increase themselves from tradesman or artisan to entrepreneur and manufacturer.

Historically the most famous, if not the most significant, technical innovation of those revolutionary years has been the method of refining saltpetre that permitted the fabrication of gunpowder by the aroused populace in the *levée en masse* of the Year II. Revolutionary martyrology represents Lavoisier offering the Republic munitions as the fruit of his science on the eve of their mutual appointment with the guillotine. It is true that the technique of refining "revolutionary" saltpetre was indeed his, but as has already been hinted it derived from the work of the Régie des poudres, from good housekeeping and not from theory, and specifically from the problem of devising an accurate test for determining the quality of the crude raw material that the saltpetremen of Paris dumped into the receiving yard of the Arsenal.

But this, though a continuation of the same story, must like Berthollet's further work on explosives await another book.

[211] Bottée and Riffault (1811), cxij; for Lavoisier's experiments on the use of woodashes, see "Expériences sur la cendre qu'emploient les salpêtriers de Paris" (1777), Lavoisier, *Oeuvres* 2, 160-173.

[212] Gillispie (1957a).

CHAPTER II

◇◇

Science and the State

◇◇

1. CLASSICAL INSTITUTIONALIZATION

During the half-century between the Turgot ministry and the Revolution of July 1830, or (to embrace the interval in dates with scientific significance) between the last years of d'Alembert and the death of Laplace in 1827, the French community of science predominated in the world to a degree that no other national complex has since done or had ever done. In its eminence, French cultural leadership in Europe reached a climax. The critical movement of classicism that first made itself fully felt in the realm of letters, architecture, and manners under Louis XIV, and that next became the system of rational ideas about nature and humanity constituting the enlightenment, issued finally in the incorporation of science into polity amid the circumstances recounted in this book. From the perspective of history of science proper, it is possible to see a second scientific revolution in what then began to transpire. Manifesting in science the displacements and renewals of revolutionary Europe, the thrust proved to be organizational rather than cognitive, as it had been in the seventeenth century. Indeed, purists may argue that only in consequence of these changes in France and later abroad does it become appropriate to speak of "scientists" and "science" in the sense of fully professional persons cultivating the modern disciplines, notably physics and biology, with their modern array of specialties and problems.

The point may at once be granted. Among the themes in this history is the transmutation of the century-or-two-old "experimental philosophy" and the age-old natural history, though it would be artificial to deprive these earlier stages of physics and biology of the name of science. For the point must not be granted simplistically, nor the relative autonomy of scientific content given away with it. It would be naive to try to map out a complete and systematic correspondence between the respective histories of the content and the organization of science. The study of chemistry, for example, went through the most profound among conceptual and the least extensive among professional transformations in these fifty years, while the study of astronomy changed but little in either respect, the gain (though large) having been in precision and volume of research. But to say that connections between aspects of an enterprise or situation are indirect and partial is not to hold that there are no relations or that they are unimportant. In instances wherein the content of science was touched by its involvement in the great events of the time, and the instances were many, and cumulative in effect when they were significant, it is to the selective influence of external factors that we must look for an explanation: the attraction of talent into certain sets of opportunity arising in the world at large; the presentation of certain problems to the scientists along with the material means for solving them; the regrouping of scientists into new forms of association. It would be misleading to take the modifications in science for evidence of some inner improvement in tone or quality. In those respects, as high a standard of discrimination relative to the state of knowledge as could be required of any body of science had already been attained in France by the late eighteenth century. Indeed, the main concern of the present chapter is to exhibit the flourishing state of the scientific enterprise in the last years of the old regime, which had endowed it with the most elaborate set of institutions in the world since the time when Colbert had founded the Academy of Science in 1666.

The attention of historians interested in scientific affairs was long ago drawn to the appearance in the mid-seventeenth century of the Accademia del Cimento in Florence, the Royal Society in London, and the Academy of Science in Paris, actual institutions emerging with the "new philosophy" amid the technical and conceptual developments comprised in the Scientific Revolution.[1] To these bodies their first historians attributed the provision of encouragement, information, instrumentation, sociability, and visibility—benefits that men of common interests might be expected to draw from one another's company under official patronage. Only in consequence of recent sociological study, however, has it begun to be understood that this gathering of men of knowledge into formal organization was somehow more than gregarious, and that throughout modern history it has been related to the very formation of science in a way that the desultory associations of writers and critics concerned with expression in the literary culture have never needed to be.[2]

The distinction derives from differences in the pattern and function of communication. It may be brought out by a comparison of phrases long in use. We say "Republic of Letters," and the image implies the intrinsic individualism of literary creation, its openness of expression. Writers have not traditionally written for each other. They write for the public. Not so the scientists, for whose collectivity the natural figure of speech is the more exclusive "Community of Science." A community is for communi-

[1] The classic works are Ornstein (1928); Merton (1938); Brown (1934); and Olschki (1927).

[2] Ben-David (1964), (1965); Merton (1957), (1961), (1965).

cation among other things, and the significant exchanges occur inside. Science is nothing until reported, or at most it is only private information, and when reported it is not judged and appreciated like some work of art. It is verified. A precept of scientific procedure imposed itself early in the seventeenth century: in principle, no experiment was to be accepted, no observation to be trusted, no phenomenon to be established, unless the work might be reproduced and the measurements or observations confirmed by scientists other than the one who announced the results and claimed them for his credit. Whether or not those rules were observed in normal practice, it tells much about the scientific enterprise that the earliest institutions should have arisen out of the necessity for communication, and not the other way about.[3]

From the very beginning, reformation of language was taken to be concomitant with the new philosophy. Bacon warned that the errors residing in words are among the more misleading idols, and looked to creation of a natural language in which names might denote things, and the rules for their combination might reflect relations in nature. Such was in fact the purpose of the nomenclature devised for modern chemistry by Lavoisier's school. That scheme owed its inception to philosophy, specifically to the precepts of Condillac, who systematically developed the notion that science is language, the only really well-made language, in which the syntax is analytic of experience as algebra is of quantity.[4] From the outset, accounting for the special success of communication in science has been a preoccupation for its philosophy. To an observer contemplating the whole panorama presented by three centuries of modern science, it does seem as if communication had succeeded there more nearly than in other concerns. The proposition involves paradox. Exponents of the literary and verbal culture might be expected to have been the more communicative. But talk is either more or less than what passes between scientists in their work, for in one form or another the restriction that makes possible the relative efficacy of those exchanges is that they be confined to matters to which people in ordinary life are generally quite indifferent. This limitation may be called literally a precondition of science, for it was clearly stated at the outset, though rather in political than philosophical terms. The rules of both the Royal Society in London and the Academy of Science in Paris, adopted quite independently, precluded political and theological discussion. Thus an ideal of neutrality for science was institutionalized

[3] For full references to the early history and pre-history of the Academy of Science, see Hahn (1971), and for references to the even more voluminous literature on the Royal Society, see its most recent historian, Purver (1967), whose actual argument is less than convincing, however. See also Lyons (1944). For the Accademia del Cimento, see Middleton (1971).

[4] Gillispie (1960), 165-170, and (1971b).

from the start, in order that men concerned with augmenting knowledge should be undistracted in their proper business by the contentions forever emanating from the forum and the pulpit. The spirit was not at all one of repressing scientists, but of liberating them from the distress of vain disputes amid the confusion that politics creates in the world.[5] There was nobility in this sentiment, and to denigrate it would distort the nature of scientific values. Yet nobility entails a certain condescension, and the implication has ever been that only scientific discussion gets results.

Beyond mere fellowship, the values operating in a community, and the norms governing the behavior of its members, are what differentiate it from other associations. In this regard, Robert K. Merton's work on the dynamics of the scientific community has taught us to understand certain motivations for the inwardness of which historians, like scientists themselves, had seldom looked beneath the pieties attesting love of truth and knowledge.[6] Scientists do wish the truth found, of course, but they also want, and perhaps more passionately, to be the ones who find it. They do not, therefore, want their truths discovered by someone else,[7] and Professor Merton proceeds to a systematic analysis of episodes like Lavoisier's secrecy, not to say deceptiveness, about the contents of his sealed note and Laplace's youthful assault upon the attention of the Academy.[8] For those familiar with science either actually or historically, that analysis made sense of what until then had seemed the trivial, if vaguely unpleasing, awareness that hardly any significant discovery has occurred unattended by a priority dispute, often involving charges, veiled and otherwise, of plagiarism. Lavoisier and Priestley over oxygen, Newton and Leibniz over the calculus, Mayer and Joule over the mechanical equivalent of heat—the list is as long as the annals of science and bespeaks an acrimony inconsistent with selfless love of knowledge. Professor Merton is himself to be credited with the recognition that science has evolved its value system in a polarity between conflicting sets of norms.

On the one hand, scientists have supposed themselves to be modest, disinterested, unassuming, and absorbed in the study of nature for its own sake and the good of humanity. On the other hand, an imperative governing the entire scientific enterprise is the high, the overwhelming premium it places upon originality, and the motor has ever been competition to achieve recognition for its exercise. True, scientists have often been personally shy and retiring, and have seldom cared greatly about worldly po-

[5] For an especially revealing passage to this effect, see Sprat (1958), 55-56; see also Hahn (1971).

[6] Merton (1957), (1961), (1965).

[7] A point that comes out clearly in James D. Watson, *The Double Helix: A Personal Account of the Discovery of the Structure of DNA* (New York, 1968).

[8] Above, Chapter I, Section 5, 6.

sition or property in the ordinary sense. But they have cared terribly about their intellectual property in their discoveries, scratching their initials upon the Parthenon of nature with eponymous passion. It is by the name of the discoverer that laws, effects, and processes are generally known in science, a practice largely responsible for the very misleading notion that it consists of an accumulation of treasure trove that, like gold coins, had lain there all along concealed under layers of ignorance until unearthed by the enterprise of the discoverer. Reflection on this phenomenon will make apparent that scientists have generally been far from selfless, and on the contrary have had to be highly ambitious—seeking, not money, to be sure, nor political eminence, nor martial glory, but fame and reputation like the artists and engineers of the Renaissance, whose successors in such respect they are. There is no reason to regret the fact except in a mythology that would make scientists greater or meaner in moral stature than other men. At all events, the precise provisions established by the research societies for reporting discoveries and establishing priorities have been intrinsic to the functioning of science from the beginning.

The waxing and waning in brilliance of great cultural galaxies appearing in history usually resist attempts at causal explanation. Nevertheless, a particular application of the analysis just summarized may cast some light on how it happened that by the end of the eighteenth century the large scientific predominance of England in Newton's time had yielded to that of France. For the Academy of Science had definitely been second to the Royal Society of London in their early decades, though by no means a poor second. With the failure of the Accademia del Cimento in Florence, those two bodies became the scientific exemplars of the shift of cultural leadership from Italy to western Europe.[9] Since then, they have epitomized in their respective relations to the state the two main institutional patterns, the official and the voluntary, that have evolved for science and learning generally in modern history. Throughout Europe and Latin America, the official mode of patronage has been adopted from France. The voluntary mode has prevailed in Britain and across the seas in this country, where support and direction of cultural, scientific, and higher educational affairs have been largely left to private endowment, particular foundations, and local authorities.

The divergence between the courses taken by the French and English states in early modern times was responsible for these differences in the patronage of science and culture. For France and England together had earlier developed the prototypes of the modern state as Europe merged from feudalism. No two systems of government and sets of institutions were more similar in the fifteenth and sixteenth centuries. Both were

[9] For an interesting discussion of the contrast, see Ben-David (1965), 30-40.

highly centralized monarchies, national in scope and deriving from a common stem of medieval kingship. The polities were similar in class and social structure. Following the Puritan civil wars in the seventeenth century, however, English affairs took the route that led to limitation of the monarchy, its subordination to the classes ruling through Parliament, and its eventual conversion into a symbol. Not only the crown was restricted. In the eighteenth century the state itself retreated to its minimal business of keeping order, enforcing law, conducting diplomacy, and fighting wars, while economic, social, and cultural concerns were abandoned to private persons and associations. The Royal Society of London benefited from its designation in name only. It received nothing from the crown beyond its charter and its mace, and English science went quite ungoverned except by the corporate arrangements of this voluntary group, in some sense a science club.

In France matters fell out more favorably for science in the long run. There, civil and religious war ended in the sixteenth century. The state having barely survived those conflicts, the policy of successive ministers in the seventeenth century, Sully, Richelieu, Mazarin, and Colbert—indeed of Louis XIV himself—was to magnify the crown in all ways. Among other respects, it began dispensing the kind of patronage once extended to arts, letters, and science by Italian rulers of the Renaissance. In London, where the promoters of the Royal Society had gone to the crown and asked for a charter, the initiative lay with science. In Paris it was otherwise. Colbert, learning that an unofficial scientific circle existed and knowing of the foundation of the Royal Society, thought to enmesh this new enterprise, too, in the machinery of state. He had before him the model of the Académie française and the example of other, lesser foundations for the arts. Thirty years previously, Richelieu had incorporated the forty leading lights of life and letters into the body of immortals to which election has since been a sovereign accolade. It was then charged by the crown with forever sharpening the edge and protecting the polish of the main instrument of French intellectual supremacy, the French language. In virtue of the same, quasi-Italianate instinct—two of the Medici had been queens of France and regents, after all, and Mazarin was Italian—for placing the state at the center of cultural panache, artists were formed into an Académie royale de peinture et sculpture in 1648; the first of the foreign academies was established in Rome, the Académie royale de France in 1655; scholars were gathered into an Académie royale des inscriptions et belles lettres in 1663; builders were provided with an Académie royale d'architecture in 1671; players were endowed with a Comédie française at a date less easy to specify so categorically; and Italian singers having been initially imported by Mazarin, an Académie de l'opéra was chartered in 1669. In effect, the extension of patronage developed through the three stages of letters (Richelieu); fine arts and theater (Mazarin); and scholarship, science, and engineering (Colbert).[10] By the charter incorporating the Academy of Science in 1666, therefore, the adjective "royal" was more than merely a courtesy to scientists. It conferred authority.

Speaking in broadly historical terms, it may very well be that the English institutional and constitutional pattern has proved the more auspicious for the protection of political liberty and the encouragement of economic enterprise. Such was the view of many literate and informed persons, French and English, who were attentive to these developments in the eighteenth century, and many historians have since agreed that the genius of modern English history resides in its combination of personal liberty with governmental stability.[11] With science, however, it was different in the period that concerns us. True, the dynamism and ferment of Puritan and Restoration England had evidently been more encouraging to scientific inquiry and talent at the start than the rigidities and artificialities surrounding Louis XIV. But once the impulse of the first generation of the Royal Society of London was exhausted, its voluntary nature neither stimulated nor even sustained amid the complacencies of the Whig supremacy a scientific enterprise that could match in vigor, and perhaps more important in rigor, the researches started and the work accomplished under the official aegis of the Academy in Paris and its counterparts, virtually its dependencies scientifically, in Berlin, Turin, and Saint Petersburg.[12]

It is a liberal's vanity to imagine that freedom, for him the better cause, has been an historical concomitant of science, either as condition or consequence. Neither a Bacon nor a Descartes ever prophesied that it would be, and the technical accomplishments of the Soviet Union are immense

[10] The standard work on the French Academy is Mesnard (1857); and that on the Académie royale des inscriptions et belles lettres is Maury (1864a). These together with the Academy of science were the premier learned academies, incorporated in the Institut de France during the Revolution. For documents and regulations governing all three, see Aucoc (1889). For the Académie royale de peinture et de sculpture, see Vitet (1880) and Lemonnier (1911, 1913); and for the Académie royale d'architecture, see Lemonnier (1911, 1913). For the Comédie française, see Valmy-Baysse (1945); for the Opera, Demuth (1963).

[11] This is the theme of Elie Halévy's great *History of the English People in the 19th Century*, 6 vols. (2nd revised ed., London, 1949-1952). For a discussion of the argument, see Gillispie (1950).

[12] For the history of the Berlin Academy, see Harnack (1900); and for the history of the Saint Petersburg Academy, together with a record of its correspondence with Paris and Berlin, see *Uchenaia Korrespontentsiia, XVIII veka . . . 1766-1728* ["The Scientific Correspondence of the Academy of Sciences of Petersburg"], ed. I. I. Lubimenko, G. A. Kniazev and L. B. Modzalevsky (Moscow and Leningrad, 1937). For a comparative history of the structure and functioning of the major national academies in the eighteenth century, see McClellan (1975).

evidence that it need not be. Scientists, like most men, may enjoy freedom but seldom require it in the way that writers do. Instead, they need standards, something more determinate than sensibility by which their work may be judged. They need support, for few and usually insignificant have been the researches that were economically profitable to the author. They need motivation, for science is hard and not the sort of thing people do to seek their ease. All this was afforded by the prospect for election to the Academy in Paris, the goal for young ambition among scientific talents throughout the most civilized country in Europe, and the model for its emulators in the other capitals, where the academies were in constant communication with Paris and always in French. Indeed, no complex of scientific institutions offers more instructive illustrations of the governance of science in its community by honorific, if not by economic norms; by desire of participants for prestige, if not for other types of profit. Perhaps the play of such ambition appeared less nakedly abroad, where the central academies, like the provincial ones in Montpellier, Dijon, Bordeaux, and Toulouse (to name the more important), associated arts and letters with the sciences in a single corps.[13] For only in Paris were the sciences distinct from the other branches of culture and given in charge to a body defined by their boundaries, a *corps d'état*, privileged for science.[14]

2. THE ACADEMY OF SCIENCE

It is natural that the scientific institutions of eighteenth-century France should have exhibited in miniature the structural characteristics of the regime that sustained them: monarchical, hierarchical, prescriptive, and privileged. The most familiar of those still surviving, officially the Muséum national d'histoire naturelle, and usually called the Jardin des plantes for its public garden, was then the Jardin du roi under the intendancy of the comte de Buffon. The most ancient, the Collège de France, founded by François Ier, was then styled "royal," as was the Observatory of Paris, officially under the direction of the Academy. Like the others, the Academy itself was an administrative dependency of the royal household illogically divided between two ministries, the Batimens du roi, which Turgot put under the direction of the comte d'Angiviller, and the

[13] On the Dijon Academy, see M. Bouchard (1950) and Tisserand (1936); and on the others see Bouillier (1879), Mornet (1954), and Saunders (1931). Roche (1965) gives a more analytical, sociological analysis, with special reference to Bordeaux, Dijon, and Châlons-sur-Marne. His recent treatise (Roche, 1978) supersedes existing literature on the provincial academies. His interest is in their relation to the class structure and their role as disseminators and forums of culture and knowledge, not as originators.

[14] For the regime of the Academy, see, in addition to Hahn (1971), the documents printed in Aucoc (1889), Gauja (1967), Maindron (1888), and Maury (1864b), the last-named being the best of the older histories.

Maison du roi, whose most active minister before the Revolution was the baron de Breteuil in the 1780s. More important for the functioning of the Academy, it was divided in its internal organization into something like estates.

The twelve honorary members comprised in its highest class were habitually chosen from the nobility and the magistracy, men of the quality of the duc de Noailles, the comte de Maillebois, Lamoignon de Malesherbes, Trudaine de Montigny, the President Bochart de Saron, and the duc de la Rochefoucauld-d'Enville. Their function was to be patrons and ornaments, although the last three named could and did contribute modestly to the work of science. Only the *honoraires* were eligible to serve in the offices of president and vice-president, to which election was for a year. In practice most of these notables seldom attended the sessions of the Academy and left direction of its affairs largely to the senior rank of serious men of science, the *pensionnaires*. In principle there were eighteen such, together with the permanent secretary and the treasurer, offices held for life. From their number a director and under-director were chosen annually to preside at meetings and oversee business in the normal absence of the president. Supplementary places were occasionally authorized, as happened for Condorcet in 1773, when he began acting for Grandjean de Fouchy before succeeding him in 1776 as permanent secretary. Only the *pensionnaires* qualified for stipends. Their presence at meetings was attested by presentation of a *jeton*, to be redeemed for cash, and their names were recorded in the ledger separately from those of their junior colleagues. They shared with the honoraires the vote in election of new members of all ranks, and with the next lower class, consisting of twelve *associés*, the decisive voice in the scientific determinations reached by the Academy. Below the associates were *adjoints*, also twelve in number, and originally called *élèves* until, in 1719, it was objected that the designation was undignified. Even so the adjoints retained the quality of neophytes, of the Academy but not participating in decisions and required to sit apart in meetings.

Such were the main categories by rank. By specialty the ordinary academicians were distributed among six sections, each consisting of three pensionnaires, two associés and two adjoints. Three sections pertained to the division of "mathematical sciences"—geometry, astronomy, and mechanics—and three to the "physical sciences"—anatomy, chemistry, and botany. Nothing ever being simple about the institutions of the old regime, however, certain memberships-at-large existed and were also called associate. Distinguished foreigners were honorary associates. *Associés regnicoles* were Frenchmen residing in the provinces. In addition, there were ten or twelve *associés libres* living in the capital who might attend sessions when they pleased, but who were ineligible for promotion to the class of pensionnaire. The emeritus status of *vétéran* had also been instituted in the hope, infrequently fulfilled, of inducing voluntary retirements.

All these devices, various though they were, had not sufficed for all contingencies, and from time to time d'Angiviller authorized a supernumerary membership in a certain class and section. An example was the provision made for Lavoisier when the then-minister overruled the Academy's recommendation in 1768 and in his stead named Jars to the section of chemistry.[15] In addition, the Academy designated non-residents of Paris to be correspondents assigned to a regular member through whom their communications were received. Not counting these last, when the Academy assembled on a ceremonial occasion, say to attend the coronation of Louis XVI at Reims; to offer a reception to some visiting dignitary, for example the Czar Paul or Prince Henry of Prussia; or to hold its public *rentrées* at Easter, and in the autumn, semi-annual features of the Paris season—at these times its full and present membership would muster some seventy persons.[16]

Of that number, only fifty-odd pensionnaires, associés, adjoints; and supernumeraries were working scientists in the regular sections. Their company was small and desirable enough to bring about intense competition for entry, and large enough to constitute a society involving a complex web of relations among its members, a set of worthy belongers. Perhaps it will not be far-fetched to compare its life to that of some single department comprising one of the major academic disciplines in a university of our own day. The scale was about the same. The Academy also combined collegiality with stratification into stated ranks and promoted highly intelligent persons up the grades from probationary standing to seniority, according them a definite and increasing measure of responsibility at each echelon. Advancement was a function of vacancy at the top combined with seniority and the evaluation that older, secure, and outwardly satisfied intellectuals made of the intelligence, the diligence, and the promise of younger, insecure, and inwardly ambitious ones. In effect, such a body makes its contribution, not through what it can do in its corporate capacity—in the case of the Academy the project of collective research had proved visionary in the seventeenth century—but through what its existence leads its members to do in order to gain in its esteem and consequently in their own.

In that respect, both the eighteenth-century Academy and the modern academic department have proved to be powerful dynamos for the generation of knowledge. Producing knowledge and communicating it have al-

[15] Above, Chapter I, Section 6, n. 171.

[16] On the regime of the Academy, see Hahn (1971), 76-83; Birembaut (1957); and the introductory "Histoire" that Condorcet prefixed annually to the successive volumes of the *Histoire et mémoires de l'Académie Royale des Sciences*.

ways been individual engagements of the scientist and scholar with his colleagues and more recently with his students. What they do collectively, when sitting together, is sit in judgment. Judging, judging—the matters other than scientific in which the Academy did so will appear in a moment—they are always judging, and quick to be jealous of the privileges that entitle them to do so. Those privileges they have won by intellectual right, not birthright. Scornful of birthright, theirs tends to be a meritocracy sheltering some of its more corrosive critics.

In the light of these considerations, it might well be thought, and sometimes it has been said, that the superior scientific efficacy of the Academy in Paris as compared to the Royal Society in London derived from its exclusiveness and consequent professionalism. It is true that the Royal Society, supported only by dues, had to admit enough people who would pay them to make ends meet, and lacked, therefore, the power to elicit scientific talent by holding out the prospect for election to its company or to reward accomplishment by advancement within its fellowship. Its membership at the end of the eighteenth century was on the order of 500 Fellows.[17] Nevertheless, the selectivity exercised by the Academy, though it certainly led in the direction of professionalism, may not, without anachronism, be identified with that stage in the development of science. Achieving a properly professional standing was one of the purposes of the protagonists of science when the Revolution dissolved the old order of things; and, as will appear, imputations were already being heard, whether justly or no, that the Academy was guilty of favoritism, sycophancy, arrogance, and corporatism, qualities inconsistent with a fully professional mode of conduct. It will be important for much that follows, therefore, to consider closely what is meant by the term "professional," and to identify which of the elements were present and which lacking in the regime of eighteenth-century science in France.

A profession—perhaps it may be agreed—is, in the first place something more than an occupation, in that its practice presupposes mastery of a body of knowledge and thereby qualifies the practitioner for the prestige that from classical antiquity to the present has attached to the theoretical and cognitive when contrasted to the crafty and commercial. In the second place, however, a profession does share the economic attribute of an occupation. It is legitimately followed for gain and is not a status held by right. For reasons of dignity, the words "profit" and "wages" are not applied to a professional man's rewards, but he does live from the fees or other honoraria that the direct employment of his knowledge brings him.

[17] For the classic indictment of the regime of the Royal Society in the late eighteenth, early nineteenth centuries, see Charles Babbage, *Reflections on the Decline of Science in England* (London, 1830), a highly polemical work, and for a modern account of its reform, see Cardwell (1972). For membership figures, see Lyons (1944), 343.

Finally and most distinctively, a profession is self-governing in that it exercises jurisdiction over the education, qualifications, and conduct of its members. In accordance with an actual or tacit delegation of regulating power by the authorities, justified by the public interest, it undertakes to police itself, claims to police all who would deal with its subject matter, and secures its members partial shelter from the interplay of external scrutiny, economic sanction, and political accountability that governs the behavior of laymen in their occupations.[18]

If this definition of professional is acceptable, it follows that the word when applied to vocations in the eighteenth century must be reserved for divinity, law, and medicine, the subjects traditionally confided to the three higher faculties of the university. Science did not yet qualify. It was coming close, closer than the humanities and social sciences, and much closer than pedagogy, architecture, or the military arts. But even the French community of science, though more advanced than its counterparts in other countries, could fully satisfy only the first of those criteria, the possession of natural knowledge.

As for the second—livelihood—scientific breadwinning was an unsystematic affair. The amount annually awarded the pensionnaires of the Academy averaged about 2,000 livres apiece, an agreeable addition to an income, but not, properly speaking, an income fit for a good bourgeois, which would have totaled five to ten times that much. For the most part the senior members, who mainly received these stipends, were already ensconced in one or another of the posts that did support a modest style. Their positions generally entailed possession of technical knowledge, but for purposes other than its advancement, and engaged their abilities obliquely to their research and well behind its frontier. In the 1780s the ministry of the Maison du roi made a policy of reserving places that could and should have been filled by scientists for members of the Academy.[19]

Of the major pieces of scientific patronage, Lavoisier's office of régisseur des poudres was probably the most lucrative. We have already seen how Berthollet, having first secured a medical income from the duc d'Orléans,

[18] For discussion of professionalism in science, see Ben-David (1972), Johnson (1972), Shils (1968), and for particular application to this period, see Hahn, "Scientific Careers in Eighteenth-Century France," and Crosland, "Development of the Professional Career of Science in France," in Crosland (1975).

[19] In a letter of 7 March 1738, d'Angiviller explains the thinking of the ministry to Cassini de Thury, director of the Observatory. The immediate subject was a successor to Maraldi in the post of concierge there. D'Angiviller explains that he cannot give it to Cassini's preference, one Vallos, because he was not of the Academy. As for those who were, Jeaurat already had the *Connaissance des temps*, which he edited well, and a pension from the Ecole militaire. The abbé Rochon was absent from Paris too much. So he would choose Méchain, who was young, married, poor, and able. (Bibliothèque de l'Observatoire, D-5, 40.)

succeeded Macquer as commissioner of dyes at the Gobelins in 1784;[20] how Bossut was given the chair of hydrography that Turgot instituted at the Academy of Architecture, which lessened his dependence on the sale of his textbooks and on his fees for examining the candidates who had to master their contents for Mézières and the Ecole militaire;[21] how Turgot had installed Condorcet at the Mint, which service associated him both with Tillet, director of the Mint at Troyes and royal commissioner of assaying,[22] and most uncongenially with Sage, whose chair and laboratory of assaying were located there in the Mint, functions that he combined with a professorship at the School of Mines in 1783;[23] and finally how d'Alembert had had Laplace named professor of mathematics at the Ecole militaire, a job he hated and quit in 1783 to succeed Bézout in the task of setting examinations for the artillery, leaving Adrien-Marie Legendre to teach the military cadets.[24]

Several academicians were themselves military men. Coulomb and Meusnier made careers in the Engineers. Borda and Bougainville were line officers, Borda rising to a captaincy in the navy,[25] and Bougainville to brigadier and squadron chief in the marine infantry.[26] Dolomieu's profession was no doubt the most anachronistic. He was a Knight of Malta.[27] Medical men generally commanded the highest incomes, however. Pierre-Isaac Poissonier, inspector of military hospitals and chief physician with the army in Germany in the Seven Years War, enjoyed pensions from the crown totaling 14,000 livres.[28] Louis-Guillaume Le Monnier, chief medical officer until 1759, received a lavish 20,000 livres as physician to the king.[29] In the Academy he was a botanist. The most eminent scientifically of the medical people were unconnected with the military and did less well from official favors. Vicq d'Azyr, besides his secretaryship of the Royal Society of Medicine, gave private courses in anatomy, physiology, and surgery, a public course in comparative anatomy at the veterinary school of Alfort, and as physician to Marie-Antoinette and the comte

[20] Above, Chapter I, Section 6; also below, Chapter VI, Section 2.

[21] Hahn (1962).

[22] On Tillet, see Wehnelt (1937).

[23] On Sage, see Birembaut, "L'Enseignement de la minéralogie . . . ," in Taton (1964), 365-418; *Archives parlementaires 13*, 523; and also below, Chapter VII, Section 2.

[24] Duveen and Hahn (1957); Hahn, "Les Ecoles Militaires" in Taton (1964), 524; Hellman (1936).

[25] On Borda, see Mascart (1919), a book that is unreliable in details.

[26] On Bougainville, see Martin-Allanic (1964).

[27] See below, Chapter III, Section 4.

[28] *Archives parlementaires 13*, 360; Jean Torlais "Le Collège Royal" in Taton (1964), 281.

[29] *Archives parlementaires 13*, 312; on Le Monnier, see Robida (1955), Paul Delaunay (1906), Yves Laissus, "Le Jardin du Roi," in Taton (1964), 326.

d'Artois was paid a pension of 4,600 livres.[30] Antoine Portal received, besides the stipend of his professorship of anatomy at the Jardin du roi, a pension of 500 livres beginning in 1773 in recognition of his publications in science, and another 1,200 beginning in 1776 "in consideration of the work he had composed on methods for resuscitating the drowning."[31]

Secondary school teaching and popular science provided more modest opportunities. Mathurin-Jacques Brisson, a naturalist and successor to Réaumur, followed the abbé Nollet as professor of *physique* at the Collège de Navarre.[32] Bailly was an inspector of military schools, even while retaining a pension derived from the curatorship of the royal art collection that he had inherited from his father.[33] Delambre, on whom was to come the main work of founding the metric system, was a retainer in the household of a financier and tutored his son.[34] More outgoing, J.-A.-C. Charles began in finance and borrowed money from his father to rent a laboratory where he opened a highly successful public course in experimental physics.[35] His *cabinet* of apparatus and instruments was famous, and even greater notoriety came to him in consequence of a spectacular flight that he made on 1 December 1783 in a balloon stabilized with a bladder arrangement of his devising, an experiment observed by all Paris, which he described to the Academy two days later.[36] The veterinary school at Alfort provided chairs of hygiene and agronomy in addition to anatomy: Daubenton taught rural economy there in addition to his duties at the Jardin du roi,[37] followed by Broussonet in 1785, who had abandoned ichthyology for agronomy and combined the duties with secretaryship of the Société Royale d'Agriculture.[38]

Of the trades connected with science, pharmacy was by far the most prosperous. Its thriving condition helps to explain why chemistry was the most active and populous sector of science just prior to the Revolution.[39] Academicians who came into science by way of pharmaceutical experience were Bayen, Baumé, Fourcroy, Macquer, Parmentier, Pelletier, Rouelle,

[30] Dufresne (1906).

[31] *Archives parlementaires 13*, 747; see also Barritault (1940), 73-83; Taton (1964), 282, 284, 330.

[32] *Archives parlementaires 13*, 651; see also René Taton, "Brisson," DSB 2, 473-475.

[33] *Archives parlementaires 13*, 379; see also Brucker (1950); E.B. Smith (1954); Hahn (1955).

[34] I. Bernard Cohen, "Delambre," DSB *4*, 14-18; Desboves (1881); Caulle (1936).

[35] *Archives parlementaires 13*, 566; Champeix (1966), 1-52; Taton (1964), 633-634.

[36] Jean-Baptiste Meusnier de La Place, "Sur l'Equilibre des machines aérostatiques," *Observations sur la physique, sur l'histoire naturelle et les arts* 25 (July 1784), 39-69, esp. 39 n. 1; see also Darboux (1910); and below, Chapter VII, Section 3.

[37] *Archives parlementaires 13*, 418; Taton (1964), 284-285, 338; Guillaume (1908-1909) *1*, 178-199; Roule (1925).

[38] Jean Motte, "Broussonet," DSB 2, 509-511.

[39] Guerlac (1959b).

and Vauquelin.[40] In the background was almost a crowd of pharmacists, some of whom—for example, Demachy, Mitouard, Trouville, and Venel—left a mark on the science of chemistry without attaining to the company of the Academy.[41] Of no other trade or science was anything of the sort then true in such ample measure. By the end of the old regime, other practical chemical arts were expanding together with pharmacy to the scale of a chemical industry. Chaptal was its first important representative in science proper,[42] though Darcet combined with his professorship at the Collège royal the post of inspector at the royal porcelain manufactory at Sèvres, where he succeeded Macquer, who had himself followed Jean Hellot.[43] While not exhaustive, such a sampling suggests the variety of possibilities available to men of science beyond the limited, though far from negligible, openings for official staff at the Observatory, the Jardin du roi, the Collège royal, and the faculties. Of those institutions, more in a moment.

With regard to the third and most decisive aspect of professionalism, scientific self-government, the Academy in Paris had attained nearer to the reality in fact than in form. In form, its identity derived from the crown, not from itself or from science, and its members enjoyed certain trivial immunities betokening status in the class structure rather than possession of knowledge. The writ of committimus entitled them to transfer litigation to the courts in Paris. In 1786 Calonne exempted them from payment of the tithe on royal pensions.[44] Such privileges pertained to minor office in the body politic and gave the beneficiary a certain cachet in those cascades of disdain through which Voltaire somewhere described French social consciousness descending.

As for serious matters, however, only in the field of education did the Academy exert little or no systematic influence over the concerns of science. In the conduct of its own business, the approximation to professional procedures was becoming close. To be sure, the crown might sometimes intervene in elections, as it did for Bailly in 1763 and for Jars in 1768, although in no such episode was lasting damage done to scientific talent through ministerial interference. In an instance in 1779, on the contrary, external pressure was the reason that one who became eminent, the chevalier Jean-Baptiste de Lamarck, was preferred for the section of botany over a rival, a certain Descemet, who was the Academy's first nom-

[40] Charles Bedel, "L'Enseignement des sciences pharmaceutiques," in Taton (1964), 237-257; Cazé (1943).

[41] Bedel, in Taton (1964), 237-257; Guerlac (1961b), *passim*. On the pharmacists, their trade and college, see below, Chapter III, Section 3.

[42] Pigeire (1932).

[43] Taton (1964), 267, 273, 284; Cuzacq (1955); see below, Chapter VI, Section 2.

[44] Hahn (1971), 72.

inee and who never made a lasting mark. Usually, however, the court confined its occasional interest to the choice of the scientifically peripheral honoraires and associés libres. Even then, the Academy might resist, asserting its freedom of choice, though it was more likely to do so when the appointment was to be a scientific one.[45]

If it were not beside the point, an argument could be sustained that the Academy made mistakes as grave as any minister's. In its selection of chemists, Turgot wrote to Condorcet in 1772, it had sometimes taken the worst candidate, and he particularly deplored its "*sottise*" in rejecting Rouelle and Darcet for Sage, who in fact did prove to be unable or unwilling to understand the new directions in theoretical chemistry, and who embarrassed the Academy in its determination to stamp out quackery by turning out to be a charlatan within the fold.[46] Not that the preponderance of control the Academy had gained over its own composition represented political victory in some campaign against the government—it was simply that the more technical science became, the less competent and usually less desirous was the crown to make its own appointments. The development is an example of the kind of unplanned consequence that the changing content of science has often had upon its organization and relation to the supporting structure of civil institutions.

For—to generalize the point—the problem of professionalism must be explored relative both to the content and the organization of science. The elements do not always correspond. A science may have attained to great technical virtuosity, as mathematics did, before it became fully professionalized. Or, like geology, it may have become professional before it was elaborated very highly. Nevertheless, history exhibits a definite secular tendency for the growth of technicality and rigor to accompany a movement toward the professional: the notion of a discipline has come to embrace both the content and practitioners of a subject. It is virtually a law of the evolution of science that its direction has been toward more sophisticated and extensive modes of quantification, and at any stage the qualitative is outranked or superseded by the quantitative. In motivational terms, the pressure is generated by the collective psychology, in which greater prestige always accrues to the exact and the rigorous. Within the scientific community, a d'Alembert was weightier than a Buffon early in our period, and a Laplace more imposing than a Cuvier at its close. Within the single science of astronomy, say, a Messier or Jeaurat, who could not calculate, suffered in comparison to a Dionis du Séjour or a Legendre, who could, however lengthily.

[45] *Ibid.*, 80-83; on the Bailly election, see below, Section 3, n. 73, and on Jars, above, Chapter I, Section 6, n. 171.

[46] Henry (1883), letter dated 21 December 1772, 123; on the Sage affair, see below, Chapter VII, Section 2.

All this was almost a naturalistic evolution, for the reason that in the signal area of education the situation was wholly unprofessional. Some few years ago a group of experts in the eighteenth century collaborated in exploring the means through which the teaching of science came about in France.[47] Like the livelihoods, it did so mainly in serving other purposes, the most notable example having been the training of cadets for the military engineering corps at Mézières. Such, however, is the only general statement that may be made about the bewildering array of formerly Jesuit schools now mostly Oratorian or Benedictine, Parisian and provincial *collèges*, military and naval schools with their cramming satellites, botanical gardens, natural history cabinets, pharmacy shops, private courses, self-education, and even universities—through which talent somehow received its education, after which the fittest were selected competitively into officialdom through the mechanism of the Academy.

For it was indeed the competitiveness set up by the regime of the Academy that favored in its effect the development of quantification and professionalism in the French mode of doing science. The intensity of that competition increased steadily throughout the eighteenth century. The most recent historian of the Academy tells us that in the 1730s the average age of election to the Academy had been just over twenty-eight, and in the 1780s it was thirty-nine. So competent if uninspired a chemist as Jean Darcet, professor at the Collège royal, having been passed over innumerable times, finally won election in 1784 at the age of fifty-nine, and then in the quality of supernumerary associé.[48] When vacancies occurred, candidates might be nominated by members of the section or they might apply on their own account. No stigma attached to the direct approach. To ask for what you want accords with Gallic realism, and Hahn reports instances of a vacancy in 1772 for which there were nine candidates, one in 1776 for which there were eight, and another in 1778 that attracted seventeen![49] Whatever the mode of application, it was the entire body of pensionnaires and honoraires who formally passed upon the qualifications and credentials of every applicant and recommended the Academy's preference to the minister.

Although it is more difficult to be categorical about temperaments than numbers, long acquaintanceship with the writings may perhaps allow an historian to recognize a certain professionalism in the scientific personalities whose generation was predominating in the Academy at the end of its old regime, a quality that had been less marked in those of earlier times. There was a commitment to hard science in the makeup of a Lavoisier, a Monge, and a Laplace that was lacking in a Cassini, a Borda, and a Buffon, older and somehow more amateurish. Maurice Daumas has convinc-

[47] Taton (1964). [48] Hahn (1971), 98. [49] *Ibid.*, 98.

ingly associated Lavoisier's theoretical taste with his admiration for the clarity and satisfactions to be found in the work of Monge and Laplace. He wanted to divest chemistry of the discursiveness appropriate to natural history, and, though it might not be ready for mathematics, still it could be invested with the exactness that marked the mathematical touch.[50] Similarly, Condorcet's election to the post of permanent secretary was taken to be a victory for the forward-looking elements over the traditionalists who supported Bailly: over Cassini *père* and *fils*, third and fourth of their line, whose qualifications to direct the Observatory were hereditary rather than astronomical; over Borda, the chevalier and naval officer who in his ingenuity started many a hare (the theory of a preferential ballot, a hydraulic model for the science of machines, the aiming circle that started the geodetic survey for the metric system), but left no body of finished science; over Buffon, the grand seigneur and last of the naturalists who ran the Jardin du roi with a high and feudal hand.[51]

In 1785 Lavoisier, serving in the office of director for that year, accomplished a reform in the Academy's regime conceived to carry the purpose of professionalism further. Its division into the six original sections no longer corresponded to the realities of science. The most obvious scandal was that to which Lavoisier himself had called attention when pressing for election in his youth: there was provision for mechanics, but none for research into phenomena of electricity, magnetism, heat, sound, and light. To accommodate investigations of the latter sort, Lavoisier proposed instituting a section of "experimental physics." A second new section for mineralogy would alleviate the pressure on that for chemistry, which was redesignated to include metallurgy. The scope of the botany section was also widened to embrace agriculture. In deference to the gathering sentiment of egalitarianism in society at large, the class of adjoints was abolished and all its members raised to the status of associé, the only practical difference being that they now had the vote in scientific matters and sat with their colleagues. In order that the Academy not become larger, the number of places in each section was reduced from seven to six. On the face of it, the old organization provided 42 places and the new 48. In fact, however, the apparent increase permitted a reshuffling to accommodate the five supernumeraries with whom various compromises had embarrassed the old structure. For Lavoisier felt strongly that the Academy must remain small. The historian is thus on sound ground in attributing the quality of scientific work in France to its competitiveness. That was also Lavoisier's view, and naturally it prevailed among others who were also

[50] Daumas (1955), 13, 160-161, 172-173.

[51] On the election of Condorcet as permanent secretary, see Hahn (1971), 134; Baker (1967b) and (1975), 35-47.

succeeding in the competition. If the Academy were to be enlarged, it would degenerate into a mere salon, a fashionable club. Lavoisier might well have adduced the inflated example of the Royal Society in London, had he known of its condition. He did not mention it, but did warn that standards would be debased, the name of academician cheapened, and the portal opened to a "half-knowledge more dangerous than ignorance and sure to be accompanied by imposture and intrigue."[52]

Anyone curious to visit in imagination the working sessions of this distinguished body, then entering upon the prime of its vitality, may find a contemporary guide in the person of a Swedish astronomer, A. J. Lexell, who in 1780 and 1781 did so in reality and sent gossipy reports back to his colleagues in Saint Petersburg, where he was of the international company of scientists sustained by its Imperial Academy.[53] The first of many possible illusions to be dispelled would be the romantic image of some forum of universal science assembling every Wednesday and Saturday afternoon to hear and discuss those classic memoirs wherein Lagrange (though he was called to Paris only in 1787) fully rationalized the science of mechanics; wherein Laplace transformed astronomy into celestial mechanics; wherein Lavoisier modernized the science of chemistry by exorcising the figment of phlogiston and replacing its futility with an exact, materialistic theory of combustion; wherein Buffon completed the description of nature in eloquent periods since anthologized in many a manual of expository style; or wherein Vicq d'Azyr, treating of comparative anatomy, inaugurated a more analytical approach to the sciences of life. For even if those epitomes were quite just in substance, and they are not, it was never thus in the flesh. The members of the eighteenth-century Academy had in common with their successors in the later twentieth century that they paid small attention to whoever might be murmuring rapidly along at a lectern in a manner to leave undisturbed the animated conversations occupying his colleagues in their places around the table or gathered in groups over at the windows, back near the door, or midway by the stove. These were assemblies in which the room served also as corridor, the papers having usually been too long to be read in their entirety,

[52] Lavoisier, *Oeuvres* 4, 567. For documents concerning this reorganization, see *ibid.* 4, 555-614; Procès-verbaux of the Academy of Science, register for 1785; and for a summary, Hahn (1971), 99-101.

[53] Published by Birembaut (1957); cf. Lubimenko (1935). The official procès-verbaux are no more informative about the private and personal inwardness of the Academy than minutes of meetings of such organizations usually are. Further detail may often be found in the *plumitif*, the rough notes kept by Condorcet, which are also conserved in the Archives of the Academy. For franker revelations of what may be called its "*petite histoire*," the so-called "*Collection*" of Lalande, an annotated copy of the Academy's regulations, is a mine of gossip and anecdote. Professor Hahn plans to publish this material, and for further indications, see his Bibliographical Note, Hahn (1971), 321-329.

and far too technical to be followed aurally, even if more than a few persons had been interested or qualified to attend to what was being presented.

Twice a week, except during the holidays, the Academy held session in the Salle Henri II, the king's old antechamber in the Louvre. One entered from the Cour Carré by the staircase to the left under the Tour de l'Horloge, mounted to the first story above ground level, and passed through the Salle des Etats. The tourist who would retrace today the route of some aspirant for academic favor in the eighteenth century will traverse a large room of Roman jewels and bronzes and find himself in a hall approximately fifty feet by thirty surrounded by Etruscan ceramics and standing under a ceiling decorated by Braque. To the right, a pair of windows at one end then gave on to the place Fromentau, now the square du Louvre, and to the left a single one opened onto the Cour Carré at its southwest corner. Dimly to be made out in the paneled obscurity was a painting by Coypel representing Minerva seated and gazing at a portrait of Louis XIV. Another canvas by Roslin portrayed the king of Sweden, and a third the constellations of the southern hemisphere. Terracotta busts of eminent predecessors—Fontenelle, Morand, the abbé Nollet, Réaumur—surrounded the academicians of the day. Some of the younger members were beginning to affect the fashions of Louis XVI. Older colleagues and all the physicians wore the black costumes and ponderous wigs proper to a learned profession in earlier, stiffer times.

The main furniture was a large table in the form of a hollow rectangle as shown in the diagram. An opening (IK) in the long side near the entrance to the chamber gave access to a smaller, oblong table GH in the center of the room with several chairs before it. At the center of the long side opposite the entrance, the places A and B were for the president and the director, the former being seldom present. Ranged on their either flank were the honoraires and the retired and senior pensionnaires, the re-

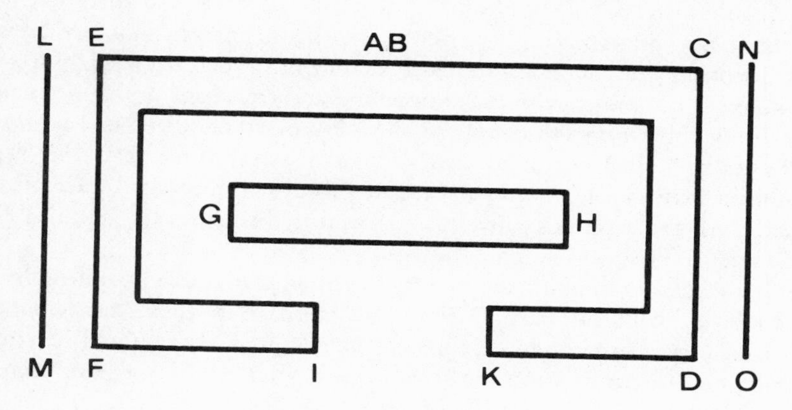

maining pensionnaires occupying the two sides EF and CD at right angles to their elders. Behind them were two benches LM and NO for correspondents who might attend. At the bottom of the table the two segments IF and KD were for the associés and adjoints (this was prior to Lavoisier's reform, which modified the seating arrangements along with the organization). The secretary, Condorcet, sat at the corner C of the upper table to the president's left, with Lavoisier at his elbow during Lexell's visits. D'Alembert would normally take a chair at the opposite end of the head table at E. Members of the Academy with a memoir or a piece of correspondence to read generally did so from a place alongside Condorcet, although they might remain in their own seat if they preferred.

After adjournment many academicians would repair for tea and further conversation to the Arsenal, where Madame Lavoisier received in the style of a salon of science. On Sundays members of the Academy might take advantage of a standing invitation to dine at the townhouse of an honoraire, a *parlementaire* of great charm, the President Bochart de Saron. Two or three were always there in the rue de l'Université, among them Lexell, as much for the excellence of the company as for the dinner. Indeed, he remarked, the noblemen in France who applied their minds to science were much less arrogant than the scientists themselves, seemed to expect nothing in return for kindnesses, and appeared, on the contrary, to feel honored that scientists would come into their company—which is only to say, though Lexell did not put it this way, that such fortunate persons had the manners of the gentlemen that scientists as a group had not had the opportunity to become.

As for the scientists, the ones who interested Lexell were mainly the mathematicians and astronomers, and before fully accepting his account of their deportment, it would be prudent to discount the pinch of malice without which there is no gossip. He implies, to begin with the most eminent example, that some resentment had developed against d'Alembert in the role of Nestor. Partly this was because d'Alembert had closed his mind to certain newer developments in analysis, notably the use of discontinuous functions and the calculus of probability. Partly it was because of his tendency to make sweeping and trenchant pronouncements on every subject and to denigrate the work of others, even of Lagrange. Physically he had begun to seem old for his sixty-three years, his head afflicted with a pronounced palsy and his always high-pitched voice grown reedy. All the same, his conversation was truly brilliant, and he never had to force it.

It was otherwise with the two who formed d'Alembert's close coterie: Condorcet, whose literary writings had merit, but whose mathematical writings were so obscure that his colleagues privately doubted he could understand them himself; the abbé Bossut, who might not have been so feeble a mathematician as Laplace said, but whom Lexell thought far from sublime. Certainly it was Laplace who had the greatest mathematical talent, an opinion in which he himself most evidently concurred. Laplace was knowledgeable in other areas of science also, even to the point of abusing his facility, for he was extremely opinionated and wanted his way in every decision the Academy took. Others thought that his abrasive and occasionally repellent manner might be a consequence of his comparative poverty, though the appearance is inconsistent with the findings of later scholarship that his family was in fact well off. He was friendly mainly with Bézout, the algebraist whom he was soon to succeed in the post of examiner for the artillery. With Monge, the third of the leading mathematicians of the Academy, Laplace was no more congenial than either was with d'Alembert. Monge was then newly arrived from Mézières. Black of brow, rough in mien, he was closely associated personally and in certain technological interests with Vandermonde, who did not look at all like a mathematician. Monge informed Lexell on their first meeting that he alone was doing geometry in France, an observation that was entirely accurate except for work by his former pupils, most notably Meusnier.

The astronomers come off rather better in Lexell's personal esteem, though less well technically. Dionis du Séjour was a small man, witty, vivacious, and amusing, whose calculations were widely known but little used because of their prolixity. He was of the *noblesse de la robe* with a seat in the Parlement of Paris, where his had been a leading part in securing the abolition of judicial torture. A propos of the senior astronomer, César-François Cassini de Thury, and of his son, Jean-Dominique, soon to succeed his father as director there at the Observatory where his grandfather and great-grandfather had been installed in their day, everyone knew how talent was declining in that family. Le Monnier was forever pretending that other people were appropriating his observations without giving him credit. Jeaurat could hardly complain of being passed over for promotion because, though of the section of geometry, he was nothing of a mathematician. Lalande, for all his bold parade of science and atheism, for all his deep, impressive voice, was in fact a charlatan, though a cheerful one, who gained popularity by suffering gladly the affronts of bigots. Bailly and Messier were two whom Lexell found congenial, as he also did Cousin, rare among the mathematicians in that he conversed with moderation and took the trouble to write with clarity. The others Lexell did not know so well. LeRoi was affable though unimpressive. Daubenton was poised, calm, and knowledgeable. Macquer appeared to be agreeable, Trudaine de Montigny cold, Cadet short and stubborn, Tillet gentle and a touch melancholy, Baumé commercial and calculating, Adanson slight and a little visionary, Brisson large of stature and a trifle stupid, Portal thin and pale, and Sage false and ingratiating.

Contrary to what might be expected, the reading of formal memoirs by these or others of the Academy's membership occupied little of its time. Lexell heard only five or six in the first three months of his visit. Most of the agenda in a normal meeting would consist of correspondence and reports on the many writings, technical inventions, and kindred enterprises submitted to the Academy for its approbation. Persons not of its membership who had something to present were placed literally in its midst at the oblong central table, where they would speak out their communications encircled by their judges. If these latter often appeared inattentive, it was because they were inconsiderate rather than unjust since the real decisions were made in ad hoc committees to which particular matters were always referred. For simple affairs, three members of the Academy would normally suffice, one acting as spokesman or *rapporteur*. For complicated or controversial business—the investigation of Mesmerism was a notable example[54]—the number might be seven or eight. Like many a deliberative body, the Academy actually exerted its authority in detail through such committees. Service upon them was an arduous responsibility, gratifying to the extent that an individual commissioner had the temperament to take inward pleasure in the dispensation of authority, irritating in that too often the duties were routine, their scientific interest slight, the projects under scrutiny trivial or absurd, and the time consumed extensive. Twelve or fifteen such assignments might befall every academician every year. In general, they lay within one of the three broad provinces in which the Academy had come to exercise the governance of this, its technical realm: first the area of science proper, second scientific publication, and third technology—a word not yet coined for knowledge of the arts and crafts, manufacturing, invention, and public works.[55]

The most significant matter touching science proper has already been considered—the preponderant influence the Academy had secured over its own composition. Only the proceedings on elections consistently aroused the intense interest of the entire membership in committee work, in part no doubt because scientists are human in being mainly concerned with their own affairs, as they also are in fancying themselves shrewd judges of the qualities of their fellow men. A few corporate enterprises were under the supervision of standing committees that oversaw finances and the secretariat. For although what the Academy performed scientifically was less influential than its manner of existing, it did sponsor and supervise certain activities. True to the original charge from Colbert, the Baconian spirit ruled its annual prize competition, instituted in 1714 and usually ad-

[54] Below, Chapter IV, Section 2.

[55] For discussion of the evolution and operation of the committee system, see Hahn (1971), 21-22; Gillmor (1971), 44-48.

dressed to some object of public utility.[56] The problem of saltpetre was an example exceptional in its urgency. The essay of 1765 with which Lavoisier won the contest on methods for lighting the streets was a more typical instance, one of several such early approaches to the Academy on the part of men afterward famous.[57] Two graduates of Mézières, Charles-Augustin Coulomb and Lazare Carnot, were among the contestants for the prize announced in 1777 and reopened in 1779 for an experimental investigation into friction and stiffness in cordage with emphasis on applicability of the results to shipboard. Coulomb's memoir won, and quickly became the point of departure in the modern study of friction.[58] Carnot received honorable mention, and though his argument remained unappreciated for a time, its analysis of the problem of transmitting power inaugurated modern engineering mechanics, which he called the science of machines.[59]

Throughout its history, the Academy maintained a tradition of support for astronomical and geodesic investigations and for geographical exploration, no doubt because the Observatory came under its official aegis.[60] Late in the century, the growth of humanitarian and reformist sentiment led the Academy to occupy itself also with public health and sanitation. Hospital reform began with a series of reports on existing horrors in the Hôtel-Dieu from an academic commission headed by Jean-Sylvain Bailly, astronomer and future mayor of Paris;[61] penal reform with a report on prisons from Lavoisier, Tenon, and Tillet in 1780;[62] and civic hygiene with a report of 1789 from Bailly, Daubenton, Laplace, Lavoisier, and Tillet calling for removal of the slaughterhouses that befouled the central market areas of the capital.[63] These prescriptions for civic betterment were serious sequels to Turgot's program for enlisting science in the service of social progress. They were not themselves science, nor meant to be. Among enlightened persons, they fortified esteem for the Academy and for scientists, thus fulfilling their mission in the relief of man's estate.

Disesteem and anger were the fruits, the bruised and bitter fruits, of the Academy's mode of discharging its responsibilities in the areas of publication and technology. Details are important, but will best await a review of the working of the other institutions important to science. Suffice it here to indicate how it came about that hopeful or ambitious persons who would publish anything pretending to scientific status, and ingenious or enterprising artisans and manufacturers who would capitalize on government patronage for industry, had first to subject their ideas and

[56] Maindron (1881).

[57] Above, Chapter I, Section 6.

[58] Gillmor (1971), 118-138.

[59] Gillispie (1971a), 11, 62-81.

[60] Below, Section 3.

[61] Below, Chapter III, Section 7.

[62] Below, *ibid.*

[63] Below, *ibid.*

productions to the scrutiny of a committee of scientists empowered to say them yea or nay.

Publication in the France of the old regime was a privilege, not a right. It had been supposed, when the Academy was founded, that its science would be done among the whole membership working in common, who would then be authorized to print the findings without further need for permissions from the crown or the Sorbonne. Anonymity and collectivity proved unconducive to writing, however, and instead of publishing as a body, the Academy early began certifying the reliability of the individual memoirs that bore its imprimatur. The practice of thus formally approving what academicians published in their official capacity was regularized in statutes drawn up in 1699, which governed the organization and procedures of the Academy until Lavoisier's reform of 1785.[64] Every such memoir in the annual volume, if not actually read in the meetings, at least had to be examined by members delegated for the purpose. A similar practice grew up for separate treatises published under the Academy's privilege to authorize printing. Writings by persons not of the Academy might also obtain its seal of approval. The permanent secretary reported many such in the "History" or prefatory review of the year with which (until 1783) he opened the annual *Histoire et mémoires de l'Académie Royale des Sciences*. The series called *Savants étrangers* amounted to a supplementary scientific journal. Of regular periodicals, the monthly *Journal des savants* was virtually an academic organ, and the daily *Journal de Paris*, edited by Cadet de Vaux, the brother of an academician, printed scientific news and memoirs prior to their official publication by the Academy after delays averaging three years. What with the logarithmically increasing volume of publication in the eighteenth century, the officials of the Librairie concerned with censorship were willing to let the Academy exercise their functions in science.[65] In the view of academicians, their doing so was in defense of knowledge, its purity and standards, a public service all innocent of dogma. Sometimes it looked different in the eyes of others, but those were not eyes through which scientists ever could see.[66]

The tangled web of privilege—literally private law—was also the nexus through which in the absence of any public patent law the Academy acquired a decisive voice in the dispensation of industrial patronage by government. It was always the mercantilist instinct of the French state that by taking thought it could add cubits to the stature of the economy, and always the instinct of French entrepreneurs to look to the state, first, for

[64] Aucoc (1889a), lxxxiv-xcii, esp. lxxxix.

[65] On the operations of the censorship, see Grosclaude (1961), 452-461; Pottinger (1958), ch. 1.

[66] Hahn (1971) gives an excellent account of the development of the Academy's control over scientific publication, 61-66.

the capitalization that their British counterparts found in banks or profits and, second, for the shelter they then required from the competition that spurred the latter on. The inventor with a new device or the manufacturer with an improved process might approach the crown in several ways. He might seek to sell his idea outright, particularly if it might be exploited in one of the royal manufactories—tapestries at the Gobelins, glass at Saint-Gobain, porcelains at Sèvres, munitions at the Arsenal. Or he might apply for an *encouragement*, a premium or subsidy to enable him to go into production on his own. Or, finally—and this alternative might be combined with a request for funds—he might petition for a monopoly of the right to make and market his product for a stated term of years. Such a grant, juridically comparable to the Academy's exclusive right to publish science, was also called a *privilège*. Whatever course or combination an inventor might choose his application was passed upon by a committee of the Academy, which made its recommendation usually to the Bureau of Commerce, more rarely to the War Office, the Marine, or another granting agency. The procedure was entirely consonant with Colbert's intention that the Academy should concern itself with the state of manufactures and their improvement.

The relationship between science and the industry that had actually developed a century later, however, is a subject requiring a chapter to itself.[67]

3. THE OBSERVATORY OF PARIS

Ever since the construction of the Observatory of Paris in 1667 and 1668, members of the staff have been wont to complain that the splendid structure where they make their professional home serves the capital better architecturally than it does their science astronomically. Seen from the gardens of the Luxembourg, it offers one of the most satisfying prospects in the city, and many a visitor has naturally supposed that the distant façade was designed to complete the perspective. In fact, that vista was opened only in 1811, but the Observatory was intended monumentally, nonetheless. Colbert confided the architect's commission to Claude Perrault and promised the fledgling Academy an edifice that would "not only surpass in size, beauty, and convenience the observatories of England, Denmark, and China, but that also, to state the whole purpose, would correspond in some sense to the magnificence of the prince who had it built."[68] The

[67] Chapter VI, below.

[68] Quoted in Wolf (1902), 4, an antiquarian history that describes the regime of the Observatory from its foundation until 1793. It may be supplemented by J.-D. Cassini's own memoir (1810), to be treated with all the caution required of an ill-tempered apologia, and by the equally partisan Devic (1851). The further volume that Wolf projected for

grand staircase remains one of the finest interior architectural sights of Paris. On the exterior, the north façade was slightly modified but not spoiled in the nineteenth century by addition of equatorial telescopic domes on either corner, the original observation sites having been platforms arranged for mural instruments. Beneath the cellar are deep caves reaching down into the limestone that had been quarried from the terrain for centuries. They are famous for their constant temperature, 11.86°C. at all seasons. The thermometer that Lavoisier installed in 1783 for monitoring experiments under those invariant conditions remained in place until the present century.[69] These subterranean reaches of the Observatory communicate with more ancient catacombs under the Boulevard Saint-Jacques, fully worthy of Victor Hugo and redolent of what medieval incarcerations no one knows. The building in its presence makes a profoundly Parisian statement.

Originally, the Observatory was imagined as a house of all science, accommodating the Academy in its prospective requirements for a chemical laboratory, an anatomical theater, a repository for apparatus and instruments, a library, a public meeting hall, and lodgings for the members. That an observatory should have been expected to fulfill all those functions bespeaks the preeminence of astronomy among the sciences in the seventeenth century. In consequence of the generality of these intentions, the installation remained the responsibility of the whole Academy until, in 1771, Louis XV formally created the post of director and appointed the third Cassini, César-François, to the function that his grandfather and father before him had exercised in all but name.[70] For the intentions had not worked out. In the event, the Academy never wished to foregather off at the edge of the city; adequate funds were never provided; the other sciences never took root within the confines there provided; and, what is more surprising, the normal observational work of astronomy itself never centered there in the eighteenth century. In the latter years of the old regime, the members of the astronomical section of the Academy got on badly with the last Cassini, Jean-Dominique, and worked mainly elsewhere, making use of telescopes at the Ecole militaire, the Collège des

---

the nineteenth century never appeared. An article by Bigourdan (1931b) recounts episodes of the later history of the establishment. Chapin (1968) in a very scholarly article discusses what he considers to be the inadequacies of the academic regime to the needs of astronomical science. Extensive documentary materials amounting to a set of archives are conserved in the Bibliothèque de l'Observatoire, where they are in good order and well catalogued. The central record of observations recorded from 1683 to 1798 occupies series D-3, 1-30, and D-4, 1-29. Administrative correspondence between the ministry and the direction of the Observatory is in series D-5, 40.

[69] For an account of these and subsequent thermometric experiments and measurements, see Lavoisier, *Oeuvres 3*, 421.

[70] Wolf (1902), 6-7, 231-232.

Quatre Nations, and the Collège de Cluny, where the navy maintained facilities.[71] Besides Cassini himself and several retainers, only Jeaurat, Le Gentil, and Méchain lodged on the premises of the Observatory.

Before we identify the geodesic tradition that did center in the Observatory, it will be well to introduce more fully those persons charged by the Academy with carrying on astronomy on the eve of the Revolution, at just the time when the planetary aspect was undergoing analytical mechanization through the calculations of Laplace. The two who had made themselves most widely known to the literary world, Jean-Sylvain Bailly and Jérôme-Lefrançois de Lalande, exemplify a distinction between notability and reputation that may often be observed in the history of science. Both were regarded as scientific lightweights by their colleagues, a judgment in which each concurred about the other. Mutual dislike originated or first appeared in 1773 when Bailly advanced as his own an explanation already given by Lalande seven years previously concerning an irregularity in the precessional motion of the nodes of moons of Jupiter. Lalande claimed priority, whereupon Bailly deprecated the discovery, now holding it to be a triviality unlikely to interest the public.[72] Their disesteem deepened in 1776 when Condorcet won election to the permanent secretaryship of the Academy. In that contest Lalande refused to cast his ballot for Bailly, despite their common membership in the astronomical section.[73]

On balance Bailly's genteel careerism appears no more disreputable than that of many a minor writer of the late enlightenment, the difference being that he chose to "*faire le piste*," or find the path to prominence through a thicket of favors, with a parade of science.[74] He showed much ingenuity, even (to do him justice) convincing himself and contriving to rise—if the expression is permissible—between three stools. He had the dexterity to win a sort of consideration from the scientific community on the grounds of modest literary talent, a certain standing in the republic of letters on the strength of early membership in the Academy of Science, and an official recognition of claims to scholarship through combining let-

[71] For an enumeration of the many observation posts, see Hahn, "Les Observatoires en France," in Taton (1964), 653-658.

[72] Documents bearing on this affair may be consulted at the Bibliothèque de l'Observatoire, B-5, 7.

[73] On this election, see E. B. Smith (1954), 447-449; Baker (1967b), (1975).

[74] On Bailly's academic career, see Seymour Chapin, DSB *1*, 400-402; E. B. Smith (1954); and Hahn (1955). The "Eloge" by Arago, *Oeuvres complètes 2* (Paris, 1854), 247-426, is unreliable in judgment and detail. See also Delambre (1827), 735-748, and the "Eloge" by Lalande, *Bibliographie astronomique* (Paris, 1803), 730-736, originally published in 1795 in the *Décade philosophique 4*, 321-330. Bailly's own *Mémoires*, 3 vols. (Paris, 1821-22), concern mainly his political career. Volume 3 has been exposed as a forgery. See Fling (1902, 1903).

ters and science with imagination in a history of astronomy. Thus, through membership in all the learned academies, did he achieve his ambition of becoming the only triune academician since Fontenelle. It is true that in no case did his election occur without intervention from ministers who had become habituated to his family's claims over four generations. In 1667 his great-grandfather, Jacques Bailly, miniaturist and member of the Académie de peinture, had been accorded a lodging in the Louvre and the post of Garde des tableaux du roi with 1,500 livres a year. The office descended *sur la tête* of his grandfather, his father, and himself. Bailly entered upon its freedom from duties in 1754 when he was eighteen and about to study astronomy with the abbé de Lacaille. A quarter of a century later d'Angiviller, Turgot's appointee to the Batimens du roi, was initiating the reforms that ultimately converted the Louvre into a museum and adopted the royal paintings for the artistic patrimony of the nation. Bailly could scarcely qualify to care for an art collection, "though he had all the artistic knowledge a man of letters could acquire."[75] Nevertheless, he protested that by rights his pension should be continued though the post be suppressed. In those years of attempted reform, the authorities acceded readily to protests, whether reasonable or not, and he prevailed.

Details of the manipulations that secured Bailly his elections to three academies are too intricate to recount here. Suffice it to record that when in 1763 he stood for the place in the Academy of Science vacated by the death of his teacher, Lacaille, his credentials consisted of a memoir on Halley's comet together with two unpublished papers on the moons of Jupiter. The minister, Saint-Florentin, chose him over Jeaurat, Messier, and Thuilier on the basis of a plurality of doubtful legitimacy, and did so despite the opposition of all the astronomers except his patron, Clairaut.[76] D'Alembert then opposed the second of his candidacies, for the French Academy. Bailly's success in 1783 after d'Alembert had died was widely attributed to his shameless flattery of Buffon (though in later years Buffon, too, broke with him).[77] Finally, the resistance of the normally timid Académie des inscriptions et belles lettres forced Breteuil, minister at the Maison du roi, to impose on it a special class of associés libres to accommodate his ambitions.[78]

[75] D'Angiviller to the king, 30 March 1783, published by Hahn (1955), 340, who points out the existence of other essential documents not known to Bailly's earlier biographers.

[76] Extract from the Procès-verbaux des séances, Archives of the Academy of Science, 19 January 1763; Joseph Delisle to the comte de Saint-Florentin, 20 January 1763, *ibid.*, 343-345.

[77] E. B. Smith (1954), 479-483; cf. Bailly, *Mémoires* 2, 150-151 n.1.

[78] Extract from the *Registres*, 18 January 1785; the baron de Breteuil to the secretary of the Académie des inscriptions et belles-lettres, 15 January 1785, Hahn (1955), 345-346.

The historian of the last body observed that, although knowledge of Latin and Greek is no measure of intelligence, a command of those languages is incumbent on any scholar who presumes to situate the origins of civilization in this or that aspect of antiquity.[79] The argument of Bailly's writings on ancient history was that astronomy derives from India and that, under the name of Atlantis, Plato was relaying the memory of an actual antediluvian land, the ultimate source of all science and knowledge.[80] Following Bailly's election to the Academy of Science, he published occasional astronomical memoirs for another ten years before turning largely to such archaeological and historical concerns.[81] By the 1780s, he was mainly chairing committees that issued enlightened reports: the exposure of Mesmer for a charlatan and his supporters for dupes;[82] the series already mentioned on the inadequacies of the Hôtel-Dieu;[83] a summons to science to complete the rationalization of industry long since begun in its *Description des arts et métiers*.[84] Somehow, the virtually unanimous judgment of superficiality that his colleagues made of his attainments never came through to the public. His manners were unassertive and his conversation flattering to the companions of the moment.[85] He was never hostile to intellect, and truly respected in others the qualities of science and scholarship to which he pretended in himself. These urbanities and his good works won him election to a final and fatal imposture: his incarnation of the spirit of science and reason amid revolutionary politics as the first mayor of Paris.

"In 1780," wrote Lalande many years later, "I did not expect that he would be assassinated and that I should be writing his éloge."[86] Not that

[79] Maury (1864a), 326-327.

[80] Bailly's historical works were *Histoire de l'astronomie ancienne depuis son origine jusqu'à l'établissement de l'école d'Alexandrie* (1775), and *Histoire de l'astronomie moderne . . . jusqu'à l'époque de* 1730, 3 vols. (1779-1782). His more speculative writings were *Lettres sur l'origine des sciences et sur celle des peuples de l'Asie, adressées à M. de Voltaire* (London and Paris, 1777), and *Lettres sur l'Atlantide de Platon et sur l'ancienne histoire de l'Asie . . .* (London and Paris, 1779).

[81] He collected his astronomical writings and papers in *Discours et mémoires*, 2 vols. (Paris, 1790).

[82] See Chapter IV, Section 2.

[83] *Rapport des commissaires chargés par l'Académie de l'examen du projet d'un nouvel Hôtel Dieu* (Paris, 1786), see below, Chapter III, Section 7.

[84] J.-S. Bailly, *Recueil des pièces intéressantes sur les arts, les sciences et la littérature* (Paris, 1810), with a biographical notice by C. Palmézeaux. See esp. "Exposé de ce que l'Académie des Sciences a fait, et de ce que lui reste à faire, pour la description complète des arts," pp. 126-146.

[85] See the "Notice" by St. A. Berville in Bailly, *Mémoires, 1*, viii-ix.

[86] Hahn (1955), 343 n.1. Lalande's éloge was written for *Décade philosophique 30 pluviose an 3* (18 February 1795) and reprinted in his *Bibliographie astronomique* (Paris, 1803), which work contains biographical notices and highly personal remarks on a number of Lalande's colleagues. Lalande's notes for the éloge are in the BN, MS Fr 12, 273, fol. 213.

Lalande was to be taken very much more seriously for the inwardness of his contribution to astronomy, for he believed that the emancipating effects of the science were more valuable than its content, and he habitually cooked his calculations. Somehow, he never aroused personal mistrust, however, perhaps because he did not take himself too seriously. He was a wisp of a man, weighing around a hundred pounds and under five feet tall, with a deep, sonorous voice and the faculty, so it would appear, of arousing protective or indulgent feelings in those about him. When a lad in Lyons, he thought to become a Jesuit like his teachers in order to be free for astronomy.[87] Sent by his father to Paris to study law, he attached himself instead to Delisle, then occupying the naval observatory at Cluny, where Lalande was the sole disciple, and attended his course, and also that of Le Monnier in the Collège royal. In a moment of indisposition the latter sent the eager student as his substitute on a mission to Berlin in 1750, bearing with him from Paris instruments needed to observe the parallax of the moon simultaneously with measurements that Lacaille would be taking at the Cape of Good Hope, almost on the same meridian.

On his own in the Prussian capital at the age of nineteen, Lalande entered upon the great world of science and letters in the company of Euler, Voltaire, d'Argens, Maupertuis, La Mettrie, Algarotti, and Frederick II himself. There he gallantly made a partner for ladies of the court upon the dance floor, carefully steered a course between Voltaire and Maupertuis in the Koenig affair, and cheerfully embraced the atheism that he afterward preached none too privately as his main philosophic stock in trade.[88]

[87] Delambre (1827), 547-621, gives a scientific biography. See also comtesse Constance de Salm, "Eloge historique de M. de Lalande," *Magasin encyclopédique* (1820) 2, 282-325, and BN, Ln[27].11, 118. Lalande papers are found in many archives in Paris and elsewhere. See Raspail (1921). Besides the manuscript "*Collections*" (above, n.53), and papers that are of interest mainly for local history, "Tablettes chronologiques pour servir à l'histoire de Bourg et de la Bresse; 1764-1806," in the Bibliothèque publique de la ville de Lyon, Fonds Coste no. 1281, there are documents in the Bibliothèque de l'Observatoire de Paris, B-4, 10; C-5, 1-12, 28-40; MSS 1022, 77-86; in the Archives of the Academy of Science, Lalande dossier; in the BN, Fr 12, 273; and in the Bibliothèque de l'Institut de France, MSS 2381, III. Bigourdan (1901, p. 239) refers to a file of 105 letters between Lalande and Zach from 1792 to 1804 given to the Observatory by Madame Laugier. Despite a search some years ago, these letters could not be found among its manuscripts. I should like to acknowledge the kindness of the late director of the Observatory, Monsieur Dangeon, who, in the course of several conversations about the qualities of the eighteenth-century astronomers, warned me of Lalande's insouciance in matters mathematical.

Mention should also be made of a collection in the British Museum (F.R. 31 [10]) of recommendations that Lalande prepared in 1789 when the cahiers of his native Bourg-en-Bresse were being drawn up.

[88] Monod-Cassidy (1967), 908-909; Jean C. I. Delisle de Sales, *Examen pacifique des paradoxes d'un célèbre astronome, en faveur des athées . . .* (Paris, 1804). For Lalande's own expression of his views, see annotations in his hand to his private copy (BN, Res Y[e].3604) of a poem by Martin de Bussy, *L'Ether, ou L'Etre suprême elémentaire; poëme philosophique et moral,*

Many years later Lalande irreverently included Napoleon in a supplement that he published to Maréchal's *Dictionary of Atheists*,[89] a work conceived in the manner of Bayle that also listed Bossuet, Pascal, and Christ. It was a measure of Lalande's capacity to disarm the severity of persons normally intolerant of levity that the emperor let him off with an admonition and did so upon the intercession of Laplace, himself grown imperious, who observed that to prohibit Lalande from publishing would be to kill him.[90]

Making up in linguistic facility what he lacked in mathematical reliability, Lalande was one of the very few enlightened spirits in France who had both German and English. In 1763, the ending of the Seven Years War drew him to London, where his good humor repeated the success of his Berlin visit. He bowed before the tomb of Newton, dined at the Mitre, was elected to the Royal Society, called on Smollett, visited Wilkes in prison, and with friends was introduced to Johnson ("un gros paysan qui boit du thé la nuit et le jour"), who insisted on talking at them in Latin. So anglicized did Lalande become in three months that he could make fun of Le Monnier for his inability to pronounce the word "strength." When La Condamine came over from Paris, he had the delight of showing his countryman a foreign city that he knew himself and paraded the great geographer about, shouting tourist information into the ear trumpet that La Condamine had had to carry since losing his hearing in Peru, where he had been leader of the expedition that measured a degree of the meridian at the Equator.[91]

Following his English visit, Lalande embarked upon freemasonry, which fellowship together with astronomy and atheism completed the trinity of his intellectual enthusiasms. He gave the inaugural address for the Grand Orient of France, and in 1776 joined Madame Helvétius in founding the Lodge of the Nine Sisters. Many leading scientists had some association with that rather nebulous company of forward-looking spirits in the years before the Revolution.[92] Gaiety, outrageousness—he held dried spiders to be nutritious and in company would pull out a snuff box and swallow a pinch[93]—and, it must be said, a conscientious command

à priori, en cinq chants (Paris, 1796). Lalande contributed a 26-page neo-Epicurean preface, and after the word *Elémentaire* in the title, he interlineated in long-hand "et Pneumatique."

[89] An edition with Lalande's additions is Sylvain Maréchal, *Dictionnaire des athées anciens et modernes . . . augmentés des supplémens de J. Lalande* (Brussels, 1833). The first edition, without Lalande's contributions, was published in 1799.

[90] For Delambre's account of this affair, see Bibliothèque de l'Institut de France, MSS 2041.

[91] Monod-Cassidy (1967), 917-923.

[92] For Lalande's freemasonry, see Amiable (1889); and Raspail (1921); and for the *Loge des neuf soeurs*, Amiable (1897).

[93] Monod-Cassidy (1967), 929-930.

of the literature of astronomy, made him a popular professor of the science at the Collège de France, where he succeeded Delisle after 1761, and taught for forty-six years. He attracted energetic young men into the subject and set them onto making and compiling observations in Paris and far parts. D'Agelet, Méchain, Burckhardt, his own nephew Michel Le Français de Lalande, and numerous foreign lads, of whom the most famous was Piazzi, were among his students.[94]

Despite voluminous publications,[95] teaching may well have been his most signal contribution, not one to be underestimated at a time when the world population of those who may properly be called professional, or proto-professional, astronomers was on the order of sixty-five persons.[96] Some fifteen or sixteen in all were being accommodated by the facilities in Paris in the last years of the old regime. Besides the people already mentioned, or to be discussed at some length, namely Laplace, Dionis du Séjour, Cassini IV, Delambre, and Méchain, and the naval figures Borda and Bory, whose interest was mainly navigational, the others who were doing the serious work of astronomy, and of whom a further word needs to be said, were Messier, Pingré, Le Gentil, Le Paute d'Agelet, Jeaurat, and Le Monnier.

The first four mentioned were all explorers in some sense, even Messier, or perhaps especially Messier, who seldom left Paris but turned his telescopes on outer space. He worked normally with a 104-power Gregorian, 32 inches in length and 7½ in aperture, an instrument far more modest

[94] Delambre (1827), 554-555, 566-567. The observations by his students at different posts in Paris are at the Bibliothèque de l'Observatoire, C-5, 2-7.

[95] Not to list his tabulations, popular writings, or numerous memoirs on particular celestial events and astronomical problems, for which see Delambre (1827), the principal works were (1) *Astronomie*, 2 vols. (1764); 2nd ed., 3 vols. (1771); 3rd ed., 3 vols. (1792) a textbook that covers all aspects of astronomy, historical, theoretical, and practical; (2) *Histoire céleste française* (vol. 1, 1801), of which no further volumes appeared since the purpose was continued by (3) *Bibliographie astronomique, avec l'histoire de l'astronomie depuis 1781 jusqu'à 1802* (1803). The latter two volumes compile and tabulate the work done in France in the last two decades of the eighteenth century. The historical part completes Bailly's treatment. In addition Lalande edited the *Connaissance des temps* from 1760 through 1775, developing that annual from a mere ephemeris or almanac into an astronomical journal that also published correspondence and memoirs. He took over the editorship again in 1794 amid the disorganization created by the Revolution and held it until the year of his death in 1807.

[96] Wolf (1902) makes the reasonable observation that a document which he prints (267-269 n) from the Bibliothèque de l'Observatoire, D-5, 33, amounts to a census of astronomers in the early 1790s. It consists of Cassini's mailing list for copies of the annual *Extrait des observations* compiled at the Observatory following his reforms of 1785 (below). In addition to various institutions, patrons, and a few members of the Paris Academy in mathematical fields, there were 9 recipients in the French provinces, 9 in the British Isles, 8 in Austria and Germany, 8 in Italy, 4 in Sweden, 2 each in America, Holland, Prussia and Spain, and 1 each in Denmark, Poland and Russia.

in scale than the famous reflectors with which William Herschel slightly later also began extending the reach of astronomy beyond the solar system. With it Messier inaugurated regular telescopic investigation of nebulae. The star catalogue that he started publishing in 1774 contains the earliest identification of these objects.[97] He noted them in the course of hunting comets, in which game he excelled all competitors. Still an assistant to Delisle in 1759, he had been thwarted by his master, who for obscure reasons forbade his claiming his first discoveries and lost him the credit for a valid and independent observation of the first return of Halley's comet, eagerly awaited for that year.[98] Only some fifty comets were then known to astronomy. There were sixty-four in 1775, when Dionis du Séjour published the *Essai sur les comètes* that set Laplace to reasoning about the spatial distribution of their orbits, and Messier had found all but two or three of the new stock.[99]

Charles Messier had come to Paris in 1751, an orphan boy from Lorraine, and attached himself to Delisle in the naval observatory wedged into the tower of the Collège de Cluny. There he succeeded to the use of the instruments in 1760 or 1761, though not to Delisle's place of naval geographer, which went to Pingré. It would appear that Messier failed to acquire mathematics, always the most inaccessible of attainments for the self-educated, for he never calculated the path of any of his comets. Prizing the discovery alone, he affected a gruff peasant's contempt for science, scientists, and theories.[100] The manner was not designed to win him favorable consideration in the Academy, and he was elected only in 1770, by which time his work had long been famous throughout Europe. According to Delambre, he took "malicious pleasure" in finding discrepancies between his observations and the tabulations or calculations of a colleague.[101] Only once did he leave Paris for astronomical purposes. In 1767 the marquis de Courtanvaux organized a voyage from Le Havre up the coast of Holland and Friesland with the object of testing rival chronometers fabricated by the leading instrument makers, Pierre Le Roy and Ferdinand Berthoud. Courtanvaux was a former military officer and honorary member of the Academy, a patron of science who interested himself in the problem of determining longitude at sea, who had learned the art of horology with his own hands, and who maintained a fine private observatory in his house at Coulombe. He retained Messier for the observations and Pingré for the calculations. They intended to make the circuit

[97] Charles Messier, "Catalogue des nébuleuses et des amas d'étoiles, que l'on découvre parmi les étoiles fixés, sur l'horizon de Paris," MARS (1771/74), 435-461. On Messier's career, see Gingerich (1953) and Delambre (1827), 767-774.

[98] Delambre (1827), 768-769.

[99] Above, Chapter I, Section 5; Gillispie "Laplace," DSB *15*, 290-292.

[100] Delambre (1827), 770-772.

[101] *Ibid.*, 774.

of the North Sea ports, but an article in a London journal warning that their ulterior purpose was espionage prevented their landing anywhere in England.[102]

Canon Alexandre-Guy Pingré, Messier's companion on that truncated voyage, plied astronomy around the world. He was one of the principal participants in its most dramatic pair of events, the expeditions organized in 1761 and again in 1769 to make observations of the transits of Venus in order to determine precise figures for the solar parallax. For the first event, his post was the island of Rodriguez, a dot in the Indian Ocean between Mauritius and Madagascar, where after a harrowing voyage, occupied with discussions about whether incarceration in a ship at sea or in the Bastille was preferable, he arrived with his party and his instruments on 27 May. They had ten days to improvise an observatory ("I think there never was a more inconvenient one," he noted)[103] and to determine its precise position. June 6th, the day of the transit, dawned cloudy and windy. When the sun did come through, Venus had already begun her transit, and they could time only the egress from the sun's disk. Eight years later he had better facilities and complete visibility at Santo Domingo, where he stopped off for the second transit in the course of a far more elaborate horological expedition than that of Courtanvaux, this one under the command of a young naval officer, Charles-Pierre d'Eveux de Fleurieu, who had also qualified himself technically by taking lessons from Berthoud.[104]

Afterwards Pingré wrote an enormous book on comets, both descriptive and historical.[105] He had had the classical education of a clergyman before turning to astronomy at the age of thirty-eight, and published excellent translations of two ancient astronomical poems, one by the Augustan Stoic Marcus Manilius and the other Cicero's adaptation of a Greek work by Aratos of Soli.[106] Pingré was an Augustinian who, when young, had been inclined to Jansenism. Being a clergyman, he could be only an associé of the Academy. It was universally said of him, as it was of another

[102] *Ibid.*, 773; Armitage (1953), 55-56; *Journal du voyage de M. le Marquis de Courtanvaux sur la frégate l'AURORE . . . mis en ordre par M. Pingré* (Paris, 1768).

[103] Armitage (1953), 51; Pingré's journal of the expedition remained in the Dépôt des Cartes et Plans de la Marine, and is described in Marguet (1917), 171. On Pingré's career, see also Delambre (1827), 664-684. Woolf (1959) gives a full account of the international effort to coordinate observations of the transits of Venus.

[104] C.-P. d'Eveux de Fleurieu, *Voyage fait par ordre du roi en 1768 et 1769*, 2 vols. (Paris, 1773). See Armitage (1953), 56-58, and for the history of all the transit expeditions, English as well as French, Woolf (1959).

[105] *Cométographie; ou, Traité historique et théorique des comètes*, 2 vols. (Paris, 1783-1784).

[106] A.-G. Pingré, ed. and trans., *Marci Manilii Astronomicon Libri Quinque; accessere Marci Tullii Ciceronis Arataea, cum interpretatione Gallica et notis*, 2 vols. (Paris, 1786).

clerical scientist, the far more notable René-Just Haüy,[107] that he was gentle of disposition and kindly in nature.

Pingré was luckier in his observations of the transits than either of the associates, Le Gentil or the abbé Chappe d'Auteroche, whom the Academy dispatched to other posts. Guillaume Le Gentil de la Galaisière, a Norman from Coutances, also began in theology but was drawn into astronomy by Delisle's lectures at the Collège royal, and thence to the Observatory by Cassini II. His mission was to observe the transit in Pondicherry. Thither he departed in 1760. What with the British navy and the weather, however, he failed to arrive and only managed to see, but not to observe, the event from the bridge of the frigate assigned him, which had had to take refuge in the harbor of the island of Réunion. When he did reach India it seemed better to stay in Asia for the second transit. After a wait of eight years, he missed even seeing it since, although the weather had been brilliant in Pondicherry for many days, a cloud crossed the sun just before Venus did. Le Gentil returned to France in 1771. By then his heirs had moved to divide up his estate, and the Academy, thinking him occupied all this time with matters other than astronomical, had retired him to the status of *vétéran*.[108] In fact, he had been traveling the Orient from Madagascar to Manila. His main book is an account of those far parts.[109] In conformity with what seems to have been almost a rule among contemporary French astronomers, he was fascinated with the astronomy of the ancients, and wrote about the origins of the Zodiac voluminously in the mode of eighteenth century quasi-scholarship, a mixture of the antiquarian and speculative.[110]

The third of the transit expedition leaders, the Auvergnat abbé Jean Chappe d'Autoroche, fared better astronomically, but in no other way. His was the most arduous of the 1761 journeys, overland across Siberia to Tobolsk. Although a momentary cloud prevented his timing the ingress of the planet, his observations of the passage itself were full and detailed. So, too, were his observations in another genre on life and government in Russia, so much so that Catherine II had a refutation published. In 1769 Chappe was sent to Cape Lucas at the southern tip of Baja California.

[107] Below, Section 6.

[108] On Le Gentil, see Delambre (1827), 688-709.

[109] *Voyage dans les mers de l'Inde, à l'occasion du passage de Vénus sur le disque du Soleil, le 6 juin 1761, et le 3 du même mois 1769*, 2 vols. (Paris, 1779-1781).

[110] It would be excessively lengthy to give the titles of the series of memoirs in which Le Gentil discussed the Chaldean Saros, the origins of the Zodiac, Hindu astronomy, Egyptian hieroglyphics and the vestiges of this remote past in medieval astronomical motifs. They will be found in MARS as follows: 1756/62, 55-69, 70-81; 1782/85, 368-456; 1784/87, 482-501; 1785/88, 9-16, 17-23; 1788/91, 390-405, 406-410, 411-438; 1789/93, 506-513.

There he managed to record full and precise data despite the ravages of an epidemic of dysentery, to which he succumbed some days later together with all the members of his party except a long draftsman, one Pauly, a graduate of the Ecole des Ponts et Chaussées, who returned with the information and the instruments.[111]

The last representative of this tradition of traveling astronomers, Joseph Le Paute d'Agelet, was younger and a student of Lalande. His combination of mathematical capacity with observational accuracy exhibited the extra increment of professionalism that distinguished the work and style of his generation from their predecessors. He began work in 1768, using Lalande's old observatory at the Collège des Quatre Nations, and left in 1773 to sail with Kerguelen's voyage to the South Seas. After returning, he was named to a professorship at the Ecole militaire, and there had the use of the Observatory until 1785. Lalande had installed for him a mural quadrant of eight-foot radius, the last instrument constructed by John Bird in London. Being the youngest and most promising member of the Academy in its section of astronomy, he was then named to accompany Lapérouse in the expedition that was intended to complete and thus outdo James Cook's exploration of the Pacific. He perished with its commander early in 1788, no one knew where until some forty years later an English sea captain came upon traces of the wreckage among the Santa Cruz islands on the reefs of Vanikoro.[112]

Edmé-Sébastien Jeaurat and Pierre-Charles Le Monnier, the two stay-at-home astronomers, need not detain us long. Given to harboring injuries, both were a trial to their colleagues and perhaps to themselves. Le Monnier had translated John Keill's *Introduction to the True Astronomy* from the original Latin of 1718, and having formed his own approach on Halley and Bradley (he was always a little "behind his century," said Delambre),[113] first constituted himself the apostle of British astronomical methods on the continent, and then had the gracelessness to open a textbook of navigation with the complaint that the British had unjustly appropriated the science of longitudes. "The perturber rather than the promoter of astronomy" (the mot was the abbé Lacaille's),[114] Le Monnier was forever accusing his colleagues of having plagiarized the data he crabbedly

[111] Delambre (1827), 621-623; Armitage (1954).

[112] Lalande, *Astronomie* (3rd ed., 1792) *1*, xxxii-xxxiii; "Notice bibliographique sur Le Paute d'Agelet," *Bulletin de la Société de Géographie*, ser. 7, 9 (1888), 293-302, special issue entitled *Centenaire de la mort de Lapérouse*. His observations made at the Ecole militaire from 1778 to 1785 are conserved at the Bibliothèque de l'Observatoire, C-2, 21-33.

[113] Delambre (1827), 179. The translation is entitled *Institutions astronomiques, ou Leçons élémentaires d'astronomie* (1746). On Le Monnier, see Robida (1955). His observations are at the Bibliothèque de l'Observatoire, C-4, 1-16.

[114] Delambre (1827), 209.

recorded in the observatory he had installed in the garden of the Capucins in the rue Saint-Honoré. The irony is that after his death his manuscripts were found to contain twelve observations of Uranus, which he had never recognized for the true discoveries they were. His brother, Louis-Guillaume, also of the Academy of Science, was a doctor, while his daughter, who could not have been over twenty-five, made him the father-in-law of Lagrange in 1792. The great mathematician had been settled in Paris for five years when he was married and was already 56.

By the time of the Revolution, Jeaurat was the oldest of the astronomers with quarters in the Observatory. Brevetted "Ingénieur-géographe" or draftsman in 1749, and thus an early graduate of the Ecole des ponts et chaussées, he had begun his career with the Cassini map, published a manual on perspective, and been appointed professor of mathematics at the Ecole militaire in 1763.[115] It was he who had installed the Observatory there in 1768. He had a certain talent for designing instruments,[116] and succeeded Lalande and preceded Méchain in editing the *Connaissance des temps*. The volumes for the years 1776 through 1787 were his responsibility.[117] His own memoirs are routine, and the archives contain his complaints of being passed over by younger colleagues.[118] A Danish astronomer, Thomas Bugge, visiting the Observatory in 1798 took him for the porter and was astonished to find that the old man shuffling papers at a table in the basement was, in fact, Jeaurat, still alive.[119]

Let us return now to the Observatory itself, where Cassini IV, nothing if not conservative by temperament, nevertheless instituted reforms largely professional in their intent. The history of science offers deeper but few more persistent examples of the operation of tradition than the fidelity exhibited by his family to the cartographic and geodesic function of astronomy. It had been for such purposes that Colbert in 1669 brought his great-grandfather, also called Jean-Dominique, from Bologna.[120] No accurate map of France then existed either for military or administrative purposes. Confection of such an instrument formed one of the charges that Colbert laid upon the new Academy. Preliminary surveys disclosed startling imprecisions, among them that the location accepted by navigators

[115] He submitted a curriculum vita to the Comité d'instruction publique, 5 pluviose an II [24 January 1794], AN, F^{17}.1065 A, doss. 4.

[116] J.-M. Faddegon and Boizard de Guise (1936).

[117] Delambre (1827), 748-755.

[118] For example, AN, $F^{17}.1344^{35}$, doss. 2.

[119] Crosland (1969), 104-105.

[120] Delambre (1912) gives a summary evaluation of all the important geodesic surveys of the seventeenth and eighteenth centuries. There is additional detail in Drapeyron (1896a) and (1896b); Gallois (1909); Laissus (1965). The Cassini map itself is the main subject of vol. 1 of Berthaut (1898). The discussion of the relation of geodesy to cartography in Brown (1949), ch. 9, is helpful but somewhat lacking in historical rigor.

for the port of Brest lay thirty leagues at sea. If cartography was thus unreliable, the general science of geodesy was no more than rudimentary. It is often said that in 1666 when Newton was making his youthful comparison of the forces with which gravity affected the moon and an object at the surface of the earth, the respective quantities did not match, one difficulty having been that he had the radius of the earth wrong.[121] It was then estimated that the degree of latitude measured about sixty English miles on the ground. The correct figure of nearly sixty-nine and a half was first approximated in 1669 when the abbé Picard adapted telescopic sights to surveying instruments, ran a pioneering triangulation along the meridian from a point just north of Paris to Amiens, and reported 57,060 toises to the degree.[122]

Installed at the Observatory, the first Cassini seconded Picard's proposal to Colbert that the meridional survey be extended on both ends until it should span the entire length of France from Dunkirk through Paris to the Pyrenees.[123] They promised in the end to provide reliable coordinates for the military and administrative map the government desired. Upon this cartographical mission Cassini superimposed the further geodesic problem of the shape of the earth. After the publication of Newton's *Principia* in 1687, it came to seem that such down-to-earth techniques of land surveying should in principle be competent to resolve the largest issue of cosmology, the choice between the Cartesian and Newtonian theories of the world. For a meridional survey ought to reveal any departure from sphericity in the surface of the globe, whether by elongation along the polar axis or by flattening. Like most academic projects, the work went slowly. Nine sheets of a beautiful map covering the Paris region were printed in 1678.[124] Colbert was dead long before anything more extensive was ready for engraving. His successors cut off funds, and although Cassini II (Jacques) managed to revive his father's undertaking in 1700 and to prepare the first full meridional survey of France for publication in 1718,[125] it proved unfortunate for his later reputation that his techniques gave his still Cartesian colleagues of the Academy the elongated earth their cosmology demanded.

With Newtonianism largely triumphant elsewhere by the 1730s, the Academy, prodded by Maupertuis, Newton's champion in France, per-

[121] I. B. Cohen, "Newton," DSB *10*, 61.

[122] Jean Picard, *Mesure de la terre* (Paris, 1671). For the instrumentation, see Daumas (1953), 70-76. The toise (or fathom) being six feet, came to approximately two yards.

[123] Gallois (1909), 292-295.

[124] *Carte particulière des Environs de Paris, par Messrs de l'Académie Royalle des Sciences, en l'année 1674*. A copy is in the map section, BN, portfolio 215, 3235.

[125] Jacques Cassini, "De la grandeur de la terre et de sa figure," HARS (1718/19), 245-256; and *De La Grandeur et de la figure de la terre (Suite des mémoires de l'académie royale des sciences, année 1718)* (Paris, 1723).

suaded the state to support yet another and more ambitious geodesic effort. Maupertuis himself, together with Clairaut, led an expedition to Lapland to measure a degree within the Arctic circle. At the same time, La Condamine and Bouguer set sail for Peru, where they surveyed three degrees within the present boundaries of Ecuador. Unfortunately, they quarreled. Recriminations compounded fearful difficulties of terrain, and the Peruvian expedition used up ten years. The two sets of results were decisive, however. La Condamine measured 56,475 toises in the equatorial degree, and Maupertuis 54,941 in the Arctic. In addition, La Condamine compared the dimensions of a simple pendulum with a frequency of one oscillation per second at Paris, where its length was 3 pieds, 0 pouces, 8.57 lignes, with those of a similar instrument at the Equator, and found that the latter measured only 3/0/7.07, the difference of one and a half lignes being between an eighth and a sixteenth of an Anglo-Saxon inch.[126] When he returned to Paris, he made use of this last determination to rectify the Toise du Châtelet, the master bar to which the royal measures were supposedly referred, and which was mortared into a pillar at the foot of the great staircase in that structure, slowly rusting and buckling with the settling masonry. What was more important, he proposed abandoning any attempt to keep such an object constant and instead adopting the equatorial second-pendulum for an international linear standard. Thus originated the idea of a scientific metric system, in which units of ordinary measurement would be related to a geodesically determined magnitude.[127]

Neither the state nor the Academy then acted on La Condamine's proposal. It was already clear from Maupertuis's results that criticism of the work of Cassini II had been well founded and that the meridional survey of France needed redoing. Officially, the work was confided to Cassini III in 1739—César-François Cassini de Thury, as he called himself from a property the family acquired north of Paris.[128] In fact the field work was mainly executed by the abbé Lacaille, who became the foremost practical astronomer in France in the midcentury, and installed the observatory in the Collège des Quatre Nations, the most active scientifically among the foundations comprised in the University of Paris. He was professor of mathematics there and taught Lavoisier, Lalande, and Bailly among many others. He was also a man of austere character who got from his instru-

[126] On these famous expeditions, see Delambre (1912), 98, 118; Maupertuis, *La Figure de la terre* (Paris, 1749); La Condamine, *Journal du voyage . . . à l'equateur* (Paris, 1751).

[127] La Condamine, "Nouveau projet d'une mesure invariable propre à servir de mesure commune à toutes les nations," MARS (1747/52), 489-514; "Remarques sur la Toise-étalon du Châtelet," MARS (1771, pt. 1/1774), 482-501. Cf. Bigourdan (1901), 1-15; Favre (1931), 33-46.

[128] Drapeyron (1896a), (1896b).

ments, as from his students, all the precision of which they were capable and verified his results in a spirit of selfless devotion to the most perfect possible accuracy of astronomical information. His great adventure was a hazardous expedition to the Cape of Good Hope beginning in 1751. Data for the southern hemisphere were to be compared with those which Lalande was recording during his year of emancipation in Berlin. Lacaille had been accorded a grant of 10,000 livres for instruments and all expenses. He had not asked for more when the navy requested him to gather additional navigational information for the Indian ocean in the Iles de Bourbon and de France. On returning to Paris in 1753 he astonished the officials of the treasury by faithfully coming around with an unexpended balance of 855 livres.

From what has just been said, it is evident that historians are mistaken who suppose that nothing had ever been attempted like the famous survey of the 1790s, intended to be the basis for the metric system. In fact, the metric survey was the third full-length measurement of the meridian of Paris in France, Lacaille's having been the second. Delambre, the one mainly responsible, knew Lacaille's work well, as indeed he did the whole history of astronomy, and admired it more than that of all other eighteenth-century predecessors. He was not given to encomiums either in his astronomical or historical writings, and his estimate of Lacaille is simple and somehow moving:

> His manuscripts when compared to his printed works attest throughout to that veracity which ought to be the foremost quality of the observer. Having through a singular combination of circumstances had occasion to redo part of his work, or to verify it with new methods, we are bound to say that, after having re-observed with greatest care all his stars, or at least all that are visible from Paris, after having undertaken lengthy research on refraction, computed new tables for the sun, surveyed the meridian of France, and had his manuscripts in our hands for several years, we have never taken a single step in his footprints without feeling redoubled esteem and admiration for a scientist who will ever do honor to French astronomy.[129]

Setting to work on the meridian in 1739, his first major commission, Lacaille had the advantage of a micrometer devised a few years earlier by the chevalier de Louville for use on sextants and quadrants. With this instrument, and with some assistance from Cassini III and J.-D. Maraldi, the second of a succession of three Cassini cousins also maintained at the Observatory, Lacaille spun a web of almost 800 triangles from Dunkirk

[129] Delambre (1827), 541-542, concluding his article on Lacaille.

to Perpignan, referring the angles to some nineteen base lines measured out on the ground. Nothing was represented except those steeples, chateaux, and terrain features that Lacaille or his associates had sighted in themselves. The result was a chart consisting of eighteen sheets at a scale of 1:878,000, completed in 1744 and presented to the Academy in the following year.[130] This was not yet the map of France. Everything beyond and between the points and lines surveyed was left blank. It constituted the geometric skeleton on which a detailed topographical map might then be fleshed out. The opening campaigns of the War of the Austrian Succession gave the impetus for its realization. In 1746 Cassini de Thury went into the Austrian Netherlands with the engineers, and the story goes that Louis XV, delighted with a handsome map of the environs of Liège prepared for ordering the battle of Rocourt, took the initiative of ordaining that his whole kingdom was to be mapped with comparable beauty and precision.[131]

The fact appears to be that Cassini de Thury took advantage of military opportunity to continue the project for a general map by recalling it to the favorable attention of the court. His efforts for cartography were nonetheless effective for being exercised more in entrepreneurship than in astronomy, and preparation of the sheet covering the region of Paris was begun in 1750. Renewed war in 1756 deepened the embarrassment of French finances, however, and the controller-general withdrew Machault's promise of a subsidy amounting to 700,000 livres over a period of twenty years, substituting for it the privilege of forming a private company to underwrite the enterprise by taking advance subscriptions to the sheets as they appeared. Cassini de Thury organized such a company consisting of fifty shareholders. He associated with himself as directors members of the Academy already encountered in this history, most notably Bochart de Saron and Jean-Rodolphe Perronet, founder of the Ecole des ponts et chaussées. Persons eminent in society and in science were pleased to have their names upon his list. Madame de Pompadour and the maréchal de Soubise, Montalembert and La Condamine, Buffon and Malesherbes, Quesnay and Daniel Trudaine were among the subscribers. Although funds were lacking in the Treasury, the Ministry of Finance did encourage the provincial estates in the pays d'états to subscribe each for its own region and itself committed the pays d'élections.[132]

Applying the techniques of geodesic triangulation in the fine detail required for topography meant sending surveying parties into every locality

[130] *Nouvelle carte qui comprend les principaux triangles qui servent de fondement à la description géométrique de la France . . . par Messrs Maraldi et Cassini de Thury . . . Année 1744*. On this work, see Gallois (1909), 304-305.

[131] Drapeyron (1896b), 246, who has the location of the battle wrong, however.

[132] Drapeyron (1896a), 3-6, 12-13.

and village across the whole of France. By dint of multiple expedients, Cassini de Thury found the money to enlarge the personnel from some half-dozen, who were all that had ever been employed on earlier, piecemeal maps and surveys, to a team of thirty-four draftsmen working with engravers and graduates of the Ecole des ponts et chaussées, all directed from a central bureau in the Observatory. Eleven engravers were required at one time or another, and three calligraphers. It was out in the countryside, however, and not in problems of finance or technique, that the most serious obstacles had to be overcome. Surveyors with their chains, stakes, and transits are always birds of ill omen in the eyes of those with a private interest in the land, and peasants resented the unexpected appearance of these strangers manipulating their strange instruments. Frequently they resisted, sometimes by threats, sometimes by force, many times by overturning or displacing markers. Parish priests and local lords themselves would take offense, complained Cassini, noting how

> The very people who would have been capable of giving us information often tried to deceive us, not looking with favor on the use that they foresaw would be made of a really detailed map in the imposition of taxes. This unfounded suspicion, which I had great difficulty in dispelling, was the reason that the part of my project concerning the natural history of each province could not be carried out.[133]

Since Turgot's proposal for a land registry presupposed just such information, local suspicion was not so unreasonable as it appeared to Cassini and his sponsors in officialdom. The work went slowly, but it went. In 1738, a year before his death, Cassini de Thury published a résumé of the operations, by which time all sheets were printed, except those for Brittany.[134] That one province, he wrote, had caused more trouble than all the rest of France.[135] Its estates never did provide funds. Only after 1815 were the last of the Breton sheets finished. With those, the 182 sheets on a scale of 1:86,400 originally projected in 1750 were completed. The work remains a triumph of the arts and science involved.[136] The plates are still in fine condition, and new impressions may even now be purchased. The most famous and ironic use that has been made of this splen-

133 Drapeyron (1896b), 248; see also Drapeyron (1896a), 6-11, for an account of the personnel.

134 C. F. Cassini de Thury, *La Description géométrique de la France* (Paris, 1783). The historical introduction is to be treated with caution.

135 Drapeyron (1896a), 13.

136 C.-F. Cassini de Thury, *Carte de la France*, 5 cartons en 8⁰ de tableaux, et 2 vols, en 4⁰ de description (Paris, 1750-1787), BN, L^{14}.11. For an evaluation, see Fordham (1929), 39-56.

did representation of the old France occurred in 1790 in the revolutionary redistribution of her manifold provinces, municipalities, and domains into the departments, cantons, and communes that constitute the subdivisions of modern France. In an earlier, and indeed the earliest civic application, the Cassini map formed the basis for the population studies published by the Academy in the final years of the old regime and already noticed in connection with the demographic interests of Laplace and Condorcet.[137]

The last Cassini, called Jean-Dominique like his great-grandfather, succeeded to the office of director-general of the Observatory in 1784 on the death of his father, whose functions he had been largely discharging for some eight or ten years. The great map done, or nearly so, his ambition was to renovate the Observatory itself, converting it from a decaying astronomical hostel into a working scientific institution independent of the Academy. The building had fallen into dilapidation, the roof leaking, the gutters overflowing, the plaster yellowing and falling, the woodwork mildewing, and the fine vaulted ceiling over the central hall threatening to collapse. Only the façade preserved the appearance of grandeur befitting the original conception of a chateau of science at the edge of the capital. Jeaurat and Le Gentil had installed telescopes in their quarters, but there was no other astronomical work. The instruments pertaining to the Observatory proper were obsolete and unusable. Cassini IV in his own apologia tells how he raised the money to start putting things right.[138] The Academy kept its distance, having lost interest in the Observatory after the decree of 1771 named his father director and ended its control. What with the importance of the map to officialdom, the association of eminent persons in ownership of its shares, and the affinity of the Cassinis themselves for fashionable society, their relations were closer with ministerial than with scientific circles. Despite the financial straits of the state in these last days of the old regime, D'Angiviller and Breteuil found funds adequate for Cassini IV to make a good beginning. Major repairs to the building were under way before the Revolution complicated matters, and Cassini appeared to be well started on the scientific aspects of his program, which consisted, first, of assembling and training staff, second, of rehabilitating or replacing instruments, and, third, of projecting the geodesic tradition he had inherited along with his post into international standardization of measurements.[139]

Breteuil secured the king's assent to an ordinance embodying new regulations for the Observatory on 26 February 1785. By its terms, the di-

[137] Above, Chapter I, Section 5.

[138] J.-D. Cassini, *Mémoires pour servir à l'histoire des sciences et à celle de l'Observatoire de Paris* (Paris, 1810); cf. Wolf (1902), 229-251.

[139] Wolf (1902), 251-263.

rector was to see to the maintenance of round-the-clock astronomical, meteorological and magnetic observations. Data were to be reported annually for the benefit of all persons concerned with astronomy and navigation in France and abroad. Every ten years an *Histoire céleste* was to be compiled and published in order that there be a permanent record of operations performed in the Observatory of Paris. Besides the director, staff was to consist of three assistants who were to work "under his eyes," and to be called *élèves*. The first was to have 900 livres a year, the second 700, and the third 600, and a further sum of 200 was to be reserved to reward whichever of them should work the best or make some discovery. The director was to select his pupils from the class of "honest citizens" with "irreproachable and spotless French family" background. He was to require of them regularity in conduct and morals. Their duties were staggered so that two at a time were always present in a working day that began at seven A.M. in summer and eight in winter and continued until three the following morning.[140]

Several prospective pupils refused these conditions. One of them, Louis-Robert Lémery, a fifty-seven-year-old assistant to Jeaurat on the *Connaissance des temps*, would consider only a "worthy (*honnête*) position, for he was apprehensive of slavery and dependence if it was too rigorous."[141] The three who did accept were a youthful Benedictine, Dom Nouet; the younger son of a soldier of fortune, Perny de Villeneuve; and a deserter from the regiment of dragoons, one Ruelle. Nouet had been sent by his superiors in the abbey of Morimont to board in the Observatory in order to qualify himself to teach science in a school they kept for the sons of poor noblemen. Of Perny, we know only that his father had commanded the artillery in the volunteer corps attached to the king's brother, the comte de Provence. Ruelle had taken refuge from prospective court-martial in the Observatory, where one of his relatives, a watchmaker called Boucher, lodged in the apartment assigned Jeaurat.[142] Obediently they minded their instruments and provided Cassini IV with the data he began circulating annually to the astronomers of Europe and America.[143] He was then styling himself "comte." All three of his *élèves* were soon to figure in the activist political company of the revolutionary Section de l'Observatoire.

When it came to instruments (for re-equipping the Observatory formed the second aspect of his program), Cassini immediately confronted the

[140] *Ibid.*, 255, 259-262. [141] *Ibid.*, 264.

[142] *Ibid.*, 263-265. Cf. Devic (1851), 207-216, who, however, thought that there were four pupils, including one whom he called only "V," who must surely have been a phantom conjured up from Perny's landed name of "Villeneuve."

[143] Wolf (1902), 266-270.

enormous superiority of English over French construction that was the counterpart in manufacturing generally to the growing French predominance in science in the later eighteenth century. The best telescopes in Paris came from London. Le Monnier had begun his career with a five-foot quadrant built by Jonathan Sisson, and he commissioned a seven-and-a-half foot replacement from John Bird for his observatory at the Capucins in the rue Saint-Honoré. The latter artisan had also made the several telescopes installed at the Ecole militaire. The secret of flint glass, required for the finest optical devices, was a British monopoly. John Dollond was unexcelled in the grinding of achromatic lenses. When in 1787 Cassini visited London in company with Méchain and Legendre, his report to Breteuil was dithyrambic in its praise for the creations of Jesse Ramsden: "The fecundity of that artist's genius, the perfection of his execution, and the consummation of his experience in his art, force me to recognize that for a very long time it will be extremely difficult, I will not say to surpass, but even to imitate him."[144]

It was necessary to try, nevertheless, since recourse to London for every important commission could only further discourage French artisans and deprive them of incentive to improve their capacities, both technical and commercial. In 1784 Cassini's initial recommendation to the crown had called for provision of three new instruments, the most considerable to be a seven-and-a-half foot mural quadrant of the English type. Endorsing these intentions, a commission of the Academy stipulated (in accordance with Cassini's own views) that all the telescopes be of French construction.[145] The crown promised funds, and Cassini set forth upon a course of scientific patronage of craftsmanship in the Observatory itself. With the approval of Breteuil and d'Angiviller, he installed a foundry in the courtyard under the supervision of one Héban, a master metal-worker who was given lodging on the premises. On the second story of the west tower he equipped a complete workshop for the fabrication of astronomical instruments. In these facilities, he proposed to have constructed the large mural quadrant which was the Observatory's most pressing requirement. It was to be modeled upon the latest instrument by John Bird at the Ecole militaire, except that the frame was to consist of a single casting, a thing never yet attempted.

The project was as unrealistic as it was ambitious, and the disappointments Cassini encountered are instructive for the limitations imposed by the mentality and scale of Parisian enterprise. The first of the artificers to whom he turned initially agreed to this commission from on high, backed by the ministry and even by the crown. This was the "sieur Charité, al-

[144] *Ibid.*, 288. [145] *Ibid.*, 274.

ready known for the construction of several instruments perfectly copied from the English."[146] His response to the opportunity, however, hardly exhibited that spirit of enthusiasm and gratitude that, so Cassini reported to Breteuil, was animating the artisans of the capital now that they had learned of the king's intended generosity in patronizing their skills in astronomical constructions. Within a few days, Charité's conduct began to mingle the grudging with the exigent. To have admitted that the job was too big for him would have been humiliating. To have come out and said he did not want to do it would have risked offense. Instead of simply declining, Charité transferred the onus of repudiating the agreement to his would-be client by posing unacceptable conditions.[147] He insisted that he and his family be lodged in the Observatory for life; that he be granted the title of "Ingénieur pour les instruments d'astronomie" by royal letters patent, which must also assure him permanent tenure of the workshop even under Cassini's successors; and that the tools provided by the establishment for fabricating this first quadrant must then become his personal property.[148]

Thereupon, Cassini turned to a second artificer, a protégé of Lalande called Mégnié, to whom he had already advanced 8,000 livres for a smaller job, an equatorial telescope. The name of Mégnié is one that the historian of technical affairs encounters frequently in the French archives, and the only identification that can be made with certainty is that there were at least two artisans who bore it. Probably they were brothers, and probably Cassini was concerned with Pierre-Bernard, sometimes designated "Mégnié le jeune."[149] Whichever it was who accepted the terms, the floor of the new workshop was strengthened; plane surfaces of marble were polished and emplaced; a small forge was furnished with hood and chimney; vessels and ovens for working glass were ranged along one wall; steel and copper rules, dividers, compasses, verniers, and micrometers were calibrated; tougher tools were purchased. Besides the considerable funds involved for all these items, Mégnié was further advanced 5,418 livres from the Treasury and another 14,000 by a moneylender to pay for additional instruments. Alas, unknown to Cassini, Mégnié was already far gone in debt before coming to the Observatory. His creditors found him there, and he fled into the limbo of the bankrupt, leaving his employees and his bills unpaid, his embarrassed patron disheartened by the ways of commerce, and the project for a shop in the observatory spoiled beyond repair.[150]

146 Cassini to Breteuil, 18 January 1785, *ibid.*, 275.

147 Daumas (1953), 358-359.

148 Wolf (1902), 278; cf. Bibliothèque de l'Observatoire, D-5, 39.

149 Daumas (1953), 362-363.

150 Wolf (1902), 282-283.

A simultaneous attempt to give out work to an established artisan fared no better. The two most highly reputed artificers in the capital were Nicolas Fortin, who made Lavoisier's finest balances and furnished the laboratory of the Arsenal with other apparatus, and Etienne Lenoir, who specialized in optical equipment. At the same time that Cassini proposed to fabricate the great quadrant on the premises of the Observatory, he commissioned Lenoir to make the second instrument authorized by the ministry, a telescope mounted on a three-foot circle of the type used in surveying. No establishment in Paris operated on a sufficient margin of capital to dispense with part payment in advance to underwrite the cost of labor and materials, and Lenoir, too, had to be credited with 8,000 livres. There was no question about his capacity or good faith. His deficiency lay in never having formally qualified himself to be a master founder. In July 1785 syndics of the corporation or guild of foundrymen searched his shop and seized his supply of copper, his tools for working metals, and the partially completed or finished instruments that fell within their competence. The magistrate imposed on him the costs of thirty-six livres they had incurred in thus enforcing their corporate rights and replied to Cassini's protests, supported though they were by the minister, that Lenoir must be admitted to the guild before the law could protect him from such otherwise legitimate and accustomed search and seizure.[151]

An instrument-maker might be equally vulnerable to visitation and confiscation by the glassmakers, the cabinet-makers, or other of the old and jealous corporations, and Lenoir's experience suggested to Cassini that the trade he thought to foster needed to be organized into a corporation of its own. He took counsel with Bailly, whose relations with ministers were also close, and they imagined an interesting device, one illustrating that the organization of a quasi-scientific trade needed a form more developed than the guilds of old with their obstructive spirit and restrictive ways. The model they followed was that of the licensed engineers who composed the Corps des ponts et chaussées. In February 1787, Breteuil drew up letters patent creating a new "Corps d'ingénieurs en instruments d'optique, de mathématiques et de physique et autres ouvrages à l'usage des sciences." Its membership was to be limited to twenty-four. It would elect its own syndics and govern itself. Other trades were forbidden to interfere with its engineers in their production of scientific instruments. Yet the new corps was not free of outside control. Election to its membership was by the Academy of Science, not by itself. Laplace soon proposed that candidates be required to satisfy the Academy concerning their knowledge of elementary geometry and mechanics before being admitted,

[151] *Ibid.*, 276; Daumas (1953), 173.

an idea that was modified to the extent of merely giving preference to artisans thus qualified.[152] At the first election those chosen were Lenoir, Carochez, Fortin, Charité, Baradelle, and Billeau; at the second, Etheret, Putois, Dumoutiez, Herbage, Tournant, Richer, Mégnié le jeune (whose skills must have quickly restored his technical if not financial credit), and Mossy; and at the third, Haupois, Lerebours, Gouffé, and Chiquet.[153] These names are worth noting, for many of them reappear in the popular societies of technology and invention that sprang into being with the Revolution.

The comparison with Britain involves more than the differences in scale and capitalization in this skilled trade, impressive though they were. In London Ramsden was then employing some fifty artisans and laborers, and when Cassini, despairing of getting his quadrant built in Paris, finally had to give him the order after all, he did not even fix the price but let it be assumed between two honorable men that the bill would be fair.[154] For the contrast was also one of social status. Scientists and their suppliers in London associated with one another as men of education and equals, partners in the development of knowledge. Cassini addressed Ramsden most courteously in their correspondence. The French artisans, on the contrary, were virtually illiterate, even Fortin and Lenoir, and were treated as underlings by those who bought their wares. It makes for a curious reflection that here at the end of the old regime, Cassini and Bailly, the most sycophantic of the scientific community in their attitude to aristocracy, should have been brought by circumstances to exemplify all unwittingly the democratic prescience of Diderot's dictum animating the *Encyclopédie*: that in the interests of rationalized industry, artisans must be raised in the world and taught to have a better opinion of their own worth.[155]

Mingled cooperation and competition with British enterprise also stimulated the third aspect of Cassini's program for the Observatory. In 1783, early in the days of his responsibility, he initiated a proposition to the Royal Society that a joint Anglo-French operation be undertaken with the object of joining the maps of Britain and France by the most precise techniques of surveying developed in each country.[156] Thus would his family's geodesic tradition be internationalized, and existing astronomical determinations of the meridians of Paris and Greenwich verified or rectified by measurements taken on the ground. In London his initiative was wel-

[152] Procès-verbaux of the Academy of Science, 13 June 1789.

[153] Daumas (1952), 89-90.

[154] Wolf (1902), 288-297, cf. Daumas (1953), 319-320.

[155] Article "Art."

[156] Bibliothèque de l'Observatoire, D-5, 7; see also J.-D. Cassini IV, *Exposé des opérations faites en 1787 pour la jonction des Observatoires de Paris et de Greenwich* (Paris, 1790).

comed by a resolution of the Royal Society of 1784, which urged upon the government the importance of covering England with a net of triangles based upon a precise determination of the meridian of Greenwich in order that mapping might attain the precision achieved in the Cassini map of France.[157] Within the Royal Society the Fellow mainly responsible for improving the occasion offered by Cassini's initiative was William Roy, an engineering officer and surveyor-general of the coasts. General Roy had made a lifetime cause of persuading government to take in hand a proper geodesic survey of the British Isles.[158] After his death in 1790 his efforts, together with the French example, bore fruit in the British Ordnance Surveys.

Both the French and English parties to the project were as concerned with developing and perfecting technique and instruments as with achieving results. Intending to exploit the capacity of the London instrument trade to the utmost, Roy turned to Ramsden for the construction of a great theodolite, to be the most accurate surveying instrument ever yet made. Ramsden's perfectionism often exasperated the very clients it attracted, for he could never keep a firm date for completing a commission. This one took three years while Roy and his French colleagues fidgeted. When done, it fully lived up to expectations, a work of unprecedented precision and monstrous size. Both the transit telescope for measuring vertical angles and the surveying telescope for horizontal angles had three-foot focal lengths and double object glasses of two-and-a-half inch aperture. The horizontal scope was mounted on a graduated circle, also three feet in diameter, and the transit scope trunnioned above it. Each was capable of 360° traverse, and all movements were equipped with two motions, the first for large traverse and the second for precise sighting. Four microscopes equipped with verniers permitted readings to 0.1 seconds of arc at the cardinal points of the circle. The feet and the plane of the circle were furnished with night-lighting devices and leveling screws. Roy in his description despaired of "entering into any detail of the minutiae: for even to have mentioned these, with the almost infinite number of little screws that serve to unite them into one entire machine . . . would have been a disgusting labour."[159] With this splendid theodolite he could determine azimuths at distances of seventy miles to an accuracy of two seconds. The instrument weighed two hundred pounds. It could not be car-

[157] Fordham (1929), 76-85.

[158] See Roy's memoir, *An Account of the mode proposed to be followed in the trigonometrical operation for determining . . . the relative situations of the Royal Observatories of Greenwich and Paris* (London, 1787). His *Account of the Trigonometrical Operation, whereby the distance between the Meridian of the Observatories of Greenwich and Paris has been determined* (London, 1790) is bound in with the Cassini *Exposé* (n. 156 above).

[159] William Roy, *An account of the trigonometrical operation, . . .* (London, 1790), 27. Roy gives a detailed account of the theodolite (27-48, illustrated with plates).

ried up the crazy ladders of ancient steeples, and required the construction of a special scaffolding on piles driven four feet into the ground. Five men using a portable crane with an eighteen-foot beam would hoist it into place, for it was not to be jarred in being mounted on its platform. The combination of accuracy and unwieldiness put a premium on finding terrain features both accessible and commanding.

While awaiting delivery, Roy had two bases measured for his survey. For the first, on Hounslow Heath, he used glass rods to tape a distance of 27,404.24 feet, almost five miles. For the other on Romney Marsh, his men used steel chains along a leg of 28,535.7 feet. The latter stretch being within sight of the French coast, Roy could link his triangles to those determining the French meridian, which junction would permit verifying the respective positions of the observatories of Paris and Greenwich within a single system of triangulation. In Paris, the Academy of Science had detailed Legendre and Méchain to join Cassini in carrying out the operations required. This same commission had taken all three to London to meet Ramsden. Their task was far less extensive than that of the English, who were surveying the Straits of Dover for the first time. The French had only to verify their own positions recorded in the Lacaille-Cassini II transit of 1739-1740,[160] sight in positions on the English coast, and provide signals for the English to do the same from their side of the Channel.

Nevertheless, their work held more than routine interest, for they too were equipped with a new instrument, the repeating circle designed by the chevalier Jean-Charles de Borda, and its employment and mode of operation were significant episodes among many others in the early development of the theory of error. Before describing that, it will be convenient to identify briefly the other two members of the first team to handle it. Méchain will be met again, putting this experience to use as Delambre's partner—a less than equal partner—in the revolutionary metric system survey, to which this junction with the English coast proved to be the immediate technical prelude. Méchain was already forty-three years old in 1787. The son of an unsuccessful architect in Laon, he had done so well in school that prominent local persons encouraged him to better his education in Paris at the Ecole des ponts et chaussées. He was admitted and began, but when his father was unable to support his expenses, he had to withdraw and take a tutoring position in a family residing near Sens. Learning of his talents and misfortunes, Lalande intervened in his life, found him a clerical position in the naval map repository at Versailles, and arranged access to an observatory at night. Méchain found two comets in 1781, and had sufficient mathematical wit combined with as-

[160] Above, Section 5.

tronomical feel to calculate observations of the body later identified as Uranus on the assumption of a planetary rather than a cometary orbit. A memoir of 1782 applied a method of Lagrange to prove that comets first observed in 1532 and 1661, respectively, were not congruent with Halley's in their orbits, and therefore would not return as some predicted in 1789 or 1790. The essay won a prize set by the Academy on theory of comets together with election the same year.[161] Still plagued by near-poverty, Méchain was given a lodging in the Observatory and in 1785 took over from Jeaurat direction of *Connaissance des temps*, an editorial chore bringing in 600 livres annually that was passed on from one impecunious astronomer to another when the predecessor received some pension or promotion. Méchain's experience at the English Channel was his first venture into the surveying field.

The third colleague, Adrien-Marie Legendre, was far more accomplished mathematically than Méchain, whose attainments were workmanlike though unoriginal, or Cassini, who had virtually none. Legendre was a very accomplished mathematician indeed, one who might be ranked a little below Monge in originality and a little below Laplace in productivity, but who could and did hold his own in such demanding company. For some reason, his work has attracted very little scholarship, and for that reason remarks about its inwardness and importance have to be somewhat more tentative than in the case of his peers.[162] His career overlaps the next generation, that of Ampère, Poisson, and Fourier, in style and subject matter rather than in age, for Legendre was born in 1752, and was thus three years younger than Laplace and six behind Monge. Recognition came to him somewhat later than to the other two. His favorite subjects, number theory and the theory of elliptical functions, were too abstract, in his view, to interest the contemporary public. He published them fully in his later years, and apart from them was of that breed of mathematicians whose problems, when more than mathematics, were beginning to be physics rather than mere mechanics. All we know of his personality is that it was said to be caustic.[163] He came from a prosperous family.

A graduate (with Bailly and Lavoisier) of the Collège Mazarin, Le-

[161] On Méchain's life and career, see Delambre (1827), 755-767, and "Eloge de Méchain," *Mémoires de l'Institut* 6 (1806), 1-28. There is a dossier of documents relating to Méchain at the Bibliothèque de l'Observatoire, E-2, 19bis.

[162] The only modern discussion is Hellman (1936). An "Eloge historique" was read to the Academy in 1861 by Elie de Beaumont, and published in the *Mémoires de l'Académie des Sciences de l'Institut de France* 32 (1864), lii ff. A translation by C. A. Alexander under the title "Memoir of Legendre" was published in the *Annual Report of the Smithsonian Institution . . . for the year 1867* (Washington, 1872), 137-157. See also Fréderic Maurice, "Mémoire sur les travaux et les écrits de M. Legendre," *Bibliothèque universelle, sciences et arts* 53 (1833), 45-82. See Hellman (1936); the DSB article by Itard, *8*, 135-143.

[163] Elie de Beaumont (1864), in *Report of the Smithsonian*, 145.

gendre made a deep impression on his mathematics professor, the abbé Marie, and in some scholarly exercise gave a definition of accelerative force so lucid that his teacher included it with acknowledgment in a textbook, whence it passed into general pedagogical usage.[164] Through the good offices of this mentor, Legendre in 1775 was appointed mathematics professor himself at the Ecole militaire, the same that supported Laplace in a like post in his early years. Rather than scorn the interest held by military problems for mechanics, Legendre submitted a memoir to the contest announced for 1782 by the Prussian Academy. The subject was one of practical artillery: determination of the trajectory of projectiles, taking account (as idealized laws of motion did not) of air resistance, muzzle velocity, and angles of projection. He won, with an analysis of the hypothesis that air resistance varies as the square of the velocity, eliciting the observation in his éloge by Elie de Beaumont a lifetime later

> that by his memoir Legendre, young as he yet was, has earned for himself a distinguished place in the series of mathematicians to whom is due the superiority of European artillery; a series which commences with Newton, in which M. Poisson occupies an eminent rank, and which is continued with so much éclat by the learned officers to whom we owe the actual precision of our artillery and the employment of rifled cannon.[165]

For the rest, the memoirs of his youth concerned problems on the mutual attractions and figures of spheroids of a sort that Laplace was discussing more cosmically and generally, and to which Legendre brought several specific mathematical solutions.[166] When Laplace was promoted to associate in the Academy in 1783, Legendre was elected to adjunct membership in his place. Nomination to the commission for joining the meridians of Paris and London led Legendre on to practical experience of surveying in the field, and that in turn to the problem of reducing observational and instrumental error to a minimum. His was the job of calculating the triangles from the data. In all the long history of surveying, the accounts that he published of those operations gave the only sophisticated analysis of the errors involved in computing the relations of lines and angles observed on a spherical surface by formulas from plane trigonometry.[167] He

[164] This was not so trivial a contribution as it might appear. For the question of eighteenth-century comprehension of the generality of the force law, see Truesdell (1960), 22-23.

[165] Elie de Beaumont in *Report of the Smithsonian,* 139. The memoir was *Recherches sur la trajectoire des projectiles dans les milieux résistants* (Berlin, 1782).

[166] See Gillispie (1979), section 2.

[167] "Mémoire sur les opérations trigonométriques, dont les résultats dépendent de la figure de la terre," MARS (1787/89), 352-383; "Suite de calcul des triangles qui servent à déterminer la différence de longitude entre l'Observatoire de Paris et celui de Green-

there demonstrated a theorem known by his name, to the effect that if a spherical triangle be of small extent, its calculation may be approximately reduced to that of a triangle in the plane by subtracting from each angle a third of the sum by which the three angles exceed 180°. He worked out formulas for reducing the calculation of triangles to the horizon under the various conditions encountered in practice, and for finding from the elements of a triangulation the shortest distance on the surface of a spheroid, which he called the geodesic line, and which always has double curvature unless it coincides with the meridian. Finally, he set out his reflections on the theory and use of Borda's circle, the principle of which was to reduce the range of error by multiplying the observations of a terrestrial or celestail feature; and certainly this was the most important consequence of Legendre's employment on these objects, since it was further refinement of the experiences of the same sort that led him in 1805 to formulate the least square rule of error as a means of resolving discrepancies between linear equations formed with astronomical data.[168]

Jean-Charles de Borda had a temperament suited to engineering. Like others of his type in the eighteenth century he oscillated between a military and a scientific career his talents hanging fire. His family were petty nobility of the Bordelais. He was schooled at Mézières; early elected to the Academy of Science on the strength of memoirs on chronometry, ballistics, and longitudes; and drawn from military engineering by way of the artillery to naval ordnance and architecture. He commanded the 64-gun *Solitaire* during the American War of Independence, in which he was

---

wich," MARS (1788/91), 747-754. See further, his *Méthode pour déterminer la longueur exacte du quart du méridien, d'après les observations faites pour la mesure de l'arc compris entre Dunkerque et Barcelone* (Paris, an VII, 1799), a critique of the metric survey, and "Analyse des triangles tracés sur la surface d'un spheroïde," *Mémoires . . . de l'Institut de France* 7 (1806), 130-161.

[168] The rule is that in a series of measurements of a single object or phenomenon, the mean value to be preferred is that for which the sum of the squares of the deviations from the others is a minimum. Legendre published it in *Nouvelles méthodes pour la détermination des orbites des comètes* (Paris, 1805), and it was excerpted and translated by David Eugene Smith as an item, "On a Method of Least Squares," for his *A Source Book of Mathematics* (New York, 1959) 2, 576-579. In this first statement, Legendre gave no justification. He elaborated it in "Méthode des moindres quarrés, pour trouver le milieu le plus probable entre les résultats de différentes observations," *Mémoires . . . de l'Institut de France* (1810/14), 149-154, read 2 September 1811. Mathematicians have never thought his treatment satisfactory and have preferred the development that Gauss gave the principle beginning in 1821, since which time the distribution it gives of values around a mean has become known as the Gaussian or normal distribution. Evidently there was some chagrin in Gauss's mind over Legendre's claims for priority, since to assert the method with merely empirical justification was in his view without mathematical merit. See the articles on Gauss by Kenneth May, DSB 5, 298-315, and on Laplace by C. C. Gillispie, *ibid.* 15, 365-366.

taken prisoner after a chivalric and unequal battle off Barbados.[169] Borda's first important navigational device was an improvement upon the reflecting circle invented about 1752 by Tobias Mayer, a German astronomer. The notion of adding a mirror to a sector goes back to Robert Hooke. The arrangement permitted measuring the angle between the directions of two distant points with a single setting of the instrument. The mirror image of one objective was made to coincide with a direct observation of the other, and the angle between the two lines of sight was then read directly off a graduated sextant or quadrant. Mayer substituted a full circle for the sector so that observations of angles greater than 60° or 90° might be multiplied without returning to zero or re-setting the instrument. Borda perfected this reflecting circle in design, though not in principle, and substituted an early model for the conventional sextant when he was navigator of the *Boussole* in an expedition of 1776 dispatched to perfect the determination of longitudes.[170] His repeating circle for geodesic purposes was in turn a development of this instrument, in which he substituted a second telescope for the mirror that navigators used to shoot the sun or a star. He had it constructed by Lenoir, the same who was prevented by the founders' guild from completing Cassini's commission for an equatorial telescope. Lenoir's success with this instrument was taken for evidence that French craftsmanship was capable of supplying the most exacting needs of science—provided it submit to the guidance of exact scientists.

Instead of straining like the Ramsden theodolite toward the maximum mechanical perfection in each angular measurement, a standard to be attained at the expense of augmenting both the scale and precision of the members, the Borda circle achieved accuracy by repeated but independent observations of the same angle. Readings made with the theodolite or with a conventional transit could be repeated only after returning the instrument to zero. Doing so might verify but not improve an observation. With the Borda circle, on the other hand, precision depended, in Cassini's words, only on the patience of the observer, "who by multiplying observations could at will eliminate all sources of error, whether in the graduation of the instrument or in observation itself."[171] In practice, the

[169] For the life and naval career of Borda, see Mascart (1919). There is also an excellent éloge by S.-F. Lacroix, read before the Société philomatique, of which a copy is in the Borda dossier in the Archives of the Academy of Science.

[170] *Ibid.*, 366-390; Daumas (1953), 243.

[171] Cassini IV (1790), 58-59. Borda (1816) gave his own account of the instrument. There is also a detailed description with plates in Delambre, *Base du système métrique*, 3 vols. (Paris, 1806-1810) 2, 160-240, which is a little harder to follow because Delambre tends to assume that the reader knows how the instrument works. Perhaps a brief summary of its mode of operation may be useful here.

Two traversing telescopes were mounted one above and one below a graduated circle. Either or both might share the rotation of the circle, and they might also turn independ-

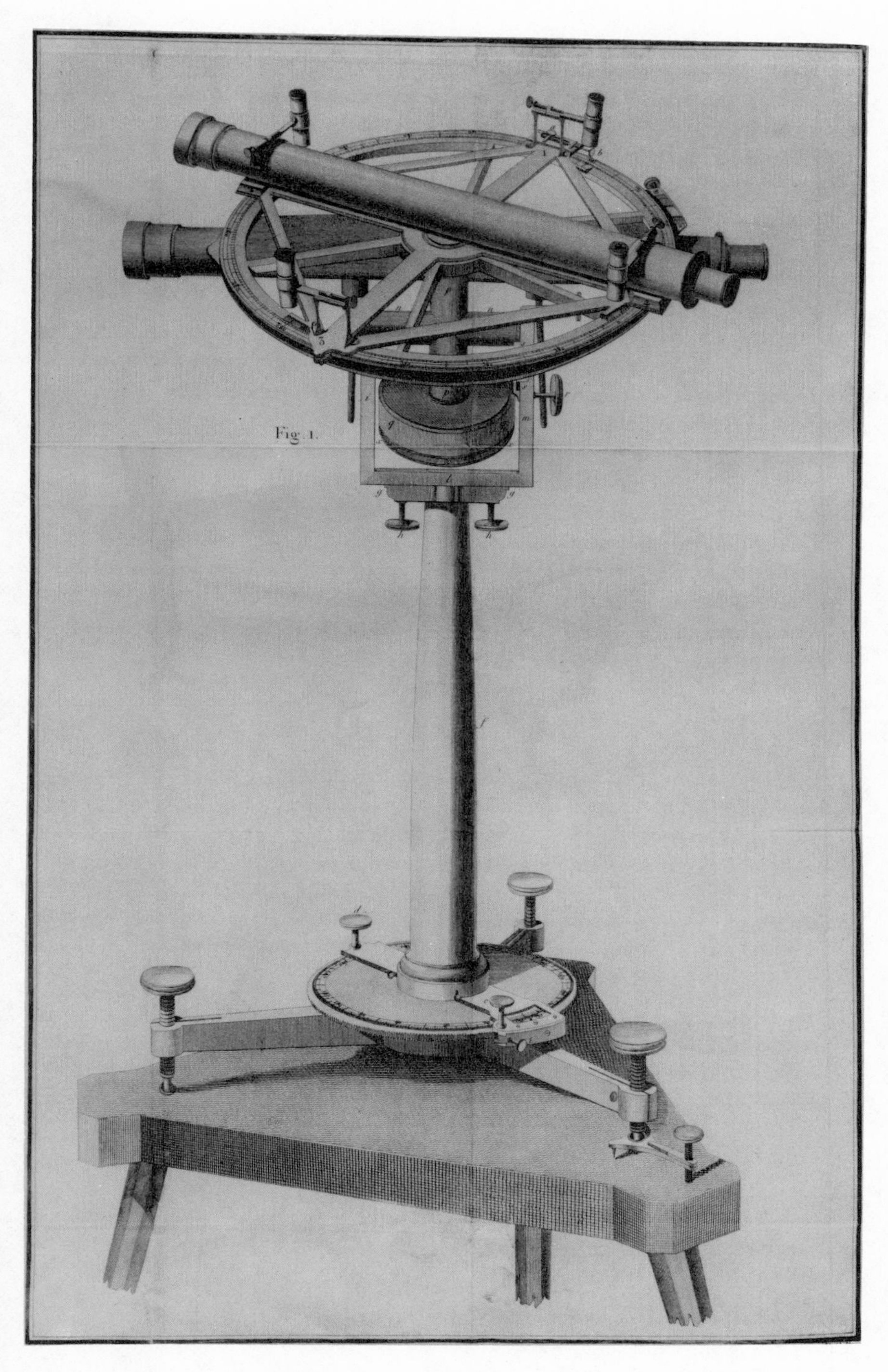

The Borda repeating circle from Delambre, *Base du système métrique* (Paris, 1806-1810), 2.

French team made from six to twelve repetitions of each angle, depending on conditions. They wrote with deference of the unprecedented accuracy of Ramsden's theodolite, "this great, this magnificent English instrument." Reduced to a plane surface, Roy's triangles closed to within 2.8′ by virtue of its precision in direct measurement of every angle. They would never have dared put themselves in competition "had not the theory, so to say, of our little circle assured to us a precision in the measurement of angles no less astonishing than the performance of M. Ramsden's beautiful instrument."[172] In the event, their own triangles closed to within 4′. What is more impressive, their instrument was as portable as a modern surveyor's transit. The diameter of the Borda circle was one foot. It stood about eighteen inches high. Even with its tripod, it can scarcely have weighed more than fifteen or twenty pounds. Rarely was a steeple so cramped or inaccessible that it could not serve both as surveying station and as object in sighting back from the next station. The circle suffered from only one important disadvantage. The repetitions that reduced errors in angular measurement were of no avail in direct astronomical readings of azimuths, and for that purpose Delambre, when in 1792 he resumed these operations in the metric survey, would have preferred a quadrant of larger radius.[173]

4. THE COLLÈGE DE FRANCE

No other foundation, wrote Lalande at the outset of his *Astronomie*, had been so useful to science as the Collège royal de France.[174] It alone among establishments for higher learning and science survived the Revolution without interruption or loss of any attribute except its regal designation.

---

ently. In taking an angle, the upper scope would first be zeroed on the circle. Then scope and circle would be turned together to take a sight on the right-hand object, and locked. Next the lower scope would be sighted on the left-hand object. But contrary to what might be expected, no reading would then be taken. Instead, the circle would be freed to rotate carrying both scopes with it, and turned clockwise until the lower scope rested on the right-hand object. It would thus have traversed the angle to be measured once. But only the positions of the upper scope might be read off the circle. The circle would be fixed, therefore, and the upper scope freed to turn counter-clockwise and back past the right-hand object until it sighted on the left-hand one. The upper scope would now have been turned through the angle twice, and the reading would be double the value to be measured. The same measure would then be repeated to give quadruple the angle, again to give sextuple, and so on around the circle as often as desired. The traditional quadrant could not measure angles over 90°, nor the sextant over 60°, without being moved. Moreover, the center of the circle was fixed and always at the same distance from the physical center of the station of observation, whereas the quadrant in turning displaced its own center with respect to that of the station.

[172] Cassini (1790), 57-85.

[173] Delambre (1912), 37.

[174] 6 volumes (3rd ed., 1792) *1*, xxvii.

Having thereby become the oldest such institution with a continuous history, it has also been the most consistently respected in modern times.[175] The edifice rebuilt for it by Chalgrin in 1774 remains the central portion of its installation. The Collège de France is situated among the *grandes écoles* in the heart of the Latin Quarter, hard by the Sorbonne but in practice independent of the University. Examinations have never formed part of its regime. The lectures, seminars, and laboratory exercises lead only to scholarly and scientific advantage and qualify no one for diplomas. Professors are appointed for their knowledge, need hold no degrees, and are expected to impart their own study and research to whoever may be interested, whether for private or professional reasons. The constraint that all teaching contribute to the advancement rather than the transmission of knowledge goes far to explain the eminence of the Collège de France ever since the lectures were inaugurated in 1530. An apology written in 1789 by the abbé Garnier, the last inspector or administrator under the old regime, remains appropriate despite the many changes since. A professor's duties are always arduous "since he has to expound the least clearly understood subjects in his art, and give an account of new discoveries before well-informed persons who have the right to ask him to clarify anything that leaves them burdened or in doubt."[176] A statement from its administrator in 1800 is fuller:

> The Collège de France is not elementary. Before the Revolution it was frequented by Frenchmen and by foreigners drawn by their interests to the cultivation of learning or who intended a career in teaching the public. They found courses there of nine months duration under accomplished masters, and completed the education they had received in the universities. Poland, Spain, and several other Eu-

[175] Lefranc (1893) is an antiquarian work concentrating largely on the foundation and early history. Altogether more informative is Sedillot (1869-1870). An anniversary compilation, *Le Collège de France, 1530-1930, livre jubilaire* (1932), even less satisfactory than such works normally are, consists of chronicles of the individual chairs written by the incumbents. The only modern study is a brief essay by Jean Torlais, "Le Collège Royal," in Taton (1964), 261-286, whose bibliography indicates the location of archival materials including the holdings of the Collège de France itself. That is useful, for many of the materials cited by Lefranc in the Archives nationales have been reclassified since his book was published. In addition to the indications given by Torlais, papers pertinent to the history of the Collège de France in the late eighteenth century and the Revolution will be found in the AN: AD VIII, 23, 26, 29; AF IV, 1289, doss. 77; F^{17}.1029, doss. 3; F^{17}.1219, doss. 3; F^{17}.1240^B; F^{17}.1337, doss. 4; F^{17}.1354, doss. 2; F^{17}.1355, doss. 3; F^{17}.1356, doss. 1, no. 93; F^{17}.1418, doss. 4; F^{17}.1457, doss. 5; F^{17}.3849-3854 (further cartons through 3879 concern nineteenth-century administration). See also correspondence of Lalande, BN, MSS Fr. 12,273, fols. 213-222; and of d'Alembert, Bibliothèque de l'Institut de France, MSS 2446.

[176] Jean-Jacques Garnier, *Eclaircissements sur le Collège royal de France* (1790), 27. BN, Rp. 5917.

> ropean states then subsidized students who remained in Paris for several years, and who became the most faithful auditors of this school. Today, a large number of distinguished scholars are proud to have been its students.
>
> It is to the concentration of almost all the branches of knowledge in a single place, as much as to the competence of the Masters, that the College owed what we have to call the fame it enjoyed in Europe and its utility for the propagation of learning.[177]

During term, professors would lecture for an hour and answer questions for half an hour three times a week. Twice a year they would announce their courses, much as their successors still do, except that the posters were in Latin. The first semester ran from All Saints Day, 1 November, until Palm Sunday; the second ran from the Sunday following Easter until 1 August. Their collegial year of nine months contrasted markedly with the six-week instructional session at the Jardin du roi, in some ways a sister institution, where courses addressed to a less informed audience consisted of fifteen or twenty lectures. Often scientists taught in both establishments, pluralism or the "cumul" having been as venerable a feature of learned life in France as the institutions that permitted it, or in a sense required it, since none paid a salary sufficient for a livelihood. A professor at the Collège de France normally began at 1,100 livres and after twenty years would have been increased to 1,500.[178]

Until 1774 the college had been housed in the buildings of the nearly moribund Collèges de Cambrai and de Tréguier. The new quarters contained six lecture halls, an anatomy theater, and a chemistry laboratory, and were designed to accommodate a curriculum reorganized in the interests of science and modernity. The changes took effect in the first year of Turgot's ministry but had been planned earlier: already in 1770 d'Alembert was corresponding about them and about a chair under the new dispensation for his brilliant protégé, Laplace.[179] Although that ap-

[177] Louis Lefèvre-Gineau, "Résumé présenté à l'assemblée des Professeurs du Collège de France." 1 germinal an VIII [31 March 1800]. Archives, Collège de France, C-XII.

[178] Garnier, *Eclaircissements* . . ., 23-30. For further detail on the regime of the college, see the memoir drawn up by Lalande, "Mémoire des professeurs du Collège de France, sur l'indivisibilité de leurs travaux," August 1793. BN, MSS Fr. 12,273, fol. 213-222. The archives maintained by the library of the Collège de France itself are not very full. There are titles of the courses taught year by year, and occasional lists of auditors. Dossiers are kept for each professor, but are seldom very informative. The most important single document is the "Régistre des déliberations prises aux assemblées des lecteurs et professeurs du roi au Collège royal de France," which records minutes of the meetings. There are three ledgers. The first runs from 7 January 1674 to 22 November 1721, the second from 9 November 1732 to 9 June 1780, and the third from 9 June 1780 through 1822.

[179] An unidentified correspondent of d'Alembert discusses (9 January 1770) the proposed changes in considerable detail, Correspondance of d'Alembert, Bibliothèque de l'Institut de France, MSS 2446.

pointment never came about, the reform of 1774 did assign the chair of Syrian to mechanics, soon renamed experimental physics, and one of the pair in Greek and Latin philosophy to "universal" (which immediately became mathematical) physics. The other chair in Greek and Latin was transferred to French literature. One of the two in canon law was turned over to the "Law of Nature and Man," and one of the two Hebrew chairs went to modern history and philosophy. Of three chairs in medicine, two were reallocated to anatomy and chemistry. Four years later the third was redesignated natural history, and Daubenton replaced Poissonier. In all, the teaching in eight of the nineteeen chairs was explicitly devoted to scientific subject matter throughout the reign of Louis XVI. Besides Daubenton and Poissonier, Lalande lectured on astronomy, A.-R. Mauduit on geometry, Le Monnier on "mathematics" (which for him meant astronomical observation), followed by Cousin on topics that later pertained to mathematical physics; Girault de Keroudou on mechanics followed by Lefèvre de Gineau on experimental physics; Portal on anatomy; and Darcet on chemistry. It is no doubt indicative of the scale of prestige that later on, when spokesmen for the College had to explain and justify its regime at various junctures early in the Revolution, these chairs and courses in science were always mentioned first.[180]

Of its company under Louis XVI, Poissonier and Le Monnier, well connected at court, were holdovers from the spirit of the previous reign. Daubenton, Darcet, Cousin, and Lefèvre de Gineau represented the dispensation of the new, in science and polity, while Lalande, Mauduit, and Portal were fixtures of the establishment, little affected by the reform of 1774. It seems more natural to consider Lalande and his career along with astronomy and Daubenton and Portal in connection with their other posts at the Jardin du roi.[181] Let us say a word of the others, and first of Mauduit, the geometer, about whom the least is known. The article on him in the *Biographie Michaud* holds that he would have been elected to the Academy of Science if his "causticity" had not been too great an obstacle. Whether the same quality was responsible for the controversy attending his election to a chair on 10 February 1768 is unknown. His future colleagues did protest some irregularity, albeit in vain, for the minister, Saint-Florentin, insisted on going forward.[182] Mauduit was already professor at the Academy of Architecture. Though occupying lodgings in the Louvre, he later demanded that his right to an apartment in the College also be respected.[183] His interests were not altogether confined to math-

[180] For the reorganization and installation in 1774, see memoir of 23 frimaire an VI, AN, F[17].3854; cf. the Résumé drawn up by Lefèvre-Gineau, 10 germinal an VIII, Archives of the Collège de France, C-XII.

[181] Above, Section 3, and below, Section 5.

[182] Régistre des déliberations 2, 107-110, Archives, Collège de France.

[183] *Ibid.* 3, 61.

ematics. In 1783 he offered the company a machine he had invented, and a little later proposed reading a memoir on the nature of fire.[184] In 1809 Poisson began supplying his teaching and became his successor. We know the subjects of the courses he offered, as we do for most of those of his colleagues, but since that is all we have for him, the impossibility of reconstructing what transpires inside an institution of higher learning from the bare bones of titles and regulations in a catalogue is nowhere more frustrating. It would be very instructive, for example, to have access to the contents of a 1771-1772 course called "Explanation of Euler: Introduction to Infinitesimal Analysis," and of another given in 1781-1782 on "The nature of curves according to the laws of central forces."[185]

About Poissonier, on the other hand, information abounds. Pierre-Isaac Poissonier, it was said, bought the succession to his chair in 1746 for some 2,000 crowns.[186] He was then a gallant young medical man-about-town. The son of a prosperous pharmacist in Dijon, he took his doctorate in Paris in 1743, and won early fame through curing a prominent but discreetly unnamed patient of a bladder ailment. Pierre Sue, a friend and associate who composed his éloge for the Royal Society of Medicine, recalled that his noble and assured countenance at the bedside supported the feelings of his patients to the end.[187] He was equally considerate of his audience when teaching, and if their attention flagged, he brought his discourse to a stop, whether he had finished the lecture or not.

In 1754 Helvétius chose Poissonier for his successor as inspector of military hospitals. He was appointed chief physician to the army in Germany in the campaigns of 1757 and 1758 and passed the rest of his life in military and diplomatic medicine. In 1758 the government dispatched him to Russia, ostensibly to consult on the health of the Empress Elisabeth, though Choiseul had also confided various political commissions to him. Apparently he became one of the sovereign's lovers. "His relations with women," again according to Sue, "were friendly and gallant. In their company he conducted himself in that agreeable manner, without the slightest sharpness, which is pleasing to most ladies, and which, while giving them the opportunity to exhibit all their graces, assures their dominion over our affection, the only form of slavery that never will be abol-

[184] *Ibid.* 3, 23, 24.

[185] Dossier Mauduit, Archives, Collège de France. According to a note in this dossier, a letter of Mauduit to Jean Hermann is conserved in the Bibliothèque municipale de Strasbourg, MS 3757, piece 61, and there is other correspondence at the Bibliothèque de Mantes, Collection Clerc de Landresse, pieces 2516-2519.

[186] For his accumulation of perquisites, see *Archives parlementaires* 13, 360. There are two modern accounts of his career, Bouvet (1936b) and Vallery-Radot (1938).

[187] Pierre Sue, *Eloge de Pierre-Isaac Poissonier* (an VII, 1799), 8. Offprint from *Recueil périodique de la société de médecine* 5, nos. 28, 29.

ished, especially in France."[188] The empress made him a lieutenant-general in order that protocol might be satisfied with his presence by her side at dinner. It was her habit to rise from the table at two in the morning. Poissonier would remain with her in the boudoir until 6:00, when she retired for the day. Eventually, he found himself "fatigued" by life in Russia, and was relieved to be called home, though Elisabeth tried to prevent his departure.[189] In the course of this mission, he became a favorite of Choiseul, who found him good company though a poor shot on hunting expeditions in later years.

Louis XV for his part preferred Poissonier's letters to the official dispatches from the ambassador. With all this, Poissonier had little difficulty in reestablishing himself in France. The post of inspector and director-general of medicine, surgery, and pharmacy in the ports of France and her colonies was created for him. The Academy of Science admitted him an associé libre in 1765, and the fledgling Royal Society of Medicine chose him for vice-director in 1776, when its patrons were first currying favor at Versailles. It is hardly astonishing that Poissonier wrote very little and taught only intermittently. In 1775 the minister—La Vrillière—authorized the last of many absences. His lectures were to be given that year by Guillotin, on the express understanding that his replacement was to entertain no expectation of the succession, which was to be scientific. Poissonier gave his final course, on diseases of women, in 1777-1778.[190] Though he then resigned his chair, he retained his membership in the College until his death in 1798 and often signed present in the register.

By contrast, the word describing Jean Darcet's approach to his duties is serious, in both its French and English meanings.[191] His assumption of the duties of the first chair of chemistry in December 1774 symbolized the new departure for the old college in that he was authorized to give his inaugural lecture in French. The plea was that Latin lacked an adequate vocabulary, for he was not ignorant of the classics.[192] Indeed, he had

[188] *Ibid.*, 37.

[189] *Ibid.*, 17-18.

[190] La Vrillière to Garnier, 21 April 1775, Archives of the Collège de France, Parmentier dossier.

[191] The most valuable account of Darcet's life is M.J.J. Dizé, *Précis historique sur la vie et les travaux de Jean d'Arcet* (an X, 1802), reprinted in Pillas and Balland (1906), 171-209. See also a notice by Antoine de Fourcroy, *Journal de Paris* (28 pluviose, an IX [1801]); the éloge by Cuvier (1819-27) *1*, 165-185; and Cuzacq (1955). According to the dossier in the Archives of the Collège de France, there are manuscripts in the libraries of Mantes (Clerc de Landresse Collection, piece 1086); Clermont-Ferrand (MSS 338-343); Versailles (Coll. d'autographes, pieces 47-48); Rouen (Duputel Coll., carton IV, piece 433; Blosseville Coll., piece 557; and Girardin Coll., pieces 45-47); and also at the Bibliothèque de l'Arsenal in Paris, MS 7054.

[192] A memorandum in his dossier, Collège de France, records the authorization for this lecture, the first ever given in the vernacular.

earned his keep as a tutor in the humanities while a medical student in Bordeaux many years before, his father having disowned him for preferring a scientific career to association with himself in the law. In 1742 Montesquieu engaged him to be preceptor to his son and took them both to Paris. There Darcet followed Rouelle's courses in chemistry at the Jardin du roi, became the great teacher's collaborator, and married his daughter in 1771, Rouelle having died in 1770. Their family affinities were always chemical. When Madame Darcet also died, her aunt, a younger sister of Madame Rouelle and herself the widow of one of the Clouet connection, kept house for him and the four children.[193] The eldest became a chemist in his turn. Darcet was also a regent doctor of the Faculty of Medicine, but never practiced that profession any more than law. When he was chosen to be professor at the College, it was because he alone among the medical men being considered had made chemistry his sole occupation.[194]

In his most important research in the 1760s, Darcet investigated the effect of very high temperatures on minerals.[195] Out of his work came techniques for the production of true or "hard" porcelains in French industry, and it culminated in his demonstration of the complete combustibility of the diamond heated in air.[196] In like manner, studies of the properties of alloys in the 1770s led to a method for making printers' plates out of a mixture of lead, bismuth, and tin that melted below the boiling point of water.[197] Because of the industrial orientation of Darcet's knowledge, the government drew him into commercial enterprises of the state. In 1784 he succeeded Macquer in the post of inspector-general of the Royal Porcelain Factory at Sèvres. Later, he also became inspector-general of the Mint.

When Darcet was appointed professor, Turgot and Malesherbes had promised to install a proper laboratory. The Maison du roi then took title to a vacant lot bordering the college where proper quarters might be built. Unfortunately, Amelot, Malesherbes' successor in the ministry, withheld the funds, and assigned Darcet instead the space under the staircase in the existing building. Darcet had to supply his own equipment, reagents, and fuel. That necessity was less difficult than it might have

[193] His protégé, Dizé, wrote a memoir of the family (23 April 1838), at the request of his son, who did not remember his mother. Pillas and Balland (1906), 68-72.

[194] La Vrillière to Garnier, 18 December 1774, Darcet dossier, Collège de France.

[195] *Mémoire sur l'action d'un feu égal, violent et continué pendant plusieurs jours sur un grand nombre de terres, de pierres et de chaux métalliques* (1766), and *Second mémoire sur l'action d'un feu égal . . .* (1771).

[196] *Mémoire sur le diamant et quelques autres pierres précieuses* (1771), and (with Rouelle) *Expériences nouvelles sur la destruction du diamant* (1773); on porcelain, see below, Chapter VI, Section 2.

[197] *Expériences sur l'alliage fusible de plomb, de bismuth et d'étain* (1775).

been since his wife had brought him Rouelle's stock and apparatus in her dowry along with an income of 20,000 livres per year. His annual outlay when he gave an experimental course came to 800 livres.[198] At a stipend of 1,300, his incentives obviously were other than economic. Recognition by the Academy came late to him. Darcet was fifty-nine when Macquer's death created the vacancy to which he finally won election. He was a hard-working chemist, and it was a matter of pride to be teaching his science on the most advanced level at the College. In 1778 he assured the ministry that his demonstrations and experiments were as full as if he had been giving private lectures for subscribers.[199]

The course alternated annually between the chemistry of the mineral and of the animal and vegetable kingdoms.[200] His auditorium was always full. In 1784 he took on an assistant, Jérôme Dizé, a nineteen-year-old pharmacist's apprentice from his native region in Les Landes. Dizé prepared the demonstrations and experiments and performed a like service for the course in experimental physics begun by Lefèvre de Gineau in 1786.[201] Darcet formed his aide's abilities into an extension of his own. The only important work related to the basic chemistry of the time done in his laboratory was an experiment that separated about a quart of water into hydrogen and oxygen and recombined the elements without loss, thus confirming its composition on a large scale.[202]

More characteristic was the research and development put in hand when a proposal for converting common salt into soda, or mineral alkali, was referred to Darcet. The author was Nicolas Leblanc, who attended Darcet's course in 1787. Forty years of age, neurotic, and at a standstill, Leblanc was a surgeon and a hanger-on in the retinue of the duc d'Orléans with a large family to support. He had some chemical training, had published on the crystallization of neutral salts, and nourished hopes of solving his problems by what he believed to be an important discovery. He had succeeded (or so he informed the duke) in obtaining pure soda by the calcination of sodium sulfate in the presence of charcoal, and he asked his patron, who was an active investor in industry, to finance the exploitation of his procedure. The duke referred Leblanc to Darcet for verification of the method. His hands full just then with commissions from the mint,

[198] There is a memoir of 1778 in his hand in the Archives of the Collège de France, "Notes sur le College roïal et sur la chaire d Chymie Experimentale que j'y possede." See also, Dizé to Darcet fils, 23 April 1838, Pillas and Balland (1906), 68-72; and the inventory of his laboratory by Pelletier, Leblanc and Darcet himself, 24 fructidor an II, AN, F^{17}.1337, doss. 4.

[199] Darcet, "Notes sur le College roïal. . . ."

[200] Course list, Darcet dossier, Collège de France.

[201] On the career and work of Dizé, see Pillas and Balland (1906), who reprinted many of his writings, and Rouquette (1965).

[202] Pillas and Balland (1906), 2.

Darcet turned the matter over to Dizé. When the first attempts to reproduce the results failed, Leblanc begged that a report be delayed and that Dizé continue working with him in order to find and overcome the source of the difficulty. Dizé agreed, with the concurrence of Darcet, and thus did the development of the Leblanc process begin in the laboratory of the Collège de France.[203] That this installation should have been a nursery of industrial chemistry, while the Arsenal under Lavoisier became the headquarters for theoretical chemistry would seem an ironic reversal of roles, except that at the end of the old regime neither laboratory belonged to the institution, but rather to the incumbent.

Developments momentous for the form of scientific disciplines may also have had their start in the Collège de France earlier than has commonly been thought. The documentation is scanty, however, and it must remain a conjecture that the teaching of mathematical and experimental physics as research subjects antedates the careers of Biot and Ampère, whose generation had clearly made the break with the old rational mechanics. Biot was named professor at the Collège de France early in his professional life, in 1801; Ampère much later, in 1823. Their respective predecessors were Jacques Cousin and Louis Lefèvre de Gineau. Cousin became professor at the Collège de France in 1766. The chair was still designated Greek and Latin philosophy, though Le Monnier had been teaching astronomy in it, and Cousin was already supplying the courses in alternate years when the subject was mathematical. Official redesignation into a chair of "*Physique générale et mathématique*" (the exact language is important here) by order of the Council of State in 1769 antedated the general reform of 1774. Lefèvre de Gineau succeeded Girault de Keroudou, who resigned in 1786, the chair now being called "*Physique expérimentale*" instead of "*Mécanique*." It may be grasping at straws to note the titles of their courses, but it is likely that they were straws in a gathering wind. In 1769-1770 Cousin gave a course on "Physique mathématique" and in 1771-1772 a semester on "Méthodes de recherches dans les sciences physico-mathématiques." In 1774-1775 (to select years in which the topics were other than the staples of rational and celestial mechanics and hydrodynamics) his subject was "Généralités sur la terre." In 1781-1782 and 1782-1783, he gave a two-year sequence on "La Précession des équinoxes, les marées, et autres phenomènes liées à la forme de la terre"; and in 1783-1784 and 1784-1785, another on "Les progrès de l'analyse, et en quoi ils peuvent servir aux progrès de la physique." As for Lefèvre de Gineau, he began in 1786-1787 with "Exposition et démonstration expérimentale des principes mathématiques de l'optique," and his next three courses were "Le flux

[203] On the origins of the Leblanc process, see Gillispie (1957a) and Pillas and Balland (1906), 5-31, 73-83, with supporting documents. See also J. G. Smith (1979).

magnétique, l'air et les autres fluides élastiques"; "La lumière"; and "Etat générale de la matière. Le mouvement. L'électricité."[204]

Unfortunately, little more is known. Lefèvre de Gineau destroyed all his papers toward the end of his life. The list of his auditors from 1808 to 1823 does survive in the archives of the Collège de France. He is said to have owed his appointment in the first place to the favor of the baron de Breteuil, minister at the Maison du roi in the later 1780s, whose children he tutored, and to have added the "de Gineau" to his name when he became a professor. Neither he nor Cousin published any science of significance. Nevertheless, Cousin was elected to the Academy in 1772, and somehow it can be felt that he had an impressive physical intelligence and was respected by his peers. He was also professor at the Ecole militaire, along with Laplace in the 1770s. He never married and instead lived with Madame Anthelmy, widow of another colleague and humanist who had translated the fables of Lessing, providing for her and her children.[205] After the Revolution, both Cousin and Lefèvre-Gineau (as he then became) took responsibility in the affairs of the Collège. The latter was its administrator from 1800 to 1823. Both also entered public service, a small fact that accords with our general argument about the interrelations of professionalism in science and modernization in polity. Cousin was elected to the municipal government of Paris in 1791, took charge of subsistence in the capital, and became president of the administration of the Department of Paris. He was a member of the Conseil des anciens, or upper house of the legislature, in 1798 and carried over into the Senate in the first two years of the Consulate. He died in 1801. Lefèvre-Gineau was a member of the legislative corps in 1807 and was reelected in 1814. Under the Restoration, he became a liberal member of the Chamber of Deputies and published writings on foreign trade and freedom of the press.[206]

In sum, the influence of the Collège de France derived, not from the stature of the incumbents, who individually were of relatively minor scientific importance, but from its regime. Taken collectively, its offerings exhibit a rudimentary sense of curriculum. According to the staff itself, a student needed three years to become a scientist, one year of elementary preparation followed by two in appropriate subjects at the Collège de France. Mauduit's courses in the elements of the higher mathematics were

[204] All that is said in these paragraphs derives from the elements of biography noted in the Cousin and Lefèvre de Gineau dossiers and from their course lists, Archives of the Collège de France. There is also a brief éloge by Charles Dupin, "Funérailles de M. Lefèvre-Gineau," Institut de France, 4 February, 1829.

[205] Minute by Lalande, on a letter Cousin had written him on some matter of collegial politics, 6 March 1788.

[206] *Opinion . . . sur l'importation des fers étrangers* (1814); *Opinion . . . sur la liberté de la presse* (1819).

intended to prepare students to follow the advanced sequences offered by Cousin. Darcet's alternation of organic and inorganic chemistry (to force those terms by several decades) formed a unit in that science. The experimental physics taught by Lefèvre de Gineau complemented the mathematical and chemical teaching and also the mineralogy and natural history of Daubenton. Portal's anatomical lectures enlarged on what was available for medical students at the Jardin du roi, in the Faculty of Medicine, and at the hospitals.[207] Absence of requirements brought variety, and standards were maintained by the imperative that teaching proceed at the forefront of research in the science as a whole, if not in the work of each professor. It is reasonable to conclude that the effect was to favor the advancement of the several disciplines, but the interpretation must not be overdone or the Collège de France made out to have been the German university straining to be born there in the Latin Quarter. Forward looking it may have been, but its tradition lay in the past, and the keynote was flexibility, not professionalism.

Indeed, in 1530, when François Ier endowed the first Royal Readers to give public instruction in Greek, Hebrew, mathematics, and Latin, his purpose, and that of Guillaume Budé, his adviser and a famous humanist, was to abate the professionalism of the University of Paris and to destroy its monopoly over learning.[208] The foundation was of Renaissance inspiration and never called a college before the seventeenth century. In the first generations Pierre La Ramée, or Petrus Ramus, personified its spirit. Not since Peter Abelard had Paris seen his like for polarizing scholarly opinion. Henri II created a chair of philosophy and eloquence for Ramus after the University expelled him for subverting logic into dialectic and denigrating everything in Aristotle that could not be made modern. Thereupon, he became a Protestant and challenged men of knowledge to make their own reformation from philosophy to mathematics. His notoriety associated the portent for a new science with the affirmation of a new religion, and in 1572 he was assassinated by hired murderers, intellectually the most illustrious victim of the massacre of Saint Bartholomew. In his will he left an income of 500 livres annually to found a chair of mathematics, the tenure to be limited to three years. In that span, the incumbent was to develop all aspects of the subject. In order to be reappointed, he had to exhibit virtuosity at least the equal of any challenger's

[207] "Mémoire des Professeurs du Collège de France, sur l'indivisibilité de leurs travaux," August 1793. BN, MS Fr. 12,273, fols. 213-222. It is true that this document was composed in order to make the best possible case for the civic importance of the institution at the moment in the Revolution when the academies were under attack and facing dissolution. The picture it presents is no doubt idealized, but at the very least it shows what the staff then thought they ought to be doing.

[208] Lefranc (1893), 101-123; Sedillot (1869), 352-356.

in a public competition to be held under the eyes of the first president of the Parlement of Paris, the Prévôt des marchands, and the first Advocate-General. Besides their eminence, the point to notice about those officials is that they were civil magistrates, not clergymen.[209]

Much disrupted during the civil and religious wars, the company looked to Henri IV for a patronage in keeping with his statesmanship. They were rewarded with a grandiose plan for an installation on the scale of the Louvre. Its realization was limited to a stunning architectural engraving by Claude de Chastillon and the laying of a cornerstone by a nine-year-old Louis XIII in 1610. In the reign of Louis XIV, learned attention centered on creation of academies. Not that the Collège de France was incompatible with the new foundations, for though Roberval was the only professor who was a charter member of the Academy of Science, tenure of a chair helped support many a later academician.[210] Notable in the late seventeenth century were Philippe de La Hire, geodesist and calculator; Joseph Sauveur, virtually the founder of acoustics; Pierre Varignon, a major though little studied figure in early analytical mechanics; and Joseph Pitton de Tournefort, botanical systematist.

Although the entire company might make recommendations, as it continues to do, on the designation of each chair, once the professor was appointed he was free to discourse of whatever he pleased. The combination of collective self-government with individual scholarly independence has, indeed, been the most remarkable feature of the tradition developed in the Collège de France, particularly when contrasted to the rigidity and resistance to change characteristic of the universities, at least in the higher faculties. In the Collège de France, incumbents never allowed the description of their chairs to constrain their interests. One professorship in mathematics was occupied in turn by Gassendi, whose subject matter was philosophical atomism; by Roberval and La Hire, both of whom actually were mathematicians; by Delisle, whose teaching founded the French school of observational astronomy; and lastly by Lalande, who openly announced conversion of the course to astronomy in 1761. We have already seen how Le Monnier was teaching astronomy in a chair of Greek and Latin philosophy. He had succeeded the most eccentric mathematician of the century, the abbé Jean-Paul de Gua de Malves, who lectured on Newtonian cosmology and mechanics, continuing in an ostensibly classical chair a mathematical tradition that reached back through the incumbencies of Privat de Molières, Varignon, and Duhamel to Ramus himself.[211]

[209] On Ramus and his chair, see Lefranc (1893), 205-225; Sedillot (1869), 388-418, 433-448.

[210] Garnier, *Eclaircissements sur le collège de France* (1790), 11-12.

[211] Sedillot (1869) 491-510, treats the incumbents of the mathematical chairs in sequence.

The reform of 1774, therefore, made less drastic a difference than might appear in what was actually done. It was initiated when the abbé Terray was controller-general. His motives were financial rather than scientific. He mainly intended to put an end to financial importunities on the part of the professors by shifting the College onto funds that also provided for the University of Paris.[212] In return for additional credits and enlarged stipends, they were to observe the ultimate authority of the University in collegiate matters. The staff had the wit to resist application of this vague provision in regard to everything except empty ceremonies like marching in academic processions. At the same time a proposal to enlarge its privilege by publishing a journal also came to naught, one of many such hares then being started in the literary and scientific world. Already in 1754 the College had been accorded the right to publish the writings of the professors under its own imprimatur, but seldom exercised it.[213]

Probably it was well for the security in which the College carried on its own mission that it abstained from trespassing on the academic preserve of scientific publication, and certainly it was fortunate for the future of science and learning in France that the professors of the College then stood apart from the University. They thus preserved themselves from the work of elementary instruction. For it is a mistake to suppose that no science at all was taught in the University. True, the emphasis of the Faculty of Arts, tradition-bound and tied to the requirements of the higher faculties, continued to be overwhelmingly literary and classical. Science and mathematics had begun to make their way, however, in the curriculum of the stronger of the twelve teaching colleges—*de pleine exercise*, they were called, as distinct from the forty *petits collèges* or hostels—that qualified students for the baccalauréat in their courses of "philosophy." A textbook of geometry published in 1732 by D.-F. Rivard of the Collège de Beauvais had become standard throughout the University.[214] The mathematical lessons of the abbé de Sigorgne at the Collège d'Harcourt were the

[212] Lefranc (1893), 255-264. The budget was not administered in a very professional manner, and when the company as a whole took an interest, which in the manner of academic bodies was intermittently, they tended to be critical of the clerical operations and to demand an overall accounting. Such a stock-taking occurred at the meeting of 5 December 1784, and is recorded in the Registre des déliberations 2, 29-30. It is there evident that the revenues assigned annually by the government consisted of 15,930 livres from the general funds of the Treasury and another 15,000 livres charged to the Messageries. In the previous six years, that had fallen short of expenses in the amount of about 2,600 livres per year, which deficit was made up from the revenues of the Collège de Cambrai. It is not clear by what authority the Collège de France drew on the remaining income of this largely moribund college, in whose buildings it had lived for over a century.

[213] Sedillot (1870), 107-109; Torlais (Taton, 1964), 266-268.

[214] Dominique-François Rivard, *Elémens de géométrie* (1732).

most highly reputed in the first part of the century. By 1750, "*Physique*" or general science occupied over half a pupil's program in his second year of "*Philosophie*," corresponding more or less to the final year of secondary school. Generally, however, science was taught historically. Students learned about the physical systems of Aristotle, Descartes, and latterly Newton. A change came with appointment of the most famous scientific pedagogue of the Enlightenment, the abbé Nollet, to a chair of experimental physics at the Collège de Navarre. He taught real science by means of demonstration and experiment, and was followed there by Brisson and Girault de Keroudou, or K'oudou as he often spelled his name. The latter we have already met teaching mechanics at the Collège de France until 1786, when he resigned. At Louis-le-Grand, "*Physique*" had become a subject in itself, separate from philosophy, in 1738. It had, however, been Mazarin's Collège des Quatre-Nations, housed in the palace that is now the Institut de France, that had given the lead in science ever since its opening in 1668.[215] Scholars in the lower faculty were still distributed among the French, German, Norman, and Picard "nations." Varignon had been its professor of mathematics, and the influence of the abbé La Caille and his successor, the abbé Marie, on the education of leading scientists of the later eighteenth century was precise and long-lasting. All this was school-teaching, of course. As for the higher faculties, that of medicine will be treated separately,[216] and both law and theology were too intimately tied to service of the traditional and conservative professions to offer the slightest opening for science or modernity in the higher reaches of the University.

Prior to the Revolution, provision of instruction in advanced science and learning remained, therefore, the function of the Collège de France alone.

5. THE JARDIN DES PLANTES

Georges-Louis Leclerc, comte de Buffon, intendant of the Jardin Royal des Plantes for half the century, died in his official residence at forty minutes past midnight in the morning of 16 April 1788. After an autopsy, his heart was enclosed in an urn of crystal and gold and bestowed, it is said, upon Madame Necker, who was much in his company in his last

[215] For the teaching of science in these forerunners of the modern *lycée*, see Lacoarret and Ter-Menassian, "Les Universités," in Taton (1964), an essay that is clear and informative on the state and organization of universities in general during the eighteenth century and that contains an admirable bibliography on the subject. For the organization of French secondary education in general, though with special attention to the Collège Louis-le-Grand, see Palmer (1975), 9-38.

[216] Below, Chapter III, Section 4.

years and beside him at his death. His brain was placed in the socle of Pajou's statue of him, which now stands in the zoological gallery of the Muséum National d'Histoire Naturelle, the name given in the Revolution to the institution he had enlarged and ruled. At his funeral the casket was drawn by fourteen horses caparisoned in black silk embroidered with silver. A crier followed by six bailiffs cleared the streets for the procession consisting of nineteen liveried servants, a detachment of the Paris guard, a contingent of schoolchildren, sixty clergymen, and a choir of thirty-six boys accompanied in their dirges by four of the old bass horns called serpents. Twenty thousand spectators crowded roofs and windows between the church of Saint-Médard and Villejuif, whither six guards from the Jardin bearing torches followed the cortège. Thence the body was transported to Buffon's native town of Montbard in Burgundy, and there buried in the chapel adjoining the fine park he had created high above the countryside on the site of the old chateau of the feudal counts of Montbard.[217]

Seldom thus did scientists or men of letters quit the world. Voltaire himself had scarcely done so, nor even Newton, from whose *Method of Fluxions* Buffon had won reputation by translating it before writing of natural history.[218] He was fortunate in the timing of his death as in the leading of his life, for he survived with his powers of work and enjoyment little diminished until the last year when such a life was thinkable.

His career belonged entirely to the eighteenth century. Naturalist and forester, landed proprietor and iron-master, bon vivant if not a rake, accomplished writer if not a great savant—those were his objectives. They were altogether unprofessional, and only as a would-be dynast did he fail in anything he thought important. His "Buffonet," the single surviving child of a drab and sickly wife, born when his father was middle-aged, turned out an ignorant wastrel and died under the guillotine. In 1771 Buffon, suffering the one grave illness of his life, sought to arrange this shaveling's inheritance of his office of intendant. The maneuver succeeded

[217] For the life and work of Buffon, see the bibliographies by E. Genet-Varcin and Jacques Roger in Piveteau (1954), 513-570, and also in Roger (1962, 1963). The starting point of Buffon biography is still Nadault de Buffon's edition of the memoir of his secretary, Humbert-Bazile, *Buffon, sa famille, ses collaborateurs et ses familiers* (1863). The article on Buffon in DSB by Roger, 2, 576-582, gives an overview of his career, as in a more discursive way does the symposium *Buffon* (1952) with contributions by L. Bertin, F. Bourdier, Ed. Dechambre, Y. François, E. Genet-Varcin, G. Heilbrun, R. Heim, J. Pelseneer, and J. Piveteau. Hanks (1966) is an admirable monograph on his career prior to his becoming intendant of the Jardin du roi; and Wohl (1960) is an original article. Much correspondence was published by Nadault de Buffon, a collateral descendant, in the most authoritative edition of his works, *Oeuvres complètes*, ed. J.-L. Lanessan, 14 vols. (1884-1885), *13* and *14*, and also in *Correspondance inédite de Buffon*, 2 vols. (1860).

[218] *La Méthode des fluxions et des suites infinies* (1740).

less well than he had planned, for although the seven-year-old boy was then promised the post, he was to have it only in reversion after Charles-Claude de Flahault, comte d'Angiviller, whom we have already encountered in the directorship of the Batimens du roi, and for whose retirement from ministerial responsibilities the intendancy was thus reserved.[219] The crown consoled Buffon both by commissioning the Pajou statue, a highly unusual tribute to a living dignitary, and by elevating to the status of a county the domain of Buffon, about three leagues north of Montbard, where he had installed a forge.[220] He had long since taken the name and attached it with the particle to his patronymic of Leclerc, the lands having been acquired by his father.

The worldliness of Buffon's financial ingenuity is instructive. His father's family and his mother's were both of the Burgundian bourgeoisie, her line having been the wealthier. She died when he was in Rome on a grand tour with an English blade, the duke of Kingston, and the latter's tutor, one Hickman, an obscure Fellow of the Royal Society. They had fallen in with one another at Angers in the course of vaguely medical studies, to which Buffon turned after taking a degree in law at Dijon. Soon after the death of Madame Leclerc, his father took a young woman to second wife. Dismayed, Buffon hastened home to sue for his maternal inheritance, carried the day in court, and won clear title to the family properties. Starting with a capital of perhaps 80,000 livres, he studied sylviculture in order to exploit the forest scientifically, installed the ironworks that returned handsome profits all his life, and, mixing purchase with litigation, acquired the seigneurial properties that made him in effect lord of Montbard, an ennobled bourgeois who, unlike an hereditary aristocrat, spent time on his lands attending to their prosperity.[221]

The canniness of a provincial landowner aggrandizing his estates was thus among the qualities that Buffon brought to the intendancy of the Jardin des plantes. The institution had just completed its first century, having been founded in 1635 on the initiative of Guy de la Brosse, physician in ordinary to Louis XIII, in order to provide medical students with the kind of scientific training in anatomy and in the chemical and botanical foundations of pharmacology that the Faculty of Medicine failed to offer in its bookish curriculum. Protected from the jealousy of the faculty by its association with the royal household, the establishment flourished under the patronage of Colbert and the administration of Guy-Crescent Fagon, chemist, doctor, and eventually chief physician to Louis XIV. In

[219] Hamy (1893), 5-8, 71-72; see also Buffon, *Correspondance inédite 1*, 388, 403, 444-447; *2*, 596-597.

[220] F. Bourdier, "La Vie et l'oeuvre de Buffon" in *Buffon* (1952), 34; Hanks (1966), 269-270.

[221] L. Bertin, "Buffon, homme d'affaires," *Buffon* (1952), 87-104.

those days Tournefort was the most distinguished figure scientifically. After his death in 1708 and Fagon's in 1718, the Jardin des plantes languished under incompetent direction until Charles-François de Cisternai Dufay was appointed to its intendancy in 1732. A recent army officer and an experimentalist in electricity, Dufay was more interested in developing a scientific program than in the courses on medical subjects.[222] After only seven years in office, he was stricken with small-pox at the age of forty-one. From his deathbed he wrote to Maurepas, his patron and an enthusiast for science in the early days of a long ministerial career (Maurepas was then at the Marine), urging that the person best qualified to carry on the program was Buffon.[223] Buffon had been a member of the Academy of Science since 1734, having submitted an original memoir on the theory of games.[224] He was further known for his translation of Stephen Hales,[225] and had good connections and the reputation of a young man ambitious in science and effective in affairs.

When Buffon took the intendancy in 1739, the Jardin des plantes consisted of modest beds, greenhouses, and uncultivated areas extending about halfway to the Seine from the frontage, or most of it, that the Muséum now occupies along the rue Geoffroy Saint-Hilaire. Below the bottom of the garden lay land belonging to the abbey of Saint-Victor. In 1782 Buffon did the monks out of this property and pushed the perimeter down to the river. Still animated at seventy-five by what A.-L. de Jussieu called his grandiose ideas, he had the head gardener, André Thouin, lay out a splendid vista there where scientific and popular Paris have met and mingled ever since.[226] The crown compensated the abbey quite inade-

[222] For the early history of the Jardin des plantes, see Bourdier (1962); Henry Guerlac, "Guy de la Brosse," DSB 7, 536-541; and Yves Laissus, "Le Jardin du roi," in Taton (1964), 287-341. There are bibliographies in Guerlac and Laissus, and the latter gives a tabulation of the personnel from the foundation until 1739. A series of memoirs by A.-L. de Jussieu is still informative, since it was drawn from his own knowledge and that of his family: "Notice historique sur le Muséum d'Histoire Naturelle," *Annales du Muséum d'Histoire Naturelle 1* (1802), 1-14; *2* (1803), 1-27; *3* (1804), 1-17; *4* (1804), 1-19; 6 (1805), 1-20; (1808), 1-39. M. Deleuze (1823) is largely superseded, but should still be consulted for the sake of completeness. There is important documentary material in the Bibliothèque centrale of the Muséum, where the current archivist, Yves Laissus, is making steady progress in the large task of classifying and cataloguing. Early in the twentieth century certain sections of the archives, especially documents surviving from the revolutionary period, were transferred to the Archives nationales, where they occupy series AJ 15. Occasional documents concerning the Museum will also be found at the Archives nationales in series AD VIII and F[17].

[223] Some obscurity surrounds the circumstances of Buffon's succession. See Hanks (1966), 135 n. 64.

[224] *Ibid.*, 35-61; for bibliographical detail, 276.

[225] *La Statique des végétaux, et l'analyse de l'air* (1735). A second edition, revised by Sigaud de La Fond, appeared in 1779.

[226] A.-L. de Jussieu, "Notice historique" *11*, 1-41 (see n. 222 above).

quately with funds from which Buffon managed to divert a portion to enriching the collections, embellishing the appointments, and increasing his own income. The expediency, not to say chicanery, of his methods forms the subject of a classic monograph illustrating the interchangeability of private and public finances in the old regime.[227] One further example will suffice. Until 1935 the present site of the library of the Museum was occupied by a gallery of natural history, originally the Cabinet du roi.[228] The exhibition halls having been inadequate, Buffon grandly vacated the intendant's quarters, in order that the rooms might serve for enlarging the display of the collection, and moved his ménage off the grounds to the Hôtel Lebrun, rue du Cardinal Lemoine, which the crown rented for him. Some few years later he bought a property bordering on the Jardin to the east. He paid 12,000 livres for it, did over the house at public expense to be an official residence (it is still called the Maison de Buffon), and seven years later sold it to the crown in return for rent-free occupancy and 80,000 livres, which he accepted in the form of an annuity at seven percent for the life of himself and his son, then aged fourteen.[229]

Once settled into the routine of his life, Buffon always passed the winter and early spring in Paris, during which months members of the staff might have audiences. For the rest of the year, from April to late November or early December, he lived in Montbard. There he wrote or assembled volume after volume of his *Histoire naturelle*, the only publishing venture of the century to rival the *Encyclopédie* of Diderot and d'Alembert in fame and magnitude. Like many another eighteenth-century sensualist, like Diderot indeed, Buffon followed a strict regime. He would rise at dawn, breakfast lightly, and climb from his house up the narrow streets to his park atop the town. There he had installed a writing room which may still be visited, in a sparsely furnished pastoral pavilion that looks northward from the ramparts over a fine Burgundian landscape. He would spend the morning hours writing, often with the assistance of aides summoned down from Paris. The afternoons he devoted to his commercial affairs, and the evenings to reading in preparation for the next morning at his desk.[230]

Although much celebrated, Buffon's work has been read mainly in excerpts, and it may be well, therefore, to describe it briefly. Publication began in 1749, just ten years after he had assumed office. The whole opus consists of four series of volumes. The form and conception were directly related to his policy in the intendancy, which was to assemble the fore-

[227] Falls (1933).

[228] For the many transformations of this structure, see Bourdier (1962).

[229] See Falls (1933), 145; Bertin, "Buffon, homme d'affaires," in *Buffon* (1952), esp. 97-101; Yves François, "Buffon au Jardin du Roi," *ibid.*, 119-123.

[230] Bourdier in *ibid.*, 38-42; Roger (1962), cxlvi.

most natural history collection in the world in the galleries of the Cabinet du roi. To carry out that mission, he created the post of Curator and Demonstrator in 1745, and appointed to it a young physician and fellow-townsman from Montbard, Louis-Jean-Marie Daubenton, diverting him from the career of country doctor into that of man Friday for zoology and mineralogy.[231] (The botanical and pharmaceutical holdings were left to the care of the brothers de Jussieu and the chemists.) The full title of Buffon's first series of fifteen volumes, on land animals, is *Histoire naturelle, générale et particulière, avec la description du Roy.*[232] Even if ever so second a fiddle, Daubenton was more than a curator. He was a collaborator, and contributed the anatomical descriptions to volumes III through XV. The author and administrator complemented each other in Buffon; he drew upon the collections for information and exploited all the fame his writings won to augment them with specimens gathered round the world by travelers and explorers and donated by them or by their patrons.[233]

In the second series of *Histoire naturelle*, Buffon turned to birds and produced nine volumes between 1770 and 1783 with major collaboration from Guéneau de Montbeillard, ornithologist and, like Daubenton, a fellow-countryman, and minor assistance from Daubenton and the abbé Bexon. A third series of seven volumes (1774-1789) supplements the first, mainly on quadrupeds, although volume IV (1777) contains an excursion into actuarial calculations of the duration of human life. The fourth series of five volumes on minerals (1783-1788) Buffon did himself in the full vigor of old age, leaving a disciple, the young comte de Lacepède, to complete the plan after his death with a final volume for series III followed by two more on oviparous quadrupeds and snakes, five on fish, and one on cetaceans.

Buffon's reputation has been kept alive in different ways by anthologists and by historians of ideas. The former have mined the mass of material for excerpts humanizing animal behavior. They provide staple items for the "dictations" with which French schoolmasters instill recognition of rhetorical style. Picasso illustrated a recent instance,[234] though not one intended for schoolchildren, and perhaps this tradition comes closer to the inwardness of Buffon's genius than does the tendency of historians to ready only the two treatises of general import embedded in the work. The first, *Théorie de la terre*, formed part of the introductory matter in 1749.

[231] For Daubenton, see below, and Cowan (1968-1971), 37-40.

[232] For full titles and an outline of the contents of each volume, see the bibliography by Genet-Varcin and Roger in Piveteau (1954).

[233] There is testimony to Buffon's success in this respect in André Thouin's "Mémoire sur le Jardin du roi, October 1788," Museum of Natural History, MSS 1934, XXX.

[234] *Eaux-fortes pour des textes de Buffon, par Pablo Piccaso* (Paris: Fabiani, 1942).

The second, *Epoques de la nature*, Buffon wrote largely between 1773 and 1778. He tried the preliminary discourse on the Academy of Dijon at a ceremonial inaugurating its new quarters in 1773 and published the whole essay in the fifth volume (1778) of this third series.[235]

Historians and scientists, writing with the controversy over Darwin's theory in mind, have occasionally looked in these two works for statements about the mutability of species, and have been wont to impute to Buffon some change of mind, some retreat from early evolutionary views under pressure from the Sorbonne. In fact, however, Buffon never thought much about the origin or the possible transformability of species. The shift he did make in his approach was toward historicism, in response to a general change of sensibility of which the writings of Herder and Turgot offer other evidence. In Buffon's early *Théorie de la terre*, the action of the seas upon the surface of the earth is a steady state of things over an indefinite span of time. Later in the *Epoques de la nature*, time is the molder of nature through seven definite stages of development.[236] Buffon was at ease writing in this vein. For by temperament he was an historian of nature, a cosmogonist in the baroque manner of Burnet, Whiston, or Woodward, and no would-be biologist or geologist seeking to hasten into specialization sciences yet unnamed.

Throughout both essays, and indeed throughout much that came from his pen, sound the overtones of sensuality and sexual naturalism that identify the medical humanism of the eighteenth century. Those notes echo more daringly (it may be) in La Mettrie, more moralistically in Diderot, more amusingly in Erasmus Darwin. It was all one literary tradition, however, and for these and other writers, Buffon was the source of what later appeared to be its biological aspects. Among those aspects the problem of generation interested him most keenly. Hostile to the notion of *emboîtement*, which he thought both artificial and mechanistic, Buffon adopted from Maupertuis the idea of organic molecules and made it serve an expansive notion that the unity of nature resides in a continuum of seminal substance immortal through all time and uniquely endowed with the power of reproducing all matter.[237] In the 1790s the conception was important for Lamarck and equally for Bichat.

To evoke either biological evolution or vitalism would be to anticipate, however. It was in Diderot's *Pensées sur l'interprétation de la nature* that Buffon's influence immediately appeared, suggesting the general strategy of

[235] Roger (1962) gives an admirable critical edition of the text; the introduction and notes constitute the most considerable contribution to the study of Buffon in the scholarly literature.

[236] Cf. Roger (1962), xxi-xxii, xci-xciii.

[237] On Buffon's organic molecules, see, e.g., Buffon, *Histoire naturelle* 2 (1749), 20-21, 425-426; Roger (1962), lxxi-lxxiii.

that very significant book, wherein Diderot presented materialism sympathetically by investing matter with sensibility instead of inimically by reducing life to matter and motion. Buffon also inspired Diderot's specific prediction that natural history would become the basis of a science centered on human interests, thereby displacing a mathematics doomed to obsolescence by its indifference to the human condition. True, Diderot, animated by a compassion foreign to the ambiance of magistracy and intendancy, adapted these ideas to a democratic vision of the future.[238] He incorporated Buffon's ideas into a model of nature as collectivity drawn rather from Leibniz, and thus invested the program for natural history with a social significance it lacked in the version of Buffon, who was nothing of a philosophe in the ideological sense.

Style interested Buffon more than humanity, and also more than science or discovery. It probably would not have displeased him to know that the *Discourse on Style*, which he gave before the French Academy on the day he was received into membership, 25 August 1753, should be the one piece still widely read two centuries later. "Style," he held, "is only the order and movement in the expression of ideas."[239] Novelty contributes nothing. Indeed, it is usually an impediment, and the writer who wants to hold his readers will stick to things they know. Buffon was consistent in his indifference to innovation. Although he was involved in many scientific disputes, none of them concerned priorities, and he is one of the very few famous figures in the history of science of whom that may be said.

Accordingly, the principle of organization throughout the *Histoire naturelle* is an order of decreasing familiarity among and within the various topics.[240] The work begins with the natural history of man, while the discussion of animals starts in volume IV with the horse and continues with lesser domesticated species before venturing into the wild. Buffon contemned all analytic systems of classification for their artificiality and particularly scorned the Linnaean scheme, its pious author, and its pedestrian adepts.[241] His immense work still repays browsing. It is enjoyable in the way that an old and handsome book of bird-fancying may be, and indeed the nine volumes of the *Histoire naturelle des oiseaux* contain the finest descriptions and illustrations. There as throughout, the habits of the creatures, their characters and animalities, form the subject matter; the

[238] For a discussion of Diderot's philosophy of nature, see C. C. Gillispie, "Diderot," DSB 4 84-90. A convenient edition of the *Pensées sur l'interprétation de la nature* is that in Vernière (1956), 165-244. Compare the passages in that edition, sec. iv-vi, 180-184, to those in Buffon, *Histoire naturelle 1* (1749), 53-54; 2 (1749), 27. Cf. Hanks (1966), 61; Roger (1963), 531-532, 535-536, 541-542.

[239] *Discours sur le style*, ed. René Nollet (Paris, 1908), 6.

[240] Buffon *1*, 36-37.

[241] *Ibid.*, 37-41.

anatomical information is haphazard, except for what Daubenton supplied, and that is largely superficial. Buffon wrote his own accounts from books, from reports of naturalists and others residing in far places, from specimens that chanced to be near at hand in the collections of the Cabinet du roi, and hardly at all from dissection or his own observations.[242]

Buffon's was a literary vein, in short, and his scientific reputation—the "French Pliny"—was a creation of his own among writers, acquiesced in by later scientists who have read them but not him. The younger generation in the Academy, who knew and may have read him, did not acquiesce (except for his protégés). They held his work in low esteem, even as he did them.[243] The President de Brosses and Hérault de Séchelles were his intimates, not his colleagues of the Academy of Science. Among them, Bailly was the only close friend who was not also a retainer. Buffon had the poor judgment to take seriously Bailly's pseudo-historical idea of a primitive Indic people versed in astronomy, and made it the motif for the seventh phase of *Epoques de la nature*, wherein the final stage of nature is readied for the recorded history of man.[244] For the rest, he treated Daubenton, Lamarck, Lacepède, Haüy, Faujas de Saint-Fond, the family de Jussieu, Thouin—in a word, the staff of the Jardin des plantes—more like the subordinates they were in status than like the colleagues they half-consciously aspired to become. If it was hardly thinkable that they should have been colleagues of himself, they could at least imagine that relationship with one another.[245]

It is time to introduce these others, for they were the ones who figured in the events that mainly concern this history, and first the family de Jussieu. In a dynastic view, the de Jussieu rivaled in botany the record of the Cassini in geodesy and of the Bernoulli in analytical mechanics.[246] Their connection with both the institution and the Academy antedated Buffon's own by a generation and thereby assured them a certain autonomy within their sphere, which was the garden itself. Antoine-Laurent, fourth of the family to serve on the staff, was just coming into his own scientifically at the time of Buffon's death. He published his *Genera plantarum secundum*

[242] Roger (1962), lxxxix.

[243] *Ibid.*, xxxv, lxxvii, lix, cxliv-cxlv; Hanks (1966), 66.

[244] Roger (1962), lxxvii; Buffon, in *ibid.*, 205-220.

[245] For the staff's own sense of its status, see below, 183-184.

[246] Two extensive accounts of the family remain indispensable, though historically somewhat amateurish: "De La Méthode naturelle et des Jussieu," in Flourens (1857) *2*, 11-173; and "Les Cinq de Jussieu: leur role d'animateurs des recherches d'histoire naturelle dans les colonies françaises," in Lacroix (1932-1938) *4*, 99-181. For more precise detail, see J. Laissus (1964, 1966). There are family papers in the dossier of A.-L. de Jussieu in the Archives of the Academy of Science. The greater part of the papers was sold at public auction, however. There is a catalogue of the sale in the Adrien de Jussieu dossier, Academy of Science.

ordines naturales disposita in a dramatic month, July 1789.[247] It was one of the last major scientific treatises to be written in Latin and might appear, therefore, to have had less in common with the revolutionary spirit than, say, Lavoisier's *Traité élémentaire de chimie*, printed almost simultaneously and deliberately addressed to the public at large. *Genera plantarum* is the classic exemplar in botany of a natural method of classification in contrast to an artificial one. For systematists in general, its importance was even wider. Despite appearances, Cuvier considered it no less epochal a book in the sciences of observation than Lavoisier's *Traité élémentaire* in experimental science: Jussieu's principle of subordination of characters was that which he himself developed into the correlation of parts in comparative anatomy.[248]

Merely to state the principle fails to convey its power, which lay in its applicability to resolving the indefinite sequence of small dilemmas encountered in classifying plants. In a word, Jussieu rejected any resort to a single discriminant arbitrarily chosen, like the Linnaean classification by organs of reproduction, which was artificial in its operation and limited in its scope to identification of genus and species. Jussieu's purpose was to group genera into natural families at the next higher level of classification. To that end he considered the whole plant and based the first choice on whatever characters exhibited the greatest generality within the floral population. Normally, the initial identification would follow from the mode of germination of seed and the respective disposition of the sexual parts, but he then made his further distinctions according to the relative importance of subsidiary properties in leaf, stem, or root systems, certain combinations of which distinguished the family.[249]

The principle of subordination of characters was a method of analysis, therefore, or, more loosely, a way of seeing and doing things, rather than the basis of a law or theory. The history of science offers no more striking illustration of actual procedures forming a tradition of research within a scientific institution. The influence of Buffon, after all, was a lordly matter of patronage and discourse; though seemingly dominant at the time, it expired with him—with him and with the mode of natural history. Not so the less obtrusive influence of the de Jussieu, which was a matter of practice and carried over through comparative anatomy and zoology into

[247] A facsimile reprint of the first edition with an historical introduction by Frans A. Stafleu is in the series, *Historiae naturalis classica*, ed. J. Cramer and H. K. Swann, 35 (New York: Wheldon & Wesley, 1964).

[248] Cuvier (1810), 305.

[249] For critical exposition of A.-L. de Jussieu's taxonomic philosophy and method, see Daudin (1926a), 204-217, and the Stafleu introduction to *Genera plantarum* (n. 247 above), xix-xxv. Jussieu's own account of his method occupies a seventy-page "Introductio" to that work.

professional biology. It styled the immense taxonomic effort conducted by successive hands within that institution from the time in 1759, when Bernard de Jussieu laid out a garden for the Trianon following the natural method, until the completion in the 1820s of Cuvier's classification of the vertebrates and Lamarck's of the invertebrates.[250] But a man had better remain in the circle to be heeded. Michel Adanson, a bolder pioneer, a ferocious eccentric, did not. His precepts in *Les Familles des plantes* (1763) were more like than unlike those of the de Jussieu; he had been Bernard's student and associate in the 1750s; and his system was based on wider exploration of the vegetable kingdom. He won little hearing in his lifetime, however, and though a member of the Academy, remained a maverick and outsider, whom Antoine-Laurent de Jussieu ignored.[251] Jussieu also ignored the early botanical work of Lamarck, about which a word in a moment.[252]

The family de Jussieu was a curious one, in that Antoine-Laurent was not the descendant but the nephew of his predecessors, Antoine, Bernard, and Joseph, brothers with a marked tendency to celibacy, all of them afflicted, as was he, with a form of progressive myopia that required poring over their plants at ever closer range as they grew older. Their father, who had had sixteen children, had been an apothecary in the town of Montrotier in the Lyonnais. There the eldest brother, Christophe, a physician and master-apothecary, had remained at the head of a family of eleven children (Antoine-Laurent was born of his second marriage in 1748); and there Antoine, who had turned from medicine to botany and established himself in Tournefort's chair at the Jardin du roi in 1710, had sent back in 1714 first for his brother, Bernard, and then about 1725 for their youngest brother, Joseph. In 1735 Joseph was named botanist with the geodesic expedition led by La Condamine and Bouguer to Peru. Funds ran out, and Joseph, without private means, was marooned in Lima and reduced to practicing medicine for a livelihood. In Peru he stayed for thirty-five years, beset by vicissitudes and faithfully remitting exotic specimens to Paris. When finally brought back, he had lost his mind and could scarcely speak French. Meanwhile, Antoine had died, and Bernard, alone in the house, rue des Bernardins, where the brothers had kept bachelor hall, had

[250] Daudin (1926a, 1926b).

[251] The work of Adanson has aroused more interest in recent years than it did in the eighteenth century. See esp. E.-T. Hamy, "Michel Adanson et A. L. de Jussieu," in Hamy (1909); Guédès (1967), containing a very full bibliography; J. F. Leroy (1967); Heim (1963); Scheler (1961a), and a publication of the Hunt Botanical Library of Pittsburgh, *The Bicentennial of Michel Adanson's "Familles des Plantes,"* Hunt Monograph Series No. 1 (Pittsburgh, 1963-64).

[252] Below, 156-164.

followed family precedent and sent home for a nephew, Antoine-Laurent, who joined him in 1765 at age seventeen.[253]

Bernard de Jussieu never cared for scientific reputation in the usual sense. His was the originating mind, but he was of retiring disposition, preferred not to lecture, and repeatedly refused advancement from the minor post of "Sous-démonstrateur de l'extérieur des plantes," which gave him charge of the botanical garden. After the death of his brother, on whom he had greatly depended, his closest associates were his nephew and a lad of almost the same age, André Thouin, whose much humbler, indeed very modest, family developed as intimate a connection with the institution as did the de Jussieu. Thouin's father, the head gardener, died in 1764 when the boy was seventeen, leaving also six younger children, three brothers and three sisters. Despite his extreme youth, André was given his father's post and provided for the family. Two of the brothers, Gabriel and Jean, worked in the garden when they grew older, as did a nephew. The sisters kept house. In later years their cottage, attached to the greenhouse, became a favorite resort for famous amateurs of botany, notably Jean-Jacques Rousseau and Lamoignon de Malesherbes, its simple hospitality in "plein Paris" according with the sentimentality of the herbarizing fad.[254]

The two young men were taught by Bernard de Jussieu—André Thouin, the gardener and subordinate, and Antoine-Laurent, the nephew and successor. They learned by watching and by listening. For Bernard wrote little. Instead, he meditated. The only record remaining of his systematic thought is the catalogue to the garden he designed for the Trianon in 1759.[255] The king had asked him to compose it of plants native to France so that it might constitute a botany school. Bernard never claimed the dignity of a principle for his method of identifying the families to which the plants belonged, but it was by the "subordination of characters"—his nephew's phrase—that he recognized the likenesses.

Like his uncles before him, Antoine-Laurent took a medical degree. In 1772 he defended a thesis comparing the vegetable to the animal economy.[256] Medicine was then the only formal road to botany, or to any sci-

[253] There is an éloge of Bernard de Jussieu by Condorcet, HARS (1777/80), 94-117; for an outline of the collective biography of the family, see J. Laissus (1964), 27-30. For Antoine and Joseph, see Lacroix (1932-1938).

[254] Reminiscenses of Ossian la Reveillière-Lebeau in his funeral discourse for Oscar Leclerc Thouin in 1844, BMHN, Doss. Doc. Biog T. On the career of Thouin, see "La correspondance d'André Thouin," in Hamy (1909), 91-99; Silvestre (1825); Thiebaut de Berneaud (1825).

[255] *Ordre des plantes établi par M. Bernard de Jussieu dans le Jardin du Trianon* (1759).

[256] *An oeconomiam animalem inter et vegetalem analogia* (J. Laissus, 1964), 29. For further detail on the career of A.-L. de Jussieu, see A.-T. Brongniart, "Notice historique sur A.-L. de Jussieu," *Annales des sciences naturelles* (Jan. 1837), 1-24; E.-T. Hamy, "A.-L. de

ence. Indeed, the successor to Antoine in the chair of botany at the Jardin des plantes was a fashionable doctor, Louis-Guillaume Le Monnier, brother of the astronomer.[257] Physician to Louis XV and more courtier than scientist, he was the companion and almost certainly the lover of Marie-Louise de Rohan-Soubise, comtesse de Marsan. His post in the Jardin des plantes recalls its seventeenth-century origin in the medical side of the royal household. In 1770 Le Monnier began exercising the functions of chief physician to a king in worsening health. Finding it awkward to travel from Versailles across Paris to give his stated demonstrations, he took Antoine-Laurent for his *suppléant* to carry out the duties in return for a portion of the salary, a normal enough arrangement in the eighteenth century.[258]

The responsibility gave Antoine-Laurent the occasion to begin his life task of articulating and generalizing his uncle's method. First he published a memoir on the renoncula, showing that the characters in this family were of varying significance, some remaining uniform in all species, others varying more or less exceptionally according to their importance.[259] Next, he published a paper explaining the new organization he was giving his course of demonstrations in the Jardin des plantes, in effect a natural classification.[260] Finally, he persuaded a slightly reluctant Buffon to replant the "botany school" of the Jardin itself in conformity with the Trianon plan, thus replacing Tournefort's classification with a natural one and his nomenclature with the Linnaean (for the de Jussieu recognized the convenience of the latter, while thinking the grouping into species and genus taxonomically inadequate).[261]

Antoine-Laurent de Jussieu cut a larger figure in scientific Paris than his recluse of an uncle had wished to do. He actively seconded Vicq d'Azyr in launching the Royal Society of Medicine. He served on the commission of the Academy that investigated Mesmerism and alone dissented from its report.[262] He faithfully attended meetings both of the Academy and the medical society, and sometimes complained that all this official business made too many claims on a scientist's attention.[263] A disappointment

---

Jussieu et Claret de la Tourrette," in Hamy (1909), 47-61; and the Stafleu introduction to the 1964 facsimile reprinting of *Genera plantarum* (n. 247, above).

[257] On Le Monnier, see Robida (1955); J. Laissus (1966); Y. Laissus (1966).

[258] J. Laissus (1964), 28; A.-L. de Jussieu, "Sixième notice historique sur le Muséum d'Histoire Naturelle," *loc. cit.* 7 (note 222 above).

[259] "Examen de la famille des renoncules," MARS (1773/77), 214-240.

[260] "Exposition d'un nouvel ordre de plantes adopté dans les démonstrations du Jardin royal," MARS (1774/78), 175-197.

[261] J. Laissus (1964), 29.

[262] Below, Chapter IV, Section 2.

[263] A.-L. de Jussieu to Claret de la Tourette, 12 April 1785, letters printed in Hamy (1909), 47-61.

saved him further distraction and enabled him to complete *Genera plantarum*. Not unnaturally, he had hoped to succeed to Le Monnier's chair of botany, for which he had been supplying the teaching since 1770. When finally Le Monnier did relinquish it in 1786, his pleas went unheeded.[264] Such matters were arranged, over the heads of the ordinary staff, at the level of ministry and court where only Buffon and Le Monnier himself had entry. The appointment went, not to Antoine-Laurent, but to a countryman and protégé of Le Monnier, originally from Brittany, one René Louiche-Desfontaines.[265]

Desfontaines did have qualifications. He, too, had studied medicine, while preferring botany. After taking an easy degree from the diploma mill in Reims, he frequented the Jardin des plantes and then departed on a botanizing and herbarizing mission to North Africa.[266] It cannot be said that his research ever made much difference to the science of botany. In the event, however, his appointment may have worked in its best interests. He was a magnificent teacher, not merely leading a few disciples from herbaceous bed to medicinal border as had Bernard de Jussieu, quietly discoursing of the complexities in a natural system, but simplifying the whole subject in the popular mode that came into fashion in the 1790s and holding spellbound the hundreds who flocked to his lectures, given at seven in the morning. Meanwhile, Antoine-Laurent, relieved however involuntarily of teaching duties, set to work assembling his great book. He had the arrangement so profoundly in mind that he sent copy directly to the publisher who printed off the sheets as they were written. The work appeared just before the Revolution began to impede normal scientific publication.

Let us turn now to persons whose careers depended at their outset on the intervention of Buffon in their lives—to Lamarck and Lacepède first, scientific retainers though of noble family, and then to Daubenton and his association with Haüy, Faujas de Saint-Fond, and Dolomieu in the mineralogical emphasis of the last years of Buffon's intendancy.

There were similarities in the uses that the lordly old bourgeois made of the necessities and aspirations of the two former, Jean-Baptiste de Monet, chevalier de Lamarck,[267] and Bernard de La Ville-sur-Illon, comte de

[264] J. Laissus (1966).

[265] Y. Laissus (1966) for their correspondence. On Desfontaines, see also Chevalier (1939), an undocumented biography.

[266] His chief work was *Flora atlantica*, 3 vols. (1798-1800), concerning mainly the flora of the Atlas range.

[267] The Bibliothèque centrale of the Museum of Natural History possesses a large manuscript collection concerning the work of Lamarck in his maturity, but little concerning his youth. For an inventory, see Vachon, Rousseau, and Laissus (1969). The same authors intend a publication of the most important documents. There has been no full biography since Landrieu (1909), but there have been many special studies in recent years, particu-

Lacepède.[268] In 1781 Buffon engaged Lamarck to be tutor for his son, then seventeen, who was departing on a grand tour to be presented to the princes and naturalists of the Low Countries, Germany, and the Empire. He designated Lamarck Correspondent of the Jardin des plantes for the duration of the journey and charged him to send back specimens for the collections. By then it was evident that Buffon could not expect quite to finish the *Histoire naturelle*. He had to provide for his work as well as his son, and in 1784 he named Lacepède to the place of Curator and Assistant Demonstrator in the Cabinet du roi on the understanding that the young man would be his literary executor and would study to write in Buffon's manner on fish and reptiles, thereby completing the master plan.

Both Lamarck and Lacepède were of the minor provincial nobility. Lamarck, born in 1744, was the eleventh and last child of a country squire in the Abbeville region of Picardy, where the Flahault de Billarderie also had their seat, the great family of which d'Angiviller was scion. Lacepède, a dozen years younger, derived his title from a childless maternal granduncle, a Gascon who left him name and fortune, the La Ville having stemmed from the Vosges and the Rhineland. He was brought up in Agen, where his youth was divided between chateau life in summer and, in winter, the literary and philosophical diversions of that spirited small city, the center of Gascony. Lamarck's education in the Jesuit college at Amiens was stricter. The benjamin of the family, he was probably intended for the priesthood.

Though far from robust, Lamarck preferred the army to the church and entered the school for pages at Versailles after his father died. He served in the campaigns of 1761 and 1762 in Germany and the Low Countries. When the Seven Years War ended in 1763, he was posted to garrison, first in Antibes and Toulon, where he began botanizing, and later in Alsace. He quit the service in 1768 at the age of twenty-four, his health threatened by an apparently scrofulous abscess of the neck. Lacepède's military associations were honorary. His father was lieutenant-general of

---

larly on the background of Lamarck's thought. See especially Aron (1957), Burkhardt (1970, 1972), Gillispie (1956a), Hodge (1971), Schiller (1969). In 1972 Franck Bourdier circulated a mimeographed "Esquisse d'une chronologie de la vie de Lamarck," which he established with the collaboration of Michel Orliac on the basis of extensive archival research into the minutiae of Lamarck's existence. I rely upon Bourdier together with Hamy (1907) for statements of biographical fact in the pages that follow.

[268] Roule (1932) is an undocumented biography of Lacepède, an expansion on the same author's earlier study (1917). See also the Cuvier "Eloge historique de . . . Lacepède," MARS *8* (1829), ccxii-ccxlviii; Villenave (1826); Van den Berg (1961); Théodoridès (1974). There are many autobiographical passages in Lacepède's two proto-romantic novels: *Ellival et Caroline*, 2 vols. (1816) and *Charles d'Ellival et Alphonsine de Florentine*, 3 vols. (1817). Hahn (1975) publishes Lacepède's recently rediscovered autobiographical sketch, and (1974) prints four previously unpublished letters.

the Senechalsy in Agen. What with his family's Rhenish connections, he arranged for a German commission in a regiment that probably existed on paper if at all: on the title pages of his first books he is qualified "Colonel au Cercle de Westphalie."

Lacepède was by then a young man about town, having come up to Paris in 1776, something of a d'Artagnan of the Enlightenment, something of a Boswell to Buffon's Johnson. Lamarck for his part had been in the capital since 1769 or 1770 in search of health, which the famous surgeon Tenon restored to him, and also of occupation. Both hesitated between music and natural history for their early allegiance. Lamarck played the cello and thought of making a career with it, although botany had been his hobby in the army, a serious hobby. Lacepède, before arriving in Paris, had written simultaneously to Gluck and to Buffon. He composed scores for the *Armide* of Quinault and for *Omphale* and did a book on operatic taste before fixing on Buffon and natural history.[269]

To conclude the comparison with its most telling aspect, Lamarck and Lacepède were both contributors, through the first books they wrote, to that eighteenth-century literature that took all nature for a subject and explained its course by the operation of some universal principle or substance of great subtlety—fire, electricity, magnetism, heat, metamorphosis, irritability, sensibility, polarity, or interconvertibility of forces. Such writing, and there was a lot of it, occupies the middle ground in a range extending from Bernardin de Saint-Pierre and Diderot on one end to publications approved by the Academy of Science on the other, or from literature to science. Lamarck and Lacepède graduated from these indiscretions, at least in part. More often only frustration and disappointment awaited authors who thus fancied to go up the garden path from speculation into science. The most significant example was Jean-Paul Marat, of whom more in the sequel.[270]

Lamarck wrote such a book in 1776 and four years later submitted it to the Academy for sanctioning. If we give the full title, the reader will know as much about its content as did the members of that body, for (if we may believe Lamarck's complaint) the commissioners procrastinated and failed to submit a report.[271] Many years later, thanks to the Revolution and freedom of the press, he could publish it, and did so in 1794, dedicating its two volumes in gratitude to the French people. The title is

> Researches on the causes of the principal phenomena of physical nature; particularly those of combustion; evaporation and vaporization of water; heat produced by friction between solid bodies; heat which

[269] *La Poétique de la musique*, 2 vols. (1785).

[270] Below, Chapter IV, Section 3.

[271] Lamarck, *Recherches sur les causes des principaux faits physiques*, 2 vols. (1794) *1*, vi.

manifests itself in sudden decomposition, in effervescence, and in the bodies of many species of animals throughout their lives; causticity; the odor and taste of certain composites; colors of bodies; the origin of composites and of all minerals; and finally the maintenance of life in organic beings, their growth, their vigor in their prime, their decline and death.

Lamarck acknowledged the work to be one of meditation, based on "physico-chemical logic" and not on experiment, and expressed the opinion that the proper mode of scientific inquiry is to move from the general to the particular.[272] The general phenomenon that interested him was the manifestation of activity and life everywhere in nature.[273] Already at the time of writing he disagreed with the phlogiston theory, though unaware of the experimental work of Lavoisier and his school, then in its early stages. By the time of publication, oxygen had largely displaced phlogiston in the theory of combustion, but Lamarck thought no better of this new version of what he called pneumatic chemistry and considered Lavoisier's experiments to be inconclusive. In any case they did not touch his own theory, which was that the activities mentioned in his title were manifestations of differing states of fire, an element incapable of causing such effects in its free and natural state. A subsidiary feature of the argument is more interesting to scholars wishing to trace Lamarck's later views on the transformability of species back to their origin. In a variation on Buffon's dichotomy between physical nature and living process,[274] he held that life is the only form of activity that produces chemical and mineral composites, the tendency of inert matter being toward decomposition and degeneration.

Lacepède's *Essai sur l'électricité naturelle et artificielle* in two volumes (1781), followed by two more of *Physique générale et particulière* (1782 and 1784), are similar: they are books written out of the author's reflections on all nature and on other books of the same sort. A passage from the former conveys the image of a scientist at his work. He is discovered at the center of one of those awe-inspiring scenes that nature reserves for the close of a stifling summer day. A vast storm advances amid flashes of lightning and claps of thunder. Clouds mass blackly. Menacing shadows overtake a terrified countryside:

> Amid this universal consternation, the Philosopher, solitary, intrepid, dares advance in order to harness the lightning bolt at the very seat of its power. All alone, he steps forth, in his hand a light and fragile instrument. Lifted on the gales, the frail device rises to the clouds to combat the lightning. No sooner has it attained the

[272] *Ibid.*, xii. [273] *Ibid.*, 5-6. [274] Above, Section 5.

> zone of thunderclaps than it is surrounded by flashes of fire. The entire storm masses and gathers round it while the Philosopher, whom all Nature now seems to be obeying, controls and directs by means of a light cord. I catch glimpses of him clearly outlined in the flares, protected from the surrounding danger by the fruits of his experiments and meditation, dominating so to say the frightful cloud laden with doom. . . .[275]

For Lacepède, too, had his theory about fire, light, magnetism, life, the annual greening of the earth, and the motions of the heavenly bodies—all in relation to the electrical fluid; but we will not stay to elaborate its fine points, nor to distinguish it from others, whether Lamarck's still unpublished system, or that of Marat, whom he cited appreciatively in several passages of temperate disagreement.[276]

Lamarck and Lacepède would themselves have been more aware of differences than likenesses in their circumstances and personalities. The most obvious contrast was economic. Lamarck was poor, and after he left the army had to contrive to supplement an income of a few hundred a year. At various junctures he studied medicine, worked in a bank, did catchpenny consulting for dealers in shells and rocks, and wrote of botany for the *Encyclopédie méthodique*. The injured tone of his preface to the work just mentioned, and the delay in publication, were instances among many of a lifelong tendency to be the victim of misunderstandings bordering on persecutions. Small of stature, frail in physique, tireless in research, stiff-necked in his parade of integrity, Lamarck had a penchant for irregularity, domestic as well as professional. He had three or four wives, and lived with the first of them, Rosalie Delaporte, who bore six of his eight children, for fifteen years before he married her on 27 September 1792, as she lay dying.

Lacepède, on the other hand, was rich and had an equally marked penchant for associating with the well-placed and powerful, in his youth with Buffon, in his prime with Malesherbes, in his maturity with Bonaparte. His letters of self-introduction to Gluck and Buffon procured him instant invitations to the opera and to the Jardin royal. Yet Lacepède's devotion to Buffon proved genuine. Not only did he learn ichthyology and herpetology, promising to finish the *Histoire naturelle*, but after Buffon's death he kept his promise; and though that was all he ever did do in science (the Empire made him into a courtier and the Restoration into a writer of Gothic novels), the volumes he produced were no inconsiderable achievement, and testimony is unanimous about his kindness and affability. It is

[275] Lacepède, *Essai sur l'électricité* I, 3.

[276] *Ibid.*, 39.

reported of his deathbed that his last words were, "I shall see Buffon again."[277]

Let us leave the comparison now, however, for Lacepède did not have an important mind, and finally Lamarck did, mingled though his thinking was with highly unprofessional ruminations. For his was a dual, not to say a split, scientific personality. Partly he was the master of detail, skilled in its analysis, and partly the visionary, gazing beyond the wide horizon. The pattern is already evident in the book that gained him reputation, *Flore française*, which, putting aside his lucubrations on fire, he composed in 1777 and published in the following year.[278] The work is a three-volume manual for identifying plants, a production of Lamarck the analyst, incongruously prefaced by a summons to a natural classification.

Lamarck never confused these two enterprises, the practical and the philosophical. On the contrary, he insisted upon the difference, deploring the vanity of botanists who claimed the dignity of natural truths for the terms of their own analysis, thus "giving orders to nature, forcing her to deploy her productions like a general his army, by brigades, by regiments, by battalions, by companies, etc."[279] It is interesting that he here took the same view of the possibilities for knowledge of living nature that Laplace did in thinking of the mechanistic universe: truth remains the goal even though the human intellect will never suffice to grasp the real relations among all the things in the world.[280] The highest purpose of natural history must, therefore, be to approximate its classifications as closely to the chain of being as ingenuity could bring them. That enterprise would bring our ideas into conformity with the structure of nature. It would not help us identify plants, however, which was the present purpose, and Lamarck promised to return to it in a future work, a "Theâtre universel de botanique," meanwhile offering this preliminary discourse to *Flore française* as an earnest of his philosophical intentions.[281] It should, perhaps, be said in passing that Lamarck made no mention of the de Jussieu principle of subordination of characters in his early thinking about natural classification.[282] He would then have proceeded by specifying *le plus vivant* or the most highly organized plant to place at the top of a scale and the most incomplete or marginally lifeless for the bottom, and would thereupon have distributed all the others along a continuum such that every species would be found between the two others to which it was most

[277] Roule (1932), 44.

[278] *Flore française*, 3 vols. (Paris, 1778).

[279] *Ibid.* 1, xc.

[280] *Ibid.*, lxxxviii-lxxxix.

[281] *Ibid.*, 4, cxviii.

[282] Clos (1896), 6-7. This excellent memoir remains the fullest discussion of Lamarck's botanical work.

closely related. At that time Lamarck imagined the direction to be one of regressive degradation rather than progressive perfection. It must further be said that there was nothing original in his handling of the idea of a chain of being.[283]

The great originality of *Flore française* lay in its practical scheme for recognizing plants. It may impress the modern reader still more forcibly than it did Lamarck's contemporaries, for the principle is that of a binary analysis, and he employed it with a brisk cogency quite uncharacteristic of his philosophy. His method was expressly artificial, ruthlessly so. It depended on progressively splitting a population by a sequence of discriminants chosen to be readily recognizable, mutually exclusive, and evenly enough divisive so that the individual species would be isolated in four or five cuts of the deck. His example applying the method to a sample collection of twelve species will make the procedure clear.[284] (The names on the chart are numbered for the sake of brevity.) *Flore française* constitutes in effect a program for subjecting the thousands of species of plants native in France to this analysis, cleverly coded for easy use. When it is said that Lamarck had an important mind in science, that sort of evidence is convincing. Explaining his scheme, he displayed the indifference of some computer to the inevitability that any separation at all would divide some plants from others with common features. The purpose of the exercise was to find out what they are, not where they belong.[285] So arbitrary an approach was unusual in the late eighteenth century. The book was a success, however, a notable success. Buffon liked it immediately, both because of its workability and because it mingled praise of himself with criticism of Linnaeus.

In 1779 a vacancy occurred in the botanical section of the Academy. Buffon then joined with d'Angiviller, the minister responsible, in arranging that the crown exercise its prerogative to prefer Lamarck over Descemet, the candidate who had the majority of votes.[286] (D'Angiviller had the double interest of his prospective succession to Buffon in the intendancy of the Jardin des plantes and his territorial association with Lamarck's impoverished family.)[287] Here we may leave Lamarck for now, entered through the favor of the great upon the life of science, composing a "Dictionnaire de botanique" for the *Encyclopédie méthodique*,[288] thinking

[283] *Flore française 1*, xci-xciv. Note, however, that Lamarck did accept the notion of families, and that he attempted an arithmetical calculation weighting the relative importance of the occurrence of the main organs of fructification in determining these relationships. *Ibid.*, cii-ciii. Cf. Daudin (1926a); Lovejoy (1936).

[284] *Flore française 1*, lxi-lxiv.

[285] *Ibid.*, lx-lxi.

[286] Hahn (1971), 81; also Bourdier, "Esquisse" (n. 267 above), 5.

[287] Above, 144-145.

[288] Lamarck wrote the first 3 volumes (1783-1789) of the 8 volume *Dictionnaire de botanique* for this enterprise.

LAMARCK'S CLASSIFICATORY SCHEME FROM *Flore Française*

1. Hieracium murorum
2. Anthemis cotula
3. Polypodium filix mas
4. Alsine media
5. Salvia pratensis
6. Agaricus campestris
7. Pyrus communis
8. Bryum murale
9. Bellis perennis
10. Anagallis arvensis
11. Boletus luteus
12. Carduus marianus

FIRST PARTITION	SECOND PARTITIONS	THIRD PARTITIONS	FOURTH PARTITIONS
Flowers, stamens, and pistils readily distinguishable: 12, 1, 10, 5, 9, 4, 7, 2	Abundant flowers in common calix: 12, 1, 9, 2	Similar flowers—all horn-shaped or all tongue-shaped: 12, 1	Flowers all horn-shaped: 12
			Flowers all tongue-shaped: 1
		Others tongue-shaped: 9, 2	Bare torus without grains: 9
			Torus loaded with grains: 2
	Individual flowers not from common calix: 10, 5, 4, 7	Corolla a single petal: 10, 5	Regular corolla: 10
			Irregular corolla: 5
		Corolla of several petals: 4, 7	10 Stamens or less: 4
			11 Stamens or more: 7
Flowers absent or with indistinguishable stamens and pistils: 3, 6, 11, 8	Leafy plants with detectable but indistinct fructification: 3, 8	Pollinating fructification on backs of leaves: 3	
		Peduncular fructification, anthers at end of stems: 8	
	Leafless plants with indistinct and undetectable fructification: 6, 11	Pileus covered with blades: 6	
		Pileus covered with pores or pipes: 11	

dark thoughts about chemists and the new chemistry, and surrounded in his cramped apartment, rue Coppeau (nowadays the rue Lacepède), by a growing brood of illegitimate children, one of them named André after Thouin. His other great friend in those early days was the abbé Haüy. Neither of them yet had their appointment to the staff of the Jardin. Haüy was then a professor of classics or "*Humanités*," priest by vocation, schoolmaster by trade, and mineralogist by taste. The second volume of *Flore française* acknowledges his participation in discussions of natural classification and attributes the style of the Preliminary Discourse entirely to his collaboration.[289] Lamarck, when left to himself, did write an abominably prolix French, and it is probable that Haüy rewrote the whole argument, eliminating many a distracting digression. Before turning to Haüy and the mineral kingdom, however, we need to appreciate the situation of old "Berger" Daubenton. With Buffon keeping so haughty a distance, the young men tended to adopt Daubenton for grandfather-figure, and one always pictures him in tremulous decrepitude, tiny, stooped, with a very long nose, a "*croulant*" supported in and out of meetings of the Academy by disciples, Haüy sustaining him on one arm and Dolomieu on the other.

Needless to say, it had not been always so. Daubenton had developed three specialties: anatomy, animal husbandry, and mineralogy. The first he undertook on Buffon's commission; the second brought him independent reputation; the third revived his collaboration with Buffon in their later years on a footing nearly as equal as that on which they had begun their lives, boyhood contemporaries in Montbard.[290] There his father had been a notary. He was schooled by the Jesuits, who smelled a recruit and sent him on to Paris for theological study. He was a slight lad, studious, accurate, retentive, and disciplined—made to be a priest, it seemed. He preferred medicine and like most of the students supplemented the bookish courses offered by the faculty, place Maubert, with the anatomy lectures at the Jardin des plantes given in the 1730s by Jacques-François Du Verney. Daubenton already knew the institution, therefore, when in 1745 Buffon proposed his giving up the medical practice he had barely begun in Montbard and collaborating instead on the scheme they had discussed in chance encounters: developing the Jardin des plantes from an auxiliary of medical teaching into a proper institute of natural history. There he

[289] *Flore française 2*, iv.

[290] On Daubenton's life and career, see Roule (1925), undocumented like all this author's biographies; Orcel (1960); Guillaume (1908) *1*, 178-197; and his correspondence with Buffon, in Nadault de Buffon's edition of the latter's *Correspondance générale*, vols. 13 and 14 of Buffon, *Oeuvres complètes* (1884-1885). There are scattered, not very helpful manuscript remains at BMHN. Daubenton is sometimes confused with his first cousin and brother-in-law, E.-L. Daubenton (1732-1785), who held the minor post of curator and assistant demonstrator in the Cabinet d'Histoire Naturelle from 1766 to 1784.

would order the vast disorder of nature herself into a domesticated *abregé*.[291]

With that commission Daubenton took on the curatorship of the *Cabinet*, a gallery some eighty to a hundred feet in extent furnished with display cases for zoological preparations, presses for botanical collections, and cases for minerals and fossils.[292] There might be twelve to fifteen hundred visitors a week. Buffon saw to enlarging the collection and expanding the accommodations. Daubenton directed the museum, making himself available to the public, cataloguing, displaying, and rearranging the exhibits. Those tasks he combined with provision of anatomical descriptions for the *Histoire naturelle*, thus advertising Buffon's program for the Cabinet in the world of letters.

The collaboration put a strain on Daubenton's self-esteem. His anatomies contributed nothing to the science. He did practice dissection, though without time to use a microscope for fine points, and what with the pressures of deadlines and limitations of space, he had to restrict his anatomical descriptions of the many creatures to superficial characteristics of the main structures: the skeleton, the digestive tract, the respiratory and circulatory systems, the dimensions of the principal organs. Even so, there were those among Buffon's readers who objected to his disfiguring the splendid panorama with his *tripailles*.[293] Thus did Daubenton pass two decades. He was being used, and he came to feel used, though less angrily than many in like circumstance might have done, for there was a vein of irony running through the gentleness of his disposition, and he did not take himself tragically.

Greater dignity and a standing of his own came in consequence of a commission from the Bureau of Commerce for improving the quality of French wool. Colbert had emphasized the advantage that would follow were the textile industry to be freed from its dependence on foreign sources of raw wool, on England for the staple grades and on Spain for the finest quality spun from the fleece of the famous merino flocks. The government took no serious measures, however, until in 1766 Trudaine resolved to support a program of research, approached Daubenton to conduct it, and promised him full expenses together with an annual pension of 5,000 livres for life.[294] Having spent twenty years on zoology for Buffon's sake, he seized the chance, installed an experimental sheep ranch near Montbard, and imported animals from Roussillon in the south, from

[291] For Daubenton's views on the matter, see "Cabinet d'histoire naturelle" in the *Encyclopédie*, in which the references to himself in the third person are spelled "D'Aubenton." His actual post was "Garde et démonstrateur des collections d'histoire naturelle du Cabinet du Roi."

[292] Bourdier (1962).

[293] Roule (1925), 106.

[294] Memoir (3), n. 295 below, 79; *Archives parlementaires* 13, 418.

Flanders, from England, from Spain, from Morocco, even from Tibet. Trudaine de Montigny continued his father's interest and support. The work took twelve years. The memoirs that report it were exactly the sort that the Academy liked to feature at its public meetings, being nontechnical and obviously in the civic interest. "The experiments of M. Daubenton have shown the way," observed Condorcet introducing one of them, "but we will need time, research and much work to win this treasure, more substantial than the Golden Fleece the Argonauts set out to conquer."[295]

Husbandry and stockbreeding were the two aspects of the research. An anatomist, Daubenton began with the mechanism of rumination, established that the level of body liquids is the critical factor in metabolism, and concluded that the inferiority of French herds was a consequence of the Gallic practice of shutting the animals in the barn throughout the winter. Cooped up in fetid warmth, they became dehydrated through sweating and were exposed to disease. When he tried the British and the Spanish example of folding sheep in the pasture the year around, they flourished even in the fiercest weather and their fleece was all the finer for it.[296] The breeding program was equally impressive and produced evidence that Darwin might well have cited, had he known it, in the chapter that opens *On the Origin of Species* with a discussion of the fancier's ability to vary the characters of a race by selecting the animals to be bred. Daubenton controlled his experiments with all the care of a geneticist, and found that rapid modification required the services of rams imported for the quality of their fleece. Merely choosing the best from an indigenous population produced a far slower and probably a less extensive possibility of improvement.[297]

When the program had been under way for some ten years, Trudaine de Montigny paid an official visit accompanied by two inspectors of manufacturing: Desmarest, then responsible for the generality of Châlons, and

[295] HARS (1772, pt. 1/1775), 9. The six memoirs were as follows:

(1) "Mémoire sur le mécanisme de la rumination, et sur le tempérament des bêtes à laine," read 13 April, 1768, MARS (1768/70), 389-398.

(2) "Observations sur des bêtes à laine," 15 Nov. 1769, MARS (1772, pt. 1/1775), 436-444.

(3) "Mémoire sur l'amélioration des bêtes à laine," 9 April 1777, MARS (1777/80), 79-87.

(4) "Mémoire sur les remèdes les plus nécessaires aux troupeaux," 3 December 1777, MSRM (1776/79), 312-320.

(5) "Mémoire sur le régime le plus nécessaire aux troupeaux," 11 December 1778, MSRM (1777 & 1778/1780), 570-578.

(6) "Mémoire sur les laines de France, comparées aux laines étrangères," St. Martin's Day, 1779, MARS (1779/82), 1-11.

[296] Memoir (1), n. 295 above, 395-397.

[297] Memoir (3), n. 295 above.

John Holker, the Jacobite Englishman who had become the leading entrepreneur in the cotton industry of Rouen and inspector-general of manufactures.[298] They judged the fabrics produced from Daubenton's ranch to approach in quality the "superfine" grades from Spain.[299] Theirs were the eyes of practical experience. Daubenton made his own investigation of the structure of textiles with the aid of microscopic analysis and measured the strength and dimensions of the filaments micrometrically before concluding that his experiments had demonstrated that the French soil and climate were capable of producing fine wool.

It remained to make the sheep-farmer mend his ways, and there was the rub. "Most country people," acknowledged Daubenton in the conclusion to his first memoir,

> appreciate neither the force of reasoning nor the authenticity of fact and can feel no confidence in any proposed innovations unless their success can be demonstrated visually and manually. Nothing but concrete examples can persuade them to follow new practices. . . . The government is doing what it can. But it is up to good citizens to cooperate. You, who have a taste for rustic tasks and who love humanity: Raise your flocks of sheep. Set such an example to the country people as will provide them with the means of bettering their lot by the profit they might draw from wool-bearing animals.[300]

For his part Daubenton composed a pastoral primer, *Instruction pour les bergers*, liberally and charmingly illustrated with scenes of enlightened peasants lavishing tender loving care upon ewes in lamb-birth.[301] The book retains an interest for the history of husbandry and also for the history of popular culture and education. It was through the catechism that villagers were accustomed to learning what to do, and for that reason Daubenton composed his *Instruction* in the form of questions and answers. The style and tone exemplify the extremely elementary, not to say childish, level in which an educated man thought he must address the working population when proceeding by the written word.

The government put more hope in educating specialists and decided to fortify the curriculum of the veterinary school it had founded in 1766 at Alfort on the eastern outskirts of Paris.[302] In the 1780s Fourcroy was assigned the course in chemistry, Vicq d'Azyr in comparative anatomy, and

[298] On Holker's career, see Fages; Rémond (1946); below, Chapter VI, Section 3.

[299] Memoir (3), n. 295 above, 85.

[300] Memoir (1), n. 295 above, 398.

[301] Paris, 1782.

[302] On the Ecole d'Alfort, see Railliet and Moulé (1908); P. Huard, "L'Enseignement médico-chirurgical," in Taton (1964), esp. 206-209.

Broussonet in zoology. For Daubenton it created a chair of Rural Economy, and he divided his teaching between that subject and his lectures at the Collège de France, where he concentrated on mineralogy, the third of his subjects, and the one in which he took the keenest interest and felt the greatest pride. His was also the teaching that introduced the science to the students who frequented the Jardin du roi.[303] There its success was such that a chair was created in the revolutionary reorganization, and in 1793 Daubenton became first professor of mineralogy, to be followed on his death in 1799 by Dolomieu, absent in a Neapolitan prison, and soon after by Haüy. Of the many whom he had attracted to the science when they were young, those two had become leading figures, one in the new science of geology and the other in the hybrid science of crystallography.

Their affection was nonetheless respectful for the contrast in sophistication between their work and his. The 1780s saw a kind of revolution in knowledge of the mineral kingdom, less famous than the contemporaneous transformation of chemistry but no less thorough.[304] Daubenton published his *Tableau méthodique des minéraux* in 1784.[305] It contains the scheme by which he had arranged the collections of the Cabinet du roi. Over five hundred minerals are distributed into orders, classes, sorts, and varieties—according to their superficial physical and chemical properties. Avoidance of the word *species* was deliberate, for Daubenton disbelieved in continuity between living and non-living beings. Already the approach seemed discursive and old-fashioned, however. In the same year Haüy brought out his *Essai d'une théorie sur la structure des cristaux*, taking a solid geometrical grasp upon this old branch of natural history, exhibiting the constancy of interfacial angles in crystals of the same type, and proposing a classification by criteria of mathematical form.[306] The technique im-

[303] Orcel (1960).

[304] Mauskopf (1970a).

[305] *Tableau méthodique des minéraux, suivant leurs différentes natures, et avec des caractères distinctifs, apparens, ou faciles à reconnoître* (Paris, 1784). See also his introduction discussing mineralogical method, *Histoire naturelle des animaux* (1782) in the *Encyclopédie méthodique*.

[306] *Essai d'une théorie sur la structure des cristaux appliquée à plusieurs genres de substances cristallisées* (Paris, 1784). The *Bulletin de la Société Française de Minéralogie* 67 (1944) devotes a commemorative issue to the life and work of Haüy. It contains essays by A. Lacroix, "La Vie et l'oeuvre de l'abbé René-Just Haüy," 15-226, who prints many of his letters and gives a thorough bibliography; Ch. Mauguin, "Conférence sur la structure des cristaux, d'après Haüy," 227-262; J. Orcel, "Haüy et la notion d'espèce en minéralogie," 265-335; and a further essay by Lacroix, "Les Deux Frères René-Just et Valentin Haüy," 338-344. Valentine Haüy devoted his life to the service of the blind and devised a type of relief script, an early stage of what became Braille. For other work on Haüy, see Kunz (1918) and more notably R. Hooykaas (1955), an article that summarizes the author's earlier studies on Haüy and the history of crystallography and draws upon manuscripts in the archives of the Academy for a close discussion of the origin of his approach; see also, Hooykaas, "Haüy," DSB, 6, 178-183; Mauskopf (1970a, 1970b).

mediately commanded the interest and respect of Laplace, Bézout, and the mathematicians of the Academy.[307] Haüy had presented his first two papers in 1781, and, since he also had the support of Buffon, he was elected two years later to a vacancy, albeit in the botany section.

This is not the place either to explore the details of Haüy's work or to deplore his failure to acknowledge indebtedness to Torbern Bergman in Sweden and to Romé de l'Isle right there in Paris. The latter's *Cristallographie* had appeared in 1783.[308] Suffice it to allow that in this one episode of an otherwise austere and blameless life, intellectual ambition does seem to have prevailed over the charity animating Haüy's vocation in the priesthood.[309] Those who relish examples of scientific prejudice will find a flagrant case in Romé's exclusion from the Academy.[310] It was Buffon who perpetrated that injustice, however, not Haüy. Romé, like Lamarck, was a scientific gentleman of little fortune, a former soldier who got into natural history after returning to Paris from vicissitudes in India and China. He had the bad luck to begin under Sage, the mineralogist and chemist who, on the losing side of the disputes in chemical theory, was misprised personally by both Buffon and Lavoisier, their entourages agreeing on little else. Romé had offended Buffon doubly, by criticizing his theory of a central fire in a cooling earth[311] and by representing the classification of minerals according to crystalline form as an extension of Linnaeus's program from the vegetable to the animal kingdom.[312] Such actually it was, and there is no doubt that it was Romé who started organizing mineralogy around determinations of the geometric plan of crystals.

Neither, however, is there any doubt that it was Haüy, a familiar of the Jardin des plantes, who, endowed with more exact measuring skills and a clearer stereometric vision, tacitly converted Romé's somewhat Platonic idea of geometric plan into the concrete, micro-architectural notion

[307] Hooykaas (1955); Haüy's earliest memoirs, presented to the Academy in 1781, were referred to Bézout and Daubenton: "Mémoire sur la structure des cristaux de grenat," and "Mémoire sur la structure des spaths calcaires," both published in *Observations sur la physique* 19 (1782), 366-370, and 20 (1782), 33-39 (Rozier's journal).

[308] J.-B.-L. Romé de l'Isle, *Cristallographie ou description des formes propres à tous les corps du règne minéral, dans l'état de combinaison saline, pierreuse ou métallique*, 4 vols. (Paris, 1783). Romé presented this work as the second, much enlarged edition of his *Essai de cristallographie* (Paris, 1772). On Romé de l'Isle, see the article by R. Hooykaas, DSB, *11*, with a full bibliography of primary and secondary references. See also Birembaut (1953).

[309] Birembaut (1953); Hooykaas (1955).

[310] J.-C. Delamétherie, "Notice sur la vie et les ouvrages de M. de Romé de l'Isle," *Observations sur la physique* 36 (1780), 315-323.

[311] *L'Action du feu central bannie de la surface du globe, et le soleil rétabli dans ses droits; contre les assertions de MM. le comte de Buffon, Bailly, de Mairan, etc.* (Paris and Stockholm, 1779); a second, revised edition appeared in 1781.

[312] Mauskopf (1970a), 188.

of geometric structure and carried the program into effect, thereby founding modern crystallography in a mathematical theory of the structure of matter.[313] Portraits usually show him holding a goniometer. His crossing of mineralogy with geometry offers an early instance of the fertility of hybrid sciences, a phenomenon that became characteristic of the nineteenth century, when boundaries were often areas where new ideas occurred. A recent essay shows how both Lamarck's evolutionary ideas and the basic laws of chemical combination had historical roots in these mineralogical discussions.[314] Although Lamarck and Haüy were great friends, Lamarck started his famous theory of transformation of animals in direct reaction against the doctrine of fixed species in minerals. As for chemistry, Haüy eventually took the view that the "integral molecule" of a mineral, by his definition the smallest particle that on subdivision still retains the properties of the substance, was characterized primarily by geometric form and secondarily by chemical composition. His mineralogical molecule was thus a direct ancestor of the modern chemical molecule: John Dalton later thought Haüy's definition of the chemical function of ultimate particles the best he had read.[315]

Haüy was a cabinet crystallographer, however, who scarcely ever ventured into the field or traveled further from Paris than St. Just-en-Chaussée, the village in Picardy where he had been born; while government was more interested in the other, economic aspect of mineralogy, that which, when combined with historical questions, developed into geology. It subsidized famous expeditions in the eighteenth century for the obvious reason that discovery might uncover new resources. In the infancy of the science, the province of Auvergne together with the neighboring Vivarais and Velay, formed a kind of nursery. They made a craggy laboratory in which theoretical geology cut its teeth on the problem of explaining the resemblance of the *puys* to volcanic cones and the origin of the dramatic columnar structures of basalt that occur roundabout. Guettard and Desmarest, both acting for government, were followed by Faujas de Saint-Fond, Giraud-Soulavie, and Déodat de Dolomieu, noblemen or clergymen.

Guettard we have already met, taking the neophyte Lavoisier on a geological mapmaking tour into Alsace, Lorraine, and the Franche-Comté in

[313] Haüy published the definitive version of his theory in the last year of his life, *Traité de cristallographie, suivi d'une application des principes de cette science à la détermination des espèces minérales*, 2 vols. plus atlas (Paris, 1822). For his place in the history of crystallography, see Burke (1966), Marx (1825), Metzger (1918).

[314] Mauskopf (1970a).

[315] *Ibid.*, 200, quoting from John Dalton in Nicholson's Journal, "Inquiries concerning the signification of the word 'particle' as used by modern chemical writers," *A Journal of Natural Philosophy, Chemistry and the Arts* 28 (1811), 85. See also Mauskopf (1969a, 1969b, 1970b, and 1976, Sections 1-2).

1767.[316] Henri Bertin, formerly controller-general and then in a lesser ministry of economic affairs, had commissioned their survey. They imagined a work of some two hundred sheets. Never finished, the project had developed out of Guettard's invention of a mineralogical map showing the occurrence of rocks and minerals in France and abroad.[317] Guettard was astringent in temperament and tended to be crusty about theories and large views. He it was, however, who started the idea that the *puys* of Auvergne are extinct volcanos. The notion came to him during a journey that he made in 1751 in company with that inevitable patron of enlightened views, Lamoignon de Malesherbes, when they noticed that the stone used for building throughout the province was largely of volcanic origin.[318]

Nicolas Desmarest we have also met, verifying the fineness of the fiber spun from Daubenton's home-grown fleece. He was then inspector of manufactures for the generality of Châlons.[319] We might, indeed, have encountered him still earlier, for he had served under Turgot in a similar capacity at Limoges.[320] Son of an obscure schoolmaster, Desmarest was educated by the Oratorians at Troyes and made an unspectacular career in public administration in the area where science, physical geography, and technology overlapped. Diderot called on him to contribute the articles "Fontaine" and "Géographie physique" to the *Encyclopédie*,[321] and in 1757

[316] Above, Chapter I, Section 6.

[317] Guettard's practice was to identify the occurrence of minerals by means of the conventional chemical symbols and to divide France into regions or *bandes* according to mineralogical characteristics of the prevailing formations. His earliest essay along these lines was "Mémoire et carte minéralogique sur la nature & la situation des terreins qui traversent la France & l'Angleterre," MARS (1746/51), 363-392. Guettard and Lavoisier completed sixteen quadrangles in their collaboration before handing the project on to Antoine Monnet in 1777, who published them in his *Atlas et description minéralogiques de la France* (1780). On this work, see Rappaport (1969), and for further bibliographical career concerning Guettard, see Rappaport's article, DSB 5, 577-579.

[318] "Mémoire sur quelques montagnes de la France qui ont été des volcans," MARS (1752/56), 27-59; for discussions of Guettard, and the importance of Auvergne in the history of geology, see Geikie (1905) and de Beer (1962).

[319] Above, 166-167.

[320] Below, Chapter VI, Section 3: His son's manuscript notes on his life are in the Bibliothèque de l'Institut de France, Fonds Cuvier, MS 3199; a collection of his own correspondence is preserved in the BN, Fonds français, nouvelles acquisitions, MS 803 and MS 10359, the latter consisting of letters from Turgot. For his importance in the history of geology, see Geikie (1905) and Taylor (1969), and for a brief life and full bibliography, see Taylor's article in the DSB 4, 70-73.

[321] *Encyclopédie* 7 (1757), 80-101, 613-626. Desmarest had first made reputation with an essay on a one-time land bridge between England and France that won a prize in 1751 from the Academy of Amiens, *Dissertation sur l'ancienne jonction de l'Angleterre et la France* (Amiens, 1753). He published a treatise on earthquakes soon after: *Conjectures physico-mécaniques sur la propagation des secousses dans les tremblemens de terre* (1756); but his most widely

Trudaine singled him out to specialize in woolens for the Bureau of Commerce. He later made himself equally expert in agronomy and the technology of paper.[322] Elected to the Academy in 1771, a plain and useful man of forty-six, he too brought to its company the quality of skepticism about theories that is often enjoined by scientists of somewhat ordinary intellect. Desmarest conveyed that attitude less corrosively than Guettard—and proved no more capable of containing his own imagination than his older colleague had been when in 1763 official travels took him in his turn south of Clermont-Ferrand into central Auvergne. Accepting from Guettard the notion that the *puys* had once been volcanic cones, Desmarest was further astonished by the great basalt prisms of the province, and took them for decisive evidence of volcanic origins.[323] He put himself in touch with Diderot again and published a fine pair of engravings representing the structures in the sixth volume of plates accompanying the *Encyclopédie*.[324]

Desmarest was sometimes accompanied in his journeys by the duc de La Rochefoucauld, even as Guettard had been by Malesherbes, for geology in its early days appealed not only to officials and economists but to noblemen and gentlemen with their feeling for terrain. The findings in Auvergne attracted the interest of a trio of investigators originally of this latter sort. About the first two, Faujas and Giraud-Soulavie, we can be brief. Barthélemy Faujas de Saint-Fond came of a family well placed in the judicial nobility of Dauphiny. Like Lacepède, he made himself and his enthusiasm for natural history known to an aging Buffon, who took a liking to him and in 1787 found a small place for him in the Jardin du roi.[325] For Faujas had a certain charm. Arthur Young was well received in his property at l'Oriol, near Montélimar, and recorded that the "liveliness,

---

read work was a translation and edition of Francis Hauksbee, *Expériences physico-mécaniques* (1754) with lengthy notes and additions.

[322] His *Art de la papéterie* (1789) is an example of the eighteenth-century technological treatise at its best and clearest. Desmarest also wrote the article on papermaking for the *Encyclopédie méthodique, Arts et métiers mécaniques* 5 (1788), 463-592, having contributed two lengthy memoirs to the Academy on the methods practiced in Holland, MARS (1771/74), 335-364; (1774/78), 599-687. For Desmarest's relations with the paper industry, see, below, Chapter VI, Section 3.

[323] Guettard disagreed. Basalt looked crystalline to him, and he thought the structures were monstrous precipitates from some former sea. Desmarest developed the detail of his argument only after a delay of ten years: "Mémoire sur l'origine & la nature de basalte . . ." MARS (1771/74), 705-775; "Mémoire sur le basalt," MARS (1773/77), 599-670.

[324] Recueil de planches . . . , 6 (1768), Pl. VII, VIII, section on "Histoire naturelle; règne minérale volcans."

[325] Laissus, "Le Jardin du Roi," in Taton (1964), 340. On the career of Faujas, see Archibald Geikie's biographical memoir in his edition, 2 vols. (Glasgow, 1907) of the 1799 translation of Faujas's *Voyage en Angleterre, en Ecosse et aux Iles Hébrides*, 2 vols. (1797). See also Freycinet (1820); Challinor (1954), 126-129.

vivacity, *phlogiston* of his character, do not run into pertness, foppery, or affectation."[326] Faujas had published *Recherches sur les volcans éteints du Vivarais et Velay* in 1778. How well he then knew the work of Desmarest is unclear. His book is an immense folio, magnificent in its illustrations and in its author's disregard for any principle of selection more systematic than whim. It quite indiscriminately recounts the history of the activity of Etna, Stromboli, and Vesuvius (known to the author from books); describes the basaltic formations of Vivarais and Velay; deplores the horrors of villainous inns in remote regions where the traveler risks falling victim to the blood feuds that preoccupy the peasants; prints excerpts from correspondence with Dolomieu and with Sir William ("milord") Hamilton in Naples; and celebrates the appearance and cult of the Black Virgin of Notre-Dame du Puy.

The abbé Jean-Louis de Giraud-Soulavie had traversed the same provinces a few years earlier, having been born at Largentière in 1752 and ordained in 1776. He is not to be ignored, for a revolutionary future awaited him, but historians of geology have taken more seriously than did his contemporaries his chronology of volcanic events and his attempt to establish a stratigraphical succession on the basis of extinct fossil forms. In a letter to Faujas, Buffon called him "a school-boy . . . who writes in the manner of a master."[327]

Of the third of these well-born geologists, Déodat de Dolomieu, as he called himself—his full style was Dieudonné-Sylvain-Guy-Tancrède de Gratet de Dolomieu—it is safe to observe that, in giving his patronymic to a mineral, dolomite, and to an Alpine range where it predominates, he was unique among the Knights of the Military and Sovereign Order of Malta, whose members, younger sons as a rule, were barred by vows of celibacy from perpetuating their names in the normal manner. His father, the marquis de Dolomieu, had entered him in the order at the age of two.[328] When he was eighteen, he killed a man in a duel and was in consequence imprisoned by the order, to be released after nine months when the pope interceded with the grand-master at the request of Louis XV.[329]

[326] Arthur Young, *Travels in France* (Everyman edition, London, 1927), 202. The date was 23 August 1789.

[327] Buffon to Faujas de Saint-Fond, 3 October 1781, in Nadault de Buffon (1860) 2, 109. Giraud-Soulavie's main work in geology is the very discursive *Histoire naturelle de la France méridionale*, 8 vols. (1780-1784). The main arguments are summarized in *Géographie de la nature* (1780). Mazon (1893) is a not very scholarly biography.

[328] Alfred Lacroix collected and published Dolomieu's correspondence and certain other papers in 1921, prefacing the edition with a carefully annotated biographical notice. The DSB article is by Kenneth L. Taylor, *4*, 149-153. It contains a full bibliography. On the naming of dolomite, see Lacroix (1921) *1*, viii.

[329] Lacroix (1921) publishes his petition to Cardinal Torrigiani, 31 October 1768 *1*, 65. See also *ibid.*, xvi, lxxi-lxxii, n. 2.

A Dauphinois, Dolomieu was a countryman of Faujas, whose correspondence and discursive book drew his interest to studying basalt and volcanos while he was aide to Prince Camille de Rohan on a diplomatic mission to Portugal.[330] He was then a lieutenant in the carabiniers. Earlier on, between 1771 and 1774, he had relieved the tedium of garrison life in Metz by studying chemistry and physics with a military apothecary there, one Thyrion, who gave public courses in science—Pilatre de Rozier was another of his pupils—and with whose niece Dolomieu had an affair.[331] He also met and became fast friends with La Rochefoucauld, then in Metz with his regiment. La Rochefoucauld encouraged him to take up mineralogy seriously, put him in touch with Desmarest and Daubenton, and in 1778 arranged his election to a corresponding membership in the Academy of Science. He was assigned to Daubenton, who thus became his official mentor.[332]

In Paris during the long intervals of freedom from duty that redeemed eighteenth-century military life, Dolomieu frequented the salon of La Rochefoucauld's mother, the duchesse d'Enville, and there mingled with the world of fashionable science, departing on occasion for mineralogical excursions. What with his villa in Malta and the intricate connections between the order, the Bourbon court in Naples, and the papal court in Rome, and given, too, a certain Mediterranean quality in his own temperament, Dolomieu felt drawn to Italy, its volcanos, its geography, its Etruscan, Greek, and Roman antiquities. His first book, *Voyage aux Iles Lipares*, published in 1783, is little more than a traveler's account of the Aeolian Islands—his "scientific novitiate" in the judgment of "tout Paris," so observed a comrade in a letter that also confided that "la marquise de M" was still "burning for him" and was impatiently awaiting his return, having been unfaithful only on three occasions, once with a seventeen-year-old child, once with another Dauphinois, and once with a not-too-elderly bishop.[333] Dolomieu published a second book in Rome in 1784, a seventy-page memoir on the effects of the earthquakes that had ravaged Calabria over a six-month period in the previous year. It goes little beyond topography and is more interesting for its account of the terror, rapacity, and sporadic heroism of the populace than for geology.[334]

Thereafter, Dolomieu did not publish anything significant for over four

[330] Faujas's *Recherches sur les volcans éteints du Vivarais et Velay* (Grenoble, 1778), prints excerpts from his early correspondence with Dolomieu, who later thought him a lightweight: See Lacroix (1921), Dolomieu's sketches of his principal friends, *1*, 49; see also Dolomieu to Picot de Lapeyrouse, 31 October 1787, *ibid.*, 190.

[331] On Thyrion, see Lacroix (1921) *1*, lxxii, n. 3.

[332] Dolomieu to Daubenton, 18 July 1776, *ibid.*, 80-83, see also nn. 1 and 2.

[333] Laqueuille to Dolomieu, 18 April 1785, *ibid.*, 139-141, n. 1.

[334] *Mémoire sur les tremblemens de terre de la Calabre pendant l'année 1783* (Rome, 1784).

years, and when he did, the work was a very different and more substantial one.[335] Indeed, there were two works, bound together, one a lithological account of the Pontine islands, a minuscule volcanic archipelago in the Tyrrhenian Sea off the Gulf of Gaeta, and the other a thorough treatise on the volcanic mineralogy of Mount Etna, in repose and in action (for in 1787 a splendid eruption had ended six years of quiescence). His earlier publications could be mistaken for the recreations of a peripatetic nobleman, like the writings of Faujas or Sir William Hamilton on the same regions. In contrast, Dolomieu's new book was geology, an analysis based on lithology.[336] His treatment of Etna bespeaks a detailed, analytic knowledge of minerals and a disciplined preoccupation with their classification. Ten years later, a Maltese vengeance landed him again in prison amid the Napoleonic wars, a Neapolitan prison where he lost his health and wrote down the reflections that had guided him into command of his science, and that he now commended to students.[337] He could not know while writing it that he was even then being elected to succeed Daubenton in the chair of mineralogy at the Muséum. He had to scrawl the draft of this *Philosophie minéralogique* onto the ample margins and end-papers of Faujas's *Minéralogie des volcans*, the only paper he had with him.[338] Reading it now, one of the classic methodological treatises that marked the emergence of new disciplines at the turn of the century, we can judge what it was he had been learning in lithology from his own teacher, Daubenton, in the Cabinet du roi; what in crystallography from his fellow pupil, Haüy, in the laboratory; and what in geology from Etna on her sulfurous slopes and at her smoking summit.

Only in the 1790s was he widely recognized to be the foremost of volcanologists, but we must not imagine him in the previous decade preparing that eminence by gradually deserting the world of fashion for the world of science. Quite the contrary; all the while he was increasing his competence, he was also leading the life of the cosmopolitan gallant, now in Paris, now in Dolomieu, now in Lisbon, now in Toulouse, now in Florence, ever and again in Malta, and everywhere taking privilege so much for granted that he no more thought to be its partisan than if it had been the air he breathed. It was in Malta that he had his property and housed

[335] *Mémoire sur les Iles Ponces, & Catalogue raisonné des produits de l'Etna; pour servir à l'histoire des volcans: suivis de la description de l'éruption de l'Etna, du mois de juillet 1787* (Paris, 1788).

[336] On the importance he attached to lithology, see Lacroix (1921) *1*, il; and Dolomieu to Gioeni, 2 April 1790, *ibid.*, 235.

[337] For a brief discussion of the importance of the work in the history of mineralogy, see Mauskopf (1976), 19-20.

[338] *Sur la philosophie minéralogique, et sur l'espèce minéralogique* (Paris, 1801). For the circumstances of its composition, see the preface, 3-9, together with his "Journal de captivité," in Lacroix (1921) *1*, 31-44, and *ibid.*, xxxiii-xliv.

his collections. There he became lieutenant-maréchal in 1781, in effect governor of the town, and there he founded an observatory and imported an astronomer, one Dangos, a pupil of Lalande.[339] The order to which Dolomieu belonged resembled nothing so much as a feudal and clerical hermit crab that centuries earlier had crawled across the Mediterranean from its crusaders' castles in the Holy Land into the old shell of the island fortress. Its labyrinthine politics now combined the spitefulness of the cloister, the brutality of the barracks, and the litigiousness of the pettiest of courts.[340] Dolomieu was forever contending against the obscure despotism of the grand-master, Emmanuel de Rohan, in defense of the equally obscure rights of his *Langue* of Auvergne.[341] Apparently he lost. In any case, Rome became increasingly the city of his choice. There he was much in the company of Monsignor Borgia, of Milady Knight and her daughter, and of the Contessa Piccolomini. He met, too, Mr. Wedgwood and Mr. Goethe, the latter "well known in Germany for writing full of grace and sensibility."[342] Angelica Kauffman painted Dolomieu's portrait in 1789, when he was thirty-nine.[343] It has the nervy quality of a greyhound in the lean and acquiline features, the high cheekbones, the strong nose, the confident eyes. Tall in stature, assured in bearing, he was one whom people noticed. "That is a Man," they would murmur when he passed—so recalled Bruun-Neergaard, a naturalist who accompanied Dolomieu on later travels.[344]

Only in the 1790s did geology emerge as a distinct science, building on a cosmogonical and mineralogical groundwork that dated back to Buffon and his theory of the earth, to Daubenton and his disciples in their youth, in short to the last years of the Jardin du roi. In those years a revival of activity in teaching also began. Although Buffon himself had never been much interested in that function, in 1787 he did order built the amphitheater that still stands, though it was not completed before he died. Chemistry and also Desfontaines's lectures on botany[345] were then going far better than were pharmacy and anatomy, the quasi-medical subjects that the Jardin had originally been founded to provide.

It is possible that the anatomical lectures may have been some help to medical students in search of the practical instruction that the faculty,

[339] Dolomieu to Lalande, 9 June 1782, Lacroix (1921) *1*, 91-96.

[340] For various works concerning the order of Malta, see *ibid.*, 65.

[341] See, for example, his letters to Dufay, 14 November 1782, *ibid.*, 100-102, and to Picot de Lapeyrouse, 10 April 1783, 3 June 1783, 30 September 1783, *ibid.*, 103-109, 112-115.

[342] Dolomieu to Fréderic Munter, 30 November 1787, *ibid.*, 191-193.

[343] *Ibid.*, frontispiece.

[344] *Ibid.*, lxvii. T. C. Bruun-Neergaard was with Dolomieu on his last geological tour and published the account in *Journal du dernier voyage du C^{en} Dolomieu dans les Alpes* (1802).

[345] Above, 156.

stiff in the rigor mortis of its Latinity, still failed to offer. Delivering them, however, had become a sideline for successful doctors who held the professorship: Antoine Petit from 1769 until 1794 (though he ceased lecturing in 1775 when he fell sick), and Antoine Portal from 1778 right through every change in politics and organization until his death in 1832 at the age of ninety.[346] They were supposed to be seconded by a lecture-demonstrator qualified in surgery, a post held successively for three generations by Antoine, Jean-Claude, and Antoine-Louis Mertrud, none of whom won much notice, and the last of whom gave ill-attended demonstrations of animal anatomy instead of assisting the professor. When Petit's health failed, he engaged Vicq d'Azyr to give the lectures. This was just the time when Turgot was also turning to Vicq d'Azyr to investigate the cattle epidemic.[347] So much the most promising was he among younger men in the medical sciences that in retrospect he almost seems their Lavoisier manqué. His expectation of succeeding to the chair was disappointed when Buffon instead named Portal, and he went on to throw his energies into the Royal Society of Medicine. Portal had powerful friends. He already held a professorship of medicine in the Collège de France and was titular physician to the king's brother, the comte de Provence. His major works have gone largely unread: an enormous chronicle of anatomy beginning in 1300 B.C., and an only slightly smaller textbook of the subject from which he omitted illustrations in the interest of economy.[348] Petit and he were not the men to continue the tradition maintained by the two Du Verneys and Winslow down to the middle of the century, when the anatomy lectures had drawn a fashionable and literate audience.

In chemistry, too, the public lectures had declined in popularity before the 1780s, the famous Rouelle having retired in 1768. Unlike anatomy, however, the subject itself may have gained scientific authority during the interval, for it was in the very capable hands of Macquer,[349] who com-

[346] On Petit and Portal, see Barritault (1940).

[347] Above, Chapter I, Section 4.

[348] *Histoire de l'anatomie et de la chirurgie*, 5 vols. (1770); *Cours d'anatomie médicale*, 5 vols. (1803).

[349] The subject matter taught by a member of the staff did not always correspond to the designation of his post. G.-F. Rouelle was succeeded as "Démonstrateur en chimie" by his son, Hilaire-Martin, who died in 1779 and was followed by A.-L. Brongniart (the first in another scientific dynasty). Macquer had the place of "Démonstrateur et opérateur pharmaceutique," in which he had been preceded by Etienne-François Geoffroy and Louis Lémery. See Y. Laissus "Le Jardin du roi" in Taton (1964), 319-341, for a tabulation of the personnel. On the career of Macquer, and the teaching of chemistry at the Jardin, see Contant (1952), Smeaton in the DSB 8, 618-624; the Eloge by Vicq d'Azyr, HSRM (1782-83/87), 69-94; and W.-J. Ahlers, "Un Chimiste du XVIII siècle, Pierre Joseph Macquer" (1969), a thèse de troisième cycle, Université de Paris, Faculté des Lettres et Sciences Humaines.

bined his professorship with consulting in two royal enterprises, the dye shops in the Gobelins tapestry works and the porcelain factory at Sèvres. When he died in 1784, applicants for the three places crowded upon the ministry, which wisely determined to divide them. The Gobelins went to Berthollet, the porcelains to Jean Darcet, and the professorship to Fourcroy, the last named being admirably suited to reviving Rouelle's tradition of magnetic, not to say hypnotic, chemical teaching in the Jardin des plantes.

Antoine-François de Fourcroy, doctor of the Faculty of Medicine of Paris, tried hard all his mature life to be a leading scientist.[350] He worked excessively. He overcame obstacles. He avoided pitfalls. He published prolifically. Outwardly he succeeded, one among many men of a generation traversing troubled times whose conduct in certain episodes often needed explanation, not to say extenuation, then and afterward.[351] It needs to be explained that, though his father was an unprosperous apothecary in the retinue of the duc d'Orléans, the family really was entitled to the particle since it was descended from a fifteenth-century Robert de Fourcroy, squire and man-at-arms from the Boulonnais, who acquired a seigneury along the Oise under Charles VII, and further that the chemist's distant cousin, Fourcroy de Ramecourt, of a more fortunate though junior branch, commanded the Royal Engineering Corps in the last years of the old regime. It needs to be explained that, though unhappily he did pretend to the title of docteur régent without having been voted it by the faculty, and though he did then join the Society of Medicine after having promised the faculty before receiving his degree that he would take no part in that reformist body, still all the regulations and obstructions of the University were such that a young man with a career to make had to find some way around them. It needs to be explained that, though it was indeed his wife's dowry that set him up in laboratory equipment for his public courses, it was only after nineteen years that he divorced her to marry a younger and more cosmopolitan lady, the widow of the famous architect, Charles de Wailly. It needs to be explained that, though in the days of his prominence he did work jointly with Nicolas-Louis Vauquelin,

[350] Two modern studies of Fourcroy have appeared. Smeaton (1962) is the fuller scientifically and is reliable on his public life. Kersaint (1966) concentrates on minute documentation and rebuts aspersions. It is still necessary to consult the Eloge by Cuvier, which was based upon notes furnished by André Laugier, Fourcroy's cousin and intimate friend. The manuscript is in the Bibliothèque de l'Institut de France, fonds Cuvier, 191, pièce 2 (See Gillispie [1956b] and Kersaint [1957]). Also to be consulted is an éloge by another intimate friend, A.-M.-F.-J. Palisot de Beauvois, *Eloge historique de M. Fourcroy* (Paris, n.d.)

[351] The biographical details in the paragraph are drawn from Kersaint (1966), 12-26, with supporting documentation in his notes and appendices. See also Smeaton (1962), 1-19.

whom he had plucked from his cousin's apothecary shop to be his laboratory assistant, the malicious remark that "Vauquelin worked and Fourcroy signed" was a canard,[352] and that Fourcroy really did contribute a wider, more sophisticated fund of chemical knowledge to a collaboration wherein the younger man was the defter and more original in technique. Moreover, Fourcroy's two elder sisters, also separated from their spouses, kept house for his inarticulate protégé and looked after him.[353] All these circumstances could be explained, for Fourcroy really did win loyal friends,[354] but they do need to be.

Fourcroy explained things very well himself. Madame Roland was at the opening of his first course in 1784.[355] He came fully into his own in the new amphitheater, however. Cuvier heard his lectures there in the 1790s and considered their quality to be an instance justifying the comparison of Paris with Athens in their complements of public and scientific spirit. In such eloquence (said Cuvier in his éloge)

> Plato and Demosthenes seemed reunited, and it would have taken the one or the other to convey an idea of it. So coherent in the organization, so rich in the elocution, so noble, just and elegant in the language, it was as if the terminology had been matured only after the most careful deliberation; but then, so lively, so rapid, and so fresh was the delivery that by contrast it was as if the words themselves were inspirations of the moment. A silvery tone, a modulated pitch, a voice lending itself to every movement and reaching all corners of the vast auditorium—nature had endowed him with every gift. Sometimes his discourse flowed smoothly and majestically, and he would then be holding his audience by the splendor of his images and the formality of his style. Sometimes, varying the emphasis, he would gradually shift over to a familiar and conversational tone, alerting attention by occasional sallies of wit. You would have seen hundreds in his audience, drawn from all classes of society and every nationality, pressed elbow to elbow hour upon hour, almost afraid to breathe, their eyes fixed on his and hanging on his words (*pendent ab ore loquentis*, as the poet has it).[356]

[352] Kersaint (1966), 165, attributing the remark to L. J. Simon. On the career of Vauquelin, see Smeaton's article in the DSB 5, 89-93, which cites the literature.

[353] In 1798 Thomas Bugge visited the ménage, then installed in the rue de l'Université in the apartment at the Ecole des mines to which Vauquelin was entitled as professor. They showed him the laboratory, and he considered that the saying "that learned females are not always the handsomest and neatest was verified in the persons of both these chemical ladies." Crosland (1969), 48.

[354] Notably Laugier and Palisot de Beauvois, the former a chemist and kinsman, the latter a botanist (see n. 349 above).

[355] Madame Roland to Roland, 21 April 1784, Perroud (1900) *1*, 351-352.

[356] Cuvier (1819-1827), *2*, 16-18.

Fourcroy discovered these gifts in himself while assisting in the laboratory of Jean-Baptiste Bucquet, professor in the Faculty of Medicine. Bucquet's early work on gas chemistry was important to Lavoisier, and he for his part was the first to teach Lavoisier's theory of combustion publicly, incorporating it in his lectures for 1779.[357] Bucquet might have become a great chemist. He had the notion of repeating all the fundamental experiments in order to rid chemistry of error and relate it to the neighboring sciences of nature. He died of a torturing rectal cancer at the age of thirty-four before carrying out his expansive idea, which was very similar to the program of research that Lavoisier set himself in the prophetic memorandum opening his laboratory register in 1772-1773.[358] Fourcroy got his start teaching in 1778. On a day when Bucquet felt too ill to meet his class the next morning, he asked Fourcroy to prepare and give the lecture. Like an understudy, Fourcroy spent the night getting up the part of resins and held the audience for two hours, a youthful virtuoso.[359]

In other circumstances Fourcroy might have gone on the stage. While a day-pupil at the Collège d'Harcourt, he liked memorizing passages from poets and dramatists and could always amuse friends by imitating famous actors. His father's affairs worsened, and he left school at fifteen to become a clerk. Fortunately he was known to Vicq d'Azyr there in the small world of the Latin Quarter. Some said that Vicq d'Azyr, whose wife (Daubenton's niece) had died young, was courting his sister.[360] In any case he rescued the boy from the writing master's and enlisted the aid of colleagues in the Royal Society of Medicine to put Fourcroy through medical training. Association with Bucquet opened the way to science, and he never practiced medicine. Pupils paid fees to attend classes in Bucquet's private laboratory and received a more detailed, more practical training, mainly in pharmacy, than they could get by hearing lectures in the faculty, or for that matter in the Jardin du roi. Bucquet died in January 1780. Fourcroy's marriage in June provided the capital to buy his late master's apparatus, and he opened his own laboratory in a shop on the north side of the Parvis, or plaza, facing Notre Dame.[361]

His industry was prodigious. He taught three private classes every day. Beginning in 1784, he had also his course at the Jardin du roi, which had to be very broad and very different since the schedule called for only

[357] Bucquet published two of his courses: *Introduction à l'étude des corps naturels tirés du règne végétal*, 2 vols. (1773). For his collaboration with Lavoisier, see E. McDonald (1966). McDonald has also written the article in the DSB *2*, 572-573, and "Jean-Baptiste Bucquet (1746-1770)—His Life and Work," a M.Sc. dissertation (University of London, 1965). Fourcroy himself published an "Eloge de M. Bucquet," *Observations sur la physique* (Rozier's journal) *15* (1780), 257-264, as did Vicq d'Azyr, HSRM (1779/82), 74-93.

[358] Above, Chapter I, Section 6; HSRM (1779/82), 79.

[359] Kersaint (1966), 16.

[360] *Ibid.*, 239, n. 1 to ch. 2.

[361] *Ibid.*, 17, Pl. IV.

twenty lectures a year, each of them a performance. What was more, in 1783 Fourcroy had been named, along with Daubenton and Vicq d'Azyr, to one of three chairs instituted at the Royal Veterinary School at Alfort. With all that, he still had energy to take on a professorship in 1781 at the Lycée de Monsieur, patronized by the comte de Provence, in the rue de Valois hard by the Palais Royal. In 1785, it was sold to subscribers.[362] Fourcroy joined under the new management. He gave two courses annually, in 1788 on chemistry and on mineralogy, and in 1789 on animal and vegetable chemistry and on zoology and botany. What with the stringencies of his childhood, money mattered to him—he even did hack reading for the censorship—and he would occasionally complain that scientists were badly paid.[363] It is natural that he should have felt some resentment at the system of society, considering how far his family had sunk in it.[364] Not that he aspired to wealth on anything like Lavoisier's scale: his ambitions were more limited in all respects, and he saw the possibilities for science in a context of service rather than authority.

He did a lot of chemistry, much of it really useful.[365] His doctoral thesis had discussed the therapeutic value of breathing various new gases,[366] and his later work had a medical emphasis. In this respect, as in the organization of his courses, he followed Bucquet's lead, and freely acknowledged doing so.[367] He studied marsh gas and other inflammable vapors and exhibited the danger of administering oxygen to tuberculosis patients, from which finding he unfortunately concluded that it was unwise to send them into the fresh air of the country or to the mountains where (he thought) the atmosphere contained more oxygen than in a closed room in the city.[368] Mineral waters interested him. The town of Enghien just north of Paris got its start as a spa from his examination of its springs.[369]

[362] On the Lycée, see Smeaton (1955a); and below, Chapter III, Section 1.

[363] Smeaton (1962), 11, quoting a doctor, F. Lanthenas, a friend of the Rolands who met Fourcroy and wrote to Roland (2 May 1784) that his new acquaintance "crie étrangement que les savants ne soient pas mieux payés." He owed his post as one of the royal censors to La Rochefoucauld (*ibid.*, 33) with whom he also managed to collaborate on a memoir, "Examen d'un sable vert cuivreux de Pérou," MARS (1786/87), 465-473.

[364] Such was Cuvier's observation, "Eloge du Fourcroy," Cuvier (1819-1827) 2, 12-13.

[365] Smeaton (1962) recounts his work in part II of his book, and summarizes its importance in his article in the DSB 5, 89-93. Both Smeaton (1962) and Kersaint (1966) give full bibliographies.

[366] "De utilitate effluviorum elasticorum gas dictorum ad tuendam sanitatem." A copy is in the Bibliothèque de l'Arsenal.

[367] A.-F. de Fourcroy, *Leçons élémentaires d'histoire naturelle et de chimie*, 2 vols. (1782) *1*, ii-v.

[368] Smeaton (1962), 136-137, and see Fourcroy, "Extrait d'un mémoire sur les propriétés médicinales de l'air vital," *Annales de chimie* 4 (1790), 83-93.

[369] *Analyse chimique de l'eau sulfureuse d'Enghien, pour servir à l'histoire des eaux sulfureuses en général* (1788); see Smeaton (1962), 19-22.

His most distinctive line of research, however, was analysis of animal and vegetable matter. Fourcroy could properly be called a pioneer in anatomical chemistry, to others a somewhat repulsive field. He made a thorough investigation of the chemical constitution of gall stones and urinary calculi, drawing samples from the ample collection assembled by the Society of Medicine.[370] It was not inappropriate, therefore, that having failed of election to the Academy in February 1784, when Macquer's death created a vacancy in chemistry, he should have made a second try in April for the section of anatomy. Portal was preferred, and Fourcroy had to await the reorganization arranged by Lavoisier in 1785.[371]

Then or soon after, he rallied to Lavoisier's theory, having learned about it when assisting Bucquet. The first of many editions of his lectures appeared in 1782. He proposed throughout the discussion to compare Stahl's account with the "pneumatic doctrine of several modern chemists," neither rejecting the one nor adopting the other and (like a good historian) reporting the facts, which alone are truly science.[372] Four years later he was collaborating with Berthollet, Guyton de Morveau, Laplace, Monge, and Lavoisier himself in refuting Kirwan's *Essay on Phlogiston* and in framing the new nomenclature, one of that entourage of whom Lavoisier later observed irritably that attribution of the new theory to "French chemists" was wrong—it was his.[373]

Here, then, was the personnel of the Jardin des plantes at the time of Buffon's death: Daubenton, Desfontaines, Faujas de Saint-Fond, Fourcroy, A.-L. de Jussieu, Lacepède, Mertrud, Portal, and Thouin, to whom must be added A.-L. Brongniart, a pharmacist, and Gérard Van Spaendonck, an artist. Brongniart, a distant cousin of Fourcroy, was demonstrator in chemistry and performed experiments to accompany the lectures.[374] He is not to be confused with his nephew, Alexandre, famous in geology in the next generation, or with his brother, Théodore, the architect who designed the Bourse and laid out the cemetery of Père Lachaise. Van Spaendonck, officially royal miniaturist at the Jardin des plantes, was successor to Magdeleine Basseporte, the only woman ever to have been on

[370] Smeaton (1962), 147-153; see, e.g., "Examen chimique de la substance feuilletée et cristalline contenue dans les calculs biliaires," *Annales de chimie* (1789) *3*, 242-252; "Expériences faites sur les matières animales," *ibid.* (1790) 7, 146-193.

[371] Kersaint (1966), 24-25.

[372] *Leçons élémentaires 1*, xxiii-xxiv.

[373] For the Lavoisier remark, see his posthumous *Mémoires de chimie* (n.d.) 2, 87, and Smeaton (1955b), 316, n. 61. It is a famous episode in the history of chemistry that Madame Lavoisier translated Richard Kirwan's book to serve as a straw man for notes refuting the thesis by Guyton, Lavoisier, Laplace, Monge, Berthollet, and Fourcroy (*Essai sur le phlogistique*, 1788). *Méthode de nomenclature chimique* (1787) appeared with the names of Lavoisier, Berthollet, Guyton, and Fourcroy. See Crosland (1962).

[374] On Antoine-Louis Brongniart, see Launay (1940), 11-15.

the staff. Increasingly, however, its members preferred employing the famous flower painter, Pierre-Joseph Redouté.[375] Lamarck was still only a correspondent officially, and neither Dolomieu nor Haüy yet had appointments.

Buffon died in lordly fashion with wages in arrears and assorted tradesmen and contractors unpaid.[376] Pulling wires on his deathbed, he sought to undo the arrangement of 1771 by which d'Angiviller was to have succeeded him, and dispatched Faujas to buy up the immediate succession for his son, now a major in the Angoumois regiment and (it will be recalled) second in line for the intendancy. The maneuver failed, though d'Angiviller did not take the post after all. With dubious legality, he instead transferred his right to his brother, the marquis de la Billarderie, head of the Flahault family, who had never had any connections with science, even administratively. Prudence may have precluded d'Angiviller's accumulating further perquisites, and there is also reason to suspect that a private understanding between La Billarderie and Condorcet entered into the change in plans.[377] If so, the permanent secretary of the Academy would himself have become La Billarderie's successor, thus superseding the young Buffon's interest in the higher interest of integrating the Jardin with the Academy.

Whatever the truth may have been, the staff was ignorant of it and was left with the immediate anxiety of how to steer the new intendant, an elderly nobleman innocent of any knowledge of its affairs, into ways consonant with its increasingly professional ambitions. Only Daubenton had the age and reputation to be spokesman. Thouin was the moving spirit, however, the gardener's boy whose whole life had been in, and indeed was, the Jardin des plantes. A memoir that he clearly intended for the education of the new intendant, and beyond him the ministry, tells how members of the staff themselves saw their mission. The positions of professor and demonstrator are not mere places (Thouin explained); they are "charges." The incumbents take an oath to carry out the duties; they register their brevets or letters of appointment at the Chambre des comptes; they observe the regulations laid down by the crown through the Ministry of the Royal Household; and (most significantly)

> they make no distinctions among themselves except those arising from merit and seniority. Since their work has no other purpose than the progress of science and public utility, each of them tries only to deserve in his own area the esteem of scientists and the affectionate regard of his fellow citizens.[378]

[375] Y. Laissus, "Le Jardin du Roi," in Taton (1964), 334.
[376] Hamy (1893), 75-76. [377] *Ibid.*, 5-12, 24.
[378] "Mémoire sur le Jardin du Roi," BMHN, MSS 1934, XXX, fol. 3.

Buffon himself in his last years (so says an accompanying memoir, not altogether plausibly) had thought certain reforms needed in order to eliminate arbitrary usages.[379] As example, Thouin cited the practice by which each member of the staff drew his stipend directly from the Treasury, the amounts varying capriciously. If all the ad hoc monies for maintenance, purchase, and salaries and wages were to be consolidated in a single fund under an annual budget, as in a rational administration they ought to be, then staff members could be paid on an equitable basis proportionally to their merit, duties, and seniority. By the same token, a uniform curriculum ought to be instituted in order that all the lecture courses might open at the same time of the year and treat their subjects thoroughly enough to serve serious students. Except in botany, the teaching (in Thouin's view) was haphazard or superficial or both.

For the rest, Thouin explained, the Jardin des plantes of the 1780s combined in its operation the functions of a public park, a botanical institute, a nursery for arboriculture, and a laboratory of acclimatization. And honorable to the capital though it was to attract foreigners together with French citizens to the study of the vegetable kingdom; profitable to the country though it was to provide landowners with the techniques and even the stock for reforestation amid a "property revolution" that was denuding the landscape; valuable to the public though it was to accustom the fruit, nut, and spice trees of Asia and America to burgeoning in France; still it was the popular aspect of the Jardin des plantes that Thouin discussed first and at greatest length. In a huge city like Paris, he pointed out, people need open spaces to be preserved where they can breathe an air that is free and pure. Such areas do exist within reach of those who live in the eastern, northern, and western sections of the capital. But to the south, there are only the Luxembourg and the Jardin des plantes. The Jardin des plantes presents the great advantage that it is accessible to a quarter of the population of the city. "Remote from the world of so-called fashion it is particularly convenient for modest citizens who have neither the means nor the desire to enhance the display of elegance in the other public gardens. But it is particularly important for hard-working people imprisoned in routine and sedentary jobs who need a chance for physical exercise as much as they do for relaxation."[380]

Thouin wrote these memoirs in October 1788.

[379] "Notes pour servir à l'histoire du Jardin du roi," BMHN, MSS 1934, XXXI, fol. 36-37, also by Thouin.

[380] "Mémoire sur le Jardin du Roi," BMHN, fol. 4-5.

PART TWO

PROFESSIONS

CHAPTER III

Science and Medicine

1. THE EXPANSION OF SCIENCE

Academy, Observatory, College and University, natural history cabinet and botanical garden—the institutions classically provided for science started in the seventeenth century or earlier, and it has been said that they had already outlived their time even before the end of the old regime. The Academy in particular (in this view) was accommodating too small a proportion of the rising population of scientists; its publications were an inadequate channel for the swelling stream of knowledge, and were insufferably tardy to boot; its organization was impeding the specialization of science into the disciplines of mathematics, astronomy, physics, and so on; its privileged status was no longer congruent with the emerging realities of society at large.[1] In retrospect these things may be said, and little doubt about their pertinence would arise if it were a question of a set of practices continuing unchanged into the 1830s. Like many similar judgments about aspects of the 1780s, however, this one risks the anachronism of reading the inevitability of developments that accompanied the revolution back into the reign of Louis XVI and earlier.

Such judgments sometimes look less compelling when reference is confined to expressions of opinion contemporary with the events. A sense of cramp here and there did provoke sporadic complaints, notably with respect to publication and scientific fellowship. That much has to be admitted. What is more, all informed observers could see how largely science was entering into polity. Indeed, the remainder of this volume will be concerned with its expansion. Part Two will deal with this theme in relation to medicine, public health, and the defense of an enlarged professional perimeter; and Part Three in relation to agriculture and the trades; industry and invention; and finally, engineering. As will appear, the regime even though aging, proved quite capable of developing institutional forms appropriate to those purposes. Before that, however, let us briefly notice a few more obvious manifestations of early growth in the form of journals, popular forums, and voluntary societies.

On the whole, the growth looks healthy. The historian who would de-

[1] Such is the tenor of Hahn (1971), chs. 4 and 5, and it is a position that can be argued.

termine the moment in the 1770s and 1780s when a new piece of science came before the public often needs to turn, not to the annual volume of *Mémoires* from the Academy, which was becoming a work of record or of reference, but to the *Observations et mémoires sur la physique* commonly known as the *Journal de physique* or Rozier's journal.[2] That famous periodical began in 1771, and continued with monthly issues until 1823. It was the pioneer in a generation of proprietary journals that flourished in France and elsewhere. Until the 1770s the annual volumes of the major academies had afforded adequate vehicles of publication. After the 1820s journals confined to a single discipline predominated. In the interval, the periodicals of first instance were those modeled on Rozier's journal: well-known examples reporting scientific research in the English-reading world were Nicholson's journal, Thomson's journal, and Silliman's journal. Older periodicals were also concerned with science, notably the *Journal des savants*, founded in 1665 contemporaneously with the Academy itself.[3] But this exemplar of the reviewing journal limited itself to reports on the literature of science and learning. The editors had no thought of publishing original reserach.

The abbé François Rozier came up to Paris from Lyons in 1771, a clergyman whose vocation was enlightenment.[4] His family having been bourgeois landowners, he already took an interest in scientific agriculture, and specially in horse breeding and viniculture. In 1765 Rozier succeeded Bourgelat briefly as director of the veterinary school in Lyons when the latter moved on to the more important establishment at Alfort.[5] Rozier had also completed a treatise on viniculture by the time he turned to scientific journalism.[6] Somehow family and friends arranged to buy him the privilège or license of the recently expired *Observations sur l'histoire naturelle . . .* , edited by Gautier d'Agoty. The initial series of the *Observations sur la physique* consists of eighteen small volumes in duodecimo published monthly between July 1771 and December 1772. The second and definitive series started in January 1773 in a generous quarto format, conform-

[2] The full title was *Observations et mémoires sur la physique, sur l'histoire naturelle et sur les arts* (1771-1823). McClellan (1979a) is an excellent, somewhat revisionist account, which gives full references to the earlier literature, notably McKie (1948), (1957), and Neave (1950-52).

[3] On the early *Journal des savants*, see Morgan (1929); on its later history, Cocheris (1860); and on the scientific periodical in general, Bolton (1885) and Kronick (1962).

[4] Nicholas-François Cochard, *Notice historique sur l'abbé Rozier* (Lyons, 1832); see also Alphonse de Boissieu, *Eloge de l'abbé Rozier* (Lyons, 1832). Portions of Rozier's correspondence are conserved at the Bibliothèque de Lyon, Fonds Coste 347, 851, 954, and 1132, and Fonds Charavay, 791.

[5] Above, Chapter I, Section 4.

[6] *Mémoire sur la meilleure manière de faire et gouverner les vins de Provence, soit pour l'usage, soit pour leur faire passer les mers* (Marseilles, 1771).

able for purposes of binding and collecting with the annual volume of memoirs of the Academy and with the normal scientific treatise. For Rozier thought to participate in the interchange of official science like a latter-day Mersenne.

The analogy was of his own making.[7] He was an expediter of academic science and not intentionally a harbinger of anything either democratic or specialized. On the contrary, the *Observations sur la physique* afforded an opportunity to publish more eighteenth-century science more rapidly than ever before. The twelve monthly issues, usually bound in two volumes, comprised almost twice as many pages as the average yearly volume for the Academy, even without the benefit of mathematics, astronomy, or rational mechanics, which sciences occupied a quarter to a third of the academic collections. Their absence from Rozier's journal is hardly to be construed as evidence of a forward-looking attitude to specialization. Rozier understood the word "*physique*" in his title rather in the old-fashioned sense of d'Alembert's distribution of the subjects of knowledge into the two main divisions of mathematical science and "general and experimental physics."[8]

The latter was the area in which Rozier's contributors wrote articles: experimental philosophy (especially electricity), chemistry, meteorology, mineralogy, anatomy and physiology, zoology, botany, agricultural and mechanical arts, and medicine. Members of the Academy of Science who worked in those fields published there, taking advantage of the most distinctive feature of the journal, which was prompt publication. The normal length of its articles was six or eight pages, and early communication of research there often took the form of excerpts or abstracts from the forty- or fifty-page memoir read before the Academy and printed two or three years later in its *Mémoires*. Academic prepublication accounted for only a small proportion of the pieces in Rozier's journal, however. Most of the authors were teachers, doctors, engineers, improving farmers, and other members of the educated classes with a strong commitment to experimental or descriptive science. How the material was selected is unclear. Much of it might appear trivial in retrospect, and controversies abound in the correspondence that Rozier printed. On the other hand, even though he dispensed with the time-consuming academic system of referees, the content seldom has the air of psuedo-science or crank literature. Nor was the mode one of popular science or vulgarization. Only the distribution and the deadlines would nowadays be called journalistic, not the scientific level. The purpose was to publish original pieces of research or observation

[7] In the Foreword to the first volume of the enlarged series, *Observation sur la physique . . . I* (1773), v.

[8] In the *Discours préliminaire* to the *Encyclopédie* (1751).

for a scientifically literate public. The circulation of 1,500 copies was divided between learned societies in France and abroad and individual subscribers. Readers serious about science relied upon it, and were correspondingly disappointed when, as too often happend, the bookseller or the mail service failed them.[9]

After 1780, the abbé Rozier returned to his first love, agronomy, and also to Lyons, and handed direction of the journal along to his nephew, the abbé J.-A. Mongez, who continued it in the same pattern. If stirrings of desire for a more differentiated and open practise of science may be detected in the 1780s, and they can, anything accomplished by the *Journal de physique* in satisfying them was incidental to its main purposes, while enterprises that might have been so intended were too fugitive or preliminary to constitute a trend. Of the two periodicals devoted to a single branch of science, the *Annales de chimie* and the *Connaissance des temps*, the former may reasonably be taken for a portent of the formation of specialized disciplines. The reform that Lavoisier brought to chemistry culminated in the virtually simultaneous publication of his *Traité élémentaire de chimie* and of the first number of the new journal, early in 1789. Besides himself, the board consisted of Guyton de Morveau, Berthollet, Fourcroy, and Monge together with three lesser lights, Dietrich, Hassenfratz, and Adet. The significance must not be overdone, however. Only restriction of the contents to the one science distinguishes the early issues from the genre of Rozier's journal. Most of the articles consist of early printings of memoirs destined for definitive publication by academies. As for the *Connoissance des temps*, it dated from 1709 and was less a journal than an astronomical almanac, edited at the Observatory under the aegis of the Academy. It contained ephemerides and other data essential for navigation and observation together with occasional memoirs of research.

Popular science, to turn to that, had an important public in the 1780s. It would be difficult to measure whether vogues like ballooning and fads like Mesmerism (more will be said of both) argue for greater real interest in science and technology than in previous decades or for a *fin-de-siècle* appetite for the sensational. Somehow the impression comes through, however, that prosperous segments of the public felt a hungrier craving for novelty, and that well-placed persons showed great readiness to join in associations for worthy causes, usually started by aspiring youth. In December 1781 the Lycée de Monsieur opened in the rue Saint-Avoye and soon moved to the rue de Valois, alongside the Palais Royal.[10] The impresario was the young scientaster François Pilatre de Rozier (not to be

[9] McClellan (1979), *passim*.

[10] Smeaton (1955a). For Pilatre de Rozier's account of himself, see Tournon de La Chapelle, ed., *La vie et les écrits de Pilatre de Rozier écrits par lui-même et publiés par M. T*^{xxxx} (1786).

confused with the abbé Rozier), whom we shall meet again ballooning in 1783, an Icarus of the enlightenment. His Musée was a Chatauqua-like enterprise of a type that became common in the later eighteenth century and the nineteenth century. It may, indeed, have been the prototype. The Royal Institution founded by Count Rumford in London in 1799 and Peale's Museum in Philadelphia were examples that are better known in the Anglo-Saxon world.[11]

Financing was by a mixture of patronage and subscription. Within two years Pilatre de Rozier enrolled over seven hundred members, about a third of whom counted as founders. Among the charter subscribers were Condorcet and Vicq d'Azyr, statesmen of science become patrons of dissemination. In 1785 further capital in the amount of 70,000 livres was infused by a group of thirteen men of means, who thereby became proprietors. Members had access to a reading room and a program of lectures offered daily from December until August. Science was emphasized at the outset, the professors being persons of useful quality: François Mitouard in chemistry, Jean-Joseph Sue in anatomy, Parcieux in physics, and Prévost in mathematics. After 1785 courses in literature and history were added, taught by people of greater note, J.-F. de La Harpe and D.-J. Garat. Fourcroy then took on the series in chemistry, improving on the reputation he had already won for his performances at the Jardin des plantes. Perhaps the Lycée was slightly in advance of its time, for membership fell off badly in 1787 and recovered only after 1790, when the pattern of joining things individually was more in keeping with the early enthusiasms of the revolution.

Overt discouragement blighted an initiative intended to move botany, the most accessible of the sciences, in the direction of democratization and specialization. On 28 December 1787 a gathering of young naturalists formed the Société Linnéenne de Paris.[12] The moving spirits were Pierre-Auguste Broussonet, Louis-Augustin Bosc d'Antic, and Aubin-Louis Millin, all in their late twenties. Broussonet, originally from Montpellier, was then in the flush of a sudden prominence, having been singled out in 1785 by Bertier de Sauvigny, intendant of Paris, to be permanent secre-

[11] On the Royal Institution, see Berman (1978), and on Peale's Museum, a paper by Charles C. Seller delivered before the American Philosophical Society in April 1979, to be published in its *Proceedings*.

[12] The most important source of information is the manuscript "Registre des procès-verbaux de la Société Linnéenne de Paris," conserved in the Bibliothèque Mazarine, MSS 4, 441. See also Arsène Thiébaut de Berneaud, *Compte rendu des travaux de la Société Linnéenne de Paris* . . . (1822), and the only number that appeared of *Actes de la Société d'Histoire Naturelle de Paris* (1792). This is not an easy publication to find. There is a copy in the Academy of Natural Sciences in Philadelphia. The BN copy has the cote Inv. S. 1333. It is not to be confused with the *Journal d'histoire naturelle*, which also began in 1792. The *Actes* was Linnaean and the *Journal* Lamarckian.

tary of the Society of Agriculture in the capital and to reawaken that comatose body. He had spent the previous five years in London, and taken a lively part in the activities of the circle that formed the Linnaean Society there in 1788.[13] Bosc came of a Protestant family prominent in medicine and also in the glass trades in the region of the Tarn. With a minor civil post in Paris, he occupied his energies botanizing in the company of a circle of friends. Madame Roland was among them, along with René-Louiche Desfontaines, C.-L. L'Héritier de Brutelle, and Guillaume Le Monnier. Bosc owed his first scientific standing to a paper on a new genus of cochineal published in Rozier's journal in 1784.[14] Millin was a man of letters at heart. He had learned natural history with a view to writing the history of the natural sciences, inspired by the example of Montucla's history of mathematics.[15]

Several factors conspired to make the Linnaean system a rallying point for feelings of dissidence from official science in Paris. The most obvious was Buffon's disdain for it. To employ Linnaean classifications and nomenclature was to fly in the face of the overlord of natural history. By a similar token, it was the system most in favor in Montpellier, where Jean-Guillaume Bruguière, Guillaume-Antoine Olivier, and Claude Riche were all good Linnaeans, carrying over into botany the dissent from institutional authority in Paris that is evident in other realms, notably in medicine and in the vitalism of philosophic biology.[16] More largely, and less logically, Linnaeus appealed to the amateur, the nature walker, who even like Rousseau found the discriminants easy to perceive and the identifications easy to make. For all these reasons, the new association had a young Turk tone to it. They cannot have been very terrible Turks, however, for the roster of senior persons who put themselves down for membership bespeaks respectability itself: Malesherbes, Lavoisier, Cels, Daubenton, Michaux, Lacepède, Fougeroux de Bondaroy, Faujas de Saint-Fond, to mention only a few of those who loaned at least their names.

The authorities at the Jardin des plantes were nonetheless displeased. Buffon was in the last year of his life, and A.-L. de Jussieu was the one who invoked academic courtesy to suppress the upstarts. According to

[13] On Broussonet, see below Chapter V, Section 5.

[14] "Description de l'*Orthezia characias*," *Observations sur la physique*, . . . *24* (1784), 171-173. On Bosc, see the Cuvier éloge (1819-1827), *3*, esp. 88-90. There is an interesting exchange of correspondence between Bosc and James Edward Smith, first president of the British Linnaean Society, who kept in close touch with the French Linnaeans. BN, MSS Fr. nouvelles acquisitions 2760, fols. 162-163. See the article by Jean-François Leroy, DSB 2, 321-323.

[15] Charles-Guillaume Krafft, *Notice sur Aubin-Louis Millin* (1818). Jean-Etienne Montucla, *Histoire des mathématiques*, 2 vols. (1758).

[16] The Cuvier éloges of Bosc, Broussonet, Bruguière, Olivier, and Riche are illuminating on this subject (1819-1872).

Bosc's recollections, confirmed by other sources, it was put about that young naturalists who hoped for election to the Academy or other privileged bodies would do well to separate themselves from the Linnaean Society. The word sufficed. The ambitious and the notable withdrew, and the society collapsed[17]—to revive less than two years later, in May 1790, under the name Société d'histoire naturelle, the climate of the Revolution again proving favorable, in this case to the echoes of Jean-Jacques as well as to the practice of voluntary association.

A further fledgling society must be mentioned, although it belonged to the old regime only in the moment of its birth. The Société philomathique, vigorous in the later 1790s and prominent throughout the nineteenth century, traces its formal origin to an inaugural meeting among six founding members on 10 December 1788. The naturalist Riche had been a leader of the dispersed Linnaean Society, and appears to have been one of the two prime movers, the other having been the agronomist, Augustin-François Silvestre.[18] In addition there were two doctors, Audirac and Petit; a mathematician, Broval; and a chemist, Antoine-Louis Brongniart. Only Brongniart and Silvestre made important names for themselves, Brongniart—with others of his family—in science, and Silvestre in scientific journalism and statesmanship. No further meetings or admissions were recorded until 9 November 1789. From the few early papers that survive, it appears that the purpose was fellowship and mutual instruction. The

[17] See especially, Bosc's recollections, incorporated in the "Registre" cited in n. 12 above, fols. 35-43; and Millin's near-contemporary account in the prefatory section of the *Actes* cited in the same note.

[18] The point of departure for the history of the Société philomathique is the centennial memoir composed by Marcellin Berthelot, "Notice sur les origines et sur l'histoire de la Société Philomatique," *Mémoires . . . à l'occasion du centenaire de sa fondation* (1888), i-xvii. In 1972 five registers containing the procès-verbaux of the society were in the possession of a bookdealer, Monsieur André Jammes, 3 rue Gozlin (6°). Among them were the first two, covering the periods 10 December 1788 to 1 October 1791 and 1 October 1791 to 23 fructidor an IV (9 September 1796). I owe this information to a private communication from M. Arthur Birembaut. The archives containing other papers of the Société philomathique were deposited in the Bibliothèque de la Sorbonne in 1900, where they occupied cartons 123 through 134 when I was able to consult them many years ago (in 1955). The present paragraph is based entirely on notes taken at that time, the papers themselves having been in a state of almost total disorder. Interspersed among them were many documents pertaining to the Société d'Histoire Naturelle. From late June 1796 the two societies shared the same quarters for some years—the current address is 24 rue Dauphine. Publication of the *Bulletin des sciences de la Société Philomathique* began only in 1797. The intention had been to bring it out much earlier. The first fifteen numbers, from July 1791 to September 1792, were prepared in manuscript and read out at meetings. They consist of reports on the meetings of the main learned societies of Paris, and amount to a series of scientific newsletters. See also, Riche and Silvestre, *Rapports généraux des travaux de la Société Philomathique de Paris, depuis son installation au 10 décembre 1788, jusqu'au 1er janvier 1792* (undated, but probably 1792). BN, [R. 16023.

members had no thought of advancing science, but only their own knowledge of it. Their jottings contain no trace of any animus against established institutions. On the contrary, the meetings heard reports of the proceedings of the Academy of Science, the Society of Medicine, the Academy of Surgery, and so on. Membership was limited to fifty, and successful candidates had to have submitted at least one memoir for the approval of an ad hoc commission passing on their applications. The Société philomatique set out to be a kind of junior academy, therefore.

However auspicious, these were small doings in the antechambers and along the peripheries of science. More convincing evidence of the capacity of academic institutions to excite emulation is to be found in medicine. Chartered in 1778, the Royal Society of Medicine was conceived by its founders as the agency through which the old profession, it is not too much to say the entrenched profession, was to be penetrated by the modernizing spirit and civic modes of eighteenth-century science.

2. THE SOCIETY OF MEDICINE

Formally the Royal Society modeled itself upon the Academy of Science, needing to be nearby, wrote Vicq d'Azyr, "always observing its research, feeling the same spirit and following the same road in order to arrive at new results."[19] Legally also the Society of Medicine was the same sort of body, a privileged corporation, established by royal letters patent authorizing it to meet regularly, to correspond, and to exercise in the field of medicine an authority over publication like that which the Academy held in science and which the Collège de France had briefly sought in scholarship.[20] Politically it had, again like the Academy of Science, the right to be consulted directly by any of the ministers or secretaries of state whose responsibilities intersected with its competence.[21] Those who most frequently had business with the Society of Medicine were the controller-general, for questions of public health and welfare; the Maison du roi, for the administration of mineral waters; the secretaries of state for war and for maritime affairs, for military and naval medicine; and Bertin's Petit Ministère for agricultural business. It published its own proceedings and memoirs, though not quite annually, and also emulated the Academy in its arrears.[22] Resident membership was limited to forty-two: thirty asso-

[19] "Nouveau plan pour la médecine," HSRM 9 (1786-87/1790), 150.

[20] Above, Chapter II, Section 4. The definitive letters patent, dated 29 August 1778, are printed in HSRM *1* (1776/79), 17-24.

[21] See a memorandum, undated but of the 1780s, on the administration of the Society, BANM, MSS 114, dossier 7.

[22] Nine volumes were published between 1779 and 1790, the first ostensibly for 1776, and the ninth for 1777-1778. The *Histoire*, or proceedings, and *Mémoires*, or submitted

ciates who were required to be doctors, and twelve lay or independent associates prominent in science or at court. The sixty provincial associates were all medical men, and there was a like number of foreigners, largely honorary, first among them Benjamin Franklin.[23]

The Society met twice a week, on Tuesdays and Fridays. At first its seat was in the rue du Sepulcre (now the rue du Dragon). In 1781 or 1782 it moved to the rue des Petits Augustins, now the rue Bonaparte, where in 1820 the restored monarchy installed its successor, the present National Academy of Medicine. In the late 1780s, however, the stated meetings were held in the Louvre, first in the pavillon de l'Infante, and later in the hall of the Academy of Science. There the Society listened to papers, judged of books and memoirs, and passed upon the remedies, cures, and pharmaceutical preparations submitted for its approbation. It also reviewed selections from its enormous correspondence, for the Society had enlisted over a thousand country doctors to be correspondents. The bylaws placed no limit on their number, stipulating only that they participate actively in order to be continued on the roll.[24] Jumbled in the library of the Academy of Medicine remain tens of thousands of their letters, reports, and memoirs sent in from the provinces to Paris.[25] In this respect, comparison with the Academy of Science becomes a contrast. There, the

---

papers, are paginated separately. A tenth volume, supposedly for 1789, was published by the revolutionary Ecole de santé in 1798/an VI. It contains papers that were being readied for the printer when the society was overtaken by the Revolution and suppressed. Further volumes were intended (*10*, vi) but never appeared.

[23] Names of the newly elected members appeared in the *Histoire* annually. There was little overlap with the Academy of Science, and the statement sometimes made that the one was a stepping stone on the way to the other is incorrect. The only members of both bodies among associates were Bucquet, Fourcroy, A.-L. de Jussieu, Macquer, and Poissonier; and among independent associates Daubenton, Duhamel du Monceau, Lavoisier, Tillet, Trudaine de Montigny, and La Rochefoucauld, the last two of whom were honorary in the Academy.

[24] Letters patent, HSRM *1* (1776/78), 21.

[25] As will appear below, the documents conserved in the Bibliothèque de l'Académie Nationale de Médecine constitute a very important collection. The procès-verbaux, or minute books, survive for all meetings between 1 December 1775 and 19 August 1793, except for those in the year 1790, which are lost. They are bound in six folio volumes, MSS 7, 8, 9, 10, 11, and 11 bis. Running notes for meetings from 17 March to 14 June 1775 are in MSS 12, and formal reports for the years 1780-1791 are in MSS 14 and 15. For the rest, the collection has never been classified or catalogued, and its state of disorganization is such that it has to be sampled rather than studied systematically, a task made easier by the kindness of the staff. The one account of these papers that has been printed emphasizes their importance for the history of climate, though in my opinion it fails to convey their full interest for the history of medicine or of institutions. (J. Meyer, 9-20 in Desaive *et al.* [1972].) The citations in the footnotes that follow refer to the sequence of cartons in which chance has placed the documents; when and if they are classified, some system will certainly have been instituted to make the existing numbering obsolete.

formal attachment of correspondents to senior members, in principle one to one, seldom produced much exchange. The great importance of correspondence within the Society of Medicine[26] offers, therefore, an index to certain differences between science and medicine continuing through the time when science was a reformist force exerting its influence on the older profession through its institutions, its personnel, its ideology, and its content.

The nature of the relationship between science and medicine may be examined through the instance of the career of Félix Vicq d'Azyr, the animator and only permanent secretary of the Society of Medicine from its chartering in 1778 until its suppression simultaneously with the Academy of Science in the Revolution. We have already encountered Vicq d'Azyr, heading Turgot's commission on the cattle plague, briefly teaching anatomy at the Jardin du roi, and helping the young Fourcroy with the expense of a medical education.[27] Together with Lavoisier and Condorcet, he was one of the small number of influential persons whom the visitor to scientific Paris in the 1780s was most likely to meet. His reputation is something of a puzzle, however. An obscure street named after him near the Hôpital Saint-Louis in the 10th arrondissement is often thought instead to commemorate some minor victory in the conquest of Algeria. A very small structure in the brain also bears his name, the thalamomamillary handle of Vicq d'Azyr.[28] Tradition credits him with having inaugurated the fine scrutiny of the cerebral cortex and more largely with having advanced the whole study of anatomy of the brain. It is evident from his meticulous *Anatomie du cerveau* that he was capable in anatomical dissection. He knew the literature. He could manipulate his scalpel. He could see what it disclosed. His sectioning of the brain improved on Haller's strategy in several particulars, and he illustrated his preparations in a series of thirty-one plates, the detailed discussion of which constitutes the treatise. It occupies over half of one of the six small volumes of his *Oeuvres*,

[26] "Of all the functions to which the Society of Medicine was called on being instituted, it has always felt that the most useful was that of gathering by an extensive correspondence observations that could hasten the progress of our art."—so wrote Vicq d'Azyr in the draft of a memorandum to all physicians elected to the Legislative Assembly in 1792 (BANM, MS 114, doss. 21). In an earlier memoir of the revolutionary period, "Observations et éclaircissements," he observed that the Society's correspondence was "sans aucune comparaison, beaucoup plus étendue que celle d'aucune autre Académie." (BANM, MSS 114, doss. 23.)

[27] Above, Chapter I, Section 4; Chapter II, Section 6.

[28] Spillman (1941); Laignel-Lavastine (1949) *2*, 305, 647; *3*, 332. The article on Vicq d'Azyr in the DSB *13*, is by Pierre Huard. The important older accounts are those of Jacques L. Moreau, "Discours sur la vie et les ouvrages de Vicq d'Azyr," in his edition of the *Oeuvres de Vicq d'Azyr* (1805) *1*, 1-88; and Dubois (1866). Dufresne (1906) is more perfunctory.

collected and published in 1805 by Jacques-Louis de Moreau, a former student and admirer.

This edition did not include the publications on epizootics that issued from his work on the cattle plague.[29] Even after making allowance for the clarity and practicality of those manuals, however, the historian who looks into Vicq d'Azyr's writings to find the record of his quality is likely to come away disappointed. Three of the six volumes are occupied by éloges composed in praise of famous men. Vicq d'Azyr had neither Fontenelle's wit nor his discernment in this genre, and was less informative than Condorcet. A certain straining after style, timeliness, and relevance dates them. The éloge of Vergennes, for example, emphasizes the American Revolution and thus associates the fellowship of the Society of Medicine with the citizens of the new nation, stalwarts all of liberty.[30] Besides *Anatomie du cerveau*, the only other evidence of research based primarily upon dissection are two short memoirs, interesting ones to be sure, on the anatomy of the auditory and vocal organs in certain birds and mammals.[31]

The rest of his scientific writing was in the vein of natural history and closer to Buffon than to biology. The most considerable piece is a general treatise on anatomy, the text of the course he gave first privately and then at the Jardin du roi from 1775 to 1777.[32] His presence and eloquence equipped him to lecture with verve. The availability of the lectures in print, however, fails to substantiate the conventional explanation of his prominence, which is that he anticipated Cuvier and was thus a founder of comparative anatomy. It is true that in this and other writings he made comparisons between human and animal anatomy, and also between analogous structures in the same organism, the hands and the feet for example.[33] Nowhere were his observations based upon a systematic correlation of parts, however. Nor was Vicq d'Azyr's interest methodological. Instead, it was professional, and the clue to his purpose will be found in the location where he offered a course actually called comparative anatomy. That was at the Royal Veterinary School at Alfort in the 1780s, in the time when Daubenton and Fourcroy were also fortifying the scientific content of its curriculum.[34]

His joining them in that small effort was consistent with the large civic purpose he thought to serve through the instrumentality of the Society,

[29] Above, Chapter I, Section 4, n. 65.

[30] Vicq d'Azyr, *Oeuvres* 2, 94-140; originally in HSRM, 8 (1786/90), 35-69, delivered in the public meeting of 12 February 1788.

[31] Vicq d'Azyr, *Oeuvres* 4, 338-387.

[32] "Discours sur l'anatomie," *ibid.*, 5-312.

[33] "Sur les rapports qui se trouvent entre les usages et la structures des quatre extremités dans l'homme et dans les quadrupèdes," *ibid.*, 313-337.

[34] Above, Chapter I, Section 4.

which was nothing less than instituting the reform of medicine—of the profession, however, not the practice, for Vicq d'Azyr did not contribute to the deeper and more intimate development of clinical medicine evident half a generation later.[35] His intention was rather to unite physician and surgeon, veterinarian and pharmacist in the responsibility for preventing epidemics, whether affecting beast or man, and for controlling the various plagues when they nevertheless broke out, as despite the most enlightened measures they were bound to do, albeit with diminishing frequency and virulence.[36] In short, the object of the new medicine was to be public health. Properly speaking, the privileges of physicians could have no other justification, and least of all prescription. Vicq d'Azyr in medicine, Lavoisier in chemistry, Condorcet in social science: they made a trinity of reformers from on high, each envisaging the whole of society from the standpoint of his own specialty (and vice versa), and all three were agents of Turgot.

As for the personal differences, they were mostly to Vicq d'Azyr's advantage. Inside and outside the scientific community, he was respected as Condorcet was not and liked as Lavoisier was not. The public had an interest in the reform of medicine that it scarcely felt in the reform of chemistry or social science. In the *Tableau de Paris*, Mercier, who rarely praised the authorities, hailed the foundation of the Society of Medicine as the "highest wisdom," and observed of the medical establishment:

> Exercising the best, the most lucrative and the easiest of trades, doctors decided, and with reason, that whoever did not wear the official fur and the scholastic gown was to be held incapable of making any discovery, and that any such claim should be protested whether fairly or unfairly (*per fas & nefas*). Thus do they immolate all humanity in the sinister interest of their honorariums: and since the dead never bring suit against their doctors, no more than do their heirs, they continue to write their own prescriptions blindly and to distribute the old poisons from the pharmacy.
>
> When will the generous and enlightened man arise who will overturn the temples of old Aesculapius, who will break the surgeon's dangerous scalpel, who will close the apothecary's shop, who will destroy that body of conjectural medicine, attended by drugs, fasts,

[35] Foucault (1963), who, however, writes of the transformation of medicine in a philosophical terminology only distantly related, if at all, to the sense of the historical reality conveyed by the sources themselves. See also Ackerknecht (1967).

[36] "Nouveau plan pour la médecine," HSRM 9 (1787-88/1790), 1-170; see also the many drafts for various similar *projets* of reform noted down at various times in the late 1770s and 1780s, BANM, MSS 115.

and diets? What friend of mankind will finally proclaim a new medicine, since the old one kills and depopulates?[37]

Although Vicq d'Azyr may not have been quite the man Mercier would have had him be, he was nevertheless charming in manner and affable in nature. Unfortunately, the few letters we have are almost illegible, but much remains of the enormous correspondence addressed to him in reply to his initiatives. We can piece together an impression of his personality from his effect on others. From the warmth of their responses, it is clear that he had the gift of leading people to feel better about themselves for being drawn into the dignified and humane concerns he took upon himself. With the well placed and powerful, his touch was light. He had the wit always to be amusing in that milieu and to be welcome when he asked for something. Although he came from provincial bourgeois stock, respectable and without a hint of fashion, his letters from the great persons at court—Vergennes, Breteuil, Calonne, La Vrillière, Malesherbes, the maréchal de Castries, Trudaine de Montigny, Villiers du Terrage, d'Angiviller, Necker, the abbé d'Espagnac, Miromesnil—are in the style of half-bantering familiarity that comes naturally to people at ease in inner circles, whether through birth or tacit invitation. "Will you be coming back to Versailles before the return of the king?" asks Villiers du Terrage in a covering note accompanying certain intendants' reports, "and will you do me the honor of accepting my soup?" (taking pot luck).[38] Of all the men of science mentioned in this book only two others, Buffon and Dolomieu, were ever thus sought after at Versailles—and Dolomieu had been born to it. Thus, among his other engagements, Vicq d'Azyr became a court doctor, official physician first to the king's brother, the comte d'Artois, and later also to Marie Antoinette, with whom he grew to be a favorite in the years just prior to the Revolution. "*Mon philosophe*," the queen would call him.[39] He was elected to the French Academy in 1788 in the place of Buffon, having tried the previous year for the seat of the marquis de Paulmy d'Argenson and been told by Poissonier that he could count on the support of the Marmontel party, Loménie de Brienne, Montesquiou, Bailly, La Harpe, and Chamfort, in addition to his "former apostles."[40]

It would be naive to suppose Vicq d'Azyr to have been politically dis-

[37] L.-S. Mercier, *Tableau de Paris*, 12 vol. in 6 (1789) *1*, 61-62.

[38] The vicomte de Villiers du Terrage to Vicq d'Azyr, BANM, MSS 108. See n. 25 above for the problems involved in consulting these papers. Much, but by no means all, of the response to Vicq d'Azyr's correspondence may be found in MSS 108, 109, and 110.

[39] Dubois (1866), 643.

[40] Poissonier to Vicq d'Azyr, 11 September 1787, BANM, MSS 109.

interested in choosing the prominent people (John Adams was among them)[41] to whom the Society offered some appropriate form of association. But having secured the use of their names, Vicq d'Azyr paid attention to their persons, remembering their anniversaries, condoling with them on deaths in the family, conveying the Society's relief on their own recovery from illnesses.[42] He was truly considerate, in a word, and if noblemen responded well to being treated like human beings, his fellow physicians among the medical correspondents in the provinces were even more strongly affected to the Society in consequence of the graceful words their permanent secretary habitually found to make them feel their observations worthy of professional consideration. "I am very flattered, Monsieur," writes Doctor Bouquet from Luçon, "that the company accepts my feeble efforts. I shall not fail to send them to you every three months," and he goes on to complain mildly that the controller-general has taxed the package of forms for reporting the weather despite its being franked, and to hope that these cross-purposes can be straightened out.[43] Writing from Dijon, Dr. Maret regrets having sent his first observations in a format that was not approved. He corrected his reporting in 1780, and then decided to spare the Society his clinical observations, thinking them probably too commonplace to serve any purpose other than fattening the files. He is delighted to learn that, on the contrary, his colleagues in the capital set store by his experience: "I wish my efforts to merit the indulgence of your celebrated company."[44]

Vicq d'Azyr's father had been a successful physician in the old town of Valognes in lower Normandy; thus he knew the life of a country doctor from his boyhood. After a secondary education at the local college, he went on to Caen for further study. There he and Laplace were enrolled in the same philosophy course. "Monsieur Adam," he would later say of their professor, "has no idea how much trouble we have taken, Laplace and I, to forget everything he taught us."[45] Poetry was then his favorite subject. For

[41] Apparently, he went to call on John Adams to bespeak his interest in arranging an interchange with the Boston College of Physicians, for there are two letters in the collection from Adams to him (20 December 1782 and 8 March 1783) and one from Adams to Geoffroy (20 December 1782), who administered the office. At the same time, Vicq d'Azyr won approbation from Vergennes for putting the Society of Medicine into correspondence with the Royal College of Physicians of London despite the war: "Divisé par des intérêts ou des préjugés politiques," wrote Vergennes in high-minded fashion, "il est beau de voir les hommes se réunir pour les intérêts de l'humanité; et l'on ne peut qu'applaudir aux savantes des deux pays qui ont opéré cette réunion." Vergennes to Vicq d'Azyr, 9 April 1782, BANM, MSS 108.

[42] See, for example, the exchange with d'Angiviller in BANM, MSS 108, doss. 8.

[43] Bouquet to Vicq d'Azyr, 10 November 1783, BANM, MSS 155.

[44] Maret to Vicq d'Azyr, 27 October 1783, BANM, MSS 156.

[45] Dufresne (1906), 13-14; for Vicq d'Azyr's early life, see also his *Oeuvres*.

a time he thought to be a man of letters and came close to taking orders with a view to providing himself with a status. His father objected, urging the advantages of a medical career. The boy agreed to try the family profession (there were also doctors on his mother's side), and went off to Paris in 1765 at the age of nineteen. Vicq d'Azyr was always one to throw himself into things. The round of courses, lectures, and visits to laboratories and hospitals fascinated him, and he sat for his licence in the summer of 1772. The following year he opened his own private course of human and animal anatomy, lecturing in the amphitheater of the School of Medicine during the summer holidays and drawing an audience among whom were well-known doctors. We have already seen that in 1775 Petit, his teacher and patron, wanted him for successor at the Jardin du roi, and Vicq d'Azyr seems not to have begrudged Buffon his preference for Portal.[46]

In the meantime, Vicq d'Azyr had been married. The story combines elements of romance and pathos more characteristic of the nineteenth century, when consumption spelled the early doom of ardent temperaments. It began in the spring of 1772, when he was still a student. A pretty girl swooned in the street outside his window. Rushing to her aid he recognized the niece of Madame Daubenton, a certain charming Mademoiselle Lenoir. They were soon married. In eighteen months she was dead of tuberculosis. There have been those who said that the liberty of a dashing young widower stood him in good stead in the world of fashion. The more immediate consequence was a threat to his own health. A coughing spell produced a minor hemorrhage, and it was thought wise that he should retire for a time to the Norman coast near his father's home. There along the quiet beaches he made studies of the anatomy of various sea creatures.[47] Communicated to the Academy of Science, they, together with his personal reputation, sufficed for election to membership on 16 March 1774. He also completed his thesis for the doctorate of the Faculty of Medicine. His return to Paris was that of a rising light, the dangers of youth behind him, the prospect for a brilliant career ahead.

[46] Above, Chapter II, Section 6.

[47] "Premier mémoire pour servir à l'histoire anatomique des poissons," SE (1773/76), 18-36; "Deuxième mémoire pour servir à l'histoire anatomique des poissons," *ibid.*, 233-262. The same volume contains a note "Observation anatomique, sur une extrémité inférieure dont les muscles ont été changés en tissu graisseux, sans aucune altération dans la forme extérieure," 301-304. The two memoirs (reprinted without illustration in Vicq d'Azyr, *Oeuvres* 5, 165-222) are drawn largely from published literature, and give a general description of the organs common to all the three orders into which Vicq d'Azyr wished to distribute fish. They are fortified with certain original observations, mainly on the swim bladder, that he had made in Normandy. A later series of three memoirs on birds is similar in conception and execution: MARS (1772, pt. II/1776), 617-633; (1773/77), 566-586; (1774/78), 489-521 (reprinted in *Oeuvres* 5, 223-294, without notes or illustrations).

Such was the still young man whom Condorcet urged upon Turgot in November 1774.[48] The cattle plague gave Vicq d'Azyr his opportunity, and he made the most of it. Although the administrative side of his activity was the more conspicuous, he minded Turgot's insistence that he also institute a scientific study and analysis of the disease with special attention to the mode of transmission. So soon as he had framed his first set of regulations on prevention and decontamination and issued them at Bordeaux, he turned to research and in January 1775 set up two laboratories in farm buildings near to Condom, one for studying cures and the other transmission. His staff consisted of a doctor, a veterinarian, and an apothecary. They soon despaired of finding effective remedies, attributing the few cures that occurred to nature rather than to medicine, and concentrated their inquiry entirely on the mechanism of contagion.[49]

Disease was commonly supposed to be spread by unhealthy conditions of the atmosphere. "Putrid air" was the ordinary phrase, and the problem that Turgot set in his instructions was to "correct and purify" it by all the means that physics and chemistry could propose.[50] It cannot be said that Vicq d'Azyr categorically abandoned this classical notion. He mentions miasmas and recommends drastic fumigations. His actual experiments, however, have the great merit of establishing that infection was transmitted, not by generalized states of the atmosphere, but only from one animal to another, and not by physical contact, but either through respiration, in which case the miasma exhaled by sick animals was the vector, or else through feeding, in which case their fodder was the source. The experiments were well designed and decisive. They established that beasts that recover are immune from the malady, and that it affects no other domesticated animals. Vicq d'Azyr tried inoculation, the first ever practiced in France, and thought it too dangerous for general use. One correct conclusion, that hides from cadavers are not infectious, was widely disbelieved, unfortunately for the economy since much usable leather was thus lost. One incorrect conclusion, that cadavers themselves remain infectious for a long period, was as widely believed, also unfortunately since much expense was incurred in overly elaborate schemes for burial and for disinfecting stables.[51] All told, the research that Vicq d'Azyr directed bespeaks a serious medical talent animating the career of one who knew how to turn his experience to account in the larger vista of medical reform.

Besides epidemics, the jurisdiction that he preempted for the Society of Medicine comprised official consultation and medical communications,

[48] Above, Chapter I, Section 4.

[49] He described the research in a 35-page pamphlet, *Recueil d'observations sur les différentes méthodes proposées pour guérir la maladie épidémique qui attaque les bêtes à cornes, sur les moyens de la reconnaître partout où elle pourra se manifester et sur la manière de désinfecter les étables* (1775).

[50] Turgot to Esmangart, 29 November 1774, Turgot, *Oeuvres 4*, 251.

[51] See Dronne (1965), 173-176, for an expert modern evaluation.

mineral baths and waters, commercial remedies and medicines, and research and publication. What provoked controversy, however, was the combination of these concerns under a new aegis. For none of them was in itself a novelty to the authorities. Already in 1772 the abbé Terray had instructed intendants to enlist country doctors in a regular correspondence about public health.[52] Vicq d'Azyr's commission on epidemics only adopted and enlarged that practice. As for mineral waters and the myriad nostrums ever being urged upon a suffering and gullible public, the state had long recognized some responsibility for regulating their distribution and sale. The incentive was greater in the former case since title to all mineral resources inhered in the crown, and the Treasury had an interest in the profits. An edict of 25 April 1772 vested these functions in a new Commission royale de médicine. It was to be administered by the dean of the Faculty of Medicine under the inspection of the king's head physician and to include representatives of the surgeons and the apothecaries.[53] The attitude of that body to its duties will be examined later for the light it throws on the inwardness of the conflict between the Faculty and the Society of Medicine over the larger question, indeed the fundamental question, of official responsibility for research and the advancement of medicine.[54]

3. SURGEONS AND APOTHECARIES

It was axiomatic in the late eighteenth century that only the academic format could serve to increase knowledge in a subject.[55] Callings far more modest than medicine had long since vested their leadership in academies with a view to bettering themselves intellectually: building and civil engineering in the Académie royale d'architecture in 1671, surgery in the Académie royale de chirurgie in 1731, seafaring in the Académie royale de marine in 1752.[56] It was surprising, wrote an enthusiast for a medical academy in 1724, that nothing of the sort had yet come about in his subject.[57] The failure was not for lack of trying. In 1673 Antoine Daquin,

[52] HSRM *1* (1776/79), 3.

[53] Delaunay (1906), 301-302. The papers and procès-verbaux of this Commission royale de médecine are in BANM, MSS 112-115.

[54] Below, Section 4.

[55] See, e.g., a memorandum of 5 January 1790, concerning possible unification of the Faculty and Society of Medicine, which takes for granted the necessity of conserving "sa forme académique, sans laquelle l'expérience a prouvé qu'une compagnie ne peut contribuer aux progrès des sciences . . ." BANM, MSS 114.

[56] On the Académie d'architecture, see Lemonnier (1911-1929), *1*, preface; on the Académie de marine, see Charliat (1934); on the Académie de chirurgie, see below, Section 3.

[57] A copy of his proposal is in BANM, MSS 114, doss. 7, together with a minute signed by the controller-general, Dodun, approving it in principle.

head physician to Louis XIV and a doctor of Montpellier, had secured letters patent inaugurating a Chambre de Médecine, designated "Royale" in 1683, which became a kind of court academy at Versailles and then survived another ten years. In 1730 one of his successors, Pierre Chirac, also trained in Montpellier, started an Académie de médecine pratique et expérimentale, to be composed of twenty-four leading doctors who would exchange therapeutic experience with others working in hospitals throughout the country.[58] The Faculty of Medicine of the University of Paris aborted all such efforts, however. Its obstruction succeeded because advocates of replicating the academic model in medicine overlooked a crucial element. Unfortunately for their proposals, medicine had already become a profession in the Middle Ages and already had its corporate guardian, its extremely jealous guardian, in the form of the faculty itself. No such obstacle impeded academies in science, surgery, or sailing.

To be sure, the faculty was constantly at odds with surgeons and apothecaries, but in its eyes these were battles with rebellious barbarians in provinces subject to their medical empire, not civil war among the rightful rulers. French surgery and surgical teaching were already famous in the eighteenth century, however, and it was because the barbarians eventually carried the day that medical reforms of the early nineteenth century could consist in adoption by physicians of the modes of clinical practice pioneered by surgeons. That they thus prevailed, to the point of ultimately entering the medical profession themselves, was owing in part to the support of the crown and in part to the visible capacity of surgeons to perform successful operations, notably amputation, lithotomy, trachaeotomy, reduction of compound fractures and strangulated hernias, lifting of cataracts, and excision of visible tumors, and thus to cure their patients in a significant number of cases.[59]

Both of the old trades of surgery and pharmacy began the transition from craft to discipline in the eighteenth century and turned from lore toward learning, the surgeons sooner, in a more decisively professional manner, and at a higher social level. The pattern was the same, however. The state authorized dissolution of the early marriages from which both guilds had been formed, that of the surgeons with the barbers and that of the apothecaries with the grocers. Colleges were then established so that the preparation of young men to practice the respective arts might consist in their being students gaining knowledge subject to an institution instead of their being apprentices imitating skills subject to a master.[60] For

[58] Corlieu (1877), 212-213; Delaunay (1906), 309-310; Steinheil (1903) *1*, ix-xi.

[59] Gelfand (1973a) is an important thesis on the teaching of surgery in eighteenth-century Paris. It gives a very full bibliography and guide to the sources, published and unpublished.

[60] Until the mid-seventeenth century there were two companies of surgeons. Surgeons

our purposes the tradition of royal favor shown to surgery need be traced no further back than the deathbed of Louis XIV. The king's head physician, Guy-Crescent Fagon, was then widely accused of negligence for having allowed the illness to become fatal, while the head surgeon, Georges Mareschal, was praised on every hand for the accuracy of his view of the medical situation. Thereafter, the latter's successor, François de La Peyronie, exercised an almost paternal influence over the infant Louis XV all during his childhood, and remained in charge of the king's health until his own death in 1747.[61] Even at those times when the king had confidence in his head physician, that official still had no authority over the regent doctors of the faculty in Paris, who always took a *frondeur* attitude to the court and its efforts to reform the practice and profession of medicine.[62] The surgeons, by contrast, recognized the king's head surgeon to be the sovereign figure in their community (Saint-Côme it was called from its patron saint and his church), which welcomed La Peyronie's initiative in securing royal authority for institution of courses in anatomy and surgery on a regular basis in 1724 and of an Academy in 1731. Differentiation of their practice from hairdressing had long been a fact, one sweetened for the barbers, at least in Paris, by the new profitability of wigmaking. In 1743 a royal decree declared the separation to be legal and, to the dismay of the faculty, recognized that surgery had attained the status of a liberal art: the degree of Master of Arts was to be required of its future practitioners.[63]

La Peyronie's two protégés, François Quesnay (of later physiocratic fame) and Antoine Louis, maintained the mutual confidence between Versailles and Saint-Côme. Physician to Madame de Pompadour, Quesnay used his position to secure letters patent in 1748 conferring the designation "royal" upon the surgical academy started in 1731. La Peyronie had made him its permanent secretary in 1740. He edited the first volume of its memoirs and consulted regularly on its affairs with La Peyronie and his successor.[64] The old guild had been in some sort a democratic body like

of the short gown were barber-surgeons, who had the right to treat minor injuries. They outnumbered the more highly qualified surgeons of the long gown, members of the Company or College of Saint-Côme, who could not be hairdressers. In 1655 the crown joined the two guilds into one, a step much resented by the more highly skilled, who felt it to be demeaning and attributed it to a plot by the faculty to reduce their standing. See Gelfand (1973a), 6-9, 36, who considers, however, that there were good economic reasons for the move. For their own account of the academy and of their prior history and humiliation, see the account by Antoine Louis, "Histoire de l'Académie Royale de Chirurgie, depuis son établissement jusqu'à 1743," *Mémoires de l'Académie Royale de Chirurgie* 4 (1768), 1-102.

[61] Gelfand (1973a), 90-99.

[62] Steinheil (1903) *1*, ix-xi; Delaunay (1906), 93-165.

[63] Gelfand (1973a), 112-115.

[64] On Quesnay and the surgeons, see *ibid.*, 125-137; Hecht (1958); Schelle (1907).

others of its kind. Not so the Academy of Surgery under its royal authority, for the 250-odd master-surgeons of Saint-Côme were now merely titular or *libre* associates without voice or vote in choosing their representatives. Only the head surgeon at Versailles, "Président-né de l'Académie," named from among their number the forty leading figures who were "counsellors" or active members.[65] Once they were installed and meritocracy was in operation, Quesnay relinquished the post of secretary to Antoine Louis in Paris, who saw four further volumes of memoirs through the press[66] and administered the affairs of the Academy until its suppression in the Revolution.

Before turning his main attention to political economy, Quesnay also arranged in 1750 that the public courses given at the amphitheater of Saint-Côme in the rue de La Harpe should constitute a College of Surgery. Comprehensive education in the art was not yet possible there. Students still needed to find education and experience as best they might by somehow penetrating into hospitals, attaching themselves to practicing surgeons, and following courses at the Jardin du roi, the Collège royal, and private centers of instruction around the capital. Nevertheless, the College of Surgery provided focus. The most persistent of the students who flocked to Paris from the provinces and abroad got access to its *école pratique* of dissections and operative technique.[67] In the declining years of Louis XV, the last of his head surgeons, La Martinière, persuaded ministers to commission a new building, a Palais des Ecoles. It combined the finest amphitheater in Paris with a small teaching "hospice," the latter under the direction of Jacques Tenon, where students could assist at the operating table and learn around the patient's bed. Tenon there acquired the experience that made him an expert sought out by advocates of general hospital reform in the next decade.[68] The government had allocated 600,000 livres to the new installation before the death of Louis XV. Despite the stringencies afflicting the Treasury, Turgot added 300,000 livres to the appropriation in the new reign, and the Academy and College of Surgery moved into their elegant quarters during the summer of 1775.[69]

The literary quality of memoirs contributed by surgeons, though noticeably below that of writings addressed to the Society of Medicine, be-

[65] For the charter members, see *Mémoires de l'Académie Royale de Chirurgie 1* (1761 edition), xxxiii-xxxiv; and for the letters patent of 2 July 1748 followed by by-laws of 18 March 1751, *ibid.* 2 (1753), vi-xx. The list of all counsellors, adjoints, and associés-libres is given on xxi-xxx.

[66] The five volumes of memoirs were published in 1743, 1753, 1757, 1768, and 1774. A second printing of the first volume was issued in response to much demand in 1761. On the career and personality of Louis, see Dubois (1866).

[67] Gelfand (1973a), 170-172.

[68] *Ibid.*, 262-268, and below, Section 7.

[69] *Ibid.*, 187-189.

speaks education in most instances. The records of the apothecaries, on the other hand, are couched in the language of shopkeepers when it sounds natural, and of servility when it does not.[70] Turgot's abortive liquidation of the guilds was the crisis that provoked conversion of the Communauté des Marchands-Apothicaires et Epiciers into the Collège de Pharmacie, opened on the site of its old garden of simples in the rue de l'Arbalète on 30 June 1777. So far as the records show, this reform was one initiated by officialdom instead of from within. True, the apothecaries had long resented the "tyrannical pretentions" of their overlords of the faculty while scorning the grocers, with whom they were trapped in an "antiquated and fatal alliance"[71]—"that gaggle of obscure artisans who know no other talent than the skill of their hands, no other way to work than blind routine,"[72] so says Jean-François de Machy, contributing his rivulet to those cascades of disdain through which Voltaire once pictured the social process descending, or condescending, from the court to the crowd. Nevertheless, no Vicq d'Azyr, no La Peyronie, and no Quesnay rose up among them.

Evidently Turgot had invited the officers of the Company of Apothecaries to submit proposals for a new organization for the "profession of pharmacy" before his edict issued suppressing all the guilds;[73] and evidently the next ministry recognized that pharmacy was somehow different

[70] The discussion that follows is based largely on the records of the Apothecaries' Company and the College of Pharmacy. They are preserved at the Bibliothèque de la Faculté de Pharmacie de l'Université de Paris, and are in four ledgers that bear the following titles and call numbers:

No. 37, "Livre des Délibérations commansé (sic) le 15^{0} janvier 1671," which date is corrected on the cover to read 1677-1735.

No. 38, "Livre pour les Délibérations des Messieurs les Marchands apothicaires—épiciers communs . . . 1736-1776."

No. A-39, "Livre des Délibérations, 1777-1797."

No. 45, "Livre des plumitifs, commence le 29 7bre 1781." The published literature on the history of pharmacy tends to be on the antiquarian side. See notably, Laugier and Duruy (1837); Philippe (1853); Bouvet (1936a); and Prevet (1940), the last being a quasi-sociological work. For the college itself, and its forerunner, see more particularly *Centenaire de l'Ecole supérieure de pharmacie de l'Université de Paris* (1903); Toraude (1904); Cazé (1942); and Charles Bedel, "L'Enseignement des sciences pharmaceutiques" in Taton (1964), 237-257. Since 1913, the Société d'histoire de la pharmacie has been publishing a journal that has undergone several changes of title. Guitard (1963) has published an index to this collection.

[71] BFP, A-39, "Livre des Délibérations," fol. 6. Address by Trevez at the opening meeting of the college, 30 June 1777.

[72] *Ibid.*, fols. 8-11, Address by J.-F. Demachy, 16 July 1777, inaugurating the courses of instruction for the first academic year.

[73] "Délibérations des . . . Marchands-Apothicaires," Session of 18 May 1776, BFP, 38. The last entry in that ledger was 14 June, when commissioners were nominated. They were obliged to seek the views of the faculty on framing proposals for new statutes.

from other trades, for it alone was exempted from the decree restoring the corporations. Instead, the crown resolved to separate the "professions" of pharmacy and grocery, and to "erect" the former into a college. Presiding over the formal installation on 30 June 1777 was the lieutenant-general of police of Paris, the chevalier Jean-Charles-Pierre Lenoir, counsellor of state, a magistrate who in his responsibility for maintaining civic order was keenly conscious of the menace of disease, pollution, and quack medicine.[74] At his instance, as one of its first acts, the college sent a deputation to inspect all the *officines* where medicines were compounded.[75] Mutual compliments at the opening ceremony make clear that the regulations governing the College of Pharmacy had been drawn to Lenoir's prescription. The task was not, he told the assembly in his address, one to be "entrusted to the whole college." He also announced who their first provosts would be: les Sieurs Trevez, Brun, Simonnet, and Becqueret. "The choice I have just made would no doubt have been yours," he observed, and there was no demur.[76] Indeed, the records also show that those pharmacists, and they were not a few, who achieved some prominence in science or in public life[77]—the younger Rouelle, Sage, Pelletier, Baumé, Mitouard, Parmentier, Bayen, Cadet de Vaux, Coste, Petit, Dizé—took little part in the affairs of the college, except for teaching in some cases, and would plead the pressure of other obligations to escape service in its offices.[78] We shall need to say something later of members of the trade in its relations with the new chemistry.[79] For the moment let us consider them in respect to their inability, even gathered into college, quite to make good, as the surgeons had begun to do, their aspiration to be accorded the standing of a coordinate branch of medicine.[80]

In addition to its regulatory functions, the College of Pharmacy offered a program of instruction and always opened its academic year on the second Monday after Quasimodo. There were to be at least twenty-six lessons of chemistry, thirty of natural history, and twenty of botany, all including practical work. Serving as demonstrators of chemistry at various times were Mitouard, Sage, Deyeux, and A.-L. Brongniart (the founder of that scientific dynasty). Botany and natural history were taught by Par-

[74] He frequently referred such matters to Vicq d'Azyr and the Society of Medicine, and besides endowing several prizes proposed for competition by the Collège de Pharmacie (BFP, A-39, fol. 84), also endowed the much more important rabies competition sponsored by the Society of Medicine. See below, Section 6.

[75] "Délibérations," 5 September 1777, BFP, A-39, fol. 15.

[76] 30 June 1777, BFP, A-39, fols. 4-6.

[77] For biographical sketches of the more famous, see Blaessinger (1948, 1952).

[78] "Délibérations." Memorandum of Parmentier, 7 July 1785, BFP, A-39, fol. 91.

[79] Below, Chapter IV, Section 4; Chapter VI, Section 4.

[80] "Délibérations," BFP, A-39, fol. 19; The Prévots to Amelot: "La Médecine est composée de trois corps."

mentier, Valmont de Bomare, Buisson, and J.-F. Demachy, who made himself into one of Lavoisier's most embittered enemies. Courses were open not only to *élèves*, the word now used for apprentices, but to masters desiring to refresh and improve their own knowledge.[81] They for their part now began calling themselves pharmacists rather than apothecaries. The old name suddenly became infra dig. (Prior to foundation of the college, only the word "pharmacy" was current. It then referred to the whole process of preparing medicines. What an apothecary who practiced it kept was a shop—indeed, *apotheke* and *boutique* are from the same Greek root.) No change in commercial organization yet accompanied this terminological shift toward elegance, however. A young man wishing to enter the business still needed to take service with the proprietor of one of the licensed officines of Paris. Chronic indiscipline and insubordination among the élèves concerned the company collectively. A deliberation of 7 August 1782 produced a tightening of regulations, and required élèves to register their name, age, native town, and local lodging at the college and to cease drifting about Paris from unknown garret to unknown cellar. The reform worked. Subjecting the neophytes to oversight by the college "brought immediate relief from the innumerable complaints that Masters of the college were making of their pupils, and at the same time furnishes the pupils a means for making known their existence and their grievances, if by chance they have any."[82]

Given the history of famous lecturing by apothecaries in their shops, a tradition reaching back through G.-F. Rouelle and Nicolas Lemery to Nicaise Lefebvre in the mid-seventeenth century, it might seem surprising that the new breed of pharmacists attached so much importance to educational reform. The opening academic exercises occurred a fortnight after the founding on 16 July 1777, again graced by the presence of Lenoir, who was accompanied this time by Joly de Fleury, advocate general, and two lesser magistrates. "To be sure," acknowledged the first provost, Trevez, in his matriculation address, "our private laboratories were already schools where we trained people in the science of pharmacy." Being enabled to accomplish the same thing before the eyes of the public and in a dignified manner was what mattered most, however. For, he goes on, "our evenings devoted to discovering, to surprising the secret process of nature, lacked incentive. Now, Gentlemen, that we are authorized to make our teaching public, with what new brilliance will Pharmacy not shine?" There had been at least three occasions in the past, in 1700, in 1759, and again in 1768, when the apothecaries had tried in an unobtru-

[81] See the regulations for the courses, 17 March 1779, BFP, A-39, fols. 25-26; Bedel in Taton (1964); Cazé (1942), 16-17.

[82] "Délibérations," 7 August 1782, BFP, A-39, fol. 80.

sive way to sponsor instruction in their quarters and garden in the rue de l'Arbalète, only to be enjoined from continuing through the intervention of the Faculty of Medicine. Both parties to the recurring dispute always saw clearly that an institutionalized, officially sanctioned education pertained to a learned profession, and that the question was more one of standing than of quality. Now at long last the pharmacists could be masters in their own house of teaching: "Today, joined together into a college, everything has changed for us."[83] Nevertheless, some ambivalence was felt about the change, for it also entailed disadvantages. The old company had been one of the six chief municipal corporations, and its representatives had sat regularly in the consular or commercial courts. Once the guilds were restored, however, all the places were immediately taken up and the pharmacists excluded. But the king had promised that nothing would be lost by their conversion into a college. Should they insist upon their rights? There were those who thought they should indeed, and their provost, Becqueret, had to put it to them that such an expectation was probably unrealistic:

> It is a commercial court. True, a number of us are in business, but many are not, although they have the right to be. The king by the declaration designating us a college appears to wish to shift our position toward the learned societies in the same measure that he moves us away from the mercantile companies. The learned societies benefit from certain privileges and special immunities that we have the right to hope for. If we ask for our place in the Consular panel, won't we be closing the door on the favors we are requesting on the model of the learned societies? Do you suppose that we can have it both ways?[84]

Only nine years later was the college persuaded that it could not. On 14 January 1786 the provosts finally placed the choice before it in these terms: Either "to ask for the place in the consular jurisdiction that they had formerly occupied jointly with the grocers"; or else "We make immediate representations to his Majesty with the object of maintaining all the privileges of notables for the members of the College of Pharmacy, and particularly the right to occupy municipal posts." Forced to choose, the college came down on the side of notability rather than profitability, but not overwhelmingly. The vote was 47 to 35, with two abstentions.[85]

In practice Lenoir, though he did refer pharmaceutical questions and preparations to the college for its judgment from time to time, never al-

[83] 16 July 1777, BFP, A-39, fol. 9; for the earlier teaching, see Bedel in Taton (1964); Planchon (1893-95), BFP, A-39 (1896), (1897), (1898).

[84] "Délibérations," 28 December 1779, BFP, A-39, fol. 45.

[85] BFP, A-39, fol. 95.

lowed it the autonomy of an Academy. The record of its deliberations had to be presented periodically for his endorsement. Decisions of its executive committee, consisting of four provosts and twelve deputies, who were seldom all present at their meetings, and also of the rarely convened general assembly of the whole college, took effect only with his approval. Its minute book bears his signature on the first and last pages, certifying that there were 187 folios to fill up. When the secretaries made mistakes, they had to cross out and initial offending passages, rather than substitute fresh sheets, lest something be slipped into the record without his having ratified it—the verb is *homologuer*. The college finances were chaotic. On one humiliating day, 3 May 1781, a *huissier* from the Conseil du roi, one Marchais, appeared at a meeting unannounced, and on orders of his general commissioners read out to the college the requirements its officers would have to meet in their accounting practices. He then sat down and wrote them with his own hand into its minute book.[86]

The college appealed to Lenoir to be dispensed from literal compliance with these rules, as they did whenever they felt threatened by forces too strong for their uncertain self-reliance. The tone is usually that of hapless supplicants. Consider, for example, a petition that the provosts addressed to Maurepas on 6 September 1778. The occasion was the design of Vicq d'Azyr and Lassone to suppress the Royal Commission on Medicine, which verified the quality of pharmaceuticals and mineral waters, and on which surgeons and apothecaries were represented along with physicians. Its responsibilities and revenues were then to be assigned to the new "Société Académique de Médecine," as a leading pharmacist called the Society of Medicine, smelling an enemy as dangerous as any faculty.[87] A copy went to their protector, Lenoir. The petition opens with a protestation that they "believed in the good faith (*croyoient la Religion*) of His Majesty and that of our Lords of his Counsel," and continues:

> My Lord, we pharmacists are lost. Unless you condescend to extend your protection to this aspect of medicine, we shall fall into a state of complete demoralization. Citizens will then be at the mercy of any charlatan for their very lives. The worthy pharmacist will either abandon his situation or be forced to become a charlatan himself in order to survive. Such will be the unhappy fate of artisans without fortune who have merited the good opinion and benevolence of all the princes, of ministers, and of all the scientists of Europe, especially in the last half-century.

[86] 3 May 1781, BFP, A-39, fols. 65-66.

[87] It was Habert, an honorary provost by virtue of his post as Pharmacist at court, who got wind of the affair. See his memorandum to Trevez, 1 July 1777, "Délibérations," BFP, A-39, fol. 18.

> The Provosts of the College humbly beg your Magnificence to be kind enough, My Lord, to take into consideration, 1. how, subject to jealousy and invasion of their rights on every side, they are reduced almost to the state of beggars; 2. that if they are deprived of the honor of meriting the confidence of the government, they will abandon their effort to improve themselves, and ignorance will be favored of the sort in which ambition or jealousy would land them. What then will become of so useful a profession, which has enriched Physic, Medicine, and the Arts and Sciences?[88]

What indeed? For this petition echoed the litany of tradesmen and skilled artisans casting themselves in self-pity upon the protection of the great. Whatever the trade, its motifs were the same: they are unappreciated, not to say humiliated; they require support, moral no less than material; they deserve it because of their honesty, fidelity, and importance to society; without it they are lost, and they threaten to down tools in despair, or, what will be worse, to abandon their standards and prostitute their skills in order to survive at all. Not thus, it must be said, did scientists, surgeons, or forward-looking physicians need to advance their interests, not by appealing all plaintively for a recognition that their actual importance failed to command. But in what measure it was subject-matter, and in what measure social and economic position, that by comparison arrested the institutional development of pharmacy; and how it was that these or other factors affected one another, are problems best left for the sociologist of occupations to resolve. As of this writing, documents that would permit a thorough social history of French pharmacy from the sixteenth century into modern times survive there in the recesses of the present Faculty of Pharmacy awaiting full study and analysis.[89]

4. THE FACULTY OF MEDICINE

Evidently Vicq d'Azyr failed to deceive the pharmacists about the academic nature of his intentions (to come back to that) even though it was almost surely in hopes of disarming the suspicions of the faculty that he had adopted the name "Society," the homonym having been the harmless Royal Society of Agriculture, still a casual association of public-spirited proprietors and writers with enlightened views about farming, and as yet nothing of an academy.[90] The modest commission on epidemics that Tur-

[88] BFP, A-39, fols. 20-21.

[89] For a guide, see Dorveaux (1893), an inventory to the papers, which were sorted and classified in 1786 by one Fessart, "Maître-maçon . . . à la réquisition et en la présence de MM Bataille et Solomé, Prévôts, et du Sieur Saintotte, écrivain-déchifreur." Dorveaux was librarian at the Ecole de pharmacie at the turn of the century, and the author of numerous articles on details of its history.

[90] Below, Chapter V, Section 5.

got had instituted by edict of 29 April 1776 first met on 13 August. By the time of its third weekly meeting on 1 September, it was calling itself the Société de correspondance royale de médecine, and Vicq d'Azyr, having begun as "premier correspondant" had already become "commissaire-général."[91] His strategy, if strategy it was, worked for a while. The faculty paid no attention. It was not much interested in surveillance of epidemics—a "*triste rôle*" one spokesman later called it, observing that no objection could have been lodged against an outside body that confined itself to anything so thankless and demeaning.[92] It was a safeguard, were any needed, that four members of the commission had to be regent doctors of the faculty.

They cannot have been very faithful or attentive. No hint of larger designs reached their parent company for many months. Throughout the year 1777 the faculty was in any case preoccupied by fearful litigation attending its attempt to discipline one of its own members, a certain shameless Guilbert de Préval, for profiteering from a fake anti-venereal preparation.[93] Hence the faculty learned what was afoot only when the self-styled Royal Society of Medicine announced a very academic sounding public *séance* for the award of prizes to be held on 27 January 1778 in the great hall of the Collège de France. The faculty's dismay over the amplitude of the occasion will best be conveyed in the words of its own remonstrance:

> In bedecking itself with the title of Royal Society of Medicine and at its own initiative arrogating to itself the duty of fulfilling the obligations that such a name implies, the former Commission set up for epidemics invades the entire domain of medicine and goes about devouring the patrimony of the Faculty.[94]

For its part, the faculty would know how to defend its dignity and privileges against interlopers as it had often had to do in earlier times. First, however, the doctors wished to propose conciliation, and they named a committee to work out a basis. Let the new society become simply an elite body within the faculty; let its members be chosen by the faculty; let the dean be its vice-president ex officio; let the course of lectures in anatomy proposed by Vicq d'Azyr be given in the faculty; and let all commissions consulting with the government on medical matters contain at least two members of the faculty who were not also in the society—if granted these conditions, the committee on conciliation would advise the faculty that its interests were protected and that it might withdraw objection.[95] Vicq d'Azyr, however, had both court and ministry, seldom thus united, sol-

[91] HSRM *1* (1776/79), 1-16.

[92] Steinheil (1903), *2*, 107.

[93] Below, Chapter IV, Section 1.

[94] Steinheil (1903) *2*, 110.

[95] *Ibid.* *1*, xiii-xiv; *2*, 106-111.

idly behind him. Successive drafts of the provisional letters patent are in his handwriting.[96] They called for the king's head physician-select, François de Lassone, to be inspector. He it was, a former surgeon, who led the action publicly, and kept in daily touch with Vicq d'Azyr behind the scenes in the rue du Sepulcre.[97] Lassone rejected this pretended compromise for the takeover it would have been, and when the faculty persisted in hostilities, took the dean to task:

> I am thoroughly informed, Monsieur, along with all the public, of the immoderate, dishonest, it might be said indecent, attacks that several Doctors of the Faculty, among them members of its committee [of conciliation] have allowed themselves to deliver in the tumultuous assemblies of that company loudly and on more than one occasion attacking the Royal Society of Medicine and all who compose it. In particular, their hostility is exhibited even more decisively by the unsparing authoritarian, pretentious and even threatening tone they adopt, an attitude that would be barely acceptable for absolute masters to take vis-à-vis their disciples.[98]

Nothing chastened, the faculty then appealed to the Parlement of Paris to refuse registration of the letters patent incorporating the society, and at a meeting of 22 June gave the twenty-eight regent doctors who had joined the "pretended" society seven days to resign or be expelled from the faculty. Lassone thereupon turned to the keeper of the seals, Miromesnil, for a royal command forbidding the faculty to give effect to that resolution or to hold further discussions of any sort touching the Society of Medicine.[99] In the face of this prohibition, the doctors of the faculty called something very like a strike against the crown and for three months refused to tend the ill unless the new Society of Medicine be disallowed by the parlement. "We would prefer to be destroyed," said one resolution, "than to survive in the status to which we would not fail to be reduced. We will have the courage of ancient Roman Senators. We will die if need be on the ruins of the faculty."[100]

Now then, Des Essarts and many of his colleagues were sensible and respectable men. It would be too simple to hear nothing in these absurd assertions beyond collective self-serving, and to conclude from such supposed sophistry that the French medical profession differed from its predecessor of the previous century more in having been spared its Molière than in anything it had learned or in any service it rendered the sick. Mat-

[96] BANM, MSS 114, doss. 2.
[97] BANM, MSS 109; Steinheil (1903) 2, 106-118.
[98] Lassone to Des Essarts, 17 March 1778, Steinheil (1903) 2, 114.
[99] *Ibid.* 1, xvii; 2, 163-168.
[100] *Ibid.* 2, 111.

ters were more complicated. For one thing, it did have its La Mettrie,[101] and, more important, it was precisely in this same profession that serious reform was germinating.[102] Moreover, certain doctors of some importance began drawing away from Vicq d'Azyr and his new, perhaps aggressive society. Most of them were older, to be sure, notably Lieutaud, officially its president in his capacity of first physician to the king, in whose succession Lassone stood; Darcet, chemist and professor at the Collège royal; Bouvard, member of the Academy of Science and also professor at the Collège royal and half a dozen others.[103] They were replaced in 1779 by little known people of doubtful quality, one of whom, Carrère, became a source of scandal.[104] At all events, something more than mere obstruction was involved in the faculty's resistance. Though we cannot here go into the history of medicine proper, the episode does exhibit the issues between prescriptive right and science-inspired modernization in institutions that have been studied less than have parlements, nobility, or guilds. For the privileges of the university were congruent with theirs, equally abusive in the eyes of reformers and equally legal in the eyes of history.

The faculty spoke in accents of desperation, for it felt control of the profession slipping from its hands. Knowing itself to be politically powerless and without friends at court, it was thrown back like many another ancient institution upon the frail defenses of law and tradition. The regent doctors still went through their medieval paces in bonnet, fur, and gown.[105] Every year five from among the 120-odd members of the faculty were chosen by lot to elect a dean and to designate which of their number were to conduct the courses. In principle all graduates were licensed to teach, though in practice it was usually the same ones who did. Seven courses were the normal complement: pathology, physiology, pharmacy, chemistry, materia medica, Latin surgery, and French surgery. The last named had been instituted in 1720 as a concession to the illiteracy of sur-

[101] No one interested in the history of eighteenth-century medicine should overlook La Mettrie's *La Politique de médecin de Machiavel* (1746), *La Faculté vengée* (1747), and *Ouvrage de Pénélope* (1748).

[102] See works cited in Section 1, nn. 19, 22 above.

[103] Steinheil (1903) *1*, xxiii.

[104] A doctor of Montpellier, Carrère had evidently committed some impropriety that led him to take refuge from the consequences in Barcelona. There he put it about that the Royal Society of Medicine was open to venality in awarding prizes, and that the Académie royale de Barcelone was itself no better than it should be in this respect. A letter from François Sanpont, M.D., to Vicq d'Azyr, 15 May 1790, protests "que Nous les Cathalans aimons beaucoup la formalité, que la veracité est notre charactere, et l'honneur notre guide." BANM, MSS 109.

[105] For the Faculty of Medicine in the eighteenth century, see H. Varnier's introductory essay in Steinheil (1903) *1*; Corlieu (1877), Delaunay (1906), 1-27.

gical students, who, however, disdained to attend.[106] Every year on Saint Luke's day (18 October), the entering students were received following an opening mass heard by the entire faculty in full regalia. After three years, if the candidates were Masters of Arts of Paris, and four years otherwise, they might sit for the degree of Bachelor of Medicine. Various exercises and disputations qualified them for the licence in two more years. Thereupon they could practice; but to become doctors they had still to compose and defend a thesis. All of that droned on in Latin while students trooped about the capital paying fees for private courses and attending lectures at the Jardin du roi and the Collège de France. Waiting out the doctorate required five to eight years in all and an average expenditure of 7,000 livres, most of it in the successive examination fees on which the faculty depended to stave off creditors. Its own income was paltry. It had 1,200 livres per year from the revenues of the University. In 1720 the crown had granted an additional 1,800 livres on condition that its instruction be free.[107] Fewer and fewer candidates enrolled. In 1789 only sixty students were inscribed, and no doctorates had been awarded since 1786.[108]

To make matters more forlorn, the quarters had fallen into decay. The old medical school had been situated at the corner of the rue de la Bûcherie and the rue des Rats (now the rue du Hôtel-Colbert), hard by the left-bank gateway to the Hôtel-Dieu. The rafters had rotted, and the ceiling of the amphitheater sagged and threatened to collapse. Rather than come to the rescue with funds from the Treasury, Turgot in 1775 provisionally allotted the faculty space in the rue Jean-de-Beauvais in a somewhat less dilapidated set of buildings that the Ecole de droit was vacating for a modern installation. Such makeshift treatment made a painful contrast, not only with the faculty of law but, more pointedly, with the Academy of Surgery, whose new school the government had handsomely installed in the Palais des écoles.[109]

All this decadence makes it the more understandable that the Faculty fought so tenaciously to defend its ancient patrimony, to use the term to which it clung. For this was a quarrel that resumed issues older by far than any separating philosophes or men of science from government in the eighteenth century. They reached back to the time when the medieval monarchy was measuring its authority against the Church in inverse proportion to the liberties made good by the University of Paris, legally the eldest and in practice one of the least obedient daughters of the crown. Since the sixteenth century (to go no further) the crown and ministers had been attempting either to assert some control over the University or, fail-

[106] Delaunay (1906), 13; Gelfand (1973a), 89-146.
[107] Steinheil (1903) *1*, v; Delaunay (1906), 7-8.
[108] Delaunay (1906), 22. [109] Above, Section 3.

ing that—and they usually did fail—to found parallel institutions that would teach modern subjects and develop knowledge in an open manner, thus serving the public interest. Consider the Collège de France, the Jardin du roi, the Academy of Science and the other academies of Colbert's time, the Academy and College of Surgery, the College of Pharmacy, and now the Royal Society of Medicine, note how the pattern between government and the reform of knowledge has persisted into our own day: witness the Ecole polytechnique, the Ecoles de santé, the Ecole normale supérieure, the Ecole pratique des hautes études, the Centre national de recherche scientifique, the many specialized "grandes écoles" that have proliferated since the Third Republic. The institutional structure of French higher learning has a geological quality and consists of strata laid down in successive eras of French history. In suspended animation during the Revolution, even the fossils have survived and all function alongside their successors, beginning with the oldest, the University itself. The spectacle suggests that in the academic world it is much easier to start things than to stop them, whatever their actual vitality.

That the question was one of function as much as form will appear in the example offered by the Medical University of Montpellier, which in contrast to the faculty in Paris exerted a consistently inceptive influence upon the development of French medicine. Montpellier was very different in its allegiances from the University of Paris.[110] Its two higher faculties, Law and Medicine, had in themselves the dignity of universities. It had no faculty of theology before 1421, when one was created subsidiary to the University of Law. Montpellier was thus always more open to royal influence, which also tended to be exercised since there it could make a difference. Charles VIII and Louis XII created four royal professorships similar in conception and modernity to those with which François Ier founded the Collège de France in Paris. The religious wars tore Montpellier into factions, inviting changes and further interventions that the University of Paris escaped. Henry IV created two more chairs at Montpellier, of anatomy and botany, and of surgery and pharmacy. He associated a royal anatomist with the former for practical demonstrations, and attached the latter to the old botanical garden, which he greatly enlarged.[111] By the time of his assassination, the chancellor, a royal official, had displaced the dean in administrative authority; and under Louis XIV the famous legist François Bosquet, though himself a bishop, reorganized the entire university under the royal supremacy.[112]

An important history is waiting to be written about how the resulting flexibility accorded, first, with an old tradition of openness stemming from receptivity to Arab and Sephardic medicine; second, with the botan-

[110] A. Germain (1890). [111] *Ibid.*, 107-109. [112] *Ibid.*, 101.

ical lore of one of the great Mediterranean gardens; third, with the Rabelaisian spirit of skeptical innovation and common sense in the sixteenth century; and finally, with the more literary medical humanism that a Diderot, a La Mettrie, and many another psychic materialist read into or out of the practice of the philosophic doctors, notably Barthez and Bordeu.[113] Those are the elements of a story that certainly transpired, and that somehow issued in the vitalistic biology of a Lamarck and Bichat. Somehow, too, it had to do with the activity throughout the eighteenth century of the Royal Society of Science in Montpellier, the one provincial scientific society admitted to regular association with the Academy of Science in Paris.[114] We do not know how all that developed in detail, but the doctors of Montpellier were certainly the most sought after in France, and specially so at Versailles. In the sixteenth, seventeenth, and eighteenth centuries, the king's head physician was almost always a Doctor of Montpellier.

In the eyes of the Faculty of Medicine of Paris, chartering the Royal Society of Medicine was thus simply the latest in a centuries-long series of attempts on the part of the crown and ministers to invade its privileges and favor its rivals, whereas Vicq d'Azyr appeared to be a schemer, a latter-day Renaudot hand in glove with government, and not really a doctor at all since he had never practiced. The question was not one of money but of dignity. It had always pertained to the Faculty of Law to be the ultimate tribunal in questions of right, to the Faculty of Theology in questions of religion, and to the Faculty of Medicine in questions of health.[115] The very fabric of civilized order depended on the University, and the Faculty of Medicine did indeed pursue charlatans and quacks and occasionally bestir itself on matters of public interest and sanitation. It heard, for example, complaints from residents of Chaillot about the prospective pollution of the air to be feared from the firing of steam pumps that the brothers Périer proposed to install by the Seine.[116] To deprive it of such jurisdiction would be a humiliation beyond accepting—an *avilissement*—for then the faculty would be reduced to its teaching functions, and they constituted the least of its claims to honor and consideration.[117]

5. POLICE OF PUBLIC HEALTH

The issue was more than merely honorific and jurisdictional, however. An essential difference in the actual procedures of Society and Faculty will

[113] See the articles in the DSB by Ruth Cowan, "Barthez" *1*, 478-479, and Louis Dulieu, "Borden," *2*, 301-302.

[114] On the relations between the Academy of Science in Paris and the Montpellier Royal Society of Science, see Dulieu (1958a, 1958b).

[115] Steinheil (1903) *2*, 163-168.

[116] *Ibid.*, 319.

[117] *Ibid.*, 168-179.

appear from a contrast of their respective approaches to overseeing the distribution of mineral waters and licensing commercial medicines. Those responsibilities, it will be recalled, were taken away from the 1772 Royal Commission on Medicine (the Louvre Commission as it was known), which was thereby dissolved, and assigned to the Society of Medicine on the authority of its letters patent in August 1778.[118] In the dozen years prior to the Revolution, the Society thereupon examined over eight hundred pharmaceutical and herbal preparations—and finally boasted of approving only four, so high were its standards, so zealous its solicitude for the public.[119] Not that its commissioners had closed minds: they were much interested, for example, in well-authenticated reports of the therapeutic properties of a type of green lizard, native to central America and similar to species occurring in Spain and Southern France, which, so the Spaniards learned from the Indians, were valuable in treating skin diseases, chronic ulcers, and certain venereal afflictions. The lizards were eaten raw after removal of the head and entrails. There were hospitals in New Spain that relied on nothing else.[120] Tests needed to be made, of course. The remedy was known to long if primitive experience, however. It had been reported to the Society by a correspondent who had it from a published source in Spanish. Therefore, the Society approached it with less skepticism than it did the private concoctions submitted by some apothecary or *guérisseuse*—healer—requesting a license, an exclusive privilege, and often a subsidy from the government.

Many hundreds survive from among applications of this latter sort that had been addressed to the former Louvre Commission. There is a sense in which the later American term "patent medicine" is appropriate. These preparations were always called secrets in these documents. Their "possession" was said to go back to time out of memory. It had been handed down under all sorts of legatary conditions from father to son, uncle to nephew, master to apprentice, and was ever being infringed by some unscrupulous rival who only pretended to the formula and scandalously promised identical benefits under a different trade-name. A certain Demoiselle Hallaire, for example, rue Saint-Denis near the rue de la Ferronerie, possessed the composition of a *topique* which cured *les maux d'aventure* and healed the scars. Its ingredients were Diachilum magnuum, Diacalcitées, Virgin wax, and Oil. She and her mother before her had the secret from "time immemorial," and she petitioned the commission to renew her privilege and deny that of one Sieur Delaroche, who falsely claimed to be possessor of the same thing under another name and was

[118] Above, Section 3. The Procès-verbaux of the Royal Commission on Medicine together with voluminous supporting papers are in the BANM, MSS 111, 112, 113.

[119] Vicq d'Azyr, "Nouveau plan pour la médecine," HSRM 9 (1786-87/1790), 148.

[120] Vicq d'Azyr to Lenoir, 12 December 1783, BANM, MSS 114, doss. 2. Lenoir, lieutenant-general of police in Paris, had inquired after hearing of the merit of the lizards.

after a license "en surprenant votre Réligion—taking advantage of your good faith." She would be infinitely grateful.[121]

Similarly the Sieur Bermen Desbarils had been willed the secret of the true Manus Dei by his cousin, LeBrun the younger, whose family, that of the artist LeBrun, first painter to the late king, had possessed this incomparable unguent for over one hundred and eighty years. Perhaps it will be well to convey its properties in the language of the handbill, since not all of its virtues are readily translatable.[122]

> Il desseche & mondifie fort, fait renaître la chair nouvelle; il est extrêmement bon pour toutes sortes de plaies, tant vieilles que nouvelles, en quelques parties du Corps où elles soient, & ne peut souffrir corruption en aucun endroit où il soit appliqué, & empêche par conséquent l'excroissance de la mauvaise chair.
>
> Il unit les nerfs désunis; il attire à lui le fer, le bois, le plomb, le verre, les esquilles, & quelqu'autres choses qui soient dans les plaies.
>
> Il guérit toutes sortes de morsures & de piqueures, attirant à soi par une vertu singulière subitement le venin.
>
> Il guérit les Panaris, le Fouchet, les Chancres, Fistules, Charbons, Ecrouelles, Ulceres, Scorbutiques, Ankilos & Cancers non adhérans, comme d'autres maux qui arrivent aux Mamelles.
>
> Il guérit les Arguebuzades, & éteint le feu Persique.
>
> Il guérit les Engelures & brûlures de feu ou d'eau bouillante; il fait des merveilles pour la rupture, & guérit du Fic, tant interne qu'externe.
>
> Il arrête les douleurs du dos, des reins, de la goûte & Rhumatisme.
>
> Il fuérit les Fluxions de poitrine abandonnées, la Plurésie & les Hémorroïdes. Si vous mettez l'Emplâtre sur la Peste, il l'empêchera de passer outre.
>
> Il fait mûrir & guérir toutes sortes d'Apostumes, Glandes & Loupes.
>
> Bref, il est encore bon à beaucoup d'autres maux que tous les jours on éprouve; & il y a eu des personnes auxquelles on étoit prêt de couper la jambe, la main, ou doigt de la main, comme à d'autres la mamelle, lesquelles en leur appliquant l'Emplâtre, sans autre chose, ont été entiérement guéries; ainsi qu'il se verifiera par les Certificats de ceux qui ont été guéris. Il a autant de vertu au bout de cinquante ans, que s'il avoit été composé le même jour.

[121] Among supporting papers, Mlle Hallaire submitted a certificate of approval signed by Sartine, lieutenant-general of police, on 31 May 1766, BANM, MSS 115.

[122] BANM, MSS 115.

In Paris the trade flourished along the Pont Neuf and in the rue Saint-Denis. By royal edict of 25 October 1728 the lieutenant-general of police had issued permissions on the recommendation of a committee of doctors under the nominal oversight of the king's head physician. It was in hopes of abating venality that the Louvre Commission was instituted in 1772. The thousands of existing licenses were to be reviewed, and the commission was to authorize the police in Paris and the provinces to renew only those that had merit. The king's head physician, his head surgeon, the dean of the Faculty of Medicine, and the permanent secretary of the Academy of Surgery—in all eight physicians, five surgeons, and four apothecaries—made up its membership. Inclusion of the latter elements disquieted the faculty, even though the administration was confided to the dean and a small staff of clerks. Though intended to reform matters, this commission does not appear to have changed anything in its execution of its task. For, nothing if not traditional in spirit, it too judged of the merit of pharmaceutical claims in the petitioners' own terms, by pedigree and not by verification of clinical consequences. Here is the form in which an authorization would be delivered, this one by Antoine-François Prost de Royer, lieutenant-general of Police in Lyons: "We certify and attest to all whom it may concern that the Sieur Charles Rousselot, brevetted by the king and the Royal Commission of Medicine for the composition and distribution of a specific called Beaume Aromatique has resided in this city for the period of two and a half months and has there compounded and distributed this preparation with success and with no complaints."[123] Not that the physicians, the surgeons, and the apothecaries, who sat together all suspicious of each other on the Louvre Commission, were indulgent of quackery. Quite the contrary, but they did approach the duty of policing on premises different from the scientific orientation of Vicq d'Azyr's Society. Tacitly they all assumed that a prescriptive privilege made the proper basis for doctoring the public, whether by book, knife, or potion.

It is unlikely that hawkers of orviétan and mithridate were more brazen late in the century than they had been earlier. What is certain, however, is that the scandal appeared to be altogether more inadmissible in the eyes of doctors animated by the new spirit of civic and scientific reform. Like many another abuse, it offered the chance for a crusade. Seizing the opportunity, Vicq d'Azyr analyzed the problem in a very revealing memoir drawn for Lassone's signature, and annotated by him, in the spring or summer of 1778. In it they developed, at the request of ministers, the reasons behind the incompetence of the Louvre Commission, and

[123] BANM, MSS 112.

proposed that its functions and its budget be assigned instead to their new Society by the letters patent even then being framed.[124]

A fundamental mistake, so they urged in confidence, had been to give surgeons and apothecaries equal voice and equal authority with physicians. The three elements of the old commission were always at odds; and (which was more serious) the consulting of surgeons about anything except external medicines violated every principle of good order, whereas including apothecaries in such decisions introduced ignorance combined with conflict of interest. Professional standards—Vicq d'Azyr's word was *émulation*—would never develop in a tribunal so constituted, and they were the essential factor in any form of technical jurisdiction. Its monthly meetings, moreover, were too infrequent for a serious examination of the applications before it. Finally, it had no systematic contact with any agencies in the provinces, where the plague of charlatans was even more murderous than in the capital. "It is impossible that a tribunal of this nature could be sufficiently imposing to intimidate and curb so vile, so bold and so numerous a species as that of charlatan."[125] That was the reason, observed Lassone in an aside, why Lieutaud and he, though heads ex officio of the Louvre Commission, had never wished to chair its sessions and had preferred leaving its administration to the dean of the faculty. But precisely there lay a deeper difficulty. Physicians from the faculty knew little chemistry and were in the wrong sort of association altogether for exercising a scientific responsibility. Faculties were scholastic bodies, apt only for transmitting to beginners the elements and relations already established in a subject, and were no good for testing or research, which functions must be kept distinct from teaching.

This document makes it clear, as indeed do many others, that the Royal Society of Medicine had become a working institution during the two-year interval between its origin in Turgot's epidemic commission and its incorporation in August 1778. All the disqualifications of the faculty were so many reasons for vesting the responsibility in the nascent and vigorous Society, and its intention to eliminate the surgeons and apothecaries offered another instance of what other trades were also experiencing as the domineering and aggressive tendencies of professionalization in science. The Society itself, on the other hand, must be a company of medical

[124] The draft memoir is entitled "Réflexions sur l'établissement de la Commission de Médecine destinée à l'examen des remèdes et à la distribution des eaux minérales par une déclaration du Roi du 25 avril 1772." According to the preamble, it had been requested of Lassone by Maurepas and Amelot. The latter, Antoine-Jean Amelot de Chaillou, had succeeded Malesherbes in 1776 as minister of the Maison du roi, and held the post until 1783. His family had close ties with Maurepas, and he had previously been intendant of Burgundy.

[125] *Ibid.*, pt. I, fols. 10-11.

peers, an elite without distinction. Within its membership the highest equality must reign. To that end, Vicq d'Azyr and Lassone in drafting the letters patent provided for only the one grade of associate and stipulated that all offices must rotate with the exception of those of president, the king's head physician ex officio (who was also to be inspector general of epidemics), and secretary (who was to be commissioner general of epidemics). Meeting twice in every week, the Society was already sustaining the intensive pace required for prosecuting charlatans. It had information from the provinces through its correspondents, all of whom would constitute so many agents of a medical police in their localities. This enlistment would fortify their sense of civic importance, an effect that Vicq d'Azyr was constantly striving to create on all personal and professional grounds. Participating in a national campaign to root out quacks and "Empiricks" could be expected to appeal strongly to physicians and medical institutions in other cities, where charlatans always repaired when Paris got too hot for them.[126] To strengthen the legal basis for such actions in the provinces, it would be well if the letters patent establishing the Society were to be registered in every parlement in the realm.[127]

Behind all these organizational factors, the decisive advantage the Society could bring to regulating pure drugs and mineral waters was scientific, for it was in a position to call on those members of the medical profession who were expert in chemistry to conduct tests. Vicq d'Azyr mentioned particularly Bucquet and Macquer. Also enrolled were colleagues with special knowledge of anatomy, botany, natural history, and physics, notably Bouvart, Lorry, Mallouët, and Geoffroy.[128] Money would be needed, of course, and the revenues of the Régie des eaux minérales, which were currently paying the expenses of the Louvre Commission, could provide an appropriate and convenient source of funds.[129] It is further evidence of Vicq d'Azyr's influence at court that the Maison du roi acted favorably on every detail of this *démarche*. The hapless dean, Des Essarts, took it hard when Amelot directed him to turn over all his books and balances, the more so when Vicq d'Azyr refused to appear at the final

[126] *Ibid.*, pt. II.

[127] "Mémoire tendant à prouver que l'enregistrement, dans tous les Parlemens du royaume, d'une déclaration du roi, portant établissement de la Société royale de médecine est indispensable, et qu'il ne peut y éprouver aucun obstacle." BANM, MSS 114, doss. 23. See also Maret to Vicq d'Azyr, 27 October 1783, MSS 156, informing him that the provisions against quack medicines could not be enforced in Dijon until the regulations were registered with the parlement there. It was through Maret that Vicq d'Azyr maintained liaison with the Academy of Dijon.

[128] The memoir "Réflexions" *loc. cit.*, n. 124 above, pt. II, fol. 5.

[129] The memoir just cited concludes with drafts of paragraphs to be added to the letters patent effecting the abolition of the Louvre Commission and the annexation of its funds and budget by the Royal Society. See also, HSRM *1* (1776/79), 17-24.

stated meeting of the Louvre Commission to receive them publicly.[130] The minister, he wrote, a little haughty in his victory, had given him no authority to attend that body, a step he could not take without knowing the wishes of the Society of Medicine, which had not been apprised of the matter. After he had consulted his colleagues, he would meet Des Essarts privately.[131] Finance, however, turned out to be the one respect in which Vicq d'Azyr disappointed the confidence of the ministry. He proved to be less than meticulous in his own accounts. On one occasion Amelot had to chide him severely.[132]

In a much later accounting to a legislative committee of the revolutionary Constituent Assembly, Vicq d'Azyr called the grant of letters patent

[130] See the "Extrait des Registres du Conseil d'Etat," 6 October 1778, BANM, MSS 114, doss. 21.

[131] Amelot to Des Essarts, 20 September 1778, and Vicq d'Azyr to Des Essarts, c. 5 October 1778, BANM, MSS 113; dossier containing the plumitif of the meetings of the Louvre Commission, No. 91 (5 May 1777) to 110 (5 October 1778), BANM, MSS 108.

[132] Amelot to Vicq d'Azyr, 22 June 1782, BANM, MSS 108, Amelot also complained that he had received no accounting for 1780 or 1781.

In the first fiscal year of the Society's operation, 1 October 1778-30 September 1779, the funds from the Eaux minérales amounted to 17,7000 livres. The most considerable items were 6,000 livres for the Secretariat and 6,000 livres to redeem the jetons or tokens to which associates, including independent and foreign associates, were entitled on attending meetings. Each jeton was worth 37 sous at the Monnoie de médailles, and the Society was awarded 3,274 to distribute annually (see the last document cited in this note, Par. D). Remaining items were in the amount of 900 livres for prizes, 800 livres for costs of correspondence, 600 livres to the two commissioners who tested mineral waters sold in Paris, and 2,800 livres in pensions for personnel of the old commission. See the "Etat" signed by Amelot, 11 November 1779, BANM, MSS 114, doss. 7.

In addition to the above expenses, the six doctors who had been appointed to the original Commission on Epidemics in 1776 were still entitled to stipends of 1,500 livres, each charged to the Treasury. In 1786 the government decided to enlarge the Society's authority over public health (below, Section 5) and at the same time laid down that in principle it wished to afford all thirty associate members a uniform stipend of 600 livres. To those ends, it increased its direct support to the sum of 23,200 livres in addition to the proceeds of the Eaux minérales. Vicq d'Azyr was to draw 7,400 livres from the increment for his stipend and that of his chief clerk (1,200 livres) and for his expenses in running the office. The goal of equal retainers for all associates had to await the death of those to whom larger commitments had already been made, and for the time being there were to be five stipends of 1,500 livres, three of 500 livres, fifteen of 400 livres, and one of 200 livres. (See "Extrait des registres du Conseil d'Etat," 24 April 1786, BANM, MSS 114, doss. 23.)

By 1790 the needs of the Society amounted to 36,200 livres annually. They are detailed in a resumé drawn by Vicq d'Azyr for the information of a committee of the Constituent Assembly ("Observations et Eclaircissements," BANM, MSS 114, doss. 2). Rising costs of printing and additional expenses incurred by commissioners in conducting tests accounted for the increase. From this document, it is evident that Vicq d'Azyr employed two clerks to conduct the business of the society, the chief of whom was an educated man, "père de famille," capable of handling letters in Latin and foreign languages. In addition the staff consisted of a Swiss guard and two office boys (*ibid.*, sections H, F).

in August 1778 the moment when "the Society founded in 1776 as the bureau of epidemics" took "a truly academic form."[133] Whatever its initial aggressiveness, or perhaps in justification of that, it also accepted its responsibilities seriously, meeting twice a week all year long, alone among the academies and learned institutions of Paris in allowing itself no vacation.[134]

The provincial faculties, which had for the most part trained its correspondents, were natural allies in the campaign against the privileges of the Faculty of Medicine in Paris, and the new Society engaged itself from the outset to form with them a "single body" of medicine.[135] Already in 1777 Vicq d'Azyr had approached fifteen of those bodies, which ranged in responsibility from the admirable institution at Montpellier to the diploma mill at Rheims, together with nineteen colleges of medicine, some of which offered instruction but granted no degrees.[136] Except at Besançon and Bordeaux, their deans welcomed the overture. The cordiality of the invitation contrasted favorably with the distance at which the Paris faculty had always held provincial medicine, and the prospect of the association was stimulating. The membership of the College of Medicine at Clermont-Ferrand, for example, resolved to meet more frequently and to exchange regular observations among themselves before selecting the more significant to send to Paris.[137] In Lille the dean was pleased to have his arm strengthened in the perpetual battle against surgeons and pseudodoctors.[138] On one occasion, Vicq d'Azyr's enthusiasm outran his tact. Toulouse was ravaged by an epidemic of uncertain etiology in the spring of 1782. Afterward the Society of Medicine resolved, in a spirit of largesse, to recognize outstanding instances of devotion to duty on the part

[133] "Observations et Eclaircissements," BANM, MSS 114, doss. 23, section A.

[134] *Ibid.*, sections D, F.

[135] HSRM *1* (1776/79), 2, preface.

[136] *Ibid.*, 34. The minutes summarizing these arrangements are in BANM, MSS 110. The Society had arranged formal association with the University of Medicine in Montpellier; with the Faculties of Medicine in Toulouse, Caen, Rheims, Perpignan, Strasbourg, Nancy, Nantes, Douai, Poitiers, Bourges, Angers and Aix; with the Colleges of Medicine in Lyons, Bordeaux, Rouen, Marseilles, Rennes, Nancy, Lille, Orléans, Grenoble, Béziers, Nîmes, Montauban, Dijon, Le Mans, Clermot-Ferrand, Abbeville, Moulins, Amiens, and Dieppe; with the medical "corps" of Troyes and of Saintes; and finally with the medical committee of the Academy of Dijon. He had also approached the faculties of Bordeaux and Besançon (MSS 110), but apparently they held aloof, for they are not mentioned in the published source. On the provincial medical faculties, see Finot (1958). It should be noted that colleges of medicine were essentially local medical associations. Some of them made arrangements for instruction to be given to candidates, who presented themselves for degrees before a convenient faculty when ready to satisfy the requirements.

[137] Duvenin (dean of the College) to Vicq d'Azyr, 25 November 1777, BANM, MSS 109, enclosing an extract from the registers of the college.

[138] Dean Boucher to Vicq d'Azyr, 29 August 1777, BANM, MSS 109.

of local doctors by awarding medals. Unfortunately, Vicq d'Azyr had omitted to consult the faculty at Toulouse first. It stiffly repudiated the announcement, not wishing the professional conduct of its members to be subject to the judgment of outside bodies, be they ever so progressive.[139]

6. MEDICAL PRACTICE

Epidemics and epizootics having provided the occasion for founding the Society, preventing and abating these catastrophes continued to be its central preoccupation in the 1780s. The published papers contain many a medical topography of some particular region or province, often very learned. The contributor who wrote on Marseilles, for example, carried the history of its climate and medical peculiarities back to Pliny and Roman antiquity.[140] In every volume the memoirs open with a lengthy recapitulation of the "constitution" of the year in Paris. First Charles Lorry and, after his death in 1783, Etienne-Louis Geoffroy rehearsed what the weather had been there month by month, and related its vagaries, its frosts, heat waves, floods, humidities, droughts, miasmic mists, unseasonable contrasts, and its rare fine intervals, to the infamous variety of catarrhs, dropsies, diarrheas, dysenteries, rhumatisms, fevers (putrid, intermittent, tertian, quartain), fluxes, scurvies, and other miseries that afflicted the populace. That long-suffering, that age-long suffering attribution of symptoms to a suspect climate has survived into the twentieth century mainly in the person of the concierge or aged farmer's wife enlarging on her aching joints. In the eighteenth century, it amounted to a meteorological folk theory of disease, a set of explanations in which the practicing doctor, like his ultimate Hippocratic predecessor, largely shared. Not much besides the weather and environmental circumstances could be known about the causes, after all.[141] That such was the profes-

[139] Dean Dubernard to Vicq d'Azyr, undated but probably April, 1783 (the prizes had been announced on 11 March): The Faculty "attend de la politesse de M. Vicq d'Azyr, et de son amour pour la verité, qu'il voudra bien expliquer par la voie des papiers publics le veritable sens de cette annonce. On pourroit en induire que les médailles ont été acceptées par la faculté, erreur de fait qui blesseroit sa delicatesse." BANM, MSS 109.

[140] The author was a physician called Raymond, MSRM 2 (1777-78/1780), 66-140.

[141] It is questionable whether this simple fact will quite bear the burden of anthropological terminology, psychological analysis, and philosophical construction that Foucault (1963) puts on it. If eighteenth-century doctors had possessed the technical and institutional resources of nineteenth-century medicine, I doubt that what he calls their "regard" would have been very different from that of the scientific medicine that they aspired to develop, and in fact contributed to developing. It puts the cart before the horse to attribute the elaboration of technique to some shift in attitude. The actual writings of these doctors in their correspondence do not exhibit a perception of the world so different from ours or so discontinuous with it as to justify the mystification entailed by calling the study of their outlook archaeology rather than history.

sional as well as the common view accounts for the otherwise surprisingly general response to the most onerous duty that Vicq d'Azyr laid on the correspondents of the Society. They were enjoined to furnish themselves with precise and indeed expensive instruments from Paris (an artisan called Mossy made the best)[142] and to record the specifics of the weather on a form provided them. Temperature inside and outside the house, atmospheric pressure, humidity, precipitation, the strength and direction of the wind, and the state of the sky and clouds—all these data were to be noted morning, noon, and evening at the same hour of the day, every day of the year. So, too, were manifestations of the aurora borealis, declinations of the compass needle, and any other incidents of meteorological interest.

Thus did correspondents of the Society become a collective weather bureau. "In order to fulfill the wishes of the Society," observed Vicq d'Azyr appreciatively, "they hastened to take on themselves a type of work that seemed hardly compatible with their manifold responsibilities."[143] The survival in its archives of much of the information they dutifully reported has excited the interest of the new school of quantitative historians who, the computer at their fingertips, now find opened to them the opportunity of determining the climate over an unusually critical period for social history.[144]

It turns out that the fifteen-year span from 1776 to 1792 knew somewhat colder winters and warmer summers than France currently experiences, that the summers in the late 1780s were very hot, and that hail ravaging the crops on 13 July 1788 was the more destructive psychologically for having climaxed a summer already blighted by excessive heat. Fear of a repetition amid the resulting dearth of 1789 compounded the apprehensiveness that unnerved the peasantry in the following summer of Revolution.[145] Such is the certainty that rewards a modern analysis of these old data. Indeed, if the forms supplied by the Society to its informants for their thrice-daily weather watch had actually been imagined with the convenience of future key-punchers in mind, they would not have needed to be much different. The facts were entered in a uniform manner on thousands of those sheets, the back of which provided space for reporting medically interesting conditions and events in each correspondent's region. The design was that of the coordinator of the project, for whom

[142] HSRM 5 (1782-83/1787), 245-246.

[143] Prefatory note to "Observations météorologiques faites par les correspondants de la Société pendant les années 1777 et 1778, et redigées par le R. P. Cotte," HSRM 2 (1777-1778/1780), 92.

[144] O. Muller has analyzed the data technically and compared the results with modern values, Desaive *et al*. (1972), 90-134.

[145] J.-P. Desaive and E. Le Roy Ladurie, "Le climat de la France (1776-1792)," *ibid*., 23-61, esp. 59-60.

the computer would literally have been the answer to prayer, Père Louis Cotte, vicar of the parish of Montmorency, a weather-wise abbé whose interest in the heavens was ever more meteorological than theological.[146]

Father Cotte was a single-minded enthusiast of a type not uncommon on the fringes of eighteenth-century science. L'Hermitage is not far from Montmorency, and he and Rousseau sometimes went herborizing together. The brothers Montgolfier and their balloons, the abbé Rozier and his journal, Duhamel du Monceau and English husbandry, Bourgelat and enlightened horse-breeding—they rode their hobbies under the banner of utility, and were attractive figures in their zeal. Without any pretension of becoming great savants, they labored to advance their causes rather than their reputations or themselves. Cotte was an Oratorian priest, educated in the colleges of Soissons and Montmorency, and bitten by the weather bug in early life. He devoted the free time afforded by an ecclesiastical career to making a science out of the sparse chaos of meteorology. For that purpose, data needed to be collected in comparable form and correlated. Like many fact gleaners, Cotte was a person of moderate intellectual attainments who eschewed any thought of theory. He would work for tomorrow, not for today, except insofar as accurate digests could be worthwhile to agriculture and medicine, the two occupations that needed information about the weather.

The *Traité de météorologie* that Cotte dedicated to the Academy of Science in 1774 is a compilation, not a book, a diffuse and incomplete one in the judgment of the author himself.[147] He regretted that Jean-André de Luc's more comprehensive treatise on the barometer and thermometer had not been completed in time for him to profit from its great superiority to his own efforts.[148] The most valuable parts of Cotte's treatise were certainly the exact description of instruments and the instructions on how to make observations, all of which Vicq d'Azyr later prescribed to his corps of medical weathermen.[149] For the rest, Cotte had assembled what observa-

[146] On Cotte, see Kenneth L. Taylor in the DSB *3*, 435-436; and Silvestre (1816). An example of the form employed by the Society is reproduced in Desaive *et al*. (1972), 28.

[147] *Traité de météorologie* (1774), parts of which were revised for inclusion in his further *Mémoires sur la météorologie*, 2 vols. (1788). Cotte wrote a large number of articles for journals, notably the *Observations sur la physique*, and composed elementary books on natural history, physical science, agriculture, and mechanics for schoolchildren, and also on health and on bread for country folk. See Taylor's bibliography in the DSB *3*, 435-436.

[148] Cotte, *Traité de météorologie*, xi-xii. De Luc's *Recherches sur les modifications de l'atmosphère, contenant l'histoire critique du baromètre et du thermomètre* . . . (2 vols. Geneva, 1772), had been ten years in the writing. Cotte thought that the work would bring about a revolution in the study of meteorology.

[149] Cotte, *Traité de météorologie*, books 2, 99-210, and 5, 517-528; HSRM 5 (1782/87), 245-246; Desaive and Le Roy Ladurie in, Desaive *et al*. (1972), 31-37. On eighteenth-century thermometry, see Birembaut (1958a) and Middleton (1966); and on barometry, Middleton (1964).

tions he could find written down by Toaldo at Pisa, Gabry at La Haye, Tully at Dunkirk, Poczubut at Vilna, and other interested parties. His most important source was the series of "Observations botanico-météorologiques" that Duhamel du Monceau had been contributing annually to the *Mémoires* of the Academy of Science since 1741.[150] In Paris Messier had been keeping up a professional weather watch in his observatory at Cluny since 1763.[151]

Beginning in 1765 Cotte himself maintained detailed observations in Montmorency, correlating the behavior of the atmosphere with the episodes of the agricultural year, the dates of planting and harvest, the yield of crops and orchards, the hatching out of insects, the migration of birds, and also with the vital statistics of his parishioners, their births, marriages, diseases, and deaths. His medical topography of Montmorency was later singled out for a gold medal by the Society and held up by Vicq d'Azyr as a model of the genre.[152] All these efforts were piecemeal, however, permitting Cotte no conclusions about the causes or sequence of climatic events from time to time and place to place. Only a lengthy series assembled in many locations over long years would ever allow of that. What a godsend, then, when organization of the Society of Medicine put at the disposal of his purpose an entire corps of intelligent volunteers able and eager to take part in such an enterprise. It entailed only a slight concession to shift his emphasis from the botanico-meteorological to the medico-meteorological generalities of his annual compilations in the *Histoire* of the Society. The summary tabulations that Cotte there published remain his monument.[153] They record the main elements of weather month by month in some fifty-odd locations, for the most part in France.

It has been said that the fidelity of the correspondents proved bootless since the Society failed either to formulate a synthetic theory of the weather or to specify the effects of climate in the etiology of disease.[154] No one expected such success in so short a time, however, and meanwhile the reputation of the Society increased in everything relating to public health. The historian reading at random in its archives (there being no other way to explore them in their present state) comes upon frequent exchanges of information with magistrates and ministers that show them treating the Society for all the world like a ministry of health. Thus, we

[150] *Traité de météorologie*, vii, on Duhamel, see below Chapter V, Section 2, note 10.

[151] The Academy of Science published a memoir on the data in Messier's journal, and elected Cotte to corresponding membership in 1769 and then encouraged him in his plans for the *Traité* (v).

[152] HSRM *3* (1779/82), 61-83.

[153] For a professional estimate of their value, see Kington (1970).

[154] Jean Meyer's introduction to Desaive *et al.* (1972), 19; cf. E. Le Roy Ladurie, in *ibid.*, 26-27, who holds the inquiry to be an example of "*conceptions irrationelles*" and false postulates fortunately producing real data.

have the detailed record from 1784 through 1787 of inoculations for smallpox that the Intendant Caumartin de Saint-Ange required local doctors in the Franche-Comté to submit. In the villages for which records survive, we may know the name and age of every child inoculated and the reaction it experienced together with the identity of each father of a family. In the year 1787, the communes in the district of Pontarlier were served by one Nicod, a physician and surgeon of Frane. In the village of Saint-Pierre, Sophie, the three-year-old daughter of Jean-Louis Fournage, had a benign reaction. Her 18-month old brother Philippe had an "abundant" one. The consequence for the six-month old Marie-Célestine Maire of Mont-Perreux was "confluante."[155]

Similarly, the intendant at Lille, Esmangart (he who had been at Bordeaux in 1774 when Vicq d'Azyr was undertaking his mission on the cattle plague) wrote to compliment the Society on a general memoir concerning the nature and treatment of epidemics, a document that Vicq d'Azyr had circulated to all provincial administrations. Esmangart had sent his copies to leading hospitals and doctors, and asked that many more be printed so that he might distribute them to enlightened persons in the garrison towns of Artois and Flanders. He would need three or four hundred, and thought the government should pay the cost.[156]

At the other end of the country, Raymond, intendant of Roussillon, requested the Society to persuade the government to disallow an action of the Bureau of Health in Perpignan. "Composed in large part of illiterate members and a number of Levantine merchants, that local body had just named an altogether uneducated apothecary to be physician at the lazar house. These people believed themselves to know more about the plague than did the doctors of the Medical College, even though that company had been fulfilling its functions for twenty-two centuries, ever since the founding of Marseilles in Roman times. (Such was the reputation of Mediterranean medicine for antiquity and continuity.)[157]

Evidently a service providing information for the prevention of epidemics, and specifying procedures for their management when they nevertheless struck, was then being tried out in several generalities. In January 1785 Raymond wrote again, this time to the controller-general, urging that the success of the program in Paris, Grenoble, Lyons, Soissons, and Poitiers justified its extension to Roussillon. His province was too poor to stand the expense, however, which he begged the Treasury to assume. Instituting those provisions would have entailed appointing a chief physician for the generality at a salary of 600 livres, designating a doctor and a surgeon in each election to treat the sick on orders by the subdelegate,

[155] BANM, MSS 156, piece 183.
[156] Esmangart to Vicq d'Azyr, 26 June 1785, BANM, MSS 108.
[157] Raymond to Vicq d'Azyr, 26 June 1758, BANM, MSS 108.

ensuring regular contact between them and the chief physician, maintaining a supply of medicaments in each subdelegation, instructing the syndics of every parish to inform the subdelegate whenever four or five persons fell gravely ill (subject to a 50-livre fine for negligence), supplying bouillon to be distributed to the poor by the curé on order of the health officers, and reporting regularly to the intendant on the course and possible cause of each infection.[158]

Already Raymond had established his own procedures in consultation with the Society of Medicine. Whenever and wherever a fever broke out, he despatched a doctor from Perpignan, accompanied sometimes by a surgeon and supplied with all the quinine to be had, and also with rice, flour, and money to buy meat of good quality to make bouillon for the poor. At all times sanitary measures were to be observed throughout the generality. Manure heaps had to be kept outside the dwelling area. Pigs had to be tied up, also outside. Streets had to be kept clean and paved where possible. Stoops and lean-to's impeding circulation of the air had to be cleared out of the streets. Stagnant ponds and marshes had to be drained. Despite occasional obstruction by the local council, Raymond flattered himself that his generality owed its relatively healthy state to his having instituted and enforced these regulations.[159]

The worst fears of the Faculty of Medicine were borne out, for ministers as well as magistrates began consulting the Society from the outset. Early in his first term in the Ministry of Finance, Necker authorized a Treasury grant of 400 livres for new experiments to determine whether "ergot" was the cause of a certain type of fever that regularly followed the harvest in regions where the important grain was rye.[160] In 1784 Calonne in his turn inquired whether there was medical justification for a regulation of 1713 prohibiting the distillation of marc, a brandy prepared from lees.[161] The previous year the maréchal de Castries, minister of marine affairs and colonies, requested the judgment of the Society on the most nutritious ration for sailors when at sea and also when in hospital. In addition, he needed copies of the adverse report that the Society had earlier returned on a mixture prepared by one Faure de Beaufort to preserve supplies of water on shipboard.[162]

The health of garrisons and civil-military relations in garrison towns

[158] Raymond to Controller-General (copy), 25 August 1785, BANM, MSS 108. For the working of this system in Brittany, see Goubert (1974), 108-118.

[159] Raymond to Controller-General, 27 February 1786, BANM, MSS 108.

[160] Necker to Lassone, 24 July 1779, BANM, MSS 108.

[161] Calonne to MM de la Société royale de médecine, 9 April 1784, BANM, MSS 108.

[162] Du Fresne (for the maréchal de Castries) to Vicq d'Azyr, December 1783, BANM, MSS 108. The Society replied with a lengthy memoir on nutrition, "Rapport sur plusieurs questions proposées par M. le Maréchal de Castries . . . relativement à la nourriture des gens de mer," HSRM 7 (1784-85/1788), 221-294.

were a constant worry for military and naval administrators. In 1779, following French intervention in the American Revolution, an epidemic broke out among English prisoners of war in the citadel of Dinan. Townspeople and the municipal administration took alarm, for the disease was more severe among the local people and took a higher toll among them than among the soldiers who first harbored it. Appealed to by the authorities, the Society sent a mission to Dinan. Two commissioners, Poissonier and Jeanroy arrived post haste in February. After inspecting the citadel, Jeanroy identified the causes: "1. A lack of cleanliness inherent in the English population, 2. the bad air resulting from that, 3. the dampness and chill of the installation, 4. a state of apathy and indifference to life difficult to imagine."[163] (A postscript to his preliminary report confirms our suspicion that Vicq d'Azyr had found consolation in his widowed state: "I embrace your delightful better half with all my heart; [illegible] be annoyed with me? I'm too far away to dread your anger.") Poissonier returned to Paris with their full account, but only after Jeanroy had himself come down with the infection. They were replaced in April by two further commissioners, Paulet and Delalouette.

Among the 800 prisoners, some 259 had by then been transferred to the Cordeliers hospital and another eight or ten were falling ill every day. The new commissioners were careful to praise the local doctor, one de Launay (a name later famous in French medical annals). His selfless devotion had won him high esteem, and they recommended him to Vicq d'Azyr for an award and corresponding membership in the Society.[164] Perhaps the modern reader will be interested in the symptoms produced by an eighteenth-century "virus," for that very word was then used to account for what was going around, which in this instance was almost certainly typhus:

> Normally the disease begins with a more or less definite chill followed by an access of fever; there is painful headache in most cases; in some the ribcage is painful; in others the limbs; in most cases pain in the chest is accompanied by unproductive coughing; in many there is diarrhea or dysentery from the beginning, and often both; but the most noticeable and constant symptom is great loss of strength. The pulse is always more or less rapid without being strong, and is indeed generally weak; the tongue is heavily coated, sometimes yellow and sometimes brown, though more rarely.
>
> The patients are not delirious. At least, we have seen only very few so affected; there is sometimes nausea, but almost never spontaneous

[163] Jeanroi to Vicq d'Azyr, 25 February 1779, BANM, MSS 109.
[164] Paulet and Delalouette to Vicq d'Azyr, 5 April 1779, BANM, MSS 109.

vomiting; the disease runs its course with little change . . . it sometimes lasts six weeks; the normal span is twenty-one days. Sometimes there is very heavy sweating at the beginning, which relieves the patients.[165]

Those afflicted had suffered more from drastic attempts at remedies than from the disease itself. Thus, Paulet and Delalouette had forbidden resort to the powerful cordials, purgatives and "amors" taken at the first hint of symptoms that these strong drugs only aggravated and that then had to be relieved by bleeding. The two doctors had found ipepacuanha helpful to relieve congestion and kermes in oil of almonds for the cough. These Englishmen were very bad patients, however, especially during convalescence. They ate too much, had unhealthy habits, and suffered frequent relapses. As for eradicating the epidemic, that would be more difficult than treating it. Some center of contagion in the citadel kept renewing the infection. There was no hope that it could be eliminated short of evacuating the structure, disinfecting it thoroughly, and reclothing all the beds and all the prisoners. To approve the expenditures required would exceed the authority of the local naval commandant. The governor of Dinan had accepted their findings in principle, therefore, and had forwarded the recommendations to the minister of marine affairs for a decision. After complimenting the governor, who had taken great pains to put into effect every measure they proposed, they closed their report: "We are doing our best to fulfill the intentions of the Society and to carry out the duties of our mission in the manner that will give it the greatest satisfaction. We trust that our zeal will make up for the knowledge that we lack, while awaiting the views of the company to which we have the good fortune to belong and which we beg to aid us with its advice and to accept our compliments."[166]

The taxonomy of diseases, called nosology, was the medical aspect of the eighteenth-century strategy for basing sciences upon classification, in this case of symptoms related to climate. In the structuralist approach to the history of medicine, which has been influential recently, the attitude has developed that, nosology having been the systematic preoccupation of this generation of doctors, the question of their actual knowledge of medicine is unimportant. Since the scientific medicine of the nineteenth century was equally an epiphenomenon of the social process, specifically of its bourgeois phase, what has changed in modern medicine is rather its "re-

[165] Paulet and Delalouette to Vicq d'Azyr, 9 April 1779, BANM, MSS 109.

[166] *Ibid*. For their published account, see Jeanroi, "Premier mémoire sur les maladies qui ont regné à Dinan en Bretagne parmi les prisonniers anglois, en 1779," and Delalouette, "Suite de la description des maladies . . . à Dinan en 1779," MSRM *3* (1779/82), 45-54, 55-60. Cf. Goubert (1974), 202-203.

gard" or outlook than its concrete capacity to heal the ill.[167] It may be so in some sense, though only in so very metaphysical or metahistorical a sense that the reader of the memoirs published by the Society on leprosy,[168] say, on venereal disease,[169] on puerperal fever[170] or on small pox,[171] or one who has studied the works of the surgeons,[172] is likely to react like Samuel Johnson kicking the chair. For when a disease could be identified objectively, and more than a few could be, the eighteenth-century doctor treated his patients like any conscientious medical man and tried to cure the condition. The limiting factors were his knowledge and skill. Whatever his outlook may have been, he would have liked to know more. He favored research, therefore, whether or not he performed it himself, and it has been through research, after all, that his successors, though never so bourgeois, ultimately did come to know more.

The vein of concreteness in eighteenth-century medicine comes out most impressively, perhaps, in the responses to the rabies competition subsidized by Lenoir, the same who was lieutenant-general of police and patron of the College of Pharmacy. The Society announced the prize in 1778, held the contest open until 1783, and published the six best memoirs and all the interesting correspondence in a special volume as substantial as the one the Academy of Science had devoted to the problem of saltpetre.[173]

Among real diseases, "rage," or hydrophobia as it was also called from its most appalling symptom, inspired a horror quite forgotten nowadays that medicine has made progress, whether in technique or structure must matter little to all those spared its ravages by Pasteur and vaccination.

[167] That seems to be the message of Foucault (1963).

[168] Vidal, "Recherches et observations sur la lèpre de Martigues," MSRM *1* (1776/79), 161-172, and "Second Mémoire sur l'elephantiasis," *ibid.* 5 (1782-83/1787), 168-195; Chamsera and Coquereau, "Recherches sur l'état actuel de la lèpre en Europe, & réflexions sur le precédent mémoire," *ibid.* 5 (1782-1783/87), 196-203; see also the separately published memoir, *Sur le mal rouge de Cayenne* (1785).

[169] Macquart, "Sur la gonorrhée virulente," MSRM *8* (1786/90), 83-97; and the separately published memoir by Lassone and Dehorne, *Instruction sommaire sur le traitement des maladies vénériennes dans les campagnes* (1787); Cf. HSRM 7 (1784-1785/87), 201-202; and drafts and correspondence, BANM, MSS 108, doss. 16.

[170] Doublet, "Nouvelles recherches sur la fièvre puerpérale," MSRM *8* (1786/90), 179-309.

[171] Lassone, "Mémoire sur quelques moyens . . . de rémedier à des accidens . . . dans les petite véroles . . . ," MSRM *3* (1779/82), 84-96; Girod, "Mémoire sur l'inoculation," *ibid.* *4* (1780-1781/85), 231-237; Dehorne, "Mémoire sur quelques abus introduits dans la pratique de l'inoculation . . . ," *ibid.* *4* (1780-1781/85), 414-425.

[172] Notably in the five volumes of the *Mémoires de l'Académie Royale de Chirurgie* (1743-1774).

[173] HMSRM 6 (1783, Pt. II/1784). On the history of rabies, see Théodoridès (1974a and 1974b), both with references to the literature.

The physician whose memoir won the second prize of a gold medal and 300 livres one Bouteille, takes the measure of its antagonism to its victims and their doctors:

> Of all the symptoms that make up the fearful picture of that disease, there is none but is sinister. In the cruellest of diseases, even in the plague, some symptoms appear through which Nature seems to have in view the salvation of the sufferer; but in rabies, they are all dire. They all accentuate each other in hastening the death of the unfortunate being against whom they conspire: atrocious pain, violent spasms, terrible convulsions, strangulation, suffocation, frenetic delirium—nothing in all that offers a means of cure, nor gives any reason for thinking that Nature is working to win over the disease. It appears that, frightened by the magnitude of the malady, she produces only turbulent moves, badly concerted, ineffective, or, to be more exact, downright harmful.[174]

A surgeon called Matthieu, whose memoir from Conze in Périgord won an honorable mention, writes to the same effect, though less Hippocratically:

> Of all the ills that afflict humanity, and that our art can help, rabies is without contradiction the most violent, the cruellest, and the one which ends the most promptly and inevitably in death.
>
> In other diseases producing convulsions as strong, for example in epilepsy, the sufferers, in spite of the lamentable state they are in, benefit in a certain sense from the advantage of having lost their senses, whereas rabies leaves those whom it attacks fully conscious. The persons whom I have seen in that pitiable condition, and whom I knew beforehand, have seemed to me possessed of even greater perspicacity, of a heightened intelligence, especially when, as is common, fever fails to appear along with other symptoms.[175]

Domesticated dogs were the greatest danger, but it was not unknown for mad wolves to come raging out of the forest. In the village street of Vaux near Metz, a she-wolf attacked seven persons, one after the other. Claude Le Roy, a roofer by trade, fought her off for three quarters of an hour in an epic combat hand to paw, his elbow caught in her jaws. Somehow, he freed himself, got an arm around her neck, and wrestled her clawing and snapping to the gutter, and though scratched and gouged by her talons and bruised all over his body, he managed to stab her again and again with his pocket knife, apart from his ruler his only weapon.[176] A

[174] MSRM 6 (1783, Pt. II/1784), 152.
[175] *Ibid.*, 295.
[176] HSRM, 6 (1783, Pt. II/1784), 133-134.

similar rampage contributed to the education of a young doctor in Thiers, Mignot de Genety. One day in February 1764, fifteen or twenty people were bitten by a mad wolf in a village near that city. Mignot was away in Paris. Friends wrote urging him to procure a supply of powder of calcinated oyster shells, which along with rubbing mercury was thought to alleviate the effects, and to hurry home. There the seven surgeons of the city had gathered the victims into a new hospital and were awaiting the worst. They invited him to take charge. The virus usually took three weeks to a month to invade the nervous system. Mignot had never yet seen a case of hydrophobia, and disbelieved the tales about

> all the singular symptoms attributed to this strange disease, which I still thought of as only a kind of melancholia, or individual obsession, taking the form of an aversion for liquids. I was persuaded that with sound judgment and a strong mind, no one would ever fall victim to such delusions, in spite of the bites of some animal or other said to be mad. In short, I took the view that hydrophobia was nothing but the consequence of a cowardly imagination . . . with the capacity of producing dire results through the sole effects of anxiety about the most trivial of accidental bites by dogs or other suspected animals.[177]

For the eighteenth century also had its skeptics, medical and otherwise. Mignot was soon undeceived, and first by a twelve-year-old patient in whom the symptoms began to "declare" the disease about a week after his return. Summoned to the bedside, he found the lad quiet enough, though from time to time he would burrow spasmodically deeper into the bed, hiding his head under the covers. His face, gentle, shy and intelligent, was little altered, except that he rolled his eyes at intervals. The doctor talked calmly with him, joking to get his confidence before asking how he felt. His throat was sore, the boy said, his gullet on fire. Mignot's suggestion of a cup of bouillon or an infusion met with a "Can't drink." So, a good children's doctor, Mignot took a silver coin, an écu or crown worth six livres, from his vest pocket, and flashed it before his patient's eyes to be the reward for taking a cup of bouillon. "Oh, Sir, you're fooling, and you won't really give me that crown." "Honest. It's yours, son, and I'll let you have it provided you drink some bouillon, an infusion, some sugared water, some water and wine—whatever you want, in other words, as long as you drink." All smiles the boy took the coin, tucked it under the covers, and reached for a cup of warm sugared water. His hand shook, and he turned his head. He tried drinking out of the corner of his mouth, eyes averted. Impossible. Every time the cup touched his lips, he recoiled

[177] *Ibid.*, 51.

and regurgitated. Grinding his teeth, he managed to take a few drops, only to go into a fearful convulsion:

> I shall never forget the striking conflict in his expression between the will to lift the glass to his mouth and swallow the water it held in order to earn his crown, and the repugnance, or rather the invincible resistance, that overcame him when that drink got near his lips.

Mignot left the lad his crown, saying he'd manage better later, but the fury seized him in the night and he died before dawn.[178]

Mignot's education was continued by a young woman of twenty-six. She had been bitten lightly on the forearm. When the "declared" symptoms appeared, she was to be transferred to the terminal ward. Her case was an example of lucidity increasing with the disease. "Sir," she said on seeing him come in, "I can see I'm done for. They're going to put me in that miserable room and no one has ever come back." It was for her own good, she would have better care therein, he reassured her, taking her arm to help her up the stairs. "Sir," she warned him, "Watch out. I'm not mistress of myself. I'll bite you." Despite herself she tried, so that, even like the roofer Le Roy wrestling the wolf's jaw away from himself, he had to hurry the girl up the staircase grasping her by the hair and forcing her head aside. As she was being tied to the bed, she snapped again all involuntarily at his face bending over her and at his hands busy with the knots, and died writhing against the cords some thirty-six hours after the telltale intolerance for liquids had come upon her.[179]

In assembling the rabies volume, the Society sought information rather than novelty and decided to print all the correspondence it might receive, provided the communications be conscientious and factual, without expressing any judgment. In only one respect did the editors permit themselves a recommendation, and that was in emphasizing the critical importance of local treatment of the wound as quickly as possible. Without wishing to say that no other therapy was any use, the Society declared that only this one was indispensable.[180] What was entailed will appear in one of the cases reported from lower Brittany by Robert de Kiavalle M.D. and correspondent of the Society in Josselin. His patient was Marguerite Le Couëdic, "*fille*" (for so did physicians speak of peasants) of some fifty years, of the remote village of Marue in the parish of Saint-Caraduc.

On 18 March 1782 she was bitten on the index finger of the left hand by a dog belonging to "*le nommé*" François Le Maux. The animal manifested all the symptoms of rabies, and in the same fit bit dogs belonging to a laborer called Alno and to the abbé Tanguy. Alno killed his pet at once; the curé killed his only when it too went mad.

[178] *Ibid.*, 52-54. [179] *Ibid.*, 55-56. [180] *Ibid.*, 1-4.

As for Marguerite, she was terrified for herself, but what was she to do, there in the depths of the country far from any professional help? Luckily, someone knew that the doctor was then stopping in a country house a league and a half away. She hurried to seek him out, thinking, poor soul, that he could just prescribe a draught of something that would save her. He told her what was involved, a sustained treatment of bleeding, baths, and mercurial massage, but that first the bite must be burned out. Three days had already passed. There was no time to lose. Lest the operation be bungled back in the village, where there was no one to perform it but the blacksmith or a neighbor, Kiavalle offered to cauterize the wound right then and there, though he had none of his instruments. She agreed. He placed her on the table in the kitchen, selected a utensil from the stove, blindfolded her, and got the strongest peasants in the village to hold her down, her arm in place. He then heated the cooking iron red hot and pressed it into the wound long and hard enough to sear the flesh beyond where the teeth had penetrated. "That cruel operation" completed, he sent the woman home, instructing her to dress the wound with lightly salted butter, to bleed her arm the next day, and not to neglect warm baths and mercury.

It is not surprising that Marguerite's appearance and a tendency to roll her eyes in the wake of that treatment gave concern to the people present. Kiavalle could do nothing more at a distance of nine leagues from his office, and proposed to her that as a last resort she come to him in Josselin, where he could give her the most refined care possible, requiring only that she find her own lodgings, since he could not put her up. She never came. He worried from time to time. On visiting her parish, five months before writing his report to the Society, he learned that Marguerite Le Couëdic had been "happy enough" with her first treatment not to come ask for another; that after the scab fell away, the wound drained healthily for some time before healing; and that "that woman has been perfectly well since, and is still well today."[181]

Such was the treatment of choice. It saved many of the cases reported in the rabies volume. Each of them comes alive as we read of his accident: Jean-Baptiste Cailleux, "fifteen years old, small, very well built, of an impassive manner, slothful habits, and dull personality;[182] "A woman named Julie Roger, the wife of Rougemont, thirty-seven years old, of a small figure, delicate, cheerful on the whole, lively and gay in personality, intelligent, very sensitive, enjoying good health and when she was bitten nursing a baby that she had borne six months before";[183] "Jeanne Bosquillon, a woman of forty-eight, extremely fragile, of a weak character and dejected personality, almost deaf, and without enough sense to feel

[181] *Ibid.*, 74-76. [182] *Ibid.*, 167. [183] *Ibid.*, 184-187.

anxiety";[184] "Le Sieur Gravan, clerk in the tax-collector's office, and in his seventy-second year, slight of stature, thin and frail, by nature dejected, anxious and melancholy";[185] "A man called Etienne Champion of Orléans, aged twenty-five, of a decidedly strong temper, given to drunkenness, . . . attacked when under the influence of wine."[186]

We follow each of their stories in suspense; are relieved to know that Rougemont's cheerful, intelligent and sensitive wife survived to bring up her baby; not surprised to learn that the fussy little old tax collector and the young drunk succumbed; and specially distressed to lose "le nommé Gervais Briquet, a child of twelve, small but husky, muscular, lively, cheerful, intelligent and enjoying good health," which peppy boy the scorching iron failed to save,[187] even though the doctor tried reinforcing the natural gaiety, harmony, and good cheer of his character with the strains of a guitar in his final struggles. Not all physicians and surgeons applied the fire, however. Le Roux, author of the memoir that took first prize (a good medal and 600 livres), preferred caustics:

> I never used the red-hot iron to cauterize the wounds: it is too frightening to the patients, is not nearly as easy to handle, and does not burn with the same precision as caustics. Among these agents, I have chosen liquid butter of antimony, because it burns more deeply and less painfully, because the scabs that it forms drop off sooner, and because it produces none of the side effects that are sometimes to be feared with the others.[188]

Among famous writers, Diderot was the one who had the most intimate familiarity with the medical world, and reading in the sources begins to make clear why it was that, in the accuracy of his sensibility, he imagined the figure of a doctor watching over patients in the night to suggest what it would be for a philosopher to transcend science in service to humanity.[189] Not the least interest in reading these papers is the vicarious experience they give of the actual meaning of enlightenment. The texts exhibit the reality behind that abstraction in the details of daily conduct of the important segment of the educated class that was made up of doctors. They make a favorable impression on the whole, do those physicians. True, their attitude to the peasantry and populace was nothing if not paternalistic, taking the lower orders to be ignorant and gullible children of a society, and perhaps of a history, that they and their kind had the natural responsibility to ameliorate. Thus, according to Geoffroy's re-

[184] *Ibid.*, 190.
[185] *Ibid.*, 175.
[186] *Ibid.*, 35.
[187] *Ibid.*, 155-167.
[188] MSRM, 6 (1783, Pt. II/1784), 66.
[189] "Entretien entre d'Alembert et Diderot," in *Oeuvres philosophiques de Diderot*, ed. Paul Vernière (1956), 293.

port of the "constitution" of the year 1780-1781 in Paris, the winter produced a type of catarrh, "to which the public gave various names, all equally ridiculous."[190] Thus, again, after Anne-Marie Baux was operated on at the age of seventy-three for a gangrenous strangulated hernia, "*cette fille*" lived another eight years, and might still have been alive had she not yielded to intemperance.[191] Thus, finally, a country doctor with affectionate exasperation calls the eighteen-year-old serving girl he was laboring to save "my imbecile patient" when she wanted to try some village remedy.[192]

For it was clear that the poor, if left to their own devices of superstition and fear, could have no other recourse than to the herbal brews of old women or the nostrums of quacks. But however worthless or self-damaging their ideas, their lives were important. Such doctors were not the ones to wall themselves off emotionally from their patients. Consider, for example, the report of a Dr. de Lavallée, who had been commissioned in June 1774 by the intendant of Tours to make an inspection of a group of ten parishes in the election (or district) of Chateaugoutiers afflicted by a verminous, putrid fever that in certain cases became malignant:

> A striking spectacle for compassionate humanity was spread before me in the course of my visits. In the various parishes there were still at least a hundred sick of both sexes and every age about to die one after another like those who had preceded them. Stretched out for the most part only on straw, or on wretched pallets, without care, with no help, they had been entirely deserted for the most part, since panic, always the deadly accompaniment of epidemics, had driven away any helpers. Water, and among the less poor milk, was their only sustenance, left beside them any old way in filthy and fetid jars. What a burden for a doctor, however prompt in responding, however zealous in his wish to help! Trying to understand the nature of the disease, and to prescribe the means for combating it, was not the greatest obstacle. It was to find someone whom he could engage and train sufficiently to carry out instructions for the benefit of patients whose need was so urgent. And how agonizing it was when a whole family was afflicted, as was not uncommon, far from the church in some remote cottage where no one dared enter, being consumed with panic. How could I have their bouillon taken to them, or have things prepared for them to drink, or have medicine administered at the appointed times, measures so necessary in order to hope for any success at all? I had to find the nearest neighbors, beg them, plead with them, soften their hearts all dejected and insensitive, pointing out

[190] MSRM, *4*, (1780-1781/85), 2.

[191] HSRM *4*, (1780-1781/85), 321-322.

[192] BANM, MSS 110.

> that they were not far removed from falling into a similar pitiable condition, and that they would be blamed and abandonned: all this was a hard and crushing task.[193]

Doctors of that stripe came when they were sent for and did what they could, and the record gives no sense that they predicated their services upon ability to pay.

Also missing from the record are marks of the rivalry that informed the life of the Academy of Science. Despite the emulation inspired by that body with respect to research, the practice of medicine was after all different from the creation of science. In their professional lives, doctors were more involved with patients, with whom the relation was one of responsibility, than they were with colleagues, with whom it might partake of competition. It was in keeping with the difference, tacitly at least, that there should have been no gradations of rank within the resident membership of the Society of Medicine,[194] that correspondence should have been more important than composing memoirs, and that even the latter were largely motivated by the need to share experience in service to the sick. The grounds for a physician's reputation usually lay elsewhere than in his publications. Priorities were less important than among scientists, and the Society made no provision for recording the precise moment at which the permanent secretary received manuscripts. The corresponding membership was the distinctive component of the Society, whose policy of admission was inclusive rather than exclusive. To be sure, practitioners of doubtful credentials were denied the cachet that use of the Society's name conferred. A certain Dr. Bouriat of Tours, for example, tried repeatedly to get in, and without success; we do not know why. "Reply in very general terms," noted Vicq d'Azyr on one of Bouriat's letters.[195] In the present state of information, it is impossible to be sure what proportion of the French medical profession the thousand and slightly more doctors who were correspondents of the Society between 1778 and 1793 represented. An estimate between one-fifth and one-third would probably

[193] BANM, MSS 157, "Histoire de la fièvre putride, vermineuse, et en certains sujets avec caractère de malignité, regnante dans les paroisses d'Athée, La Noë, St. Michel, Fontaine couverte, Brain, Ballots, Cuillé, La Selle, Livré, Gastinnes, Circom voisines de Cruôn, confiées à mes soins, par commission de Monseigneur l'Intendant." This memoir was received by the Society at one of its early meetings, 5 November 1776.

[194] Lassone insisted particularly on this point in a letter to Lorry, 5 July 1779, BANM, MSS 109, in which he freely "abdicates" for himself and his successors as chief physician to the king the Permanent Presidency of the Society: "Je me retrouve avec la plus grande satisfaction l'égal pour le rang de touts [sic] mes confrères: je n'auray jamais que l'ambition de les surpasser par mon zèle et par mon dévouement." Ministers whom he had consulted approved the step.

[195] Bouriat to Vicq d'Azyr, 25 January 1789, BANM, MSS 156.

not be grossly wrong, however:[196] not so much an elite, then, as an active, progressive minority, self-selected by their own interest, the conditions of membership being competence and participation.

However the patients may have fared, the economic and social historian has equal reason with the intellectual historian to feel grateful to these doctors two centuries later, for their writings open a window wider than other sources often do onto the intimate lives of ordinary people. Before the doctor's eyes the normal barriers were down, and what with his personal involvement, and the prevailing conception that health and illness depended on factors of temperament, climate, and surroundings, many a medical man left accounts of particular cases and social conditions more circumstantial than those of his successors in more scientific medical milieux have usually been.

Near Carcassonne we meet a sensitive "*brassier*" (the word for laborer in those parts, explains Dr. Gallet Duplessis) whose elder brother made him abandon his own vines to work off an obligation in the family fields and then so castigated him for idleness and incompetence that he fell into a state of mortification and chagrin imitating the symptoms of hydrophobia.[197]

In a nearby village, a young lad, already on terms of intimacy with his girl, failed to satisfy her needs as often as at the outset of their affair. Contemptuously, she took another lover, spurning his renewed advances; he died of a wound inflicted on himself in his frustration and mortification. In remorse, she never married.[198]

A certain Jacotin of Villers Saint-Frambourg, seventy years old, tall and thin, frequently gave himself over to drink, depending mainly on brandy, for which he sacrificed everything and even deprived himself of food. He had a cheerful disposition and never worried.[199]

Quite undone by the effects of intense anxiety, on the other hand, was Jean-Joseph Soleilhet, a twelve-year-old ship's boy of Martigues on the coast of Languedoc. A fearful gale overtook the fishing tartan on which he had embarked in March 1779. Compass and lateen rig were lost. The bark

[196] Goubert (1974) is an exact study of the medical profession between 1770 and 1790 in one province, Brittany, using modern techniques of quantitative analysis. He finds (p. 82) that there were approximately 142 physicians and 429 surgeons in the province in that 20-year period. It had a population of 2,300,000, or approximately 10 percent of that of France. Unlike the country as a whole, however, the population of Brittany was declining, and the province then as now was one of the poorest and most backward. There is reason to suppose, therefore, that the ratio of doctors to population was a good deal lower in Brittany than in France generally—hence my guess that there may have been between 3,000 and 5,000 doctors in the country, numbers that do not seem beyond the capacity of 16 medical faculties to produce.

[197] HSRM, 6 (1783, Pt. II/1784), 57-58.

[198] *Ibid.*, 59-60.

[199] *Ibid.*, 204-205.

drifted helplessly. In a panic for fear of foundering, starving, or being overtaken by Barbary pirates, the crew assembled before the mast to cast themselves on the mercy of the Almighty. Vowing that if they escaped, one of their number would make a barefoot pilgrimage to the Church of Notre-Dame de Miséricorde, they drew lots in ritualistic accordance with some sailors' superstition. Terrified, Jean-Joseph imagined that they were selecting the one to be thrown overboard to appease the elements, and that he, the youngest and most helpless, would be sacrificed. When instead they did blow ashore at Barcelona, he was so shattered that he soon fell victim to elephantiasis, to which affliction his family had a tendency.[200]

Dr. Bougirard, one of many surgeons who also had a medical degree from Montpellier, sent in a memoir reporting cerebral accidents he had encountered in his practice in Saint-Malo. One patient was a middle-aged gentleman, stricken in the very act of fulfilling his conjugal duty, who expired on the spot. Another was a lady of about the same age who liked to eat well, undiscouraged by a history of digestive trouble. After one gastric episode she neglected to purge her system, but the application of leeches having restored her well-being, "she sat down at the table at noon, and enjoying a very good appetite, she dined extremely well." She then went out into the sun to move some laundry hanging up to dry, and three hours later was dead of a massive sunstroke.[201]

Running observations that correspondents volunteered on the verso of their meteorological returns provide a more general sort of information on local agricultural and economic affairs. For example, an unseasonably cool August in 1785 made Dr. Housset of Auxerre, who collaborated in recording data with his lawyer son, Housset de Fortbois, "fear for the quality of the wine and doubt that there would be an early or a good ripening." By the same token, it had been an unusually healthy summer month, and he had little medical news.[202] Still more informative are speculations concerning the cause of epidemics called for on forms designed to furnish data on such catastrophes. Dr. Pierre reports to the intendant of the Franche-Comté, Caumartin de Saint-Ange, on the adjoining villages of Celeux and Villers-Farlay in the bailliage of Arbois, administered from Poligny, both infested by a bilious and putrid fever in the months of April and May 1788. A rainy winter, frequent changes of temperature, uncleanliness, an unhealthy diet, and bad ventilation in the domiciles were the apparent causes. The villages themselves seemed to be fortunately situated on the bank of a small stream. The houses were separated by gardens and built on sandy, well-drained soil, though there were a few with damp clay

[200] MSRM, 5 (1728-1783/87), 171-173.

[201] BANM, MSS 155.

[202] Report dated 31 [sic] September 1785, BANM, MSS 154.

for foundation. The fault, therefore, had to lie elsewhere. The inhabitants were almost all laborers. Their normal diet consisted of porridge, vegetable greens, and sometimes pork. They made their bread from cornmeal. When that crop failed, as it had last year, they substituted a mixture of barley and "*perette*" (whatever that was). There being no spring, each house had its own well. The water seeped underground from a little river called La Louve, and was clear and good.[203]

Altogether more understandable was a catarrh that spread through the village of Abbans Dessus in March and April of 1786. The cause was "The lack of every essential. You see miserable paupers, almost half-naked so ragged is their flimsy clothing, trembling in all their limbs while huddled round a fire barely sufficient to cook the wretched food they use. Uncleanliness, cramped quarters, all damp and airless, are the most obvious causes of this epidemic, and of its progress. I should also add the contact of healthy people with the sick."[204] The inhabitants of Boussière, on the other hand, had avarice to blame for a fever that struck in October and November 1787. Situated in a plain bordering the Doubs and surrounded by mountains, their location could scarcely have been healthier or more agreeable. Woods, vineyards, the cultivation of wheat and rye—all flourished. But the desire to prosper led them to sell their best produce in nearby Besançon and to keep only the worst for themselves.[205]

7. SANITATION, PRISONS, AND HOSPITALS

A distinguished series of investigations into public hygiene and urban sanitation, prison conditions and hospital care, occupied the attention of the public in the last decade of the old regime.[206] The famous architects Claude-Nicolas Ledoux and Bernard Poyet drew plans respectively for model prisons and model hospitals.[207] It was the French counterpart of the English crusade, led and made famous by John Howard, for humanitarian reform of institutions incarcerating the old, the mad, the sick, the indigent, and the criminal, the difference having been that in England the motivation came from evangelical religion and in France from the philanthropy of enlightened statesmen and civically minded scientists.[208]

[203] Report of 6 June 1788, No. 273, BANM, MSS 154.

[204] Report of 10 May 1786, No. 85, BANM, MSS 154.

[205] Report of December 1787, BANM, MSS 154.

[206] Below, notes 214, 215, 222, 231, 232, 236.

[207] For Ledoux's schemes, see his *L'Architecture considérée sous le rapport de l'art, des moeurs, et de la législation* (1804), and Haug (1934); Poyet's designs for a new hospital system are printed as an appendix to "Troisième rapport des Commissaires chargés . . . de l'examen des projets relatifs à l'établissement des quatre Hôpitaux," HARS (1786/88), 13-42; see below, note 232.

[208] John Howard, *The State of the Prisons in England and Wales with Preliminary Observations and an Account of Some Foreign Prisons* (Warrington, 1780); Cf. Halévy (1949), 85;

Throughout, the government turned primarily to the Academy of Science, and the Society of Medicine took part less prominently than might have been expected, particularly in proposals for the Hôtel-Dieu. There is no hint of competition or jurisdictional dudgeon in the record, however, and the explanation, if any be needed beyond the prestige of the Academy, is probably that the prime mover on the scientific side was Bailly, who had by now quite transcended astronomy in his commitment to civic projects. In any case, Vicq d'Azyr and his colleagues had their hands full with their main responsibility, which continued to be surveillance and abatement of disease and provision of professional assistance whenever and wherever epidemics and epizootics struck.

Certainly the statesman who in the 1780s gathered the threads of public health administration into his own hands, the baron de Breteuil, secretary of state for the department of Paris and minister of the royal household from 1783 to 1788, gave every mark of confidence in Vicq d'Azyr and in the work of the Society. Early in 1786 the controller-general advised the intendants that the Society had more than fulfilled the hopes the king had held for it ten years before when the original commission on epidemics had been created, and that he intended further support in order that the medical topographies might be completed for all regions. To that end, they were enjoined anew to require their subdelegates to report relevant medical information from the physicians and surgeons in the service of the generality.[209] The government for its part had decided to grant the Society funds from the Treasury to supplement its proceeds from the administration of mineral waters.[210] At the same time, the Council of State ordered the creation of a standing committee within the Society, consisting of the officers and four associate members, two of them to be elected annually for two-year terms. They were to oversee compilation of topographical and meteorological information on a national scale, relate the data to the history and treatment of epidemics, and systematically encourage publication of research.[211]

The Society did, moreover, contribute in passing to the movement among scientists for sanitary reform. Lavoisier began his famous research

---

W. H. McMenemcy, "The Hospital Movement of the 18th Century and Its Development," in Poynter (1964).

[209] BANM, MSS 114, doss. 23. "Copie de la lettre écritte par M. le Controleur-Général à MM les Intendants des Provinces, le 6 février 1786," from "*plumitif*" of 14 February in the office of Villiers du Terrage. The document also calls for a census of physicians and surgeons in each generality, together with recommendations on the number actually needed. I have found nothing to indicate that such an enumeration was ever put in hand.

[210] For financial details, see above, Section 5, n. 132.

[211] BANM, MSS 114, doss. 23. Printer's proof of the "Arêt du conseil d'état du Roi, portant règlement pour la Société royale de médecine, concernant les epidémies, du 24 avril 1786," over Breteuil's signature and corrected in his hand.

into the physiological chemistry of breathing with determinations of the respirability of air in theaters and hospital wards, and published the findings in its *Mémoires*.[212] He also collaborated on a piece about the adulteration of cider.[213] Breteuil entrusted one considerable commission to the Society rather than to the Academy, and most unpleasant it was, exhuming the contents of the Cimetière des Saints-Innocents and disinfecting the terrain. Just off the rue Saint-Denis, the graveyard was the only considerable burying ground in Paris, the only place where rich and poor were equal. Already ancient when Phillip Augustus had it enclosed in 1186, it had long since become a scandal and source of disgusting emanations when Jean Fernel and Houllier of the Faculty of Medicine advised removing it from the city in 1554. There was no telling how many millions of cadavers it contained. Recent records showed over 90,000 burials in the previous thirty years, mostly in common ditches dug twenty-five to thirty feet deep into the deposits of previous centuries. The fill had raised the level of the enclosure eight to ten feet above that of the surrounding streets, and gases had begun seeping into cellars along the rue de la Lingerie immediately to the west, making the houses uninhabitable. Removing such remains presented a delicate problem. The Society placed Michel-Augustin Thouret in charge of the operation. A single false move would have touched off trouble in the quarter, he recognized, the more so as a few squatters had made this charnel house their refuge, living wretchedly amid the dead. He brought it off with consummate tact, however, aided by a surgeon called Marquais, in the presence of clergymen of all the sects, and in the winter months. Thereupon, the emplacement was paved over to accommodate stalls for vegetables and herbs in the general market system which adjoined it. As of this writing, the ground has once again gaped open for some years awaiting the decision of government of the Fifth Republic to lay out a park to replace Les Halles, now that their time too has come.[214] A little later in 1788 Thouret had an equally un-

[212] "Mémoire sur les altérations qui arrivent à l'air dans plusieurs circonstances où se trouvent les hommes réunis en société," read 15 February 1785, MSRM, *5* (1782-1783/87), 569-582. Cf. Smeaton (1956a), 231-232; Storrs (1966, 1968); Greenbaum (1972), 656-657.

[213] Lavoisier, Thouret, and Fourcroy, "Rapport sur la falsification des cidres," HSRM, *8* (1786/90), 159-166; a further report on the same subject follows immediately, 167-172. Lavoisier also collaborated on the report to Castries on the nutrition of seamen in hospital (above, Section 5, n. 162), for which he analyzed the composition and preparation of bouillon. His experiments were left in manuscript, and are published in his *Oeuvres*, "Mémoire sur le degré de force qui doit avoir le bouillon, sur sa pesanteur spécifique, et sur la quantité de matière gélatineuse solide qu'il contient," *3* (1865), 563-575; cf. Smeaton (1956a), 234-236.

[214] Thouret, "Rapport sur les exhumations du cimetière et de l'Eglise des Saints-Innocens," HSRM, read 5 February 1788, *8* (1786/90), 238-271. The commission was appointed in October 1785, at the instance of DeCrosne, Lenoir's successor as lieutenant-

savory commission: an investigation of the sewage plant at Montfaucon, where fecal matter from all the public conveniences of Paris was carted and spread out to dry for conversion into fertilizer or other disposition. Until 1781 there had been a second such installation for the southern part of the city, the "voierie de l'Enfant-Jésus."[215]

The movement for civil improvements in the 1780s was led by a kind of partnership in good works between Bailly and Breteuil, reciprocally each other's men of confidence in the academies and the government.[216] Bailly (it may be recalled) owed his third academic membership, in the Académie des inscriptions et belles lettres, to Breteuil's intervention.[217] Nephew of Voltaire's (and Newton's) Marquise du Châtelet, Louis-Auguste Le Tonnelier, baron de Breteuil, led a military and diplomatic career culminating in Vienna, where as ambassador from 1778 to 1783 he skillfully freed French diplomacy from a certain dependence on the dynastic connection with the family of Marie Antoinette. Returning to Paris, his credit high, especially with the Queen, Breteuil was rewarded with appointment to the department comprising the Royal Household and the affairs of the capital. As minister responsible for science, he arranged permission for Lavoisier and his associates to begin the *Annales de chimie*,[218] supported Cassini's rehabilitation of the Observatory,[219] and brought Lagrange to Paris from Berlin.[220] As minister for Paris, he threw himself into urban renewal like a Turgot père after his time or a Haussmann before it, clearing the slummy old tenements off the bridges, seeking to clean out and close down the slaughterhouses along with the cemetery and the practice of interment in churches other than the Innocents, pressing forward with projects for amelioration of prisons and hospitals, asking and taking the advice of the experts he patronized, championing Beaumarchais in the affair of the *Marriage of Figaro*, attacking the cardinal de Rohan in that of the diamond necklace, and regulating the employment of the arbitrary *lettres de cachet* that came under his jurisdiction. For Breteuil was a high-handed reformer, a patrician large in gesture, inevitably an emigré in the Revolution, and in the 1780s an opponent both of Necker and Calonne. In all accuracy, however, it has to be acknowledged that such prideful enmities diminished his effectiveness, and in all fairness that

---

general of police, and consisted of La Rochefoucauld, Lassone, Poulletier de La Salle, Geoffroy, Desperrieres, Colombier, Dehorne, Vicq d'Azyr, Fourcroy, and Thouret himself.

[215] "Rapport sur la voierie de Montfaucon," HSRM, *8* (1786/90), 198-221. A "Supplement" and "Second Rapport" occupy 222-237. This request came from DeCrosne. Besides Thouret, the commission consisted of Fourcroy, Dehorne, and Hallé.

[216] Greenbaum (1973) treats the association between Bailly and Breteuil.

[217] Above, Chapter II, Section 3.

[218] Above, Chapter I, Section 6.

[219] Above, Chapter II, Section 3.

[220] Above, Chapter I, Section 5.

the movement for ameliorating prisons and hospitals had already received much impetus from the first ministry of Necker.

Or more precisely from Necker's wife. In 1779 Madame Necker founded a hospital for the poor in the parish of Saint-Sulpice, intending it to be a model; and hers was almost certainly the initiative that led the government to request the Academy of Science in 1780 to name a commission on replacing the ancient dungeons constituting the central prison system of Paris, the Châtelet, the Petit Châtelet, and For-l'Evêque.[221] Lavoisier took charge of this body. His report held that, though many considerations relevant to prisons were outside the competence of science, still "everything concerning the circulation and renewal of air, methods for preventing putrefaction or guarding against its effects, in short whatever related to the sanitation of the installations and the protection of those who inhabit them, these objects were within the purview of the Academy."[222] Those three gaols (there were others in Paris) had been constructed to be fortresses in the Middle Ages, and were now all under the administration of the Châtelet. Having disarmed or overcome the reluctance of their keepers to allow the commission even to enter on the premises,[223] its members had the shock experienced in many other times and places by secure citizens on learning of conditions in penal institutions:

> The three prisons . . . together cover a surface of only 522½ toises (about 20,000 square feet) and generally hold in all some six or eight hundred, and sometimes up to a thousand, people. To go into the detail of their arrangements, they present the following features: very small courtyards and area ways, very high buildings that interfere with the circulation of air, very small rooms with low ceilings in which too many prisoners are placed; rooms arranged, moreover, so that air and light reach them only with difficulty, and so that they draw from each other air already vitiated and infected; narrow and badly placed openings; pallets, on which the prisoners, rather than lying down, are piled up; straw ticks, often rotten, that serve them for a bed; latrines and gutters for urine that run through most of the cells; sewers from which the foul gases pass into the living quarters;

[221] Besides Lavoisier, the commission consisted of Duhamel du Monceau, Trudaine de Montigny, Leroi, Tenon, and Tillet. D'Alembert engaged Lavoisier to send an advance copy of their report to Madame Necker (d'Alembert to Lavoisier 23 March 1780), *Oeuvres de Lavoisier 3* (1865), 461-462, and it is apparent in an exchange of letters between her and Lavoisier that in these matters she was acting for her husband, or with his authority (462-464).

[222] "Rapport fait à l'Académie Royale des Sciences, sur les prisons, le 17 mars 1780," MARS (1780/84), 409-424. The report was reprinted with additional details and supporting documents in the *Oeuvres de Lavoisier 3* (1865), 465-498.

[223] Lavoisier to Mme Necker, 25 March 1780, *ibid.*, 462-463.

> dungeons where water drips from the vaults, where the prisoners' clothing rots on their bodies, where they take care of all their needs; earthen floors and flagstones swimming in stagnant water for lack of drainage; everywhere muck, vermin, and filth. Such is the horrible spectacle offered by the three prisons which it is proposed to destroy or to reform, a spectacle which it would have been difficult for us to imagine, if we had not actually seen it.[224]

Lavoisier and his associates also deplored the system by which prisoners with means to pay the gaolers were incarcerated less miserably and fed less badly than indigent inmates (though only in the Bastille could persons of substance be confined with some small decency). Not surprisingly, the commission found that these ghastly old piles could never be renovated, and recommended their demolition and the adoption of a plan submitted by the architect Moreau for a new, enlightened prison to be constructed on the site of the Cordeliers. That was not done, of course. A new debtors' prison, observed Condorcet in his comment, was all that humanity could obtain.[225] Thus did it fall out that some fourteen years later Lavoisier saw those same cells again and passed his last days in one of them, there in the dankness of the Conciergerie.

As for hospital reform, Necker's plans for the Hôtel-Dieu entered into its history in the aftermath of a fire that destroyed its western part in 1772.[226] In that institution, even older than the prisons, every ward was no less a chamber of horrors than every cell bank. The main part of its complex of buildings, dating largely from the time of François Ier, stood on the opposite side of the Ile de la Cité from the present hospital and straggled along the quay downstream from Notre Dame, whose western portal they much encumbered on the south side near the smaller arm of the Seine. An umbilical cord of corridors and wards built across the Pont-au Double connected these structures with a secondary wing of the hospital, which actually contained more patients. It reached along the left bank, on the location of the present quai de Montebello, and joined onto the Petit Châtelet just above the Petit Pont. The fire of 1772, subject of a canvas by Hubert Robert, had been preceded by earlier conflagrations in 1737 and in 1718, to go no further back. The first proposal of reformers, incorporated in letters patent in 1773, was to demolish the surviving sections and move the hospital away from the center of the city. Its two dependencies, the Hospitals of Saint Louis in the Faubourg du Temple and Saint Anne or la Santé near the Gobelins, would have been expanded im-

[224] MARS (1780/84), 410-411.

[225] HARS (1780/84), 9.

[226] Coury (1969), 18-19; on the history of the Hôtel-Dieu, see also Fosseyeux (1912); Richmond (1961); Greenbaum (1971, 1973); and for published documents and a guide to archival resources, Brièle (1881-1887) and Bordier and Brièle (1877).

mediately, and additional facilities would later have been constructed elsewhere. Adopted by Turgot, that policy encountered fierce opposition from the Augustinian sisters, who provided nursing, and from the administrators, who agreed with the nuns on little else.[227]

Soon after Necker came to office in 1777, he appointed a commission to mediate the dispute. Together nurses and staff prevailed on him to abandon his previous plan in favor of an enlargement and renovation of the Hôtel-Dieu on its ancient site. Reformers were quick to attribute the retreat to budgetary cowardice and capitulation to vested interest, but that may have been unfair. A recent historian of the hospital feels that it is in the municipal genius of Paris that medical care for its citizens should always be provided there, under the shadow of Notre Dame and in the cradle of the city, and it is to be remembered that, appalling though the care may have become, the move was opposed by the great Desault himself, who was even then conducting the surgical clinic where Bichat and many other reformers of French medicine learned their craft.[228] Moreover, Necker did retrieve the hospitals from ecclesiastical control and subjected them to civil authority. He appointed a magistrate, Antoine-Louis de Chaumont de La Millière, intendant of finance, to head a new hospital service under the Ministry of Finance, and a physician, Jean Colombier, to be his chief inspector.[229]

Whatever the merit in Necker's approach to the hospital, Breteuil shouldered La Millière and Colombier aside and made an investigation of the Hôtel-Dieu the object of one of the blue-ribbon commissions on dangers to public health appointed on his request by the Academy of Science. Mesmerism and slaughterhouses were the other abuses, and Bailly chaired all three groups. It was in rejecting Mesmer's claims that Bailly came before the eyes of Paris a champion of medical propriety, and of that we shall say more in the next chapter.[230] As for the abattoirs, the advisability of moving them to the outskirts of the city might seem obvious, although the commission did have to meet the argument from butchers that members of their trade were rosier of cheek and huskier of frame than any other class of artisan. They also lived longer, and thus the proximity of their work could not possibly be unhealthy.[231]

The campaign to replace the Hôtel-Dieu with hospitals worthy of the century was a *succès d'estime* crowning a *succès de scandale* in the very last years of the old regime. Serving under Bailly on the Academy's commis-

[227] Greenbaum (1973), 265.

[228] Coury (1969), 52, 150-151.

[229] Greenbaum (1973), 263.

[230] Below, Chapter IV, Section 2.

[231] "Rapport des mémoires & projets pour éloigner les Tueries de l'intérieur de Paris," HARS (1787/89), 19-43, esp. 26. The commission consisted of Bailly, Lavoisier, Laplace, Coulomb, Darcet, Daubenton, and Tillet.

sion were Lavoisier, Laplace, Coulomb, Tenon, Darcet, and Lassone.[232] Its cause profited from the piquancy of obstruction by the officials of the hospital, who, though bound to admit free any person claiming care, male or female, old or young, rich or indigent, French or foreign, drunk or sober, nevertheless refused to allow the commissioners to set foot on its premises or to inspect its records. They visited other important hospitals, the Charité, the Pitié, Madame Necker's new installation, and the infirmaries at the Salpêtrière and the Invalides. Everywhere else they were well received and shown every fact and courtesy. They called on the archbishop of Paris, under whose titular jurisdiction the Hôtel-Dieu still officially came, on the premier president of the Parlement of Paris, and even on some of the administrators or trustees. Finally, they asked at the office "that we be allowed to see the Hôtel-Dieu in detail, accompanied by someone who could guide us and explain things. We wanted the plans of the whole institution, the dimensions of the wards, the number of beds they contained, the number of patients admitted, and the number of those who died, on a monthly basis for the last ten years. We needed all those items; we asked for them; and we got nothing."[233]

Whatever this attitude cost the lay members of the commission in first-hand exposure (and gained them in public sympathy for their recommendations), their exclusion did not vitiate their reports, which drew upon the experience of over twenty years accumulated by the only commissioner who was an expert, and also from private information leaked to him by a confidante, Marie Dugès, chief midwife at the Hôtel-Dieu.[234] For though Bailly was impresario and spokesman, the colleague who alone really knew whereof they spoke was Jacques Tenon, the surgeon who had cured Lamarck of his scrofulous abscess some years before.[235] Tenon had already

[232] HARS (1785/88), 2. The commission brought in three reports. The first was submitted on 22 November 1786. The ostensible occasion was a request by Breteuil for an opinion on a proposal by Bernard Poyet, "architect et contrôleur des bâtiments de la Ville," who had composed a *Mémoire sur la nécessité de transférer et reconstruire l'Hôtel-Dieu de Paris, suivi d'un projet de translation de cet hôpital*. The body of the report contains the findings on the actual condition of the Hôtel-Dieu, HARS (1785/87), 2-77, followed by an "Examen" of Poyet's proposal and plans, 78-110. The second report, 20 June 1787, HARS (1786/88), 1-12, concerns locations in Paris suitable for new hospitals. The third, 12 March 1788, HARS (1786/88), 13-41, is an account of foreign hospitals, mainly in Britain, where Tenon and Coulomb had been despatched on an inspection (below n. 249). It is followed by a fold-out plate exhibiting the revision that Poyet made of his original plan to conform to the recommendation of the commission. For a contemporary criticism of the commission's work, and a rival scheme proposed by Jean-Baptiste LeRoi, see Greenbaum (1974). LeRoi was a veteran member of the Academy and originally a member of the commission. He was replaced by Coulomb in the early stages of its work, for reasons that are unclear.

[233] HARS (1785/87), 4.

[234] Greenbaum (1971), 322.

[235] Above, Chapter II, Section 5.

come forward with the supplement on infirmaries appended to the Lavoisier report on prisons.[236] He is another example of the eighteenth-century expert who was self-made in two respects, in his own rise from obscurity and in his creating a technical specialty out of some set of problems, namely what the sick endured in hospitals, which had previously been suffered rather than studied systematically and practically. His father having been a surgeon, and poor, in the region of the Yonne, Tenon had made his way to Paris when a boy of seventeen to study with Winslow.[237] Exposed to the agonies inflicted on patients cut by the knife, awaiting the knife, and surviving the knife in the surgical wards of the Hôtel-Dieu (which had no separate operating rooms), the young Tenon resolved to devote his life to the relief of fear and pain in hospitals, to the abatement there of crowding and of filth. For six years he was resident surgeon at the Salpêtrière, the enormous hospital for women in the Faubourg Saint-Victor that included the prison incarcerating criminal and delinquent females. He it was who proposed the idea of the tiny teaching hospital instituted in 1774 in the College of Surgery, and who got La Martinière, Louis XV's last head surgeon, to prevail on the Treasury for funds.[238] He operated and taught in it himself and was succeeded in the post of chief surgeon by Desault.

Tenon's *Mémoires sur les hôpitaux de Paris*,[239] though not a notably literate work, is much more than a bed check and body count. It is an analysis that specifies the very factors to be considered in any evaluation of hospital care. Those elements derive from the anatomy and pathology of the individual human organism, and are a function of the requirements of the body when most in need of rest. Its stature is to determine the length of the bed, its weakness the height of the steps, its need for air the ratio of the number of patients to the volume of the ward. Institutionally, the quality of a hospital will depend on maintaining an optimal proportion

[236] "Mémoire sur les infirmeries des trois principales prisons de la jurisdiction du Châtelet de Paris," MARS (1780/84), 425-447.

[237] On the career of Tenon, see Greenbaum (1971); Huard (1966); and the Cuvier éloge, Cuvier (1819-1827) 2, 269-304.

[238] Gelfand (1973c); on the practice of Desault, see Gelfand (1973b).

[239] The work was published in 1788. It contains five memoirs: The first, "Tableau des Hôpitaux de Paris," 1-25, is an overview of the hospital system of the capital, and of the care it afforded; the second, "Description abregée des principaux Hôpitaux de Paris," 26-108, gives a description of each of the major installations; the third, "Où l'on détermine ce que l'Hôtel-Dieu de Paris occupe de terrain . . . ," 109-117, discusses the location of hospitals in a major city; the fourth and most considerable, "Description de la Maison de malades de l'Hôtel-Dieu de Paris," 118-348, gives the full detail of the regime of the Hôtel-Dieu; the last, "De la formation & de la distribution des maisons destinées à remplacer l'Hôtel-Dieu," 349-451, contains recommendations. So thorough was the analysis, and so relevant were the recommendations, even a quarter of a century later, that the work was reprinted, apparently from the original plates, in 1816.

between medical and administrative capacity and the number of beds, on assigning an adequate number of wards to each type of disease, and on arranging the housekeeping for convenience and economy. Socially, a hospital is to be judged by the accessibility of its services, by the degree of specialization in the several wards and departments, and by the adequacy of its protection against contagion and infection inside and outside the walls. "Hospitals are in a sense the measure of the civilization of a country: they are better adapted to its needs and better run in proportion to its qualities of integration, humanitarianism, and education."[240]

It was not by the standard of the Hôtel-Dieu, then, that a Parisian would have wished the city judged, for though its daily average of 2,500 patients was almost half the hospitalized population of the capital at any given time, their situation was the least eligible on every count. Their number would rise to 3,500 in the unhealthy seasons, and might go as high as 4,800 in epidemics. Its twenty-five wards contained 486 single beds and 733 multiple beds ranged three feet apart. The latter might better be called pauper-sized than king-sized, for when the hospital was full they would contain six or more patients stretched out head to foot, several of whom might have been dead for hours or even days. Four per bed was the normal complement. Many of the single beds were always preempted by the orderlies, who had no quarters of their own. Opening every which way off steep staircases and into each other, the wards lacked air and ventilation, privacy and discipline. Madmen spilled over into the surgical wards. Sufferers from small-pox and syphilis mingled with expectant mothers. Everywhere everything had to be done right there in fetid company: calls of nature answered, bandages changed, infections drained, veins bled, meals provided, operations performed (trepanning was almost always fatal at the Hôtel-Dieu because the air infected the *Dura mater*), and straw changed in beds soaked through with the sweat, pus, and secretions of the sick. Sanitary facilities, if such they can be called, opened right into the Seine, which then carried the outflow down through the middle of the city.[241] "It is obvious, therefore, that there is nowhere any hospital as badly located, as cramped, as unreasonably overcrowded, as dangerous, or which combines so many causes of unhealthiness and death as the Hôtel-Dieu. There does not exist, no not in the whole universe, a house for the sick which, as important as this in its purpose, is however so harmful in its results to society."[242]

In only one respect did the main report of the commission go beyond a rhetorically effective abstract of Tenon's *Mémoires*, and that was statisti-

240 *Mémoires sur les hôpitaux de Paris*, i.

241 "Rapport des commissaires chargés par l'Académie . . . ," HARS (1785/87), 6-55, summarizing information drawn from Tenon's fourth memoir, n. 239 above.

242 Tenon, *Mémoires sur les hôpitaux*, 345.

cally. Laplace was one of the members, and this was just the time when he was applying the calculus of probability to the vital statistics of the city of Paris and to the problems of population.[243] For these and other demographic purposes, the Academy of Science had assembled in its library the annual summaries of births, baptisms, and deaths in the capital and the faubourgs for the years 1720-1785. The reports included the number of admissions to the Hôtel-Dieu and of deaths there. Thus, even though denied access to the records of the hospital itself, the commission had the information it needed to make comparisons. The comparisons were damaging, not to say devastating. The capacity of the Hôtel-Dieu was less than twice that of the Hôtel-Dieu of Lyons, a city one quarter the size of Paris. A table of mortality exhibiting the probability of death in various hospitals gave the following results: at Edinburgh one patient died out of every 25½ admitted; at Saint-Denis, one out of 15⅛; at Lyons, one out of 11⅖ or 13⅔ (there were slightly inconsistent reports); at the Holy Spirit in Rome, one out of 11; at Versailles, one out of 8⅖; and in Paris itself, one out of 7½ at the Charité, one out of 6½ at Saint-Sulpice, and one out of 4½ at the Hôtel-Dieu. The death rate there of almost one-quarter was even worse than it seemed, for the figures included the maternity wards, which were proportionally larger than in the other institutions,[244] and even though its nativity record was also bad, fatalities in childbirth were nowhere near twenty-five percent, so that they improved a figure that would have been yet more alarming otherwise.

When all allowances were made, the commission estimated that in a period of fifty-two years, the Hôtel-Dieu was responsible for the deaths of 99,044 citizens, or 1,906 every year, who would have been saved had they gone into La Charité instead, or for the deaths of 81,318 or 1,564 a year, that Saint-Sulpice would have saved.[245] The former figure amounted to one-tenth and the latter to one-thirteenth of the annual mortality in Paris. "Preserving that hospital . . . produces, therefore, the same effect as a kind of plague perpetually afflicting the capital. It is a cause of depopulation that can be removed, and we believe that the Academy should put its consequences before the eyes of the Government."[246]

The commission brought in this first and main report on 22 November 1786. In fact its purpose was to expose existing conditions in the Hôtel-Dieu. In form it was a reply to Breteuil's referral of a design that the architect, Bernard Poyet, had proposed for a new Hôtel-Dieu of over five thousand beds to be built to the west on the Ile des Cygnes just downstream from Paris proper. Plans called for a circular structure three stories high. In the event, the commission did not endorse them. Instead of a

[243] Above, Chapter I, Section 5.

[244] HARS (1785/87), 44-50.

[245] *Ibid.*, 55-71.

[246] *Ibid.*, 71.

single hospital, the size "of a city, and a city more populous than three-fourths of the cities of France,"[247] which would be bound to develop many of the faults of the old Hôtel-Dieu and would also be inaccessible to most Parisians, they recommended decentralizing the installation into four hospitals in different quarters. Two of them were to be realized by reviving the plan to enlarge and modernize the hospitals of Saint-Louis to the north and Sainte-Anne to the south, and the other two constructed on the property of the Célestins to the east and beyond the Ecole Militaire to the west.

The crown accepted the recommendation, and issued a decree ordering the work begun under the continuing eye of the Academy's commission.[248] But first it would need information on building requirements and on the internal organization of hospitals abroad. A questionnaire went the medical rounds in Holland, Germany, and Italy. The French always took Britain more seriously than other countries, however, and at the request of the commission, Breteuil and the Academy dispatched Tenon together with Coulomb on an intensive tour to see with their own eyes how matters were arranged beyond the Channel.[249] No trace remains of Coulomb's part in the inquiry, though it seems obvious that an engineer, and scientifically the most eminent of the engineers, would have been wanted for his experience with the practicalities of construction.

The Tenon-Coulomb mission gave occasion for one of those feasts of international reason and flows of civic soul that men of science and of letters liked to celebrate when reaching out their hands across frontiers. An initial skepticism in London needed to be overcome. Tenon's name was unknown there, and though Condorcet had recommended both colleagues to the Royal Society on behalf of the Academy, Blagden put Sir Joseph Banks on guard: "Certainly what the Secretary says of Mr. Coulomb's electrical and magnetic lucubrations is most overcharged."[250] A countryman resident in London, the marquis d'Herbouville, took them in tow, arranged appointments, and translated. Tenon combined seriousness with enthusiasm for things English and soon disarmed reserve. His reports, sent back by diplomatic pouch, offer a medical chapter in the eighteenth-century history of French self-criticism in the guise of Anglophilia. Between 1 June and 3 August 1787, Tenon, Coulomb, and the faithful d'Herbouville visited no fewer than fifty-two hospitals, prisons, and eleemosynary institutions. They inspected all the important installations in

[247] *Ibid.*, 85.

[248] "Deuxième rapport des Commissaires . . . ," HARS (1786/88), 1-2.

[249] Greenbaum (1971) is an excellent account of this mission. Tenon's journal of their inspection trip is among others of his papers at the BN, MSS, nouvelles, acquisitions françaises 11359-60, 22744. Cf. Huard (1966).

[250] Greenbaum (1971), 335.

London and in provincial centers as far as Birmingham to the north and Exeter to the west, observing and even measuring every aspect of accommodations, diet, hygiene, therapy, housekeeping, personnel, administration, and financing.[251] Tenon was impressed by the British practice of philanthropy through annual contributions, and the consequent involvement of the community, a voluntary system unknown in France, where reliance was on corporate endowments and on the state. Returning to Paris in the late summer, they reinforced in their idealization of British provision for the poor and sick ("Concern is universal. . . . Welfare in proportion to need") a public appeal for funds for the new hospitals.[252] The commission now recommended also that the new structures be built on the British model of wards dispersed in pavilions connected by walkways.

The appeal to charity, launched by Breteuil himself in January 1787, was not at all characteristic of the French way of financing such things, and yet it worked. The times were exceptional, and were felt to be; and fashionable Paris was nothing if not emotional in the twilight hours of the old regime. Over 2,000,000 livres were collected. The Treasury added another 1,200,000 from the proceeds of a public lottery. Alas, Breteuil fell out even with his old friend, Loménie de Brienne, archbishop of Toulouse, who became chief minister after the failure of the Assembly of the Notables and the fall of Calonne, and resigned himself in July 1788. When Necker then returned to office, his former aides, La Millière and Colombier, took their revenge on the scientists of the hospital commission and resumed the policy, though not the practice, of rehabilitating the Hôtel-Dieu. No ground was ever broken for the new hospitals, thus abandoned by a government itself sick to death, and the money subscribed by the public in that expiring gasp of humanitarianism disappeared down the chasm of deficit like a raindrop in the Grand Canyon.[253]

[251] Gillmor (1971), 67-68.

[252] The words are from the third and final report of the commissioners, HARS (1785/88), 13-42.

[253] Greenbaum (1973), 269-277.

CHAPTER IV

Scientists and Charlatans

1. A MEDICAL QUACK

The word "charlatan" comes from the Italian "*ciarlatano*," one who clothes ("*ciarlare*") himself in the colors, bells, and tassels of the medicinal peddler crying his nostrums and potions from a hastily movable platform set up on the street corner, in the marketplace, or at the fair, and heaped with phials and sachets. His type has been legion throughout history, and although it may never be known whether charlatans, who have seldom kept records, were in fact more numerous or brazen in one time or place than in another, what is quite definite is that intolerance for their operations is a measure of the significance that official masters of skill or knowledge in a subject attach to their own institutions, to their trusteeship of the public interest, and to their custody of truth and rigor.[1] The relative importance of these factors varies with the discipline, but persons held to be charlatans have in general given greater offense the more medical or mathematical the domain they are thought to adulterate. It is difficult to imagine historians reacting to the invasion of their fields by an Immanuel Velikovsky with the severity of the astronomers and physicists who forced a publisher to drop his book.[2] It is equally difficult to imagine writers responding to the forgery of Ossian like the geneticists and zoologists who drove a Paul Kammerer to suicide for having (so it was said) faked the nuptial pads on a midwife toad.[3] Practitioners of the more verbal indisciplines try to ignore rather than to suppress the flood of nonsense or falsehood ever being visited on a public which, in respect of their subjects, must protect itself from being gulled or vulgarized.

If, therefore, hostility to charlatans is a function of some collective sense of professional identity, strong and rigid internally, punitive and anxious externally, then the reform of science and medicine at the end of

[1] On the derivation, see Bouvet (1936a), 289; and on charlatanism and professionalization, see Johnson (1972) especially p. 57, cf. Jean Meyer, in Desaive *et al.* (1972), 185; Foucault (1963), 72-74.

[2] For an account of the reaction to Immanuel Velikovsky, *Worlds in Collision* (New York, 1950), see Juergens (1963), and for a review of the whole controversy see North (1976).

[3] Koestler (1971), who thinks that Kammerer did not falsify his specimens.

the old regime may indeed be taken for a conscious effort of professionalization. For its leaders saw charlatans pullulating everywhere and posing the clearest and most present of dangers to the very fabric of civilization. Thus (it will be recalled) Lavoisier urged the Academy of Science to remain small and selective upon its reorganization in 1785, lest the door be opened to imposture and intrigue and the standing of academician be debased.[4] Thus Vicq d'Azyr, inviting colleagues to correspond with the new Society of Medicine, praised the enlightenment and humanity of the medical profession, attributing its urbanity to the happy effects of an education that accustomed its members "to living with people who are cultivated and well-born, [attributes] for which genius alone can furnish a substitute, and of which a vulgar and disgusting charlatanism tries to give the appearance in the persons of those who would usurp our place and function."[5] Thus, at a humbler level, the pharmacists warned Lenoir that, were they to be deprived of recognition, the citizenry would be at the mercy of charlatans for their very lives, and they themselves would face the choice of abandoning their livelihoods or joining the competition.[6] Thus, later, Condorcet, in the most fully developed of his schemes for a national system of education, would place it under the administrative authority of scientists, for "It is not only ignorance that is to be feared: it is charlatanism, which would speedily ruin public education as well as the arts and sciences, or which would at least apply to their ruination everything that the nation might devote to their improvement."[7]

Let us have a look, therefore, at three examples of the breed. Two of them, Franz-Anton Mesmer and Jean-Paul Marat, are famous. It must be acknowledged at the outset that deciding in what degree they were self-deceivers, and in what degree deceivers of others, is as difficult as it is to know whether anyone was actually damaged by believing in their pretensions. But those questions do not need to be answered. More important, and also more interesting, are the manner in which scientists responded to their claims and the damage science thereby did itself politically. The third of our specimens, Guilbert de Préval, was a simpler case. Of another of his type, someone once said, "Of course he's a fake, but at least he's a real fake." Préval's case being the more obvious, we shall begin there in order to show how the authorities made themselves unpopular through identifying the public interest with protection of their own privileges. In these matters, the differences that otherwise divided the Faculty of Med-

[4] Above, Chapter II, Section 2.

[5] Above, Chapter III, Section 2, n. 136 (BANM, MSS 110).

[6] Above, Chapter III, Section 3, n. 88.

[7] "Rapport et projet de décret sur l'organisation générale de l'instruction publique, . . . présentés à l'Assemblée nationale . . . les 20 et 21 avril 1792," *Procès-verbaux du Comité d'instruction publique de l'Assemblée législative*, ed. James Guillaume (1904), 188-246, 220.

icine, the Society of Medicine, and the Academy of Science disappeared before the threat posed by the common enemy, the outsider or the renegade, allied with his victims.

In the year 1771, Préval was dispensing from his own premises at a louis per dose an "*Eau fondante anti-vénérienne*," which he claimed to have perfected from a Scotch receipt. According to the prospectus, "This medicine is so antipathetic to the disease that it serves as a detector: it changes color; it becomes disturbed; from its naturally limpid state, it becomes turbid, chalky, milky, at the mere proximity of the disease, and effects vary in proportion to its severity."[8] Not only was it a venereal detector, it was an infallible prophylactic. Challenged in the faculty, Préval did not reply there and instead exhibited its efficacy by personal demonstrations before prospective clients. On 6 May, according to the *Mémoires de Bachaumont*, he anointed himself and had connection with a hideously diseased prostitute in the presence of the duc de Chartres (the future Philippe Egalité) and the prince de Condé. On 6 June he repeated the experiment with an even more ravaged partner before the eyes of the surgeon of the comte de la Marche, eldest son of the prince de Conti. He suffered no ill effects, and Sartine, lieutenant-general of police, ordered that further trials be made.[9] Later, when Tessier and Bucquet analyzed the preparation at the request of the faculty, they found that it consisted of a suspension of chalk with a little salt and mercury and an odorant.

Thus defied, Préval's colleagues put their disciplinary machinery into its customarily slow motion, and after a second hearing on 8 August 1772 required that his name be eliminated from their roster. He might still publicize and sell his quack medicine, but without the endorsement of a doctor of the faculty. At once Préval got a lawyer and appealed the expulsion to the Parlement of Paris. On 19 October the dean, Le Thieullier, countered this move with a resolution from the faculty stating that its procedures precluded appeals prior to a third and definitive hearing. Only after three years did the case come up. Judgment on the procedural issue then boomeranged against the faculty. On 4 May 1776 parlement ordered that Préval must be given his third hearing, and that, since he had not had it, he must be paid all the fees due him for attendance at stated meetings of the faculty retroactively to 8 August 1772. Moving fast now, the faculty held a formal assembly on 5 June and confirmed the earlier two acts of expulsion. Nothing daunted, Préval's lawyer opposed that outcome and got a further order from the court requiring that his client be paid the emoluments due since June, and that they be continued pending ultimate decision in parlement.

[8] *Journal de médecine* 48 (1777), 16.

[9] *Mémoires secrets pour servir à l'histoire de la république des lettres en France* 5 (London, 1777), 305-306, 317-318.

Alleaume was dean by now, a peaceable man who did not inform his colleagues of this order, hoping that the affair would simply subside. On 23 September, however, Préval turned up for the official vesper service and signed the register, thus qualifying for a jeton or token to be redeemed. Astonished to see him there, Des Essarts, a physician highly respected inside the faculty and out, took Alleaume aside and asked by what right Préval was thus resuming his regency? Alleaume mumbled something about an order from the court, upon which Des Essarts reproached him for negligence and for allowing the faculty to appear in contempt when it was only in ignorance. Préval was nearby, and whether he heard the words, he gathered their drift. He had taken the precaution of being accompanied with a *huissier* or marshal, whom he now instructed to serve Des Essarts with a copy of the order. Des Essarts refused, on the ground that he was neither an officer of the company nor otherwise responsible.

Just six weeks later, however, on 2 November 1776, Des Essarts was himself elected dean. Informed now of the earlier order, the faculty charged him to instruct Préval that he was excluded from its assemblies but would be paid the value of his tokens of attendance. Préval was waiting in the anteroom in the company of his lawyer and another marshal. Taking the new dean for the one who had hardened the faculty, he now brought criminal charges against Des Essarts for having insulted and mistreated an officer of the court, and also for personal offenses. The parlement, ever touchy about the encroachment of other bodies upon judicial privilege, suspended Des Essarts from the deanship pending adjudication of the complaint. At the same time the faculty refused Préval entry at every session. Finally, on 13 August 1777, after an interregnum in the faculty's leadership of almost a year, all the regent doctors were summoned to attend on parlement in full regalia. At these last accounts, Des Essarts was vindicated and restored; Préval was condemned, fined, and made to pay costs, and professional justice was thus finally rendered. But at what a price: for the many months during which the faculty was preoccupied with this affair were precisely those when Vicq d'Azyr was organizing the Society of Medicine right under its unseeing eyes.[10]

A further price was the resentment of those who wished for easy protection from venereal disease. It has often been thus in attempts to force facts upon a public that prefers miracles. First the quack wins notoriety and gathers testimonials to his success. Exposed and condemned by the professional authorities, he appeals over their head to those needing relief. He is sure to find patrons among persons of fashion and promoters among journalists, ever eager for thrill and novelty, ever suspicious of science and

[10] For the Préval affair, see Steinheil (1903), Documents, 1, 133, 160, 227, 311, 384; and Notes, 2, 60-65.

organized intelligence. In this situation the responsible parties are almost bound to put a foot wrong, for they really are jealous of their prerogatives. The issue then becomes one of equity and a fair hearing. Substance is increasingly lost to view in the discussion of procedures, and it is the authorities (since the procedures are theirs) who find themselves on the defensive, not the quack, who scores a moral victory before the bar of opinion, however badly he may fare before that of officialdom or law.

2. MESMERISM

Just such a fashionable triumph embittered by official rejection lay behind Franz Anton Mesmer in Vienna when he arrived in Paris in late February 1778.[11] He was accompanied by a student-lackey, Antoine, and a French military surgeon, one Leroux. They took lodgings in a hotel kept by the brothers Bourret in the place Vendôme, hard by the present location of the Ritz. Having been preceded by his notoriety, they began receiving patients almost immediately. Mesmer later described himself as "assailed by the people who came to consult me."[12] Most of them were suffering from melancholia, rheumatism, paralysis, menstrual disorders, obstruction of the liver and mesentery, failure to sweat, or an eye disease called "*goutte sereine*," probably glaucoma. Mesmer was then forty-four years of age, imposing of mien, penetrating of eye, imperial of forehead, mobile of lips, Roman of nose, columnar of neck, statuesque of carriage, deep of understanding. He had the ability to right emotional imbalances by merging the instability of his patients into the cosmic equilibrium, accomplishing that effect by magnetism, normally aided by music, in which art he was also gifted. He played the cello and the clavecin. Mozart owed

[11] There is an enormous literature on Mesmerism. Contemporary pamphlets have been collected and bound in 14 volumes quarto by the Bibliothèque nationale, 4° Tb[62].1. Two near-contemporary historians are Deleuze (1813) and Bertrand (1826). The most careful and complete medical history is Tischner (1928), with a full bibliography. More recent writings are mentioned in Darnton (1968), 173-174, whose book is concerned with public opinion rather than with Mesmer and his work. The most interesting biography is Vinchon (1936), republished by Privat in the collection "Rhadamanthe" and edited by R. de Saussure (1971). Walmsley (1967) is a biography in English, more reliable than Buranelli (1975), a partisan work to be treated with caution. Amadou (1971) is a carefully annotated publication of the principal writings issued in Mesmer's name, together with salient items in his correspondence. There is a valuable critical bibliography, 337-344. The notes are from Vinchon, Amadou himself, and Frank A. Pattie, the author of an important article on Mesmer's debt to predecessors (Pattie, 1956). The library of the New York Academy of Medicine has an important collection on Mesmerism.

[12] In "Précis historique des faits relatifs au magnétisme animal jusqu'en avril 1781," Amadou (1971), 105. The original edition under the dateline London, 1781, bore the legend "Par M. Mesmer, Docteur en Médecine de la Faculté de Vienne. Ouvrage traduit de l'allemand."

the inspiration for certain passages in the Magic Flute to their youthful intimacy when members of the same Masonic lodge.[13]

A doctor of the Faculty of Medicine in Vienna, Mesmer had written of the influence of the stars on the human body in the thesis that qualified him for the degree in 1766.[14] Further reading and reflection in the early years of his practice revealed to him how the world is subtly lapped in an infinite ocean of fluid that, all imperceptible though it is, may be known by its effects in the phenomena of nature, of which it is the underlying agent. Alexandre Bertrand, one of the earliest and still perhaps the fairest historian of Mesmerism, considered that the notion derived ultimately from the universal fluids imagined in antiquity, and more immediately from sixteenth- and seventeenth-century theories of universal magnetism.[15] He had in mind the writings of Paracelsus, van Helmont, Gilbert, Santinelli, and most notably William Maxwell, a seventeenth-century physician in Edinburgh, with whose doctrines Mesmer's physical ideas were virtually identical.[16] The resemblance did not escape his detractors, though Mesmer himself vehemently denied ever having heard of Maxwell.[17] However that may have been, the magnetic fluid suffuses everything in nature. In living beings there is only one illness, which is an excess or defect or maldistribution thereof, and only one cure, a restoration of the balance.[18] (In practice, however, Mesmer always made an exception of venereal disease, and refused to treat its victims.)

Mesmer opened his magnetic clinic in Vienna in 1773. Like Sigmund Freud, another healer in that city a century later, he developed much of his technique from the treatment of a paradigm case, a young woman twenty-nine years of age called Oesterline. For several years she had been suffering from convulsive seizures in which the blood rushed to her head producing excruciating toothache and earache followed by delirium, fits of violence, vomiting, and fainting, after which the pain was usually relieved. The cyclic rhythms of these episodes led Mesmer to wonder whether a form of reciprocal action among terrestrial bodies might not exist similar to that among celestial bodies, and whether, if only he could

[13] Amadou (1971), 17.

[14] "Dissertatio physico-medica de planetarum influxu (1776)," trans. in Amadou (1971), 32-45; cf. Pattie (1956).

[15] Bertrand (1826), 1-18; Cf. Pattie (1956).

[16] *De medicina magnetica libri iii* (Frankfurt, 1679). An edition in German, translated by Georg Franck, was also published in Frankfurt.

[17] Notably Michel-Augustin Thouret, *Recherches et doutes sur l'existence du magnétisme animal* (1784), 5 and *passim*; Mesmer to Vicq d'Azyr, 16 August 1784, in Amadou (1971), 245.

[18] For a relatively succinct statement, see the 27 propositions with which Mesmer concluded "Mémoire sur la découverte du magnétisme animal," which he published at Geneva in 1779, Amadou (1971), 76-78.

identify it, he might not be able to produce the healing climax in his patient artificially. Of magnetism (he later said) he then had nothing more than common knowledge, but having heard of its employment for relieving stomach ache and toothache, he thought it might be the agent to try. He turned to an astronomer, Father Hell of the Observatory at Vienna, a Hungarian who had experimented with magnetic painkilling. Hell was a friend (though they soon quarreled), and he supplied Mesmer with several magnetized rods fitted to the human body.[19] Mesmer prepared his patient by prescribing a tonic laced with salts of iron. When next the fit came on her, on 28 July 1774, he applied the magnets to her stomach and her legs. Painful currents surged through her body almost immediately. At first they were random, but soon took their direction downward. Within six hours, Fräulein Oesterline was well, and though further treatments were needed, eventually Mesmer did succeed in ridding her of the recurrent syndrome.[20]

Experience with other patients soon taught Mesmer to dispense with magnets and tonics, for he came to recognize that ferro-magnetism is only one manifestation of the universal fluid. Electricity is another. Far more effective than either, however, was the true instrument of his therapeutic success: animal magnetism, emanating from his own person and acting on the nervous fluid of his patients. In the routine he developed for regulating its effect, he would begin by putting himself in harmony with his patient. He sat himself before her (for he always had greater success with women), feet and knees touching, his back to the north in conformity with the universal law of polarization. He then (the literal translation from the French accounts will convey the ribaldry of skeptics) made passes at the patient. He placed his thumbs like pivots upon the hollow of her stomach and, without exerting pressure, massaged the thorax and abdomen with parabolic strokes of his fingers, his eyes all the while boring calmly yet deeply into hers. When matters went well, the affected organs would be traversed by sensations of pain, of shivering, and finally of warmth. Further manipulation would vary according to the location of the trouble, in the head, in the eyes themselves, wherever. Making these passes, Mesmer's hands became opposite magnetic poles, one transmitting and the other drawing off the current until equilibrium should be restored.

In a more generalized therapy, high-current magnetization, Mesmer joined his fingers into the form of a pyramid and ran them all over the body, head to foot, front to back, until the patient was saturated and

[19] On Father Hell, see Sarton (1944).

[20] Mesmer's account is in "Mémoires sur la découverte du magnétisme animal," Amadou (1971), 63.

fainted, whether with pain or pleasure made no difference to the cure. In yet another approach, the longitudinal, he never touched her, and discharged or drew off magnetic fluid at a distance by means of rods of iron or copper, and sometimes merely through his fingers. As treatments continued, the fits became less and less intense, the patient came more and more under the control of the magnetizer, and eventually he obtained a cure or at least a remission of the distress sufficient so that the patient could once again live effectively in the world.[21]

In the later months of his practice in Vienna, Mesmer developed a method of group therapy, which he also transplanted to Paris after moving to larger quarters. While paying court to the Academy of Science and the Society of Medicine, he took a house in the suburb of Créteil; when he failed in finding official favor and the location proved inconvenient for his clients, he returned to Paris and leased the Hôtel Bullion, rue Coq-Héron, near Les Halles. At the height of his popularity there, he was operating four "*baquets*" or magnetic tubs while reserving individual magnetization for patients of great wealth and importance. Three of the installations served clients who could afford the fees. Mesmer supervised these sessions himself. The fourth, for charity cases, came under the charge of Antoine, the former valet, who, having been magnetized himself, had the touch and the insight, and alone among Mesmer's disciples never gave the offense of aspiring above the role of assistant. The *baquets* were shallow oaken tubs, one-and-a-half feet deep and four-and-a-half feet in diameter, with a cover pierced by holes through each of which protruded an iron stem on a flexible mounting of some sort. Inside the *baquet* bottles of magnetized water were arranged in concentric circles in a bath of iron filings or bits of scrap and ground-glass submerged in water or damp sand.

While a séance was in session, doors and windows were kept closed, curtains were drawn, and atmospheric conditions were monitored by a thermometer, a hygrometer, and a barometer. Patients were seated cross-legged on the floor around the *baquet*, knees touching. They were joined by a cable, and each also grasped the left thumb of his neighbor between his right thumb and index finger. The magnetic fluid thus circulated counterclockwise through several channels, kept in motion by the patients themselves, each of whom pressed the thumb to his right when his own left thumb was pressed. From time to time the chain was interrupted in order that patients might recharge themselves from the *baquet* by passing the iron stem that communicated with the interior over the affected part of their anatomy. A pianist played throughout the séance, alternating martial with soothing music, major with minor key, according to Mesmer's perception of whether the mood needed to be intensified or calmed.

[21] Vinchon (1971), 67-68.

He himself moved majestically among the patients, clothed in a suit of lilac silk with accessories of finest lace, attending to their state. Now he would pass his wand over the chest of this one; now he would join his hands and run them down the vertebral column of that one; now he would gaze deeply into the eyes of some other, bringing her safely through a healing frenzy, or modulating a salutary onset of hysteria.[22]

It enhanced the mystique that Mesmer never learned to speak French properly. He was not himself incommoded, for he did not need words in order to think or communicate, having learned when a lad to merge his sensibility directly into reality during long walks in the woods near lake Constance in his native Swabia.[23] The consequence for the historian, however, is that we have nothing in the way of documentation of his theories or practice directly from his own pen and, like the Parisian public, must make what we can of utterances and manifestos drawn up and issued in his name by the disciples who succeeded one another, first in his favor and then in his disfavor. The most important were a doctor, Charles Deslon; a lawyer, Nicolas Bergasse; and a nobleman, the marquis de Puységur. All three were eventually accused of betrayal, even as his wife and former associates in Vienna had been. Psychoanalytically—if we may believe a diagnosis by the editor of a classic biography who is conversant with Freudian categories—Mesmer was both paranoid and obsessive. Obsessive personalities, according to this account, inevitably repudiate even, indeed especially, the strongest of personal commitments, and events impinge on them in an isolated manner, unrelated except by the effect on themselves at the moment.

> With Mesmer that isolation was manifest in certain contradictory attitudes. . . . Thus he received rich and poor around his baquets, but at the same time he was a snob who above all else wanted entrée at Court. He claimed to be disinterested, but grasped desperately at his pupil's money and at the wealth that his discovery could bring him. He wished to be the savior of humanity, but at the same time to keep to himself the secret that gave him power. He wished to communicate his knowledge to his disciples, but to be the only one to know the truth. Being sincere in each of these attitudes, he did not perceive them to be contradictory.[24]

But whatever his psychological requirements, Mesmer also needed disciples for the practical purposes of advancing his interests in French society and gaining acceptance by the learned institutions of the capital. That

[22] Vinchon (1971), 66-67.

[23] "Précis historique," Amadou (1971), 100-101.

[24] R. de Saussure, "Le Caractère de Mesmer," in Vinchon (1971), 11.

was why he had come to Paris, after all, the seat of learning, where he would win accolades from the academies and be avenged on their provincial counterparts in Vienna. For, consistently contradictory, he always wished both to be acclaimed by faculties and academies and to humiliate them, both to be a physicist who had discovered new phenomena and to be a healer transcending science. Inevitably, therefore, he approached such bodies, as he did every manifestation of authority, in a manner that insured his failure, which then forced him to appeal instead to humanity at large.

Mesmer began with the loftiest, the Academy of Science. He availed himself of the good offices of one he thought well placed to introduce him to the company, its director in 1778, the physicist Jean-Baptiste LeRoi, who at first had appeared sympathetic. In fact, LeRoi was a wheelhorse of electrical science with little or no influence, though Mesmer could not know that. A practical joke made it urgent to retrieve reputation almost at the outset. Antoine Portal, anatomist and professor at the Jardin du roi and the Collège de France, had got himself up in the guise of an eminent jurist and consulted Mesmer in the place Vendôme, concocting a complicated case history and feigning a farrago of symptoms. He then dined out all over Paris on his story of the magnetic trumpery that dissipated his pretended complaints.[25]

According to the recital that Mesmer published of his encounter with the Academy, edited we know not by whom, LeRoi offered to get him a hearing for a memoir on his discoveries. It is more likely that Mesmer importuned LeRoi to do so, but however that may have been, he drew up his paper, communicated it in proper form, and presented himself at the Louvre bright and early on the appointed day. As the members arrived, they gathered in small groups for conversation. Ad hoc committees, Mesmer supposed these to be, and imagined that at a given moment they would convene into a coherent assembly, and proceed through the agenda. How wrong he was, with his Teutonic sense of seemliness. When LeRoi called them to order, everyone continued talking. He persisted, and a colleague told him impatiently to leave the first memoir to be presented on the table. People could read it if they liked. LeRoi got no better reception for a second item (dismissed without a hearing for banality) or a third (for charlatanism). Losing the thread, Mesmer here began thinking how little this collection of arrogant gossips resembled the great Paris Academy he had venerated from afar. When LeRoi announced that his paper was to be next, Mesmer came to and objected vigorously, begging that it be put over until a time when attitudes might be more propitious. After adjournment, ten or a dozen of the younger and more curious did stay

[25] Vinchon (1971), 54.

behind and pressed him to show them some experiments (childishly enough, since they knew nothing of the matter). Mesmer then relented, unwisely he realized even at the time, and tried magnetizing a certain Monsieur A****, a sufferer from asthma. He even had a small success in spite of the unsympathetic surroundings.[26]

Reflecting on his treatment at the hands of the Academy, Mesmer began threatening to abandon an unappreciative Paris, some six months after his arrival. He had soon perceived how "superficial curiosity was a dominant taste in that capital."[27] Having fled his native land to escape the injuries visited on him by jealous colleagues in revenge over his triumphs in treating certain maladies, he had not settled abroad in order to expose himself to identical chagrins. The scientists of England and Holland awaited him. It was, he said, going off on his scientific tack, in order to move among physicists and prove the existence of an unknown physical truth that he had come. What interest could he have in provoking useless medical disputes, in again arousing the hostility of mere doctors against his discovery, and even against his person?

Nevertheless, if even the scientists would judge of such a truth only by its medical utility, he would make one more attempt to convince them, persuaded to be magnanimous by the urbanity of an honorary member of the Academy, the comte de Maillebois, who had visited the place Vendôme several times and who now intervened to allay Mesmer's resentment over LeRoi's having exposed him, "a foreigner and without support," to the incivility of his colleagues. He would even overlook LeRoi's apostasy and puerility in observing, what indeed Mesmer had heard said before, that the effect of his manipulations on his patients derived from their imagination. Against his better judgment, therefore, even against all his principles, and thanks only to Maillebois, he accepted the challenge of treating a certain number of patients. Oh, he ought not to have acceded, for there is no way to prove conclusively that what actually cures an illness is the medicine or the doctor. "Though when, for example, I put my finger on a localized pain caused by an indisposition, and when I move it at will from the head to the stomach and from the stomach back to the head, it is only complete madness or outright bad faith that can explain the refusal to recognize the author of such sensations."[28] Still, he would oblige, and for this purpose he gathered his handful of sufferers and retired to Créteil, in the country just east of Paris. From there he wrote to LeRoi. The letter is dated 22 August 1778. He had as yet received no judgment

[26] Mesmer's account of his relations with learned bodies is the burden of the "Précis historique," published in Amadou (1971), 89-202. The episode with the Academy of Science is recounted on 105-110.

[27] *Ibid.*, 105.

[28] *Ibid.*, 109.

from the Academy concerning his propositions about magnetism, and he now invited its representatives to attend on the treatment of his most inveterate cases.

He never had a reply. LeRoi did communicate his letter, so he learned, but its reading was interrupted by two persons who prevailed with a motion that the Academy take no cognizance of his discovery. Both were members also of the Society of Medicine. The identity of the first surprised him. It was Daubenton, who ought not to have forgotten that, "if he was striding toward posterity with firm steps beside Monsieur de Buffon, it was in order to be seated there on a throne of undoubted marvels."[29] The hostility of the other did not surprise him in the least. For it was Vicq d'Azyr.

> It must be neither dissimulated nor forgotten: the difficulties between the scientists and myself arise only because I asked for nothing more than their testimony, urging them merely to determine and acknowledge openly the existence and reality of my discovery. They for their part wished to be its arbiters, its judges, its dispensers. The only thing that matters to them is their tribunal, and the truth not at all unless they can exploit it for their glory or their fortunes. Humanity can perish before they will abandon their pretentions.[30]

Mutual intolerance reached a higher pitch in Mesmer's relations with Vicq d'Azyr than in any of his other encounters, and inevitably so, given their respective aspirations. The spring and summer months of 1778, Mesmer's first year in Paris, were also those when Vicq d'Azyr and Lassone were paying court for adoption of the letters patent that gave the Society a watching brief over public health and vested in it responsibility for the reform of medicine by the infusion of science. "The Royal Society of Medicine of Paris," wrote Mesmer, or rather had someone write in his name, "was so new an establishment when I had dealings with it that it wasn't even born yet. It was only conceived and in the womb. I was present, so to speak, at the delivery."[31] It was to Vicq d'Azyr that he addressed his most bitter reproaches about being expected to hand over "to the examination of a committee of doctors a doctrine that arouses all their prejudice, or, if you prefer, that is not in accordance with their knowledge."[32] After all, he had

> in hand a truth essential for human welfare. It's not enough that I should wish to be the benefactor of mankind. They also have to accept the boon, and first of all they must believe in it. To this end, I have sought out those persons whose opinion carries some weight

[29] *Ibid.*, 110. [30] *Ibid.*, 112. [31] *Ibid.*, 112.
[32] Mesmer to Vicq d'Azyr, 16 August 1784, Amadou (1971), 244-247.

> with the public. I invited them to witness the salutary effects of my discovery, to make the truth appreciated by paying tribute to it, and in this easy way to deserve the gratitude of nations. The Royal Society of Medicine of Paris has not thought this role worthy of it.[33]

What actually happened is not quite clear.

> Eager [according to Mesmer] to increase its prestige by novelties of every sort, the Society had the kindness to take notice of me. It held out to me the scepter with which it rules over charlatanism. I did not bow down. That was my misfortune. Possibly, however, things would have worked out amiably if the rigid permanent secretary of that company had not ruthlessly overborne the protective measures of his flexible colleagues.[34]

Apparently, Mesmer either asked for the approbation of the Society in addition to that of the Academy or was told that for his purposes he should have it (to which advice, however, he opposed the objection that it was less a question of approving some therapy than of verifying a physical truth). In any case, he refused the appointment of a commission by the Society. He would agree only to exhibit his procedures to commissioners. The distinction may seem a nice one, but Mesmer made it all important. For what purpose did the Society name commissions? Why, to examine the virtue claimed by some apothecary who possessed a powder, or a liquid, or some other concoction, and sought a license to sell it. Not so Mesmer, whom nature had favored with a hitherto unsuspected power. He had never harbored the low thought of extracting profit from it. Indeed, the very comparison was inadmissible (not that the Society's tyrannizing over these trifling shopkeepers offered the public any protection from the poisons dispensed under the name of traditional medicine).

He would, however, welcome in his clinic at Créteil commissioners from the Society, even as he would from the Academy, provided they come in the spirit of participating in the truths that awaited them. Only there must be no more talk of their examining some artificially or randomly selected group of patients whom Mesmer himself had not yet treated. They must come as invited visitors, not judges, free to see or, if they wished, to experience the effects of animal magnetism on the patients already assembled, and to interrogate them about their cures. He would himself, and he did, supply the Society with his own case histories describing their condition when they came to him. On 6 May 1778 Vicq d'Azyr returned those documents unopened, and when Mesmer persisted, in the pretense that his willingness to be reasonable must not have been transmitted to the Society, Vicq d'Azyr replied shortly, "That company,

[33] "Précis historique," *ibid.*, 111.   [34] *Ibid.*, 113.

which had no knowledge of the anterior condition of the patients subjected to your treatment, can make no judgment about it."[35] With that, a "definite dismissal," Mesmer abandoned hope of the Society.[36]

Quite independently of Mesmer, the Society together with the Academy of Science had been encouraged by the government to commission an investigation of therapeutic applications of electricity and magnetism,[37] a circumstance that intensified mutual disdain between Mesmer and official medicine. Physicians had begun experimenting with shock treatments soon after the invention of the Leyden jar. In 1747 one Jallabert, practicing in Geneva, gave electrical relief to a patient partially paralyzed by a stroke. Other doctors made similar attempts, with varying success, mainly in Germany and northern Europe, where, according to the report prepared for the Society, work was published "in that diffuse and obscure style which serves to retard for other nations the progress that they make in science."[38] The judgment was by Mauduyt. He had undertaken the mission with great interest, though he confessed that he had only the educated man's ordinary knowledge of electricity at the outset. He had first to make himself expert, therefore, and he started with the work of Benjamin Franklin, which in his view had turned the subject from a curiosity into a science.[39]

Being a medical man, Mauduyt naturally began his own research by studying the effects of meteorological conditions on charged bodies and comparing the observations to those made on organic bodies.[40] Thereafter, he published a series of important memoirs surveying the literature, describing the techniques, and reporting his experiences in the electrotherapy of eighty-two patients.[41] Some fifty were victims of paralysis, mostly following strokes. Thirty-one of them showed marked improvement after electrification, and several recovered completely. The remaining cases involved hormonal imbalance (*épanchements laiteux*), menstrual irregularity, constipation, rheumatism, gout, deafness, and loss of vision

[35] Vicq d'Azyr to Mesmer, 27 August 1778, *ibid.*, 117.

[36] *Ibid.*, 113-117.

[37] Pierre-Jean-Claude Mauduyt de La Varenne, "Premier mémoire sur l'électricité considérée relativement à l'économie animale et à l'utilité dont elle peut être en médecine." 7 October 1777, MSRM (1776/78), 461-513. On this research, see Schenk (1959); other important contemporary treatises are J.-A. Sigaud de La Fond, *Précis historique et expérimental des phénomènes électriques, depuis l'origine de cette découverte jusqu'à ce jour* (1781), and J.-H. Van Swinden, *Analogie de l'électricité et du magnétisme; ou Recueil de mémoires couronnés par l'Académie de Bavière* (La Haye, 1785).

[38] MSRM (1776/78), 463.

[39] *Ibid.*, 464.

[40] *Ibid.*, 509-513; "Seconde mémoire sur l'électricité médicale," *ibid.*, 514-528.

[41] The memoirs cited in note 37 were followed by "Mémoire sur le traitement électrique, administré à quatre-vingt-deux malades," 18 December 1775, MSRM (1777-1778/80), 199-431.

in "*goutte sereine*," a serous discharge. For most of these complaints, the ratio of full or partial relief was more favorable than in hemiplegia. All the patients were started in an electric bath, that is to say a field of electrostatic charge in proximity to a generator. The time of exposure increased with each treatment. When appropriate, Mauduyt went on to apply the electric spark from a generator to localized ailments, and generalized "commotion" or shock from a Leyden jar to systemic dysfunction.[42]

Mauduyt was cautious in his treatments and his claims. It might easily be observed, by the breed of critics who are wont to combine cynicism about the actual with fantasy about the possible, that his cures must have owed as much to nature and imagination as ever Mesmer's did, and no doubt that would be correct. The difference was that Mauduyt never set up to be an advocate. He accepted patients only on referral by their regular doctor, whose presence he invited at the treatments. He advanced no conclusions about the efficacy of electrification and put his confidence in the medical free market. If physicians found electricity helpful, they would prescribe it, and it would win a place in the armory, whatever the fulminations of opponents. If they found it useless, it would be forgotten, whatever the enthusiasm of panegyrists. "Time and experience will thus decide the true value of electricity, as indeed they reduce everything to its proper price."[43] Such an attitude of caution never commended itself to Mesmer. He had heard of Mauduyt's work while still in Vienna, and had paid a call as soon as he arrived in Paris. He was undeceived when Mauduyt told him that no certain cures could be claimed for electricity.[44]

Compared to electricity, magnetism was a generally familiar phenomenon. Attempts to employ the lodestone for healing go back to antiquity. Nothing sure had emerged from centuries of desultory curiosity, however. Such at least was the opinion of François Andry and Augustin Thouret, commissioners who reported for the Society of Medicine in 1780.[45] Andry had been one of La Mettrie's targets in the *Politique du médecin de Machiavel*. He was the original of the character "Verminosus" in that satire.[46]

[42] The apparatus and techniques are described in "Mémoire sur les effets généraux, la nature, et l'usage du fluide électrique, considéré comme médicament," 29 December 1778, MSRM (1777-1778/80), 432-455.

[43] "Mémoire sur le traitement électrique," MSRM (1777-1778/80), 427.

[44] "Précis historique," Amadou (1971), 113.

[45] "Observations et recherches sur l'usage de l'aimant en médecine, ou, Mémoire sur le magnétisme médicinal," 29 August 1780, MSRM (1779/82), 531-688. Originally, the Society had commissioned Mauduyt to undertake this investigation also, in conjunction with Andry. He was too busy, however, what with his research on medical electricity, and Thouret took his place (582). The footnotes to this memoir constitute a full bibliography of the literature on medical magnetism.

[46] (Amsterdam, n.d.), ch. IX, 38-40.

Thouret we have already met exhuming the Cimétière des innocents at a slightly later stage in his career.[47] He became interested in medical magnetism during a visit to Rouen. A merchant there had been suffering years of agony from an exposed dental nerve. When Thouret saw him, he was following local advice to carry a lodestone and apply it to his jaw from time to time. It worked. He could feel the skin being drawn and the nerves shifting position inside his face. Other patients confirmed the effect. A considerable literature attested to the virtue of the lodestone in relieving generalized afflictions of the nerves, even epilepsy. Most promising were the results achieved with a line of increasingly powerful steel magnets fabricated by the abbé Le Noble, canon of Vernon-sur-Seine, whose chef d'oeuvre weighed fifteen pounds and was capable of lifting 230 pounds. Like Father Hell, Mesmer's onetime associate in Vienna, he fashioned medical magnets in various forms for different parts of the body.[48] All things considered, it seemed possible to Thouret that magnetism might prove to be as valuable a resource in medicine as its directive properties made it in physics.[49] Thouret devoted several paragraphs of his review to Mesmer's work in Vienna, though without distinguishing him in principle from other practitioners at this juncture, or acknowledging that by animal magnetism Mesmer meant something deeper and more cosmic than the ferro-magnetism he had been commissioned to investigate.[50]

That is getting ahead of the story, however. We left Mesmer in August 1778, returning from Créteil to rue Coq-Héron, after his rejection by Academy and Society, reduced to a state of irresolution and yet continuing out of the abundance of his loyalty to treat his patients, and "lonelier in Paris than if I had never known a soul. . . . Oh Heaven! What immense solitude! What a desert peopled by beings insensitive to Good."[51]

Enter now the disciples, and first Charles Deslon, who came to him in the depths of that September. Little is known about Deslon's life or career prior to that moment. He was over forty, having been born in 1739 in the region of Toul. He had been admitted a regent-doctor of the faculty in 1746, had a practice in Pont-à-Mousson, and (even like Marat) was a physician-in-ordinary to the comte d'Artois.[52] Deslon was first told of

[47] Above, Chapter III, Section 7.

[48] MSRM (1779/82), 581-582.

[49] *Ibid.*, 532.

[50] When that difference was borne in, Thouret undertook a long and careful survey of everything ever written on the subject of animal magnetism, *Recherches et doutes sur le magnétisme animal* (1784). The committee appointed to examine the book for the Royal Society consisted of Geoffroy, Desperrieres, Jeanroi, Fourcroy, Chambon, and Vicq d'Azyr himself, who no doubt set the tone. Their report (printed on pp. xix-xxiv) was far more severe in its condemnation of animal magnetism than the book itself, which held to a reasoned and slightly condescending skepticism.

[51] "Précis historique," Amadou (1971), 123.

[52] Steinheil (1903) 2, 81.

Mesmer by a client, a cool level-headed gentleman who had been eased by magnetization after long years of getting little help from the resources of traditional medicine. One day Deslon was paying a professional call on this person when Mesmer arrived. Their patient introduced them, and to Deslon's astonishment went into a violent convulsion at the touch of Mesmer's hand. Deslon remained to observe the treatment, began frequenting the circle himself, and was a speedy convert.[53] Judging from his deportment once he became the medical apostle of Mesmerism, we may suppose that he was of a naive, enthusiastic, and consistent character, largely innocent of guile. In his very openness lay a source of trouble, however. He soon began urging the master that the way to convince the medical authorities along with the public was to reveal every aspect of the secret that nature had entrusted to him, and Mesmer never proved willing to reveal it all, even to Deslon, though often promising to do so one day when it might be safe.

In order to hasten the day, Deslon set about persuading his colleagues on the faculty of the reality of the experiences that had converted him. He got Mesmer to release a *Mémoire sur la découverte du magnétisme animal* in late 1779 (having probably put it into French for him).[54] He published his own *Observations sur le magnétisme animal* in 1780. He reported on the séances rue Coq-Héron at stated meetings of the Faculty of Medicine. He invited twelve senior members of the faculty to dine and to hear Mesmer read a draft of his 1779 memoir. The occasion was a disaster. In this instance, Mesmer's doughy Viennese accent did hurt the cause. Deslon then persuaded three doctors whose good faith he trusted—Mallouët, Bertrand, and Sollier—to observe Mesmer's treatment of four patients, and (what was harder) got Mesmer to admit them. For a period of seven months into the late summer of 1780, they attended a séance every fortnight. Their skepticism (prejudice?) proved invincible, however.[55] Meanwhile, critical articles were appearing, by Paulet and by Dehorne in the *Gazette de la santé*, by Bacher in the *Journal de médecine*. Undaunted in the face of this gathering hostility, Deslon took Mesmer's cause to the floor of the faculty at a general assembly called for 18 September 1780. He had transmitted to his colleagues a set of nine propositions signed by Mesmer on 2 July challenging the faculty to select a group of twenty-four patients (venereal disease to be excluded). He would treat twelve, and the faculty would arrange conventional treatment for the other twelve. Should comparison of the results be favorable to animal magnetism, he would expect

[53] Deslon, *Observations sur le magnétisme animal* (London, 1780), 16-22; for Mesmer's account, see "Précis historique," Amadou (1971), 125-126.

[54] Amadou (1971), 59-88.

[55] "Précis historique," Amadou (1971), 127-137; Vinchon (1971), 72, 79-88.

the faculty to take his part in making application to the government.[56] "My discussion with the faculty," wrote Mesmer of the hearing Deslon received, "had the one agreeable feature that it lasted only one day and that everything passed in writing between us."[57]

Outraged, the defenders of legitimacy had chosen one of the youngest regent doctors, Roussel de Vauzesmes, to be their champion. He opened the September meeting with a contemptuous harangue reviewing the whole record of this "German mountebank" in the form of a diatribe against Deslon's book and conduct.[58] Among other offenses, Deslon had announced in the *Journal de Paris* that he gave free consultations at certain hours in his office in the Temple, "an indirect and indecent way of giving his address, and unworthy of a real doctor."[59] In reply Deslon defended Mesmer rather than himself. He distinguished between the Academy, which had been guilty only of inattention, and the Society of Medicine, which had refused a new truth and a new therapy out of vanity and prejudice. Deslon reminded the faculty of its own recent humiliation at the hands of that arrogant Society (perhaps he was not entirely guileless after all), and emphasized Mesmer's intention of confiding his discovery to the government, which would naturally need to take counsel with "real scientists."[60]

The bait was dangled in vain. After due deliberation, the faculty formally (that is in Latin) rejected Mesmer's propositions and less formally (in French) resolved that the dean should warn Deslon to be more circumspect. He was to be suspended from a voice in its assemblies for a year and required to disavow his book on penalty of being eliminated from the catalogue. Thus began the process that, after the three hearings required in as many years, ended for Deslon as it had for Préval, with the medical equivalent of disbarment or unfrocking.[61] For in the event he continued in his practice faithful to the error of Mesmer's ways, even though Mesmer proved anything but faithful to him.

Mesmer had indeed begun sending signals to the government. "I ought to have patronage," he observed in the "Précis Historique" of 1781,

> I should like to have it, but it must come from the Monarch who is a Father to his people, from the Minister who has his confidence, from a system of law favoring the just and useful man. Any patron

[56] Steinheil (1903) *1*, 565-567.

[57] "Précis historique," Amadou (1971), 137.

[58] Steinheil (1903) *1*, 545-563; excerpts, with Mesmer's comments, are in Amadou (1971), 142-162.

[59] Steinheil (1903) *1*, 551.

[60] *Ibid.*, 567-572.

[61] *Ibid.*, 573. A decree of 20 August 1782 excluded him from the assembly for a further two years (*ibid.*, 943-944); and one of 28 August 1784 sealed his expulsion (*ibid.*, 1157).

> worthy of the name will never see me blush over the standing of a protégé. But I will never be the protégé of some swarm of petty, self-important types who understand nothing of the value of patronage beyond the sordid price it cost them to acquire it.[62]

That message began to reach the court so soon as Mesmer had clients sufficiently well placed to carry it. Once installed at rue Coq-Héron, the baquets became fashionable. Among his regular patients in 1781 were the duchesse de Chaulnes, the first friend Marie Antoinette had found on her own arrival from Vienna, and also her current favorite, the princesse de Lamballe, grand-mistress of the distaff masonic Mère Loge écossaise.

French freemasonry had welcomed Mesmer once he became notorious, and it was almost certainly these ladies who appealed to the queen for intervention by government when, early in the year 1781, Mesmer let the word devastate his patients that he must abandon them. The threefold rebuffs from the Academy, Society, and Faculty had discouraged him with France. He must go where his discovery would be appreciated. Her sensibility touched, and in mourning for her mother, Marie-Antoinette turned to Maurepas to find a way to keep Mesmer in France over the heads of the learned societies. Early in March emissaries waited on him in confidence. For the queen's sake, he would give the government until 15 April. At first, he even agreed to accept a government commission. Should their report on his procedures be favorable, the king would grant him a pension of 20,000 livres for life together with title to a property where he might treat patients and train associates, in effect an Institute for Animal Magnetism.

But then Mesmer drew back: not even his regard for the queen could enable him to swallow the indignity of preliminary examinations of his patients by outsiders. People must begin by accepting the reality of his discovery a priori. Very well then, Maurepas would drop that, and still the king would grant the pension and pay 10,000 livres in rent for a suitable installation, the only condition now being that Mesmer accept three pupils, to be chosen by the minister. Again Mesmer balked. These so-called pupils would be government agents spying on his work. For himself, he would not haggle over the stipend. He was confident it would be worthy of the French nation and its monarch. Twenty thousand a year and the deed of the chateau he had chosen, that was all he needed. But he could see that "conviction is a plant foreign to French soil, and the simplest thing for me is to cultivate some less ungrateful ground."[63] Maurepas, who knew how to lose his patience and keep his self-control, now told him that the agreement Mesmer had already signed was final.

[62] "Précis historique," Amadou (1971), 123.

[63] Vinchon (1971), 101; cf. "Précis historique," Amadou (1971), 172-187.

Meditating that word, Mesmer returned to his clinic and put his name to what would surely be one of the most extraordinary letters ever written to a queen of France even if he had sent it privately. Instead, he had it printed, rating her in public about the offer made in her name and giving her an ultimatum. He had abandoned hope of the government. He would remain in Paris tending his patients until 18 September, only because of his regard for her. Four or five hundred thousand pounds were nothing to her. He had no more interest in money than she did. What mattered was the welfare of mankind. Balanced against that, the twenty or thirty patients he was caring for in Paris counted for nothing, and so did the duchesse de Chaulnes. He would be as blameworthy to neglect the suffering of humanity to look after the queen's friend as he would be to seek gain. Once before he had known how to abandon patients who were dear to him in order to carry his discovery to the world. That had been in his native land, which was also that of Her Majesty. Ill-disposed persons there had poisoned the minds of her august mother and her august brother, even as the same sort now sought to do in France. Public opinion would vindicate him, however. The date he was fixing for his new departure was the anniversary of the rejection of his propositions by the Faculty of Medicine of Paris, and of the dishonor it brought upon the one colleague to whom he owed everything. He had counted on her intercession with her husband and her brother, the King of France and the Emperor of Austria.[64]

That this letter did not land its author in the Bastille or worse was one symptom among many of the loss of nerve that was progressively paralyzing the French monarchy. In August, well before his deadline, Mesmer went to take the waters at Spa. He hurried back as the time came on for the second hearing of Deslon's case before the faculty, not so much out of the solidarity expressed to the queen as out of alarm that his disciple was pretending to a mastery adequate for establishing an independent practice. His chagrins and clientele increased throughout the autumn, winter, and spring of 1781-1782. In July 1782 he accepted an invitation from the marquise de Fleury to remove again to Spa, there to install a bosky clinic surrounded by the most grateful of his patients. Among the company of thirty-odd, the leading spirits were Nicolas Bergasse and Guillaume Kornmann.

Bergasse was a neurotic lawyer thirty-three of age and originally from Lyons, a perpetual patient now become Mesmer's man of confidence in succession to Deslon, whom Mesmer disowned once he had set up on his own.[65] Kornmann was an Alsatian banker. Mesmer had treated his

[64] "Précis historique," Amadou (1971), 187-190.

[65] Bergasse published his own account of Mesmer, *Considérations sur le magnétisme animal* (La Haye, 1784). There is considerable documentation on his political career in the Revolution, AN, F^7.4595; Cf. Louis Bergasse (1910).

infant son for incipient blindness and, so the father believed, saved the boy from the life of helpless dependency to which the doctors had condemned him.[66] Kornmann's peace of mind was being further eroded by the adultery of his wife with Lenoir, lieutenant-general of police, whom we have met as patron of the pharmacists. The scandal threatened Kornmann's business since his wife's dowry furnished much of its working capital. Bergasse endeared himself to Kornmann by publicizing à la Beaumarchais the injuries of the worthy citizen at the hands of the philandering magistrate. Both further ingratiated themselves with Mesmer by taking his part in Spa against the pretensions of Deslon, who had openly set up for himself in Paris. Casting about for ways to institutionalize and propagate legitimately the secret capable of doing so much good, they pooled their legal talents and banker's instincts and hit upon the notion that became the Société de l'harmonie universelle, in their original conception a joint-stock company for animal magnetism.[67]

Publicity and promotion occupied the winter of 1782-1783. In March, 100 shares were offered at 100 louis each. (The louis was worth twenty-four livres.) The opportunity was seized. Besides Bergasse and Kornmann themselves, the marquis and the comte de Puységur were high on the list. It included two eminent clergymen, Père Gérard, superior of la Charité, and Dom Gentil, prior of Fontanet; certain grands seigneurs, a Noailles, a Montesquieu, a Choiseul-Gourrier; the marquis de Lafayette; and even two scientists, Berthollet and the young Cabanis. Berthollet, it is true, resigned almost immediately, claiming to have been deceived.[68] Lafayette, however, carried the news to George Washington on his triumphal return to America in June 1784. The minute book of the Philosophical Society in Philadelphia records that on 12 August Lafayette "entertained" the company "with a particular relation of the wonderful effects of a certain invisible power in Nature called Animal Magnetism discovered by a German Philosopher, M. Mesmer," but was not at liberty to say just how it worked.[69] In Paris a momentary reconciliation between Mesmer and Deslon threatened the enterprise, but its rupture threw Mesmer back upon Bergasse, whose turn it now became to suffer the master's failure to impart the essence of his doctrine as he always promised. Other shareholders, notably the brothers Puységur, persuaded Bergasse that the cha-

[66] See, "Cure opérée par M. Mesmer sur le fils de M. Kornmann," *Recueil . . . de tous les écrits pour et contre le magnétisme animal* 3, pièce 35, 38-42. BN, 4° Tb[62].1.

[67] For the background, see Vinchon (1971), 107-117, together with his note, "La Société de l'harmonie universelle," in Amadou (1971), 203-206; also Darnton (1968), app. 3 and 4, 180-186.

[68] His disavowal is printed in Bertrand (1826), 62-63.

[69] The entry is conserved in the minute-book, Library of the American Philosophical Society.

grins of dealing with the healer were trivialities to be absorbed in winning through to the alleviation of suffering and the regeneration of morality by magnetism. The Hôtel Bullion had been bought during the Spa interlude by Freemasons for the Loge du Contrat Social. Bergasse and his associates now took a neighboring townhouse, the Hôtel de Coigny, and there installed Mesmer's tubs and clinic in company with their headquarters.

Through the medium of their Société de l'Harmonie Universelle, animal magnetism expanded from the scale of Mesmer's practice to that of a national movement. The network of masonic lodges provided the organizational model, and there was much overlap in membership, in Paris and in the provinces.[70] The statutes comprised some sixty articles of agreement between Mesmer, the senior "Société de France" in the capital, sister societies in the provinces, and their respective members and pupils. Mesmer would be president for life and confide his discovery to the society. He would initiate the original subscribers into the mystery, and they would train pupils in the affiliated societies. These correspondents would be authorized to take patients privately, though not to give treatments in public without the consent of the local group. All moneys paid were to be for the benefit of humanity, and every member agreed to dispense magnetism only for the relief of suffering and never for personal gain.[71] The funds thus raised were considerable. The membership of the society in Paris increased to 430, and Bertrand estimated that it paid better than 340,000 livres to Mesmer himself.[72] Provincial societies started in Strasbourg, Bordeaux, Lyons, Grenoble, Dijon, Montpellier, Marseilles, Douai, and Nîmes; in ten or fifteen lesser towns; and overseas in Cap Saint-François. Mesmer proposed setting those dues at 50 louis per member, half of which would be his portion.

By Chapter II, Article IV, of the regulations Mesmer undertook to deposit in the archives of each society a signed copy of his principles, to be supplemented by further insights from time to time. Those documents would constitute his true doctrine and determine the training of students. He never did, of course, and when Bergasse and his fellow promoters forced the issue, Mesmer avoided coming to grips with his now too legalistic followers by vanishing from France in a puff of recrimination to settle near Lake Constance in the Swabia of his birth.[73] We need not follow Mesmerism proper (or improper) further. The movement no longer required its founder's presence, and continued through the decade becoming ideological and literary rather than medical in emphasis, one among

[70] For detailed discussion of the relations between the Society of Universal Harmony and Freemasonry, *see* Amadou (1971), 361-399.

[71] The statutes are printed in Amadou (1971), 209-224.

[72] Bertrand (1826), 52-53.

[73] Vinchon (1971), 142-159.

many flirtations with the marvelous titillating the world of fashion and of letters.[74]

Meanwhile the government, which for this purpose is to say Breteuil, had taken a hand, less alarmed over animal magnetism itself, a six-year-old story by 1784, than over the attendant financial and institutional excesses. There is confusion in the literature about the commission of that year, often said to have been appointed in the Academy of Science. In fact two commissions were appointed both at the instance of Breteuil, who had the reports published in August. The first originated in the Faculty of Medicine. Members of the Academy were joined to it at the request of the four doctors originally named, Guillotin (whose blade replaced the hangman in the Revolution), Darcet (the chemist), Sallin, and Borie, who died and was replaced by Majault. The five academic members dominated the proceedings. The Academy named Lavoisier, Benjamin Franklin, LeRoi, Bory, and Bailly. Bailly took the chair and magnified its mission into one of the blue-ribbon investigations concerning public health on which he and Breteuil, representing science and government respectively, were collaborating in those years.[75]

The second commission emanated from the Society of Medicine and proceeded separately.[76] It consisted of Mauduyt, Andry, Caille, Poissonier, and Antoine-Laurent de Jussieu. Poissonier was formally chairman, no doubt because of his position at court, but took no part. Jussieu did, and wrote a dissenting opinion much celebrated in the annals of Mesmerism. It has sometimes been taken for a minority report of the Bailly commission, however, which is often called "the Mesmer commission." That denomination brings up one further element of confusion. By the letter of their charges, both commissions were to examine the practice and procedures of Deslon, who, if not precisely the official, was at least the French protagonist of animal magnetism, still a regent doctor until his final expulsion by the faculty in August, and a subject of the crown.

In the eyes of the government and the public, the Bailly commission was by far the more important. It enlarged its purview to encompass the whole practice of Mesmerism, on the plea that Deslon accepted and practiced all its principles. Deslon for his part welcomed the investigation and was the soul of cooperation, thus in Mesmer's eyes compounding impos-

[74] Darnton (1968); Amadou (1971), 361-375.

[75] The *Rapport des commissaires chargés par le roi de l'examen du magnétisme animal* (1784) was separately published. It has been reprinted many times, and appears in company with the other reports in Bertrand (1826). Bailly read a brief summary before a public meeting of the Academy held on 4 September 1784 in honor of Prince Henry of Prussia, who was present. HARS (1784/87), 6-15.

[76] Published "par ordre du roi à l'imprimerie royale 1784," reprinted in Bertrand (1826), 482-510, an abstract was presented before the Society of Medicine on 24 August 1784, HRSM (1780-1781/85), 257-258.

ture by betrayal.[77] He invited the commissioners to his clinic and sat them down to observe the patients around his tubs. Here is Bailly's account of a *crise*:

> Nothing could be more astonishing than the spectacle of these convulsions. If you have never seen it, you can have no idea; and when you do see it, you are as surprised by the profound repose of some of the patients as by the agitation of the others. Accidents occur repeatedly. Deep sympathies form. You see the patients seek out an exclusive partner, and, throwing themselves into each other's arms, smile, talk affectionately, calm each other's fit. They are all entirely subject to the magnetiser. It is no use their falling into what appears a doze—his voice, his glance, his gesture brings them out of it. From these unvarying effects, it is impossible not to recognize that a strong power is agitating and mastering the patients, and that the magnetiser seems to be its repository.[78]

Was animal magnetism this power? There was the question, in the view of the commission, which laid down one cardinal rule of procedure at the outset: the answer could be affirmative if and only if this fluid could be shown to exist, this ineffable substance that in principle escapes detection by the senses. Thus, before ever they began, they disallowed Deslon's plea that the procedure be evaluated by its success in the treatment of patients over a period of time. Herein they noted a rare moment of agreement with Mesmer: curing diseases proves nothing, since nature alone does much of that. No, no, existence was the issue, and the commissioners must have physical proof, unequivocal evidence of the immediate effect of animal magnetism upon human organisms in discrete instances.[79]

In their search the commissioners designed a series of experiments to be

[77] On the brink of departure from France, Mesmer reacted to the appointment of the commission by transmitting to the editors of the *Journal de Paris* (20 August 1784) a public letter addressed to Benjamin Franklin, whom he took to be its head, saying that Deslon had broken his word of honor to keep secret what Mesmer had confided, but that anyway Deslon and his followers did not really possess his system. He also brought suit against Deslon before the Parlement of Paris (the text of Mesmer's "Requête à Nosseigneurs du Parlement en la Grand'chambre," August 1784, is printed in Steinheil [1903] *1*, 1252-1256). The *Journal de Paris* failing to print these communications, Mesmer brought them out as a pamphlet, *Lettre de M. Mesmer à Messieurs du Journal de Paris*. Deslon, patient man, permitted himself only to rejoin "M. Mesmer est un homme bien inexplicable. Il me fait assigner, . . . renvoie à son assignation, . . . et puis m'accuse au Parlement. On reconnoît bien le même homme qui dit et répète qu'il ne m'a rien appris, et qui cependant m'accuse de faire un mauvais usage de ce qu'il m'a appris; qui dit et répète que je ne connois rien à sa méthode, et qui cependant m'accuse de lui avoir volé sa méthode." Charles Deslon, *Observations sur les deux rapports de MM les Commissaires nommés par sa Majesté pour l'examen du magnétisme animal* (1784).

[78] In Bertrand (1826), 74.

[79] *Ibid.*, 85-86.

performed on all sorts and conditions of men, women, and children, beginning with themselves. For try it out they did. They gathered weekly for a séance of two and a half hours, making the chain of thumbs around a baquet in Deslon's clinic. He and his pupils alternated in magnetizing them, now by the finger, now by the wand, and again by massage. One of the commissioners had a slight stomach ache after the first session. In the next few days two others experienced neurological twinges to which they were subject anyway. Otherwise nothing, not even when they tried it three days running. Infirmities of old age had prevented Franklin from attending, and Deslon obligingly took his equipment out to Passy and magnetized him at home, along with a throng of the curious. A few habitual patients went into their customary commotions, but Franklin, his friend Madame de B., two members of his family, his secretary, and an American officer felt nothing at all, though the officer had a fever and one of the relatives was convalescing.[80]

There is no point in recounting the remaining tests. They established that susceptibility to magnetism was a function of suggestibility, poverty, and ignorance. Children, commissioners, and stable persons experienced nothing. No shred of physical or physiological evidence could be adduced to indicate the existence of the fluid. Since that had been defined to be the crucial question, the commission's finding was that animal magnetism was in the literal sense imaginary, that is to say its effects were produced by the imagination sensitized and amplified by the touch of the magnetizer's hands and instruments and by imitation of the behavior of others in a group. It is not clear how far Deslon disagreed. In a discussion at Franklin's house he readily acknowledged that imagination was largely responsible for the influence of magnetism, and speculated that the faculty of imagination might actually be animal magnetism. Who cared? If Mesmer had done nothing more than enlist imagination in the restoration of health, would not that itself have been a marvel? And the commission had to allow that a patient's belief in getting well could often make him so. The question remained whether the fits and gyrations produced by Mesmer were beneficial or harmful, and here the commission did part company with the obliging Deslon. Violent storms of emotion were always dangerous. They became habitual. They were substitutes for serious treatment of disease.[81]

[80] *Ibid.*, 95.

[81] *Ibid.*, 141-142, 145. Evidently Deslon had succeeded in keeping his feelings under control throughout the months when he and his patients were subjected to the investigations of the two commissions. When both reports were out, however, he did publish a pamphlet expressing his indignation (n. 77 above). Despite the kind words for him personally, he felt that the commissioners had been unfair and less than candid. They had concealed the salutary effects they had observed (so he said), misrepresented the techniques

But that was not all. There was worse: an aspect so sensitive that the commissioners omitted it from the published report and conveyed their alarm in a confidential note for the eyes of the king alone.[82] In so doing they acted on the instructions of Breteuil. The issue concerned morals. In the séances they had attended, the great majority of patients who became hysterical were women. Women's nerves are the more mobile, wrote Bailly, and make them more susceptible to the triple effects of touching, imagination, and imitation:

> In touching them anywhere, it could be said that you touch them all over. . . . Women, as someone has said, are like musical strings perfectly tuned in unison. Let one begin to vibrate, all the others share the motion. . . . As soon as one woman falls into a fit, the others immediately go into the same state.[83]

A further factor produces in female patients emotional tumults readily to be confused with magnetic transports. "That factor is the dominance that nature has given one sex over the other in order to attach and arouse it."[84] Men magnetize women, and though the relation is that of patient and doctor, still the doctor is a man. Moreover, many women who seek out magnetism are only bored and idle, and those who really do feel some indisposition are not thereby deprived of charm and sensibility, nor their physician of masculine instincts.

The magnetizer normally places the woman's knees between his own. The lower parts of their bodies are in contact. He places his hands on her abdomen, and sometimes over the ovaries. Often he passes his right hand behind her back. They lean closer. Their faces almost touch. Their breath mingles. Judgment is suspended. Attention wanders. Her face becomes flushed and her eyes ardent. She lowers her head and passes a hand across her brow. Her modesty leads her to conceal what is happening, when suddenly her eyes appear almost pained: an unequivocal sign, the significance of which she is past appreciating, but it has not escaped the doctors on the commission:

> So soon as the sign appears, the eyelids become moist; breathing is short and gasping; the chest rises and falls rapidly; convulsions set in, together with jerky and sudden movements of the limbs or of the whole body. In lively, sensitive women the tenderest of these emo-

---

of touching and stroking, and exaggerated the incidence of violent convulsions. In any case, animal magnetism had progressed beyond their power to suppress it: he himself had trained 160 doctors who were practicing, and Mesmer over 300.

[82] This "Rapport secret sur le mesmérisme ou magnétisme animal," also drafted by Bailly, is printed in Bertrand (1826), 511-516.

[83] *Ibid.*, 511.

[84] *Ibid.*, 512.

tions often ends in a convulsion. That state is succeeded by languor, prostration, a kind slumber of the senses which is the repose required after a violent disturbance.[85]

Lenoir, well versed in gallantry, put the question straight to Deslon: "As Lieutenant-General of Police I ask you whether, when a woman is magnetised and in a fit, would it not be easy to take advantage of her?"[86] And Deslon admitted that indeed it would. For that very reason, he made it a rule never to magnetize a patient except in the presence of others. Doing him justice, the commission expressly acknowledged his respect for the decencies. But could other doctors always answer for their self-command? And what of Mesmer?

Here we may leave the Bailly commission, prey to its chairman's own imagination, and turn to the differing opinion of Jussieu.[87] It is much more interesting than the contemptuous report with which the majority of the commission from the Society of Medicine dismissed Deslon's practice. Indeed, Jussieu's remarks have made that otherwise not very vivid botanist something of a hero in the eyes of the medical and scientific outsider. His colleagues, he felt, had taken too narrow a view of their charge and simply rendered a verdict, when what the public and government needed was a methodical exposition of many and various facts in order to form their own opinions. For his part, good taxonomist that he was, he classified those facts into four categories: (1) general and positive facts, the cause of which remains undetermined; (2) negative facts, which exhibit only the nonaction of the fluid in question; (3) facts, positive or negative, that can be entirely attributed to the power of imagination; (4) positive facts that appear to require some outside agent. He went beyond observing in committee the gyrations of Deslon's patients, and, being a doctor, tried his own hand at magnetizing people individually. The experiment succeeded. He often got effects, most of which he could easily range under one of his first three headings. Most but not all, for he had four cases that he could not thus explain. One was blind, one was in a coma, one was in spasm, and one had her back turned. None of them could have seen his gestures or even perceived his presence in any known manner. Still, they responded, and with some consistency, to movements of his rod. Here, then, were four episodes not to be attributed to imagination, touch, or imitation. Not that Jussieu concluded in favor of animal magnetism—he took what he had observed to be instances of animal heat acting through the electrical effluvia that surround all bodies, and he was as severe as his

[85] *Ibid.*, 513. [86] *Ibid.*, 514.

[87] "Rapport de l'un des commissaires chargés par le roi de l'examen du magnétisme animal," Bertrand (1826), 151-206. On the Jussieu report, see "A.-L. de Jussieu et Clouet," in Hamy (1909).

colleagues in condemning the group hysteria induced around the baquet along with the whole notion of curing by convulsion. He comforted the magnetizers by an open-minded hearing, nothing more.

Mention has been made of Alexandre Bertrand's history of animal magnetism, published in 1826. Its subtitle promises "Considerations on Ecstatic Manifestations in Magnetic Treatment." The author was a medical man trained scientifically at Ecole polytechnique. He believed that among Jussieu's patients at least two were instances of a psychic phenomenon older by far than Mesmerism. Both of them had appeared to function consciously in a trance and afterward recalled nothing of what they had done or said.[88] Instances of comparable episodes abounded through history—speaking in tongues, clairvoyance, walking on coals, and possession by devils, notably in the famous case of Loudun. Bertrand's generation preferred the word somnambulism for this state of altered consciousness, and attributed its identification amid the trumperies of animal magnetism to the marquis de Puységur, the third of Mesmer's most prominent disciples. A word must be said of that to complete this phase of the story and to mark the branching of the path that led from Mesmerism through Charcot and hypnotism to Freud and psychoanalysis.[89]

Armand de Chastenet, marquis de Puységur, grandson to the first marquis, who had been a marshal of France under Louis XIV, was commissioned in the artillery in 1768, and was colonel of the Strasbourg Artillery Regiment and commandant of the Artillery School at La Fère on the eve of the Revolution. A congenial, enthusiastic nobleman, he fell in with progressive movements of opinion in Paris, was a playwright and minor man of letters,[90] and spent more time than was common even among the more liberal aristocracy in looking after his estate at Buzancy, near Soissons. His dependents held him and his two brothers in high esteem. The younger, Maxime, used the title of "comte," was commissioned in the regiment of Languedoc, and had property in the vicinity of Bayonne. All three were early adepts of animal magnetism, though it appears that only Armand and Maxime actually practiced after subscribing to the Society of Universal Harmony.

Many years later, in 1807, the marquis recalled the ambiance of sitting at Mesmer's feet. In the interval, he had welcomed the Revolution in its early phases, traversed the Terror safely thanks to the solidarity of the population in his region, and been named mayor of Soissons when Napoleon came to power. He was thirty-one years of age back in 1783. He then

[88] Bertrand (1862), 211-212.

[89] Ellenberger (1965) has discussed the significance of Puységur and somnambulism from the standpoint of modern practice, and has enlarged upon it in his history of dynamic psychiatry (1969). See also Barrucand (1967), a slighter work historically.

[90] Two of his political comedies were published in the 1790s, *L'Intérieur d'un ménage républicain* (an II) and *Le Juge bienfaisant* (an VIII).

believed somehow in attraction, electrical fluid, igneous matter, mephitic gas, phlogiston, dephlogisticated air, and the four elements, deriving his vague notions from dim recollections of courses on physics by Charles and chemistry by Sage that he had followed in his youth. All the names had changed, however, and in reality he knew nothing at all, merely supposing that any novel effect ought to pertain to some aspect of chemistry or physics. At first he inclined to be skeptical of Mesmer, and suspected his brothers of thrill-seeking or complicity with charlatanism. He paid over his hundred louis mainly to go along with them and with others of their circle. The confession that he was no wiser when the course was finished would have been equally appropriate on the part of his ninety-nine fellow students. He had learned, indeed, so little that, only a day or two before the lessons ended, he went to Mesmer to ask how to magnetize a tree, and also how to go about emanating his fluid. "I recall that I got precious little clarification from him, whether through my fault or his."[91]

Puységur returned to Buzancy in March to relax on his lands before rejoining his regiment. There he found the daughter of his registrar suffering from toothache and thought in a desultory way to try his magnetism on her, ill-assimilated though the technique was. To his surprise, he succeeded, as he also then did with some vague complaint of the guardian's wife. Not a little mystified, he next went about to alleviate the ailments of a peasant called Victor, a husky lad of twenty-three who had been wheezing and coughing in bed for several days. Puységur got the young man up and magnetized him for a quarter of an hour or so, when all of a sudden Victor went to sleep in his arms like a child and began talking calmly and lucidly about aspects of his life and work. Astonished, and fearful lest his patient become upset, Puységur led him quietly on to look down cheerful vistas and found that he could implant in Victor the belief that he was dancing at a festival, and then firing at a target, a favorite diversion for he was a good marksman. Of it all Victor remembered nothing when recalled to normalcy. Further sessions brought out somnambulistically that Victor was repressing deep resentment against a sister who had taken possessions that their mother had promised him when he cared for her in mortal illness, and Puységur put it to him that he would do well to stand up for what was due him.[92]

In a word not yet coined, Puységur had all unwittingly hypnotized his ailing peasant. He was astounded at discovering these powers in himself, and wrote of what he had just done to his brother Maxime, who (as mag-

[91] A.-M.-J. Chastenet de Puységur, *Du magnétisme animal, considéré dans ses rapports avec diverses branches de la physique générale* (1807), 28-31.

[92] A.-M.-J. Chastenet de Puységur, *Mémoires pour servir à l'histoire et à l' établissement du magnétisme animal* (1784), 30-37; extracts from Puységur's reports are also given in Bertrand (1826), 213-220.

netism would have it) was enjoying an equally unexpected success near Bayonne. On joining a command there, the younger Puységur had no thought of making himself known to his officers and men in the guise of a doctor. Still, he could scarcely resist magnetizing a guardsman, when having been braced to attention, the soldier toppled like a plank on the command "Forward March." The new colonel then found himself constrained to form a Society of Harmony there in town and garrison.[93] At home meanwhile the marquis was healing sufferers in the marketplace of Buzancy. Word spread all across the district and as far as Paris. He magnetized a great and ancient elm in the center of the square where a limpid spring gurgled at its roots. Cords were wound around, and this pastoral baquet proved more effective than any artificial model, emanating health from every leaf. Circulating among the throng were three or four somnambulists, surrogate magnetizers healing sufferers out of a waking sleep. An eyewitness tells of it:

> Among his patients Monsieur de Puységur . . . selected several subjects whom he put into a trance (en crise parfaite) by extending his baton and touching them with his hands. Complementing their state is the appearance of slumber, during which their physical faculties seem to be suspended, though in favor of the intellectual faculties. The eyes are closed. The sense of hearing is cut off. The subject wakens only at his master's voice. It is important not to touch a patient in a trance. . . . It causes distress and produces convulsions that only the master can alleviate. These patients in a trance, who are called "doctors," have supernatural powers, through which in touching a sick person brought to them, or in running a hand underneath his clothing, they can sense what organ is afflicted and where the illness is located. They identify the affected part, and prescribe the appropriate remedies.[94]

The account is that of a visiting Parisian, one Clocquet, who goes on to tell how one of these sleep-walking "doctors," a fifty-year-old woman, put her hands on his brow and correctly informed him that he suffered from headaches of a certain pattern and from ringing in the ears. Others were equally successful with various members of the crowd. At the end of a session, Puységur would awaken his assistants by touching them on the eyelids, or instructing them to embrace the tree, whereupon they would come to themselves, a peaceful smile on their lips, and remember nothing.

Such was the recognition of a mental state that, as psychiatric experi-

[93] Maxime de Puységur, *Rapport des cures opérées à Bayonne par le magnétisme animal, adressé à M. l'abbé de Poulouzat, avec les notes de M. Duval d'Esprémenil* (Bayonne, 1784), 1-10.

[94] Bertrand (1826), 222-223.

ence has since made clear, really can be induced in certain susceptible persons. There was a certain innocence in Puységur and his brother, however, and it is that quality, rather than later vindication of his variant on Mesmerism, that makes them the most sympathetic figures in the early history of animal magnetism, whether among enthusiasts or critics. The selflessness of the pleasure he expressed over relieving the sufferings of others rings true. He claimed no credit, and indeed marveled all his life that it had been Victor, in his normal state an extremely limited peasant, who gave his master a magnetic education by perceiving and articulating the requisites so clearly when in trance.

> I have only one regret, which is that it is impossible to touch everyone. But my man, or to put it better, My Intelligencer, reassures me. He teaches me the conduct I must observe. According to him, I do not have to touch everyone. A look a gesture, an act of WILL—they suffice. And it is a peasant, among the most ignorant in the region, who teaches me that. When he is in a trance, I know no one more profound, more prudent, or more clear-sighted.[95]

Puységur was often gullible, to be sure, but that is not the same as deception, or even self-deception, and he recognized his limitations. He learned that people differ in susceptibility, and that they could not be hynotized against their will. He appreciated the impossibility of evoking knowledge of things altogether outside the subject's experience: a peasant who normally spoke only in patois could be made to talk French but not Iroquois. He sensed that he must treat patients only so long as their need lasted, and never experiment with Victor or anyone else to satisfy his own vanity. Though he practiced and wrote on magnetic somnambulism all his life, he kept up his military, literary, and civic interests and never became a fanatic. He agreed that Mesmer might have been wrong about a cosmic fluid. For his part, he had no explanation of how he produced the phenomena of somnambulism. Perhaps it was by electricity instead of magnetism. Perhaps they were the same thing. What the agent be called signified little, provided only it was recognized that the human will has the power of acting on matter. The motto "*Croyez et Veuillez*" appears on the flyleaf of his writings down to the third and last edition of *Magnétisme animal* in 1820.[96] It never surprised or angered him, however, that many friends remained unconvinced, or even that people in society occasionally made fun of him.

[95] A.-M.-J. Chastenet de Puységur to Maxime, 17 May 1784, *Mémoires . . . du magnétisme animal* (1784), 33.

[96] *Mémoires pour servir à l'histoire et à l'établissement du magnétisme animal* (3rd ed., 1820). A second edition had appeared in 1809.

Only in one connection did he express a sentiment near to rancor. The insouciance of scientific bodies did disturb his tranquility. He had learned, he wrote in 1820, that magnetism can be perceived only by persons in whom the sensibility to it is sufficiently developed, just as light affects only those with sight. He would suppose, therefore, that if almost all scientists and doctors fail to experience its effects, the explanation must be that in them the appropriate physical or intellectual organ of perception has been obstructed, rigidified, or even completely paralyzed.[97] Thirty-six years before, his brother's reaction had been the same when handed a copy of the report of the Bailly commission. Maxime was sitting then in a grove of trees he had magnetized near Bayonne among a throng of 300 grateful patients, for each and every one of whom he felt penetrated by a sentiment of sincere attachment. At first he could not take the scientists seriously, such was his sense of well-being. Then indignation overcame him at the thought of our "good Mesmer" thus persecuted for having tried to do good. "The arrogant philosopher is disdainful of anything that does not feed his vanity." These reflections depressed him. Then he got hold of himself and found consolation in the pleasure of serving others, and in that "truth proved physically by Mesmer: how powerful is the influence that man exerts on man, and also the mutual need that we have of one another."[98]

Let us draw a veil over this flow of benevolent soul in Bayonne, shift our attention to Languedoc, and conclude with the story of how Mesmerism conquered the ancient city of Castres in September 1784. The account is found in a long and reflective letter of 24 October written to Vicq d'Azyr by the local correspondent of the Society of Medicine, a Doctor Payol.[99] Six weeks previously he, together with his fellow correspondents throughout France, had been informed of the formal judgment reached by the Society on Mesmerism. Its opinion accorded perfectly with his own since, though he had no detailed information, what he had heard of Mesmer's inflated fantasies had revolted him. Still not everyone in Castres took the same view, and it was nothing wonderful if news of an agent for curing any illness without bitter medicines had aroused hopes. Then, when the crown issued the two learned reports on its futility and danger, people lost interest. For it was not to be imagined that all the scientific bodies of the capital had deceived themselves, the public, and the king.

Within the month everything had changed, in consequence of the return of a local physician who, despite Payol's most earnest dissuasion, had gone to Paris six months before "to buy the secret of these convulsions." Now he was back, with all sorts of pamphlets and testimonials, among

[97] *Ibid.* (1820), iii-iv; cf. *Du magnétisme animal* (1807), 6-7.

[98] Maxime de Puységur, *Rapport*, 15 (n. 93, above).

[99] BANM, MSS 115.

them a copy of Mesmer's fanatical appeal to the Parlement of Paris, and word flew around that two or three invalids were already better for his touch and wand. Suddenly, enthusiasm turned the coolest heads. People talked only of magnetism, of the ferocity of Mesmer's enemies, "of the black jealousy felt of him by all the doctors of the capital, of the emotionalism of the Franklins, the Lavoisiers, and all the royal commissioners who had signed the two reports." In no time the mesmerist had more patients than he could magnetize, and initiated two assistants into the mysteries. True, Castres had no baquet just yet, but the temple was all prepared, the sacred cords were spun, the arch constructed that would span the altar.

Payol reflected that it was not, after all, so surprising that Castres had thus easily been possessed of the "demon of magnetism." Look what had happened in Paris, where sound physics was so widely cultivated, whereas there was no science at all in their small city. He himself had watched several sessions. He had also submitted to being magnetized. And he had to acknowledge, even as had Sigaud de La Fond of the Faculty of Medicine in Paris, that despite his best incredulity, he could not remain unmoved when the magnetizer stood over him "making faces and going through his monkey business." Fellow skeptics admitted to the same failure of self-command. If that was how they were affected, what must be the tumult in the nervous systems of the sick, the credulous, the ignorant, the desperate? No wonder all the raptures and convulsions, the sudden sweats and unforseen evacuations of bowel and bladder.

Thinking on these matters, Payol began to wonder, and he put the question to Vicq d'Azyr, whether the two commissions had perhaps been a little hasty in asserting so categorically that animal magnetism could have no medical effects merely because it does not exist. For certainly the belief in it exists, and the emotional effects of that are real, however they may be produced. It is not at all impossible that arousing in patients a desire to be cured and a belief that a cure is occurring may lead them back toward health. Medicine men in ancient times and in foreign parts, like sorcerers and witch doctors among the common people of the countryside, did and do accomplish similar purposes. "All these things, Monsieur, are in no way extraordinary for a doctor who has carefully considered the dominion that our soul has over our body and our emotions over our maladies." Indeed, throughout the history of medicine, the first care of the physician has ever been to encourage the patient, to implant hope, to inspire confidence. If Mesmer has accomplished that in certain cases, why not give the devil his due, instead of attributing to nature alone the cures with which his patients credited him? It would then be time to emphasize the undoubted dangers of credulity, illusion, and charlatanism.[100]

[100] The immediately foregoing passages are all quoted from the letter cited in n. 99. For accounts of Mesmerism in two far more important provincial cities, Toulouse and Lyons, see respectively Tournier (1911) and Audry (1922).

3. MARAT

Pretenders to scientific favor have seldom made common cause in normal times. They start out with no wish to lessen the value of the recognition to which they aspire, and when rejected their reaction is to transcend the system rather than to legitimate its jurisdiction by joining with fellow victims. Thus, it is scarcely surprising to find Jean-Paul Marat, doctor in medicine and sometime physician to the bodyguard of the comte d'Artois, equalling either of the Mesmer commissions in the contempt he, too, expressed for animal magnetism in 1783, even though his own recent experiences at the hands of members of the Academy of Science and the Society of Medicine had also entailed humiliation.[101]

In treating of Marat, his claims and his unhappiness, the historian meets with difficulties different from the obscurities enveloping the acts of Mesmer. Scholars have recently found political significance in Mesmerism, and who peers with them into the darkness of its mirror may indeed discern it there, or perhaps anywhere. With Marat, on the other hand, the major problem, when trying to reconstruct his life "as it really was" in all but his last four years, is to abstract attention from the all engulfing political intensity of its culmination. Popular champion or monster? About none of the famous figures of the Revolution, not Robespierre, not Bonaparte, have opinions been more emotionally divided. For historians indeed, even as for participants, reactions to Marat are a touchstone of feelings about the French Revolution in all its inwardness, such deep feelings as color judgment and perhaps determine it.[102]

Nevertheless, as a recent and sympathetic biographer points out, Marat was already forty-six in September 1789, when he brought out the first number of *L'Ami du peuple*.[103] His age was fifteen to twenty years above the average of the Revolutionary leaders. Born in 1743, he was a contemporary of Lavoisier and Condorcet and older than Vicq d'Azyr or Laplace. Although he had published two highly political books,[104] both angry, the career he had failed to make was that of philosophic physician turned

[101] Marat, *Mémoire sur l'électricité médicale* (1784), 110-111. Marat submitted this memoir to the Academy of Rouen in the summer of 1783.

[102] The most recent biography (Massin, 1970) has a selected bibliography. Indispensable among modern writings are Gottschalk (1967) and Walter (1960), and among older studies, Bougeart (1865) and Chèvremont (1880), who also published the standard bibliography for *L'Ami du peuple* (1876). Marat's works have never been collected. The best selection is Vovelle (1963). Vellay has published unsatisfactory editions of his pamphlets (1911) and his correspondence (1908), neither of them complete. Further letters have been printed by Vellay (1910 and 1912), Payenneville (1919), Gottschalk (1967), Bonno (1932), Darnton (1966), and Birembaut (1967).

[103] Massin (1970), 10.

[104] *The Chains of Slavery* (London, 1774); *Plan de législation en matière criminelle* (1790). The latter piece was composed in 1777 and first printed in 1780 (see below, n. 146).

would-be scientist. He presents us, moreover, with an advantage denied by Mesmer. Marat was no mystifier. He published detailed accounts of his therapeutic practices and his experiments on heat, light, and electricity. The trouble is, so far as an historian of science can see, that though many views on how he was treated have been expressed, none of the colleagues who have written of him pro or con has really read these books.[105] At least, anyone who has done so has refrained from reporting what they contain. Let us do that, therefore. Let us try very hard to forget about revolutionary politics, however artificially, and confine ourselves to exposition. Let us try to recover how Marat must have appeared in his own eyes and in the eyes of others up to the end of the year 1788, when he published the last of his scientific books, *Mémoires académiques*, following hard upon his translation of Newton's *Opticks*. Only afterward will we venture a few reflections on the latency of the connection between his frustration and his radicalism.

His name began being mentioned in the world of French letters in 1777. Voltaire then annihilated his book *De l'Homme* in the 5 May issue of La Harpe's *Gazette de politique et de littérature.*[106] Of that work, which dealt with the relation of body and soul, and its initial publication in England, more in a moment. Later in the year a series of letters appeared in the *Gazette de santé*, at first over the signature of a man of confidence, the abbé Filassier, later over Marat's own, telling how he had saved the marquise de Laubespine from the ravages of pulmonary disease after the doctors had despaired of her case.[107] Thereupon, Marat put on sale bottles of the "Eau Anti-Pulmonique" with which he had worked the cure, claiming for it the virtues of the acidic waters of Harrogate in Yorkshire. On 1 January 1778 the *Gazette de santé* published a report on Marat's medicine by the abbé Tessier, regent doctor of the faculty and member of the Royal Society of Medicine. The preparation was found to consist of limewater in which the lime had been precipitated from previous solution by a small quantity of fixed alkali, for the commission had detected the presence of a little saltpetre when they treated it with nitric acid. A pint contained about four grains of calcerous earth and two grains of fixed alkali. (Tessier had the assistance of Bucquet and two pharmacists, Laplanche and Lelong.)

[105] Certain writings purport to deal with Marat's scientific work, but their authors never make clear what he actually did. The most frequently cited are the successive editions of Cabanès (1891, 1911). See also Juskiewenski (1933), Rozbroj (1937), Dauben (1969).

[106] In the Kehl edition of Voltaire's *Oeuvres* (70 vols., 1784-1789), the Marat review appears under "Mélanges littéraires," *48*, 226-234.

[107] Marat's communications to the *Gazette de santé* are reprinted by Vellay (1910); cf. Cabanès (1911), 101-113.

Tessier began this report by recalling how disagreeable his experience had been when called upon to evaluate the "*eau fondante*" of Guilbert de Préval. He had been motivated in undertaking that chore by the will, first, to undeceive the public about a pretended prophylactic; second, to awaken skepticism about those who announce mysterious discoveries; and third, to convince reasonable people, and especially the magistracy, that the Faculty of Medicine had been right in proscribing Préval. Unfortunately, all he accomplished, and this quite independently of Préval, was to antagonize a number of persons who ought to have been grateful for his pains. Disenchanted, he resolved then and there to abstain from ever again taking on such an analysis, and instead "to let the charlatans profit undisturbed from their exclusive privilege of persuading whom they could, and to agree to being the detached observer of some portion of the damage they do." Why should he play at Hercules scotching that hydra? That, at any rate, was his intention when an older and much respected colleague sent him a sample of this "*eau minérale factice*" of M. Marat, asking him to examine it chemically, and he yielded, out of a sense of the public good, out of the wish to oblige an eminent senior.[108]

Such was Marat's reception at the hands of the mandarins. He had set up in Paris a little over a year previously, in late 1776. Behind him lay eleven years in Britain, in London for the most part, where he kept a dispensary in Church Street, Soho, practiced medicine, and wrote of psychology and politics. Probably he knew and certainly he admired John Wilkes. In later years his style of political writing was very like that of the Junius letters of Philip Francis. There is no foundation for the story that Marat was a veterinary in Newcastle, though probably he did spend a little time there, or that he fled to Ireland to escape prosecution for a theft from the Ashmolean Museum in Oxford, though he himself said that he passed a year in Dublin.[109]

We are equally ill-informed about Marat's childhood and youth. He was born in 1743 at Boudry, a small town on the lake in Neuchâtel, which was then a principality subject to the king of Prussia. The house still stands. His father, Jean-Baptiste Mara, had been a priest born of Spanish stock at Cagliari in Sardinia. After conversion to Calvinism in Geneva, the father married Louise Cabrol, daughter of a Huguenot family exiled from Languedoc after the revocation of the Edict of Nantes. He then added the Gallicizing "t" to the name. Both Marat's father, a draftsman and designer in a textile shop, and his grandfather, a wigmaker, were artisans and well-to-do by the standards of their class. Marat was sent to

[108] Tear sheets containing Tessier's report are bound in with Marat's *Découvertes sur la lumière* (2nd ed., 1780) in the BN copy bearing cote Inv. R. 42996.

[109] Darnton (1966). The canard was given great currency by Phipson (1924), whose book on Marat's sojourn in England is untrustworthy in other respects.

a good school in Boudry and to a college in Neuchâtel. He always claimed to have been an eager student, and it is evident in his writings that he had read very widely indeed in science, in the classics, and in history, and that he retained a ready command over a large stock of information. At the age of seventeen he was rebuffed when he applied to be a member of the party that the abbé Chappe d'Auteroche was assembling to journey to Tobolsk for the transit of Venus. In that same year 1760, he took the post of tutor in the household of Paul Nairac, a naval contractor in Bordeaux, and spent two years in the city of Montaigne and Montesquieu.[110] The latter was one of the only two writers for whom he consistently expressed respect.[111] The other was Rousseau.

Apparently Marat studied medicine in the time he had to himself. He left for Paris in 1762 and passed the next three years there, living no one knows how, attending courses, and beginning the practice of medicine before removing to London in 1765. Only ten years later did he take a medical degree, however, by which time he was preparing his return to France. The University of Saint Andrews admitted him to the grade of Doctor of Medicine by a diploma of 30 June 1775. It is known that Marat was in Edinburgh in the summer of that year, but not that he ever set foot in Saint Andrews. The award required nothing of him beyond the payment of a fee and certificates of competence signed by two other doctors. His sponsors were Hugh James and William Buchan, both of Edinburgh.[112]

These facts and little else remain when we ignore the embroidery that Marat and the denigration that his enemies placed upon them. Let us turn to what we can know, which is the content of his books, and write their history from what is in them, occasionally reporting what Marat later said of them and of himself, but never relying on his testimony unless it is corroborated independently. It will be convenient to begin with two medical pamphlets. Though he published other writings earlier, they are the only vestige of his practice in London. *An Essay on Gleets* is a twenty-one page pamphlet dated 21 November 1775. A note to the reader apologizes for the author's imperfect command of English. There are, indeed, many Gallicisms—"pression" for pressure, and "Long I had not seen bougies employed for curing gleets, without finding them often ineffectual."[113]

[110] Massin (1970), 10-21, summarizes all that is known of these years.

[111] In 1785 Marat submitted an *Eloge de Montesquieu* to the Academy of Bordeaux in competition for a prize. It was published with introduction and notes by Arthur de Brésetz (Libourne, 1883).

[112] James B. Bailey's introduction to the tracts cited below in note 113, viii-ix. Cabanès (1911) gives a translation of the text of his diploma, 65-66.

[113] *An Essay on Gleets: Wherein the Defects of the Actual Method of Treating the Complaints of the Urethra Are Pointed Out and an Effectual Way of Caring for Them Indicated* (November 1775), 8. Only one copy of this tract was known when James B. Bailey reprinted it, to-

It has to be said, however, that Marat's meaning is generally clear and concrete. He dedicated the pamphlet to the worshipful Company of Surgeons of London, since both in London and Paris surgeons had long since assumed responsibility for treating venereal disease. For his part, he failed to understand why physicians regularly refused such cases, for these maladies did affect the whole system, at least in their later stages, and seldom required the services of an "operator" at any stage. He would not "strive against the torrent," however. Were he animated by "mercenary principles," he would keep his method a secret, "but a liberal mind is above such interested procedures. To promote the good of society is the duty of all its members. . . . Thus, not satisfied with relieving the patients who come to me, I wish I could relieve many more by your hands. Happy, if in this respect, the fruit of my labour is not lost!"[114] That last construction, "Heureux, trop heureux . . ." became habitual with him in commending his writings to the reader.

A "gleet" was any purulent discharge. Marat's concern was with the miseries of chronic gonorrhea. A standard treatment, devised by a famous French surgeon, Jacques Daran, provoked suppuration and drainage by repeated insertion in the urethra of a wax bougie or taper impregnated with various oils, herbs, and chemicals.[115] Purges and supportive medicines were also taken orally. The content of Marat's essay was unknown to biographers in France until its translation in 1912 by a medical scholar, who implies that Marat plagiarized Daran's method while also accusing him of profiteering.[116] There is no basis whatever, in the text or elsewhere, for either of these imputations. Marat simply criticized Daran's method for being excessively drastic and for creating scar tissue by too brutal an application of dessicating tapers in the wake of suppuration. His own technique was a variation on that procedure. First, he located the lesions by exploration with an unmedicated taper. Then he prepared further bougies by applying suppurative agents only at those points that would be in contact with the sore spots, leaving the healthy tissue unirritated. He used milder pharmaceuticals, mucilage of marshmallow instead of olive oil for lubrication, and Diachilum cum Gummis instead of litharge of lead for provoking drainage. As the infection decreased, he lessened the strength of these agents. At a certain point he substituted soothing irrigation with dilute sal ammoniac and a French "Onguent de

---

gether with Marat's *Enquiry into a Singular Disease of the Eyes*, under the title *Two Medical Tracts by J. P. Marat* (London, 1891). The pagination is continuous. Only 84 copies were printed.

[114] *Ibid.*, 6.

[115] Daran, *Composition de M. Daran* (1780).

[116] J. Payenneville, *Marat spécialiste des maladies vénériennes*: "An essay on gleets" traduit de l'anglais (Rouen, 1912).

la mer" unknown in England, and finished off with a very gentle suppurative composed of gold litharge, olive oil, Venetian turpentine, and Bol Armoen. Light cases required twenty-five to thirty days to cure, stubborn ones ten weeks, but in principle "*there is no gleet incurable*."[117]

Not that Marat specialized in venereal disease: he had got into the subject accidentally in Paris, so he said, many years before. A bosom friend, engaged to a rich and beautiful young lady whom he loved, had caught an infection, gone to Daran, and suffered a relapse shortly before his wedding day. Marat had followed the treatment, and reflecting on its failure, had thought that its faults might be repaired in the manner just outlined. His success in curing his comrade then confirmed the analysis. Several similar cases were referred to him in London. The *Essay* gives the history of two of them. Both clients were prominent persons who came to him after noted doctors abroad had failed. They did not want their names made public, "but will not decline to appear in support of truth, if a private interview was desired by patients. I have their word for it."[118]

Marat directed his second pamphlet, *An Enquiry into the Nature, Cause, and Cure of a Singular Disease of the Eyes*, to the Royal Society. This time he refrained from dedicating it: "Such a matter of form I have ever thought beneath the Dignity of Philosophy." He simply wished to communicate a "singular Phaenomenon, which has hitherto escaped the attention of Physiologers." Verifying his observations "might not, perhaps, be a regrettable Employ of Time."[119] Here, too, his subject matter, eye disease, fell to surgeons to treat. Addressing the Royal Society, however, Marat now took occasion to reproach these practitioners with their ignorance of the science of optics and the anatomy of the eye. Indeed, one of the interests that the booklet holds for the historian is the evidence it affords that Marat himself was already well versed in optics and also that he had had experience of the medical use of electricity.

The condition in question was a loss of accommodation in persons too young to develop natural far-sightedness. In each instance, the symptoms appeared following the administration of mercurial preparations (calomel, panacea, or corrosive sublimate) for some other ailment. Here, too, the first case had come to him in Paris, the eleven-year-old daughter of a merchant called Blondel. She was experiencing pain in the eye when touched, a sensation of internal pressure or stiffness, difficulty of glancing from side

[117] *Essay on Gleets*, 21.

[118] *Ibid.*, 9-10, 17-21.

[119] *An Enquiry into the Nature, Cause, and Cure of a Singular Disease of the Eyes, Hitherto Unknown, and yet Common, Produced by the Use of Certain Mercurial Preparations* (1 January 1776), 25-26. See note 113 above. Only one copy of the first printing of this tract was known when it was reprinted in 1891. It was then in the library of the Royal Medical and Chirurgical Society of London. A French translation by Georges Pilotelle was published in Paris in 1891 under the title *De La Presbytie accidentelle*.

to side, failure of near vision, and loss of some perception at a distance. Marat considered that the eye muscles needed relaxing, and tried emollients and bathing for some weeks. Then he stimulated functioning electrically by drawing sparks from the eyeball. (Blondel had to overcome his wife's reluctance before that was authorized.) At the same time he bled the patient from the foot to diminish swelling. He succeeded with four such cases in the course of his entire practice, and expressly says he never treated any others. On the basis of that experience, he asserts that most ailments—he goes so far as to say seven out of ten—wrongly diagnosed as *gutta serena* were in fact this "Accidental Presbyopia." The condition, also called amaurosis, was a common one leading to loss of function in the optic nerve. So far-reaching a conclusion, together with his strictures about the wanton use of mercury among physicians and the ignorance of optics among surgeons, are the only hints of the grandiose and the scornful in this pair of clear and otherwise creditable medical papers.

Traits of temperament are more evident in his earliest publication, *An Essay on the Human Soul*, a hundred-page booklet that appeared in London in 1772.[120] If it is true, as someone has said, that the clue to a thinker's motivation is given by his view of human nature, then what shows through the naiveté of this small effort is important for the study of Marat. It would be unreadable if one did not know who the author was, consisting as it does of commonplaces about the formation of ideas from sensations and about the faculties—sensibility, judgment, memory, will, and so on—which are only to be known by their effects. There is nothing of Locke, however, nor of Condillac, and of Helvétius only the contemptuous refutation of a sophism in *De l'Esprit*. To Helvétius's observation that every passion derives from the body and its needs, Marat retorts, "But he never will deduce therefrom the love of glory."[121]

Love of self is said to be the only motivation, and Marat takes philosophers as a class bitterly to task for their soft-headed views on compassion and reason. For "Nature formed not man compassionate," and "those that fare sumptuously and live in perfect resignation to pleasure, are not anyways affected by the sufferings of others; their sensibility is fixed on themselves, they are negligent of all beings besides, and are steeled against the sufferings of humanity."[122] As for reason, "the so much boasted resource of the wise," it is equally absurd of philosophers to suppose that it is ever capable of governing the passions. The appearance of rational conduct in a Socrates, in a Seneca, in a Zeno is only a deception, a mask for spiritual pride. What appears to be self-command is at bottom arrogance. "It was not force of soul that prevented Socrates from revealing his trouble, and

[120] This booklet, too, is exceedingly rare. Only five copies are known to exist, and there is none in the Bibliothèque nationale.

[121] *Ibid.*, 37.

[122] *Ibid.*, 11, 13.

venting his tears—it was pride." For, and this is the peroration, "those sages so greatly renowned, those who pretend to possess this force of mind, are really the weakest of men. During the time they believe themselves to be superior to every passion, and are boasting of their victory, they are subject to the most imperious masters; for reason can never counter-balance one sentiment but by an opposite one, or repress a weaker passion but by a stronger; that is to free the soul from one kind of servitude, and subject it to another more severe."[123]

Should this initial venture on the soul meet with favor, Marat promised his readers a further work containing discoveries. Evidently he already had *A Philosophical Essay on Man* under way, therefore, since he published it the next year, in 1773.[124] It is subtitled "An attempt to investigate the principles and laws of the reciprocal influence of the Soul and Body," and it contains an instance earlier than any given in the *Oxford English Dictionary* of the word "psychologist." Montesquieu is said to have been "the first that despised the unintelligible jargon of Psycologists, and reduced the study of Man to that of Nature."[125] Except for this, and one favorable reference each to Locke, La Rochefoucauld, and Winslow, philosophers and scientists too are contemned out of hand. "We may conclude that except a few scattered rays of light diffused over some particular phenomena, the science of Man is hitherto entirely unknown." For evidence of the "pompous inanity" of the literature, a footnote refers to "the works of Hume, Voltaire, Bonnet, Racine, Pascal, etc."[126] The subject, indeed, has awaited the attention of physicians, "whose profession qualified them for making such observations; and who, being called to relieve the sufferings of mankind, can contemplate the soul in all its various situations, and surprize it, if I may be allowed the expression, in every degree of misery or greatness."[127] For himself, he would begin by considering the body "as a hydraulic machine," relating its various mechanical arrangements to its functions while "carefully avoiding a minute and disgusting display of anatomical erudition."[128] He would then turn to the soul, and finally—perhaps his literary erudition included the Timaeus—after having handled the body and soul independently of each other, he would "consider the two substances as united."

Among the structures treated in the first part of his book, Marat was mainly interested in the nervous system. The nerves appear compact under the finest microscopes and are evidently continuations of the membranes coating the spinal marrow. They cannot themselves, therefore, be

[123] *Ibid.*, 107, 113.

[124] *A Philosophical Essay on Man, Being an Attempt To Investigate the Principles and Laws of the Reciprocal Influence of the Soul and Body*, 2 vols. in 1 (London; 1773).

[125] *Ibid.*, xiv.

[126] *Ibid.*, xx.

[127] *Ibid.*, viii-ix.

[128] *Ibid.*, xxiv.

the intermediaries between body and soul, and for this function Marat immediately resorted to the usual agency of an eighteenth-century theorizer, a subtle fluid invested with the needed properties, in this case the power of conveying sensations to the soul and instructions to the body. It must have a dual nature, a spirituous and extremely subtle part called animal spirits, and a gelatinous juice, the nervous lymph. In voluntary movements, the nervous fluid transmits messages from the soul; in involuntary movements, it is itself the principal motor. As for the location of the soul, if the nerves be traced to their junction with the membranes of the brain, it is clear that the *meninges* must be its physical seat, and not the cerebrum, nor the cerebellum, nor the corpus callosum, much less the pineal gland as Descartes in his ignorance had thought.[129] This solution to the mind-body problem was the point on which Voltaire focused his scorn.[130] It is somewhat unfair, therefore, that it should be the one physiological statement Marat made that is remembered. He never claimed that it was original.

When all allowances are made, however, it has to be confessed that this book requires a greater effort of attention on the part of the historian trying to enter into Marat's mind than anything else that came from his pen. It is verbose, empty, and boring in a way that none of his other writings is, be they never so chimerical or polemical. Book II on the soul is simply an expansion, somewhat rewritten, of the 1772 essay. There is a single quotable phrase in a new section on "Regular Thought," here said to be "a painful and irksome state of mind."[131] Book III, finally, on the reciprocal influence of soul and body, may be epitomized by quoting the enunciation of the law with which it concludes:

> Corporeal sensibility, the regular or disordered course of our fluids, primitive or organic elasticity, the rigidity or relaxation of the fibres, the force or volume of the organs, are the causes of the surprising diversities in souls, and the secret principle of that great influence of the soul on the body, and of the body on the soul, hitherto deemed an impenetrable mystery.[132]

Marat himself, however, said later that this was his best book.[133] If we ask ourselves how he can have thought so, it may be that he left several clues. At one point, he remarked that the independence of body and soul is an instance of the wisdom of nature, for the intermediary role of the nervous fluid prevents us corporeally from committing suicide in moments of despair and discouragement of soul. At another juncture he ob-

[129] *Ibid.*, 49-50.

[130] Above, n. 106.

[131] *Essay on Man*, 218.

[132] *Ibid.*, 216-217.

[133] ". . . Ouvrage fort au-dessus de tout ce qui est depuis sorti de ma plume . . ." *Mémoire sur l'électricité médicale* (1784), 48, n. 1.

served that he knew from lifelong experience what physical consequences mental distress might have.[134] This might well appear to be an honest book, therefore, written out of preoccupation with himself and with his own states of mind and body, and nothing is more boring to others when it is untouched by those qualities of genius and imagination with which some few souls are able to make their torments universal through art or through philosophy. Only through political journalism did Marat one day manage that.

Marat was thirty and had been in Britain eight years when his book on man came out. Among his papers was the manuscript of an earlier piece of writing, *Les Aventures du jeune comte Potowski*, which there is no evidence that he ever tried to publish.[135] It is a very pallid imitation of Rousseau, in the form of an exchange of letters between young noblefolk about the delights of flirtation and the raptures of love in a preposterously pastoral Poland. Of his private life, we know only that he later described himself as so absorbed in his studies that he was still a virgin at twenty-two,[136] and that Brissot, with whom he was friendly for a time after his return to France, tells in his *Mémoires* (which are far from reliable) that Marat boasted of having enjoyed the favors of Angelica Kauffmann in London, where she was then painting in the first flush of artistic success.[137] Evidently, he also followed British politics. On the eve of the general election of 1774 he visited a foretaste of the political Marat upon his English hosts by publishing *The Chains of Slavery*. The full title will convey the flavor:

> A Work wherein the clandestine and villainous attempts of princes to ruin liberty are pointed out, and the dreadful scenes of despotism disclosed. To which is prefixed an Address to the Electors of Great Britain, in order to draw their timely attention to the choice of proper Representatives in the next Parliament. Vitam impendere vero.

The preface inveighs against Lord North's government and the Parliament whose support it has bought, "a profligate faction, . . . a band of disguised traitors," and summons voters to choose virtuous men in their stead.[138] In the body of the treatise Marat collects and classifies examples of despots traducing peoples into subjection throughout all of history. Among the many authors he had read was Machiavelli. Indeed, the work

[134] *Essay on Man* *1*, 96; cf. *Mémoire sur l'électricité médicale* (1784), 34, 48, n. 1.

[135] *Les Aventures du jeune comte Potowski, un roman de coeur*, ed. Paul L. Jacob (1848).

[136] Marat printed a brief autobiographical essay in *L'Ami du peuple* (14 January 1793). Passages are quoted in Brissot (n. 137 below) *1*, 365-368, and also in Massin (1970), 14-15.

[137] Brissot, *Mémoires relatifs à la Révolution française*, 2 vols. (1830), *1*, 337.

[138] (London, 1774), vi.

might best be described as an "Anti-Prince." One of the devices by which rulers reduce their subjects to dependence and slavery is the patronage of arts and science.[139] Fellow enemies of tyranny are enjoined to employ a "grave and animated style" and warned to eschew satire. Satirical writing is counterproductive. It attacks only the tyrant, not the condition, and merely exasperates him, while amusing the public instead of arousing it. The French in particular are taken to task in a footnote for lightening the mood about public misfortunes "with songs and epigrams."[140]

Typographically, *Chains of Slavery* is a lavish book of xvi plus 259 pages, printed in royal quarto on fine linen paper. Marat made a similar show with certain of his scientific books. Apparently he wished his books to make a statement by their physical appearance as well as by their content, and he must have sacrificed a lot to achieve that effect. Certainly the price of twelve shillings was out of all proportion to the appetite of the British public for yet another political polemic, this one composed in the stilted English of an anonymous foreigner. In 1793 Marat published a much expanded French version in the heat of the political and constitutional conflicts of that spring.[141] Its preface gives an account of the original book, telling of efforts by the British government to suppress it, of the "fermentation" it caused in public opinion, of the government's being forced in consequence to bring in a conflict-of-interest bill against its will, of the stimulus he afforded to the movement for parliamentary reform, of his being driven from London by ministerial persecution and taking refuge first in Holland and then in the welcoming arms of the patriotic societies of the North of England, where a second edition was called for and widely distributed throughout the three kingdoms. In fact—it has been shown—this story was a fabrication. Four journals did notice the book. One called it "spirited." There is no evidence that the authorities took any action whatever. The second edition was only a remaindering of unsold copies in Newcastle.[142] Amid these alarums, real or fancied, Marat had begun meditating his return to France, to be prepared by the publication in French of his book on man. It is clear from his style that he had written his drafts in French and translated them into English, with help or without. He certainly had the text largely ready when he went to Holland in 1774. Only after visiting Holland—it is worth reminding ourselves—did he take his medical degree at Saint Andrews and publish his pamphlets on gleets and presbyopia, souvenirs for the English. Rousseau's publisher, Marc-Michel Rey, issued *De l'Homme* in Amsterdam in 1775.[143] The title page identifies the author as J.-P. Marat, Docteur en

[139] *Ibid.*, 20-22.
[140] *Ibid.*, 71.
[141] *Les Chaînes de l'esclavage* (1793).
[142] Catterall (1910).
[143] *De l'Homme, ou des principes et des loix, de l'influence de l'âme sur le corps, et du corps sur l'âme*, 3 vols. in 2 (Amsterdam, 1775). Volume 3 was printed later and consists of anatom-

Médecine. Except for the two medical tracts just mentioned, it is the first thing printed with his name on it. The text is very close to the English version, the existence of which is nowhere mentioned. Two or three references to the deity are omitted. The footnote dismissing the works of Voltaire and others is included. The seat of the soul in the *meninges* is given a section to itself, and featured more prominently in the sequence of topics. A peroration is added invoking the model of Rousseau: "Sublime *Rousseau*, prête-moi ta plume pour célébrer toutes ces merveilles."[144]

In later years he said that the police in Paris ordered the customs officials to impound shipment of his book at Rouen. There is no independent evidence to confirm this suspicion, though the packages may have been returned to Amsterdam.[145] Whatever the misadventure, Marat arrived in Paris from London in April 1776 at about the time when copies were going on sale. His next literary venture, *Plan de législation en matière criminelle*, carries into the juridical and social realm attitudes already evident in *Chains of Slavery*.[146] Indeed, it incorporates several passages from that work into a more violent denunciation of the civic condition, and constitutes the only further expression of political opinion that Marat published before the Revolution. It was also his first entry for an academic competition of the kind regularly set by learned societies, national and local. The occasion was a prize announced on 15 February 1777 by the Société économique of Berne for a detailed criminal code. Scattered assertions in his essay go beyond the vague transvaluation of values among enlightened writers who in the mode of the century attribute immorality to society rather than to individual persons. Certain passages in Marat make society

---

ical and physiological considerations that Marat had not included in the English edition. It would have been more logical, he acknowledges, to incorporate them in Book I, but he did not wish to disturb the original organization, and they were independent enough to stand alone: ". . . c'est comme un nouveau champ qu'il offre à la curiosité des connoisseurs qui pourront en s'égayant [sic] rebattre les buissons."

[144] *Ibid.* 2, 378-379.

[145] Marat to Roume de Saint-Laurent, 20 November 1783, Vellay (1908), 26-27. On this correspondence, see below, nn. 184, 188.

[146] (Neuchâtel, 1780). Apparently, no copies of the original printing survive. Brissot reprinted the text in the collection he published in Berlin, *Bibliothèque philosophique du législateur* 5 (1782), 109-290. The prize, of 100 louis, had been established by Voltaire, who preceded its announcement with a brochure of his own, *Prix de la justice et de l'humanité*, which Brissot placed just before Marat's in the same volume. It was, said Brissot, in Voltaire's usual vein. "Il y verse le ridicule sur nos lois, et s'égaie sur leurs atrocités. Mais au surplus, point de discussion, peu de raisonnement, point de chaleur sur-tout, de cette chaleur qui doit faire naître la vue de l'humanité souffrante. Voltaire ne la connut que sur le théatre" (5). The next piece, by implicit contrast, "sort de la plume d'un écrivain célèbre. Sa modestie m'ordonne de taire son nom." It had not had the success he had had a right to expect for "la grandeur, l'énergie des idées, et par sa simplicité" (111). Marat published a revised and expanded version in 1790. On this work, see Gottschalk (1967), 22-24; Günther (1902).

out to be a criminal conspiracy, property a theft, government a usurpation, and the judge an assassin when a poor wretch is condemned for taking some small part of the common store of humanity from a so-called owner in order to feed, clothe, and shelter himself.[147] His entry failed to win the prize, and he had it printed on his own account in Neuchâtel in 1780. He later said that all the offending pages had been removed before copies reached Paris and that he then destroyed what the censorship had spared lest his essay circulate with its heart cut out.

Marat composed this indictment during the first year or so of his medical practice in Paris. Nothing is known of his professional situation in its early months. He later implies that he was already attached to the civil list of the comte d'Artois when he was called to Madame de Laubespine in her extremity. It seems more likely that the marquis or his friends arranged the appointment in gratitude for his success. When denigratory letters in the *Gazette de santé* said that his wife's disorder had been merely nervous, Laubespine came forward to testify in print that Marat had indeed saved her from pulmonary disease after their regular doctors declared her state hopeless.[148] Only malice informs Brissot's story that Laubespine was a rake who had infected his wife with venereal disease, though it is true that marital fidelity never constrained their ménage.[149] However mixed his medical press, the episode drew Marat from obscurity and brought him other clients. His position along with some twenty-odd other doctors in the retinue of the comte d'Artois was worth 2,000 livres annually, a competence that together with the high fees he acknowledged charging permitted his beginning work in physics. Apparently Laubespine merged gratitude into complaisance when his wife took her doctor for her lover. The Hôtel de Laubespine was in the Faubourg Saint-Germain, at the corner of the rue de Bourgogne and the rue de Grenelle. There Marat installed a laboratory, and there he invited the commissioners of the Academy of Science to foregather along with Benjamin Franklin in order to verify his experiments demonstrating, for all with eyes to see, the nature of fire, light, and electricity.

Though laying claim to the sobriquet "Médecin des Incurables,"[150] Marat appears to have largely shifted his ambition from medicine to physics in the course of the year 1778. We do not know how he started. The experimental situation he created was so similar to Newton's, however, that he may probably have begun with an adaptation of the prismatic experiments in the *Opticks* to a favorite eighteenth-century problem, the nature of fire and heat. *Découvertes de M. Marat . . . sur le feu l'électricité et la*

[147] See *Plan de législation*, esp. 142-154.
[148] *Gazette de santé*, 4 December 1777.
[149] *Mémoires 1*, 347.
[150] Marat to Roume de Saint-Laurent, 20 November 1783, Vellay (1908), 28.

lumière is a pamphlet of thirty-eight pages.[151] Let us review the content before the circumstances. In place of Newton's prism, Marat worked with an instrument not yet much used in medicine or physics, the microscope. He made all his observations in a darkroom by means of a "solar" microscope, which he fixed to a shutter in such wise that an aperture admitted a single beam of sunlight to pass through the instrument. The apparatus was equipped with an objective lens only, seven inches in focal length. Instead of magnifying an already enlarged image to a higher degree by observing it through the lens of an eyepiece, Marat intercepted the divergent beam from the objective on a screen of cloth or paper. He then inserted the flame of a candle in the diverging cone of sunlight at a distance of several feet from the focus. He thus produced a blown-up shadow of the flame on the screen. The effect was a ruddy cylindrical form, a kind of undulating shuttle. Its diaphanous fringe surrounded a distinctly less brilliant zone, at the center of which glowed a tiny bright jet, white to incandescent. Small puffs or streamers of shaded incandescence licked upward from the top. When he substituted other objects for the flame—a glowing charcoal, a red-hot piece of iron, fragments of gold and silver, rock crystals, Japanese porcelains—he found that everything heated hotter than the surrounding atmosphere gave variants of the same effect.[152] The booklet recounts one hundred and twenty discrete experiments.

The conclusions are less clear than the descriptions. "It seems to me," Marat stated at the outset, "that the theory of fire is today in the same state as that of color before Newton. Fire is taken for a material, whereas it is only a modification of a particulate fluid, just as color is only a modification of the light reflected by bodies."[153] Rays from the sun produce the effects of heat, as do other sources, simply by exciting motion in the igneous fluid contained in bodies. Its particles have the properties of transparency, tensile strength, weight, mobility, and extreme hardness. What Marat had done was to surprise that fluid escaping visibly from bodies. For the moment, he insisted only on the novelty of his technique. His method was "absolutely new." He urged other physicists to try it, submitting his memoir forthwith to the judgment of the Academy of Science.

Marat communicated it toward the end of the year 1778, through the good offices of the comte de Maillebois, the same who a year later took up Mesmer's cause with all the urbanity of honorary membership. Besides the sponsor, the reviewing committee consisted of another honorary member,

[151] The full title reads, *Découvertes de M. Marat, Docteur en Médecine et Médecin du Garde des Corps de Monseigneur le Comte d'Artois, sur le feu, l'électricité, et la Lumière, constatées par une suite d'Expériences nouvelles, qui ont été verifiées par MM les Commissaires de l'Académie des Sciences.* A copy annotated in Marat's hand is in the BN, MSS Fr. 14734.

[152] Plates illustrating the effects are included in *Recherches physiques sur le feu* (1780).

[153] *Découvertes . . . sur le feu, l'électricité, et la lumière* (1779), 1-2.

Trudaine de Montigny, more knowledgeable (it is true) than Maillebois, and two active members, LeRoi and Sage. There is no evidence that Condorcet named such dim lights out of adverse prejudice, but it would not be inconsistent with the behavior of learned bodies if he had preferred to spare important people the drudgery of accommodating an unpromising proposal brought in by a well-meaning and ignorant patron. Marat never betrayed awareness that Sage and LeRoi were held in low esteem by their colleagues. Evidently, however, he did half-expect treatment that could later be called unfair, and took precautions lest he should need to ingratiate his work with a wider public.

On 13 December 1778 Benjamin Franklin recorded in his diary the receipt in Passy of a memoir "on the subject of elementary fire, containing experiments in a dark chamber." It came from an "unknown Philosopher," and Franklin goes on to say, "It seems to be well written, and is in English, with a little Tincture of French idiom. I wish to see the Experiments, without which I cannot well judge of it." Now, Marat must have known that Franklin read French. Did he give himself the trouble of translating his memoir with a view to an English outlet, in case he should be refused in Paris? It may be so, though no trace nor any other mention of this English version has ever been found. At all events, Franklin's reply must have given some encouragement, for he was thereupon importuned throughout the winter and spring to dine at the Hôtel de Laubespine in company with the commissioners: May he soon recover from the indisposition that at first prevented his accepting; several additional trials have been imagined solely in hopes of his presence; would not his grandson like to come too? It lends a conspiratorial tone to the exchange that these letters purport to come, not from Marat himself, who remains unnamed, but from "le Représentant de l'Auteur." In like manner had he reported the cure of the marquise de Laubespine to the *Gazette de santé* over the signature of the abbé Filassier. Finally Franklin did attend, on one occasion at least. The commission reports an experiment that successfully substituted his bald pate for the candle and other incandescent objects in the solar microscope.[154] That may well be the only light touch in the whole chronicle of Marat's existence. More characteristic is the letter of 12 April 1779:

> I again beg earnestly you would be so good as to be present then to give your opinion, which will be requested by M. le Comte de Maillebois.

Was it not so material a point to the Author, that a candid judgment should be pass'd upon his work, we would trust to time alone. But he is certain that many a academical gentleman do not look with

[154] *Ibid*. Their report is bound in with the memoir as a preface.

pleasure upon his discoveries, and will do their utmost to prejudice the whole Body. Let the cabal be ever so warm, it certainly will be silenced by the sanction of such a Man as Doctor Franklin; and how far a judgment passed by himself and the Royal Academy can influence public opinion is well known. If I prove troublesome, Sir, my consciousness of your benevolence, and my respect for your candour and understanding are my apology.

The Representative.[155]

Five days later, on 17 April, the commission brought in its report. Whether the author had proved the existence of an igneous fluid escaping from heated bodies, they felt unable to decide. That was a very far-reaching question. They had learned, however, that the author specifically wished them to withhold judgment on his theory and to confine themselves to attesting the truth and accuracy of his account of his experiments. A footnote by Marat confirms this request. He is "persuaded that the facts which form the basis of his theory had been entirely unknown before him."[156] They were difficult to produce and required special equipment. Once they were confirmed, any enlightened physicist could draw whatever consequences seemed appropriate. There is no way of knowing whether Marat started with that restriction, or limited his losses when he felt uneasiness about his conclusions building up in successive sessions with the commissioners throughout the spring of that year. The latter seems the more probable a reconstruction. If it was a trap, the commission fell into it. They did not hesitate to affirm that the experiments "on which the author rests his theory, and several of which were repeated for us a number of times, appeared very precise to us, and that we have verified them all as far as possible," and concluded:

We regard this memoir as very interesting in its object, and as containing a series of new and precise experiments, performed by a method which is as ingenious as it is suited to opening to physicists a vast field of research, not only on the emanations of heated bodies, but also on the evaporation of fluids, whether left to themselves or provoked by fermentation, solution, etc.[157]

(Signed) Maillebois, de Montigny, LeRoi and Sage

Condorcet certified the report on 25 April 1779, and Marat might surely have been excused if he thought that he had taken the first, large step on the way to approbation by the full Academy. He added a postscript ex-

[155] The exchange has been published by Vellay (1910).

[156] *Découvertes . . . sur le feu*, etc. Report of the commission.

[157] "Extrait des registres de l'Académie Royale des Sciences," 17 April 1779, printed *ibid*.

plaining that, though the commissioners had limited their report to his work on heat, they had also verified his experiments on electricity and light. These he reserved for future development.

He lost no time. On 16 June Marat told Maillebois that his memoir had a sequel, perfecting Newton's theory of colors, or rather establishing a new theory. The faithful Maillebois duly informed the Academy on the 19th, which now named the same commissioners as before, with the addition of Lalande. The smell of a rat must have been about, however. Lalande, no giant in any respect, bowed out in favor of Cousin, his colleague at the Collège de France.[158] Cousin, it will be recalled, was then professor in a new chair. His courses treated the application of the calculus to physics, and were among the influences from which the modern discipline emerged in later years.[159] Cousin himself was rigorous, but not prolific. This time, according to Marat, Maillebois and Trudaine de Montigny were unable to attend. Sage was also much absent, deferring to his more qualified colleagues, though he did sign the report.[160] That left LeRoi and Cousin, and effectively this meant Cousin.

Marat's *Découvertes sur la lumière* is a treatise of 141 pages reporting the results of 202 experiments. According to his account of its reception, he began demonstrating them before the commissioners on 22 June 1779. He met with his judges at intervals for a period of seven months through the summer, autumn, and into the winter.[161] Even before their final session on 30 January, Marat began asking LeRoi when the report would be ready. As soon as he had done another for the navy, said LeRoi on 7 January; but I am very busy, and this is no ordinary report, he said on 17 January; but I had to go visit the Châtelet for the commission on prisons, said LeRoi on 28 January; it will be done next week, he said on 13 February—and early in May, he was asking Marat to return the copies of Newton's *Opticks* that he had loaned him so that he could check a few points! Marat got no better satisfaction from Cousin, and in April began importuning Condorcet, requesting replies by return on the bottom of his letters.[162] Finally on 10 May, almost a year after submitting his memoir, he got his judgment. The experiments it contained were very numerous, ran the finding and since

> We were for that reason unable to verify them all with the requisite precision (in spite of all the attention we gave them); since, moreover, they do not appear to us to prove what the author imagines

[158] *Découvertes sur la lumière, constatées par une suite d'expériences nouvelles* (London, 1780), 4-5.

[159] Above, Chapter II, Section 4.

[160] Sage to Marat, 8 October 1779, Vellay (1908), 58.

[161] *Découvertes sur la lumière*, Avis aux Lecteurs, 5.

[162] The correspondence is printed in Vellay (1908), 58-65.

> that they establish, and since they are in general contrary to what is most fully known in Optics, we believe that it would be useless to enter into any detail in order to make that [his theory] known, for we do not regard them as anything . . . to which the Academy can give its sanction or accord.[163]

Thereupon, Marat published his memoir on his own account. If he has to be judged, let it be by "an enlightened and impartial public: it is to its tribunal that I appeal with confidence, the supreme tribunal whose decrees scientific bodies are themselves forced to respect."[164] For his part, he would always be happy to have the approval of distinguished scientists. No learned society, however, can make true what is false or false what is true. Of course his theory ran counter to existing optical doctrine. How could it be otherwise when it was new? He will not comment on the disposition that the commissioners had so kindly made of him. He had never asked for a judgment of his theory, but only for a verification of his experiments, about which they chose to remain silent. He did think he had had the right to expect them to testify to their novelty and precision.

Marat imagined that these experiments proved conclusions of three sorts (though he did not put it thus categorically): first, that the colors that Newton attributed to refraction in Book I of the *Opticks*, and also those attributed to interference in Book II, are in fact produced by diffraction, which Newton treated only in the fragmentary Book III; second, that the primary colors are red, blue, and green, and not, as Newton held, the seven shades of the prismatic spectrum; and, third, that Newton had erred in his determination of the distinctive refractive indices of those seven colors, and might even have falsified the values. He closes with a hint that Newton had misled generations of astronomers and instrument makers in insisting on the inevitability of chromatic aberration in lenses. It was most probably the first of these assertions that Cousin (for the commission) found gratuitous. If so, his judgment has on the whole been borne out by the subsequent history of optics (not that this is a standard to be applied in deciding the interest or importance of a statement, though neither is it to be dismissed as simply whiggish). It further seems likely that a scientist in Cousin's position would have felt impatient with Marat's fantasies about diffraction and failed to give the remaining contentions the attention that others like them have turned out to merit. For the matter of primary colors and the question of Newton's measurements

[163] Printed in *Découvertes sur la lumière*, 3-4, and also Vellay (1908), 64-65.

[164] *Ibid*., 6. Evidently Marat planned a second, enlarged edition of this work. A copy at the BN, MSS nouvelles acquisitions françaises 309, is full of interlined and interleaved corrections and additions in his hand. He has here crossed out the passage quoted above about appealing to the public.

of specific refrangibility have proved far from empty, though the discussion has owed nothing to anyone's ever having read Marat.[165] Indeed, he stated the latter criticism so vaguely that it is unclear precisely what he meant. Ten years afterward he himself acknowledged that this first book on optics was only "a slight sketch, the result of a fortunate insight and an easy piece of work."[166]

All Marat's theories about light, heat, and electricity started out of his solar microscopic observations in the dark room. The first optical experiments demonstrated how the shadow of an object on the screen is always surrounded by a luminous nimbus, the brightness and extent depending on the distance from the object to the screen. What can be the cause? Not double refraction, for the glow fails to appear in the steel mirror. Not the surrounding air, for the effect appears as clearly around extremely thin bodies as around spherical ones (a footnote here excuses himself for having left a remark indicating the contrary in his book on fire, printed a few months earlier, for if he had admitted the true cause there, he could have been anticipated on the discovery he was now publishing). To what, then, is this shining fringe to be attributed? Why, to the attraction of bodies, the attraction by which rays are drawn in toward the surfaces they bathe in light and correspondingly thinned out in the surrounding shell or layer of space, which therefore appears dimmer. The phenomenon is lost in spaces the size of ordinary experience, however, and is to be observed only when the "field of light" is formed in slits and pinholes.[167]

The novelty in Marat's experiments consisted in his technique for observing these fringe effects through his solar microscope, and only in that. The effects themselves were well known. As for his conclusions, he did indeed put them together in a way that was unheard of, though there was no element in them that can itself be called original. In thinking about color, Newton had shifted back and forth between the attraction of light by matter and the density gradient of aether in the space circumambient about refracting surfaces.[168] All that Marat did, whether consciously or not, was to adopt both those models, which, in Newton's mind, had been

[165] Lohne (1961), (1965), (1968), and Bechler (1975a), (1975b), are the most important items in the recent intensive and often very critical scholarship bearing on Newton's optical work. For the question of the validity of his determination of specific refractive indices, see esp. Lohne (1961). Cf. also Shapiro (1980).

[166] *Mémoires académiques* (1788), vi.

[167] *Découvertes sur la lumière*, 2.

[168] His first printed reference to the aether model, and to "inflexion," appears in his second paper on light and colors submitted to the Royal Society in 1675, and printed in his *Correspondence*, H. W. Turnbull, ed., *1* (1959), 362-386. In section 14 of Book I of the *Principia*, however, he derives the sine law of refraction from the attraction hypothesis. For a guide to the discussion of these changes in Newton's views, see I. B. Cohen, "Newton," DSB *10*, 88 n. 81, and Bibliography.

partial and tentatively held alternatives, and promote the combination into an all-inclusive theory of color in light.

The further one reads in Marat, the more complex becomes the evidence for the mingling of dependence and hostility in his attitude to Newton's work. Thus, he employed Newton's word "inflexion" although contemporary French usage preferred "diffraction," the term initially supplied by Grimaldi, who had discovered this class of effects.[169] Marat now proposed to develop it into a new branch of optics, to be called perioptrics. It was to be distinct from dioptrics (the science of refraction) and catoptrics (the science of reflection) and would treat instead of the bending of rays in their propagation through space. Where dioptrics speaks of refrangibility and refraction, perioptrics will say deviability and deviation. To suppose that light travels in straight lines is illusory. In reality it is ever being deflected by the force of attraction residing in bodies. Newton had been mistaken in attributing the prismatic colors to the differential refrangibility of specific rays. The dissociation of white light that he located within the prism had already occurred by virtue of attraction, in part at the edges of the aperture in his window blind and in part externally to the first surface of the glass, whether prism or lens. Actually, all rays are equally refrangible. It is their angles of incidence that are different in Newton's experiment. Thus, inflexion is the true origin of colors in light, not merely of those produced as fringe effects at knife edges and in slits, but of all color manifestations whatsoever, including the phenomena that Newton attributed to refraction and to the passage of light through thin plates (about which Marat here had much less to say).[170] Such was the assertion that Cousin considered unproved by whichever of Marat's experiments the commission had actually observed and by the others of which it only read the descriptions.

Meanwhile, Marat had brought out *Recherches physiques sur le feu*, the treatise he had promised in the last sentence of the 1779 essay. It was the first to be printed of the full-scale memoirs on fire, light, and electricity, all the three topics he had broached in the earlier work. His permission to publish was registered on 11 April 1780. Apparently, he never submitted this book to the judgment of the Academy. He must, on the contrary, have been working on it throughout the months of waiting, fretting, pleading for the commission's report on his memoir on light. It opens in a more truculent manner than anything he had yet printed about science: People used to think that fire is an element. Many still do. Some readers

[169] Grimaldi's *Physico-Mathesis de lumine* (1665) is an enormous and inchoate book, which also discusses certain of the most important matters, notably the interaction of light and matter, broached by Newton when he broke off experimentation in Book III of the *Opticks* and turned to speculations in the Queries.

[170] *Découvertes sur la lumière*, 30-44.

who learn that he is going to attack that notion will stop right there. Certain opinions are so sacrosanct that one makes oneself ridiculous by setting out to destroy them. That this error has grown venerable along the centuries is hardly his fault, however. If persons who prefer to nurse their prejudices refuse to read him, well, he is interested only in the small number who know how to think. He will preface the account of his discovery with a history of the opinions of physicists on the nature of fire. When a subject has been treated speciously by people who carry authority, you have to destroy the whole structure.[171]

Repeating word for word his 1779 statement to the effect that fire, instead of being a substance, is produced by the movement of an igneous fluid, Marat now gives more detail on the genesis of his idea. Such a fluid would surely be less subtle than light and much less rapid in its propagation. Accordingly, the igneous fluid should be visible under the right conditions. From something Brissot once wrote, it seems probable that Marat in one of their frequent conversations alluded in this connection to the effect of shimmering above a hot stove or over roof-tiles in the sun of a summer day. He says nothing so naive in his book, however. What he does say is that he could think of no way to detect its presence until it occurred to him that if the igneous fluid is indeed different from light and thus visible in principle, it should cast a shadow. Thus did he hit upon the idea of adapting the microscope to the examination of shadows instead of images. He tried it with the candle flame, and it worked. The cylindrical penumbra shimmering on the screen that he had substituted for the eyepiece could only be the shadow of the igneous fluid itself. It quite surrounded the area blocked by the visible flame and varied in texture with the density. All excited, Marat tried again, replaced the flame with other incandescent bodies, re-examined all the phenomena of fire there in the darkness of his laboratory, multiplied experiments, worked like mad. He compared the weights of cold and red-hot bullets. He tried the shadow of a jet of steam. He measured rates of cooling in the atmosphere, in other mediums, and in vacuums of varying degree, together with the variation of rates with temperature. He determined that the ratio of temperature to distance from a source of heat fails to serve the inverse-square ratio. By dint of these observations, and many, many more, he established the properties of the igneous fluid. In contrast to the luminous and electrical fluids, it affects only the sense of touch and produces the sensation of heat by exciting the particles of gross bodies to intestine motion.[172]

In preliminary remarks Marat comes to the defense of phlogiston, which does exist in nature though it is not to be confused with the igneous

[171] *Recherches physiques sur le feu* (1780), 1-2.

[172] *Ibid.*, 41. Figure 2 reproduces Plate I, facing p. 20.

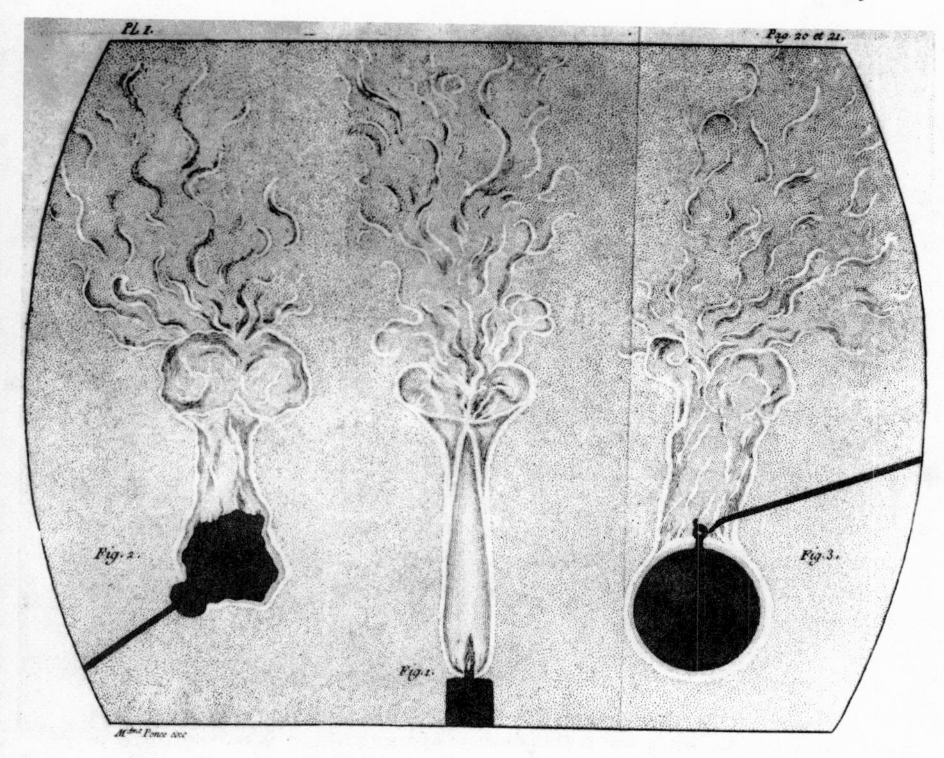

Effects produced by the solar microscope. Plate I from Marat, *Recherches physiques sur le feu* (1780).

fluid. Representing the "inflammable principle," it gives a clearer idea of that which burns in bodies than does "oil" or "sulfur" or "bitumin." In the same passage appears his first published reference to Lavoisier, whom he came to hate more than any other of the Academy. For certainly it was he whom Marat had in his sights in writing how "a famous author clamors against the nomenclature of chemistry."[173] Not that Lavoisier is identified. Literary manners precluded mentioning living writers by name, and Marat still had himself largely under control, at least in print. Even denunciation of predecessors, who enjoyed no such protection from convention, is pretty much confined to the introductory material. As for the body of the memoir, if a browser knowledgeable about eighteenth-century physics were to happen upon this treatise, or its companion on light, on the shelves of some large library, and were to open it and begin reading *in medias res* without knowing the identity of the author, he would be un-

[173] *Ibid.*, 8 n. 1; cf. 13-14.

able to distinguish the work from any of scores of other titles in the literature of minor experimental science.

Marat affected a Baconian mode in all these writings. The argument is developed through the arrangement of "experiments" numbered in the margin and recounted in italics. These passages are interspersed with theoretical commentary and general remarks about nature and science. To be sure, many of the experiments are only trivial observations of commonplace phenomena—the word "*expérience*" covers both meanings. That there are some 202 experiments in his book on fire is less impressive, therefore, than it would be if all of them had required the contrivances of sunbeam, candle, lens, and microscope that marked the more ingenious, not to mention the conventional vacuum chamber and exhaust pump, or the even more elaborate instruments that he designed for his electrical researches. Nevertheless, he did imagine and perform such experiments, and they are impossible to distinguish in principle from others by means of which persons then and now recognized to be scientists have explored the world. His early books were, in fact, more substantial than some others by persons who duly became members of scientific societies and academies, for example those already mentioned of Lamarck and Lacepède.[174]

The solar microscope did discover effects not previously seen, even if Marat exploited it for more than the technique was worth, and even though they led to no further work. In themselves, they might have done, for the more striking are well illustrated, and Marat was careful to give instructions on the method of making observations. He listed all his instruments with full specifications: microscope, lenses of various focal lengths, precision vise, pneumatic receiver, exhaust pump, spirit lamp, thermometer, scales, and many others. Together with his electrical apparatus, he had assembled a considerable physical laboratory. Most of his pieces were made to his requirements by a leading instrument maker, one Sikes, to whose shop in the place du Palais Royal Marat commended readers wishing to repeat his experiments and to perform their own.[175] How can he have paid for it all, and for the publication of his books? Did patrons help, perhaps the marquis de Laubespine? Did he wring it all out of his medical practice and his own hide? For he never cared anything about comfort or amenities. All we know is that he went from publisher to publisher—Jombert fils, Clousier, Didot, Mequignon, and Leroy—to get his books printed, constantly fussing over the texts, revising whole pages and replacing whole signatures even after the sheets had been printed off. It is not rare to find copies in libraries around the world with marginal or in-

[174] Above, Chapter II, Section 6, nn. 271, 275.

[175] He gives specifications for his instruments in *Recherches sur le feu*, 198-200.

terlined corrections in ink in his own hand and with pages out of sequence.

Brissot once remarked that the trouble was less with Marat's science than with his behavior.[176] What that behavior was becomes more evident in the third installment of his trilogy of *Recherches physiques*, the memoir on electricity. It was published two years after those on light and fire, in 1782, and is much longer. An anonymous copy of this work would be less likely than the others to be mistaken for a minor item in the literature. One passage, for example, takes issue with Franklin's opinion that non-electrics, or as Marat prefers to call them "indeferens," are completely impermeable to the electrical fluid. A footnote reads:

> In my writings I certainly don't at all set out to attack the opinions of great men, as my adversaries (of whom there are many) like to say in print. If I contest those that seem to me ill founded, it's only when they are closely related to my own subject. But then, what writer hasn't got that right? Otherwise, no one is more respectful than myself of the wise men who have devoted their meditations to enlightening the times in which they lived. That statement is the simple truth, and I don't mind saying so to still the clamor of the whole crowd of stupid people, who, jealous of the success kindly applauded by the public, make it their duty to denigrate me.[177]

Despite this disclaimer, the treatise on electricity includes many passages girding at famous authors, all of whom are wrong, stupid, and blind. Instead of being confined mainly to the preface, these expressions of contempt for existing work and theory now introduce almost every topic. The history of the subject is hardly worth noticing. Marat must remodel the whole science. The trouble is that the electrical fluid has generally been confused with the luminous, igneous, and magnetic fluids, and they with each other. He has rendered it, too, visible in the solar microscope and found that its properties are different from what is commonly supposed. Its corpuscles have generally been invested with forces of attraction for all other forms of matter and of repulsion for each other. He has confirmed the truth of the former assertion and exposed the falsity of the latter. For he had made still another crucial discovery: particles of electricity *attract* each other. What fixes the electrical fluid on surfaces is not the attraction of gross matter overcoming the mutual repulsion of its particles. Instead, it is atmospheric pressure.[178] The terms "conductor" and "non-conductor," "electric" and "non-electric," are utterly inappropriate, therefore. The correct distinction is between "*déferens*" and "*non-*

[176] Brissot, *Mémoires* 1, 339.

[177] *Recherches physiques sur l'électricité* (1782), 84-85, note.

[178] *Ibid.*, 13-17. His instruments are carefully described, 17-23.

déferens," bodies that do or do not have the capacity to transmit an excess charge of fluid and to produce a shock. Both the one-fluid theory, with its conventions of positive and negative charge, and its counterpart, the two-fluid theory, "recently warmed over by a 'savant academicien,' " are thus equally false.[179] (In his section on applications, Marat even dared decry Franklin and the lightning rod.)[180]

There would be little profit in reporting the detail by which Marat thought to sustain these propositions. They were meant seriously: no electrical fun and games for him or for his readers. As always, he knew the literature. However far-fetched his theory of the Leyden jar, he knew its workings and characteristics[181] and had mastered the electrical properties of bodies and the various methods of electrifying them. His attack upon the subject becomes an onslaught, however, and the sheer inadmissibility of his very approach to doing science shows clearly through the whole configuration of this work, though it merely amplifies the pattern already evident in the others. In each of the subjects he took up—psychology, heat, optics, and electricity—he had read voraciously. In each of them, he dismissed existing knowledge as radically inane. Into each of them, he introduced (no other word will do) a *truc*, a gimmick—the seat of the soul in the meninges, the visibility of heat in the solar microscope, the deviability of light, the attractive force in electricity—with which he would reorder the whole subject. In each, an ad hoc fluid—nervous, igneous, luminous, electrical—is invested with properties and made to bear effects. For in the last analysis, or rather in the first, Marat had no physical imagination. His models were naively Newtonian—corpuscularity, attraction, affinity, intensity gradients in the surface layers of a medium. His ideas were introduced, not so much originally, as unconformably to the state of theory and understanding. Whether science be considered methodologically or institutionally, it has never operated like this. But of course Marat's explanation of the failure he thus prepared so diligently, so painfully, so angrily was at bottom personal and political.

It is clear that already in 1780 Marat was perceived as a pest, one among many, if not a menace, by members of the Academy of Science. Their opinion was confirmed, no doubt unalterably, by 1782, when the electrical memoir appeared. Brissot was then an intimate friend, himself scrambling and scribbling on the fringes of literature in those years. He had entrée, however, and was acquainted with important scientists and writers. A heated dispute he had had with Laplace over the optical claims of Marat is reported in a dialogue on "Academic Prejudice" that he interpellated in his book *De la vérité*, begun in 1779 and published in 1782.

[179] *Ibid.*, 158-159. The reference is to LeRoi, on whom Marat had turned.

[180] *Ibid.*, 367, 406-411.

[181] *Ibid.*, 116.

All arrogant in his "*fauteuil*," the mathematician there dismisses the aspiring physicist contemptuously and justifies refusal to observe his experiments or to read his memoir on the grounds that calculation shows Newton to be right. To which the skeptical interlocutor retorts that Newton's critics also have their calculations: "What to do in this chaos of figures? Return to nature; look at the facts. . . . I would rather believe nature and my senses than all your volumes of figures."[182] To the following March (1783) belongs the legend of Marat's drawing a sword on another of his oppressors, J.-A.-C. Charles, in the middle of a public lecture in an amphitheater in the Louvre. What does appear authentic is that Charles's scornful remarks were repeated by a candid friend to Marat, who went round to Charles's lodgings to demand an accounting, and that the two men came to blows.[183]

Marat's own view of his situation appears most fully in a lengthy correspondence that attended the disappointment of his hopes of being named director of a Spanish Academy of Science to be inaugurated in Madrid. Among his friends was one Philippe Roume de Saint-Laurent, of whom little is known except that he must have been prosperous and experienced in colonial affairs with interests in Santo Domingo and also in Spain. In April 1783 Roume was in Madrid, where the chief minister, Floridablanca, in office since 1777, was striving within the rigid limits of the possible to introduce enlightened despotism. The project that brought Roume there was the design for a model colony to be planted in Trinidad. Apparently Floridablanca also talked with him on other matters, among them institutionalization for science and creation of an academy. What with all the doors closing on Marat in Paris, Roume thought to commend his friend's qualifications for heading the new body to the Spanish minister. Marat rose to the possibility. Having rejected, so he said (there is no other evidence), a feeler about entering the service of some "northern Sovereign," he set about learning the language. He approached the heraldic authorities in Paris about a noble lineage on his father's Spanish side, adopted the particle unilaterally, and began signing himself "de Marat." He put into the mouth of Roume assurances about his being a man who has "always respected the government, the laws, and the customs of the countries he has traversed." But he is apprehensive lest the Spanish ambassador in Paris might get wind of the "clamor of our philosophes, for whom it is a crime to believe in God." The minister must be warned of their hostility to one who had refused to join their "criminal sect," and had dared oppose their "pernicious errors." Why, "How many respectable

[182] Brissot, *De La Vérité* (Neuchâtel, 1782), 335. For Brissot's second thoughts on this episode, see *Mémoires 1*, 346-347.

[183] Chèvremont *1* (1880), 71-75; Massin (1970), 61.

clergymen could I not supply as reference!"[184] In a word, Marat was desperate.

As bad luck, or perhaps vanity, would have it, Marat had simultaneously allowed Brissot to reprint his *Plan de législation criminelle* in a series he was editing, thus giving the lie to his own protestations.[185] His forebodings about what the ambassador might hear also proved all too well grounded. Twenty letters, so Roume had to tell him, depicted his character in the blackest tones.[186] As hope guttered out, Marat unburdened himself in an enormous and impassioned letter accompanied by forty supporting documents attesting his merits in philosophy, medicine, and science. Among them is an encouraging letter from LeRoi, written at an early stage in his approach to the Academy, urging upon him the importance of Lavoisier's attending a demonstration of his experiments since the slightest air of wishing to exclude Lavoisier would wear the appearance of fearing his presence.[187] The faithful Roume made and kept a copy of this unhappy *apologia pro vita sua*, which was first printed in the incongruously Victorian *Miscellanies* of the Philobiblon Society of London.[188]

Envy is responsible for all his misfortunes, envy and cowardice on the part of the "rabble" (*tourbe nombreuse*) of modern philosophers skulking in anonymity. They could not make him regret having preferred truth to the boot-licking that would have won academic honors, or having saved the lives of fellow beings given up for incurable, or having advanced useful knowledge and been a man of good will. As a child he had studied hard. No sooner was he eighteen than our "pretended philosophers" had tried to take him into camp. Repelled by their principles, he held out and resolved to live in England for a time to avoid dissipation and learn more science. In London a French man of the world, a certain Monsieur de La Rochette, advised him to publish his book on body and soul anonymously in English. Thus would he escape the malice of the philosophic cabal, which otherwise would be roused by his attack on materialism. English statesmen and scholars welcomed his book. It created a sensation in the journals. Unfortunately, however, certain philosophers saw advance copies of the translation, and instigated its interception by the customs authorities.

When he returned to Paris, these same enemies incited the jealousy of the Faculty of Medicine against his near-miraculous cures. Frustrated in his medical career, he turned to physics and submitted his experiments to

[184] Marat to Roume de Saint-Laurent, 20 July 1783, Vellay (1908), 19-21.

[185] Above, n. 146.

[186] Marat to Roume de Saint-Laurent, 20 November 1783, Vellay (1908), 24.

[187] LeRoi to the Marquis de Laubespine (for Marat), undated, but sometime in 1779 or early 1780, Vellay (1908), 52-53. Vellay (1908) reproduces these letters.

[188] *Miscellanies 8* (London, 1863-1864), 1-97.

the Academy. Several philosophers among its company spoiled the initially favorable disposition of his commissioners, even though his discoveries made a great stir all over Europe. For over six months, the court and the city, princes of the blood and eminent statesmen, thronged his laboratory. Scientists came to observe from Stockholm and from Germany. Journals throughout the continent reported his work, with the exception of the one organ controlled by philosophers, the *Journal des savants*. When he consummated his findings on fire with the discoveries on light that dared call Newton into question, and once it was clear that he would not appease his critics by joining the Academy, they enveloped him in a conspiracy of silence. The Academy could not acknowledge that it had been basing its science on false principles for forty years and more. Until his work, there had been no progress in optics since Newton, but only blind repetition. Until his work, science had spent 2,000 years vainly asking what fire is, and he had shown the Academy; until his work, electricity had been a chaos, and he made the fluid visible and formed the science around a coherent theory. And was it an ignoramus who had accomplished all that? If he was not of the Academy of Science, it was because he had not troubled to be; if he was not of the Academy of Berlin, it was because he had not asked to be; if he was not of the Academy of . . . but enough! His enemies feared, not so much what he had done, but what they knew he could yet do, under the auspices of a great king.

Let that king be warned, however, and let his counsellors take heed, for the importance of the question far surpassed the dark maneuvers of philosophers in the small matter of Marat. The moral philosophy of these persons, corrupt to the core, holds many attractions for the young. Their proselytes are multiplying. Every day they make converts. Theirs is a confederation the more dangerous that there is no external sign by which to recognize them. Thus do they infiltrate all orders of society and all institutions—universities, learned bodies, judicial tribunals, royal councils. They have conceived the "horrible project" of abolishing all religious orders and of extinguishing religion itself. They poison the wells of useful knowledge and fill up all the teaching posts. If one day, they turn toward politics "and act through their agents informed of everything that happens in the offices of ministers, who can prevent them from stirring up the government itself and overthrowing the state?"

"I can see only one way, my friend, to prevent these misfortunes. That is to engage all the great writers to heap ridicule upon these apostles of modern philosophy.

"Let's come back to me."[189]

Thus Jean-Paul Marat in November 1783. A reader unversed in the

[189] Marat to Roume de Saint-Laurent, 20 November 1783, Vellay (1908), 38.

vagaries of abnormal psychology might reasonably suppose that such a letter bespeaks a mind undone and incapable of further coherent effort. Nothing of the sort, however—although Marat's disappointment from Madrid was compounded in Paris by his being eliminated from the civil list of the comte d'Artois, his only known source of income, he was even then or soon afterward composing a simplified abstract of his theory of light, *Notions élémentaires d'optique*.[190] Moreover, he had just scored an academic success, albeit a provincial one, with a memoir on medical electricity that carried off a prize. It had been set by the only learned body that ever crowned his efforts, the Academy of Rouen. The subject was the therapeutic value of magnetism and electricity, "negative as well as positive."[191]

Though Marat refrained from discussing magnetism, pleading lack of experience, the secretary considered that with respect to electricity, he had entirely satisfied the conditions. The Academy did feel bound, however, to deplore the author's language in refuting the opinions of an estimable person who was a member of nine learned societies.[192] The target of Marat's contempt here was the abbé Bertholon, whose *Traité de l'électricité du corps humain* (1780) had won a prize from the Academy of Lyons.[193] Having been thus chided, Marat (so he tells his readers) went back over his memoir before publication lest some regrettable expression might indeed have escaped his pen. He could find nothing, no not a single phrase, that any really self-respecting author need have denied himself. His criticism was harsh. He admitted it. But he would have been wanting in zeal for humanity if he had been gentle in refuting a system of therapy that had become dangerous in consequence of the ill-considered praises heaped upon it.

> For the rest, the author of this system is himself too concerned for the public good to refrain from applauding the manner in which I have opposed his opinions. Actually, they furnished me with material for additional observations, but since these did not pertain di-

[190] (1784). The front-matter says that two printings of the *Découvertes sur la lumière* (1780) having been sold out, he has been urged to prepare this abstract. In 44 pages, it gives a summary of his disagreements with Newton, and recounts one, as Marat thought, decisive experiment.

[191] *Mémoire sur l'électricité médicale, couronné le 6 août 1783 par l'Académie Royale des Sciences, Belles Lettres et Arts de Rouen* (1784).

[192] *Ibid.*, 6.

[193] Bertholon was a minor scientific figure who often wrote under the pseudonym the "abbé Sans." Among his other main works were *De L'Électricité des météores*, 2 vols. (1787) and *De L'Électricité des végétaux* (1783). He had made known his dismay over Marat's attack on the lightning rod in particular, and over his hostility toward Newton, Franklin, and great scientists in general. See Vellay (1912a) and Duval (1912).

rectly to my subject, I have suppressed them, and the author ought to be grateful to me for that.[194]

Unpleasantries apart, the actual content of Marat's memoir is marked by good sense and skepticism. He had been studying electricity and experimenting with its effects on selected patients for some six years. He tells of exposing cases of hypochondria and asthma to an electrified atmosphere in his office for periods up to three hours a day. The results were nil. Nor did the electrification of his consulting room have the slightest effect on his own ailments (which Marat very rarely mentioned). He concluded, in short, that the electrical fluid diffused throughout the atmosphere has no appreciable influence upon the human body, and that only local applications in the form of friction, shocks, or sparks can afford modest relief in cases of cramp, sprain, and inflammation of the joints, and in reducing certain tumors.[195] Angry leaflets and letters to periodicals prolonged the vendetta between Marat and Bertholon through the years 1784 and 1785.[196]

A pamphlet under the pseudonym "L'Observateur Bon-Sens" is also credited to Marat at this period. Its author criticizes the design of the balloons in which the aeronauts Pilatre de Rozier and Romain met their deaths. It was disputed whether a fire or a rupture in the bag caused them to crash in their attempt on 14 June 1785 to cross the Channel from the coast near Calais to England. Marat was confident that an explosion was responsible. He had warned Pilatre de Rozier that igneous fluid rising from the brazier would dangerously increase the expansibility of the hydrogen. Perhaps that adventurer did not understand the physics of the thing well enough to take the point, or perhaps he was simply headstrong about novelty. Whatever the reason, "he was deaf to my voice, and like another Cassandra, I cried in the wilderness."[197] These were but trivialities, however, if indeed Marat was even the author of this brochure.[198]

[194] *Mémoire sur l'électricité médicale*, 7-8.

[195] *Ibid.*, 106-110.

[196] The editors of *L'Esprit des journaux* abstracted the direst of Marat's strictures from the Rouen *Mémoire* and reprinted them (June 1785, pp. 157-168), to which reiterated attack Bertholon replied in the same journal (November 1785, pp. 343-358). See Duval (1912). Bertholon further criticized the Marat *Mémoire* in a "Lettre de l'abbé Sans . . . ," published in *L'Année littéraire*, No. 16 (1785), to which Marat replied by reprinting it in a pamphlet adding his own polemical footnotes under the title, *Observations de M. l'Amateur avec à M. l'Abbé Sans, sur la nécessité indispensable d'avoir une théorie solide & lumineuse, avant d'ouvrir boutique d'électricité médicale. . . .* (1785).

[197] *Lettre de l'observateur Bon-Sens à M. de * * * sur la fatale catastrophe des infortunés Platre* [*sic*] *de Rosier & Romain* (1785), p. 19; BN, 8°, Ln^{27}.16320.

[198] I am not altogether confident that he was. The preachy and smarmy tone is unlike him, and so is the solecism of putting Cassandra's voice in the wilderness.

What principally possessed him in the later 1780s was the extension of his work in optics. In 1787 he published a translation of Newton's *Opticks* in two volumes. In 1788 he brought out a collection of four pieces, *Mémoires académiques: ou nouvelles découvertes sur la lumière*. The title page bears the motto, "They will swim against wind and tide."

Comparison of these, the last of Marat's scientific writings, to what has been said about them, to his earlier work, to each other, and to certain recent findings of Newtonian scholarship, contains surprises on all counts. Let us begin with the translation. In his memoirs Brissot says that Marat had undertaken the translation and (he made no doubt) falsified it in order to damage its principles.[199] The author who purports to deal with Marat as scientist devotes a single sentence to it, which says that Marat tricked the Academy into giving its approbation by having the book issued under the auspices of Nicolas Beauzée, a grammarian and member of the French Academy, who was secretary to the comte d'Artois, and who composed a preface dedicating it to the king.[200] The most recent biographer, calling the publication "a good farce," says that through a ruse Marat got his academic enemies to endorse a restatement of his own theories in the form of notes to the translation.[201] A widely read biographer in English, to go no further, considers that the translation and the *Mémoires académiques* that followed are evidence of a decline in Marat's scientific interest and powers.[202]

Except for Brissot, who pretended to disenchanted impartiality, these writers were largely favorable to Marat. On this score, it will appear, he needs protection from his friends among historians no less than from his enemies among philosophers. For his is an excellent translation. A careful comparison with the fourth edition of the *Opticks*, from which Marat worked, requires it to be said in all fairness that he was nowhere unfaithful to Newton's meaning. It is a free rendering, somewhat abbreviated in wording though not in content, and it gains thereby in readability, for Newton could be prolix. In regard to Newton himself, the tone is impeccable, respectful without being adulatory. The only criticism partaking of hostility, though in nothing like the degree of Marat's other publications, is directed at the original translator, Pierre Cotte. The editor's preface charges him with servility to the literal wording and with ignorance of physics and of English.[203] Actually, Cotte was an excellent linguist and a good writer who had also translated Locke. The definitive edition of his translation of the *Opticks* appeared in 1722 in a beautifully

[199] Brissot, *Mémoires 1*, 358.

[200] Cabanès (4th ed., 1911), 185.

[201] Massin (1970), 67.

[202] Gottshalk (1967), 30.

[203] *Optique de Newton*, Traduction nouvelle, *Faite par M * * * sur la dernière Edition originale, ornée de vingt-une planches, et approuvée par l'Académie Royale des Sciences* (2 vols., 1787), *1*, x.

printed and illustrated edition. It is indeed a great deal stiffer than Marat's, who had no way of knowing that every detail had been jealously overseen by Newton himself through the good offices of Varignon and De Moivre.[204]

Marat did append a critical apparatus to his translation, but it is not true that he smuggled his own theory into the notes. They refer the reader in every instance to passages in the memoirs of the abbé Rochon.[205] Rochon was a practical physicist, an experimentalist and instrumentalist of the older school, with an abiding interest in lighthouses, lenses, and navigation. He is remembered mainly for a travel book, the first European account of a journey to Madagascar.[206] It holds sufficient anthropological interest for the English translation to have been reprinted in 1971, and is written with disarming modesty. "I am not a literary man," he acknowledged; "this may be readily perceived by the want of order and method which is observed in whatever I write."[207] Rochon himself was on the commission that recommended the Academy's approbation for Marat's translation, and since the other two were Bailly and Condorcet, it is clear that he was the effective member.

Reservations about the *Opticks* are restricted in these notes to certain experiments or inferences, some nine in all. Most of them are justified with reference to the history of the science since Newton's time. For a qualification of the proposition that differently colored rays are differently refrangible, the reader is referred to Rochon's observations of starlight, which had detected no elongation of the spectral band.[208] For criticism of Newton's distinction between heterogeneous and homogeneous rays, he is sent to Rochon's argument, also based on analysis of starlight, that it is impossible to isolate a monochromatic ray, the spectrum being a continuum.[209] The most considerable discussion concerns limitations of the reflecting telescope, which Newton had held to be the only possible method for overcoming chromatic aberration. Newton is said to have mistaken the one experiment that could have undeceived him about the invincibility of chromatic aberration in lenses. Marat here develops a topic he had just mentioned in the last sentence of his *Découvertes sur la lumière*.[210] The only hint of the extreme is the statement that, in consequence of Newton's

[204] On the Cotte translation, see Lohne (1968), 191-196; Hall (1975).

[205] A.-M. de Rochon, *Recueil de mémoires sur la mécanique et la physique* (1783).

[206] *Voyage à Madagascar et aux Indes Orientales* (1791).

[207] *A Voyage to Madagascar* (Johnson Reprint Corp. edition of the English translation of 1792), xliii.

[208] *Optique de Newton* 2, 282; Rochon, *Recueil*, 28.

[209] *Optique* 2, 283; Rochon, *Recueil* 29-32.

[210] *Optique* 2, 285-287; *Découvertes sur la lumière*, 141. In question here is Experiment 8, Part II, Book I, *Opticks*.

mistake, opticians for half a century had abandoned all attempts to find a way around the obstacle. Marat follows Rochon in recounting the history of the achromatic lens. In 1747 Euler had suggested combining two transparent substances of different refractive index. John Dollond, having first taken issue with Euler, then disproved Newton and himself by successfully coupling crown glass and flint glass in 1757.[211] The one significant reservation not derived from Rochon (and this is referred to Euler) is Marat's remark that the notion of fits of easy reflection and transmission in Book II "is not nearly as satisfactory as it is ingenious."[212] It is true that all these criticisms could be contained in Marat's most extreme views on light and colors. There is nothing here, however, that goes beyond the bounds of permissible scientific discourse. The evidence is that Marat was in command of all these issues substantively, and also that, when the occasion was right, he could be in command of himself. What made that occasion right remains a mystery: perhaps the necessity of translating and producing a salable edition imposed restraint.

In this respect, the contrast with the *Mémoires académiques* published in the next year, 1788, is startling. Marat's last work of science, it is far from empty physically. Anger rules the tone, however, and confusion reigns over dates and circumstances. A copy in the Bibliothèque nationale, accompanied by a covering letter to Billaud de Varenne, has a tipped-in frontispiece portraying Marat in revolutionary garb.[213] The content consists of four memoirs, all composed for academic competitions. The first two had been submitted to the Academy of Lyons for a prize announced in 1784 on the question whether Newton's experiments on differential refrangibility were definitive or illusory. One could almost believe that Marat had set the topic himself, given his insistent denial of the phenomenon in his previous writings on optics. He could take little consolation in the judgment of the academy, however. It held that his two papers were the only negative contributions deserving of consideration or comparison with the affirmative pair that took the prize and honorable mention, respectively.[214]

It is unclear when he actually wrote any of these four memoirs. The flyleaf announces "Oeuvres de M. Marat." In the preface he addresses his readers in his own name and recalls his earlier work. He had not revised passages in the memoirs themselves, where, in accordance with academic

[211] *Optique* 2, 286-295; Rochon, *Recueil*, 45-52.

[212] *Optique* 2, 299-300.

[213] *Mémoires académiques, ou nouvelles découvertes sur la lumière, relative aux points les plus importans de l'optique* (1788).

[214] The author of the winning essay was a provincal physician, one Flaugergues, a correspondent of the Society of Medicine. The honorable mention went to a Dutchman, Antoine Brugmans, of Groningen. See *Mémoires académiques*, 4-5.

convention, the author is referred to in the third person and anonymously. As if to compound the confusion, the author is said to be a friend of M. Marat, one who has assisted him in the experiments that have made him a famous critic of scientific orthodoxy.[215] There must have been some revisions in the original texts, however. Although three of them were due before the end of 1786, notes refer readers to the 1787 translation of Newton's *Opticks*, but without identifying Marat as its translator.[216] The references to Newton are equally ambivalent. Newton was a great man—but then his greatness consisted partly in the skill with which he was able to put over a false theory. It is Newton's inferences, not his experiments, that are at fault—but then Marat goes on to impugn the experiments too. Mathematical virtuosity was the secret of his power—but then it is specious to call the *Opticks* mathematical just because it is full of lines and diagrams.[217] About the epigones, Marat expressed himself without ambivalence. The learned academicians of Lyons, unable to decide on the merits of the various memoirs, had awarded the prize by counting pages.[218] The instrument makers and lens grinders of Paris are ignorant of the simplest elements of optics and do not even know how to make spectacles properly, and those of London are not much better, slavishly copying Dollond.[219]

Marat devotes the initial memoir to criticizing the first ten experiments in the *Opticks*, those that Newton adduced in arguing the differential refrangibility of heterogeneous rays in white light. Only the second experiment is direct, he says. Newton had there taken a slip of stiff paper colored red on one half and blue on the other and illuminated it by a candle in a dark room. He mounted a four-and-a-quarter-inch double-convex lens just over six feet from the object, tried to catch its image on a white paper an equal distance beyond the lens, and found that the focus for the red rays was one and one-half inches further from the lens than for the blue rays. It is now known that this was one of Newton's earliest experiments, dating back to his undergraduate days.[220] By calling it the only "direct" trial, Marat meant that in the other nine, Newton exhibited the spectral dispersion of colors on a screen, and inferred by induction that differential refrangibility is the cause. He then roundly asserts that this one comparison of focal lengths "is not only illusory. It is false."[221] Newton should have repeated it in daylight. When he, Marat, first exposed such a card to the sun's rays, and afterward tried it in a dark room, the red and blue images converged at the same point. The rays of the sun had soaked into the colors so that they behaved naturally. It has to be said that

[215] *Mémoires académiques*, 17-19, 117.
[216] *Ibid.*, 9-10, 14, and *passim*.
[217] *Ibid.*, 8, 35, 89.
[218] *Ibid.*, 6.
[219] *Ibid.*, 13.
[220] Westfall (1962).
[221] *Mémoires académiques*, 23.

this objection is specious. If Marat did perform this experiment, the negative results he reported probably resulted from his failure to accustom his eyes to the dark or to achieve the definition that Newton produced by winding a fine black thread around the card to create a pattern of parallel hairlines on the red and blue halves of the background.

Beyond that, it would be tedious to follow Marat retracing Newton's footprints and impugning every step. First, he transcribes the essential passages of each experiment. Next he attacks the conclusions. In Newton's most famous experiment, a beam of sunlight was admitted through a hole in the shutter, passed through a prism, and projected onto the opposite wall where the colors of the spectrum appeared along a band. Newton takes its elongation for evidence that the heterogeneous rays of white light had been refracted each a specific amount according to its color, and thus spread into the oblong shape of the image. Very persuasive, replies Marat, but look: even allowing Newton his arrangement of the apparatus, there is a difficulty. He supposes that the rays of the original beam are parallel when they impinge on the incident face of the prism. In fact they are not, for they have already been inflected in their passage through the aperture of the shutter. The elongation of the spectrum results from that initial decomposition, and owes nothing to refraction in the glass.[222]

In the final stage of refuting each experiment, Marat set about varying its conditions in order to impeach the legitimacy of the conclusions in the general case. The labor he performed is staggering. The pages of these four memoirs report hundreds, it may well be thousands, of observations, each requiring a distinct manipulation of lens, prism, screen, and light beam. He intercepted the refracted colors closer to the prism than Newton had done, right up to the emergent surface of the prism. He altered the inclination of the incident surface to the beam. He moved the screen far back and observed that the colors became less brilliant instead of appearing more sharply differentiated, as by Newton's theory they should have done. Other modifications produced circular images, wedge-shaped images, oval images, and elliptical images, often attended by, and sometimes breaking up into, varicolored crescents, each one a partial spectrum. From those effects he could argue that colors are produced by the mingling and not the separation of the rays. He illustrated the phenomena in a series of water-colored plates pasted into the volume. The state of graphic art was such that the five or six figures on each copy of each plate had to be colored in by hand. These watercolors retain their quality today, at least in several copies of the work. In a prefatory note Marat apologized

[222] *Ibid.*, 37-38. Lohne (1961) objects to the same assumption, though on different grounds—namely that the rays come from all points of the disk of the sun and are not parallel.

for the price, and the printer warns binders to protect the plates with tissue paper before hammering the sheets.

Having thus claimed to destroy Newton's system of differential refrangibility in his first Lyons memoir, Marat turns in the second, "purely physical, but spicier, tighter, peppier,"[223] to establishing his own theory. The experiments here are said to be all new. He has performed them with lenses and prisms made to his order and tested by his method of observing in a dark room, the only infallible way to select glass fit for optical instruments. There are five classes of experiment, all demonstrating that heterogeneous rays are equally refrangible and are separated only by their passage in proximity to the edges of bodies. To this end, he took prisms and viewed plane surfaces through them—clouds in the sky, the wall opposite, a white card applied to the prism—and pointed out that colors appeared only along boundaries of light and dark, the frame of a window for example. (These were observations that Goethe later singled out for admiration.)[224] But if a prism dissociated light, it must always do so. On the other hand, hairlines and thin slits invariably do have this effect. Thus, we are back at Marat's favorite positive assertion, which attributes the appearance of color to inflection and not refraction.

Increasing querulousness and repetitiousness mark the final pair of memoirs, which are printed in inverse chronological sequence. The third, on Newton's theory of rainbow, was sent to the Royal Society of Montpellier in October 1786, whereas the last, on Newton's rings and colors in thin transparent media, had already carried off the prize from the Academy of Rouen on 2 August 1786. Let us discuss them in their probable order of composition. The chief interest attaching to the Rouen memoir is the suspicion, even stronger than in the case of the Lyons memoirs, that Marat himself may have had a hand in framing the terms of the competition. That Academy had crowned his memoir on medical electricity two years previously, and some of its members must have been favorably disposed to his work. What is more telling, the *programme* or announcement is couched in phrases that appear elsewhere in Marat's writings on optics and invite the restatement of objections to Newton and of counter-experiments that Marat had already adduced. It reads:

> Colors that appear in plates of glass, soap bubbles, and other extremely thin diaphanous bodies pre-suppose the doctrine of differential refrangibility and that of fits of easy reflection and transmission. The former of these doctrines having been called in question,

[223] *Ibid.*, viii.

[224] J. W. Goethe, *Zur Farbenlehre*, in *Gedenkausgabe der Werke, Briefe und Gespräche*, ed. Ernst Beutler, 24 vols. (Zurich and Stuttgart, 1950-1971), *16*, 764-765; for Goethe's summary view of Marat's science, see *ibid.*, 657-660.

and the latter in no way satisfying the mind, the Academy proposes for the subject of the Physics Prize, TO DETERMINE THE TRUE CAUSES OF THESE COLORS. But it warns authors that by the same token it will reject all hypotheses and that as proof of their assertions it will admit only simple and unvarying facts.[225]

Thereupon, Marat reiterated the criticisms of Newton's prismatic experiments contained in the first of his Lyons memoirs. His discussion of Newton's rings is still more scornful:

> This doctrine, of little interest in itself, becomes still less so because of the way it is treated. There is to be found in it neither an exact exposition of the phenomena nor a just explanation of them. Out of a heap of badly made observations, piled up pell-mell, several formulas are deduced which are then erected into a principle.[226]

Marat here had in his sights those measurements of the separation of the rings of color produced when a slightly convex lens is pressed on a plane glass plate from which Newton computed the depth of the air film corresponding to each ring and color.[227] It was from these figures that Thomas Young calculated and in 1802 published wavelengths of the colors in the visible spectrum.[228] Marat may, therefore, be thought unfortunate in having fixed on just this target for his contempt. No doubt it would be anachronistic to reproach him, since the wave theory of light, itself deduced from the hypothesis that these are interference phenomena, lay the same distance in the future. Certainly he was correct at the time of writing in fixing on the notion of fits in light as the one of Newton's optical ideas that was least pleasing to eighteenth-century successors. In its place he advanced the proposition that thin-plate effects are produced by compression and distortion of the surfaces in contact.[229] The irises in soap bubbles he held to be altogether different. They derive from the properties of the dissolved particles, of which there are three distinct species, each of them reflecting only the rays of one of the primitive colors. They group themselves in regular patterns in service to laws of affinity and equilibrium.[230]

In the Montpellier memoir on Newton's theory of the rainbow, Marat set about impugning what has generally been taken, then and since, to be one of the most beautiful propositions in the *Opticks*. Newton had accepted from Descartes, though crediting it largely to Marko Antonije

[225] *Mémoires académiques*, 253.

[226] *Ibid.*, 285.

[227] Newton, *Opticks*, Book II, Part I, Observations 6 and 7.

[228] "On the Theory of Light and Colours," *Philosophical Transactions* (1802), 12; reprinted in *Miscellaneous Works of the Late Thomas Young*, 3 vols. (London, 1855) *1*, 140-169, see esp. 160-163.

[229] Marat, *Mémoires académiques*, 295-301.

[230] *Ibid.*, 302-321.

Dominis, the physical model in which two refractions and one reflection in each droplet produce the inner arc, and two refractions interspersed by two reflections produce the outer one with the order of colors reversed. Newton's contribution is a mathematical confirmation of his theory of refracted colors. He derives the angular dimensions of the two arcs and their apparent locations in the sky from the indices of refraction of the several colors of the spectrum, and then verifies the deduction by measurements.[231]

Marat acknowledges that attacking this, a universally admired chef d'oeuvre, risks the appearance of bravado. All we really find in Newton's proposition, however, is the hand of the great mathematician, "whereas the Science and the Dialectic of the Physicist are often faulty."[232] To bring out those defects, Marat told of watching many a storm from the tower of Saint-Sulpice. Sometimes the rainbow would form against a clear sky instead of in the receding curtain of rain, where the received theory required it to be. On occasion he has even seen right through it and made out the trees on the plain of Ivry beyond. That would be impossible if indeed the rainbow consisted of reflected light. Moreover, from the Newtonian model it followed geometrically that the arc of color seen from on high, say from a balloon looking down upon a sunlit shower, should become a ring. At least, it should approximate more closely to a circle the higher the eminence. Though few physicists have ever been in a position to make such observations, he himself some eleven years since had looked down on a storm from the summit of a mountain in Wales, and the form of the rainbow fell far short of what Newton's theory required.[233] (This passing mention of a visit to Wales seems to have escaped Marat's biographers).

Marat's central physical objection was that which he brought against the general theory of differentially refracted colors: it depended upon the false assumption that the sun's rays are parallel when they impinge on the refracting surface. He could not here invoke inflection to work some prior deviation; this time it is the finite breadth of the sun that is responsible. The rays strike each droplet, criss-crossing each other at every angle from $0°$ to $32'$.[234] Marat was correct, of course, except that the distance from the rainbow to the eye is relatively so short that the effect is negligible. Another objection was specious in principle as well as practice. He contended that the spherical inside surface of the droplets would distort the reflections supposed to occur there. In one important respect, that each observer sees the arc of color within a narrow angular range subtended at

[231] Newton, *Opticks* (4th ed., 1952 reprint), Book I, Part II, Proposition IX, Problem IV. For the history of the rainbow, see Boyer (1959), especially chs. 9 and 10.

[232] Marat, *Mémoires académiques*, 163.

[233] *Ibid.*, 176.

[234] *Ibid.*, 172.

his eye, Marat seems to have misunderstood Newton's model. He complained that if the theory were correct, the whole curtain of rain should be a sheet of color. In another respect, he misrepresented the theory, saying that the explanation of the rainbow presupposed fits of easy reflection and transmission, which he then attacked again in his last, and for this purpose altogether gratuitous section. Misstatements of this kind are not characteristic flaws in Marat's scientific writings, of which this memoir is probably the weakest example. In all its many pages there is only one of those observations, much less infrequent elsewhere, that strike his reader as having been shrewd, however partial in substance and destructive in motivation. He accuses Newton of having chosen experimental conditions so that the values would fit, "for in this much touted system, all the art of the author consists in adapting formulas to the observations, and appearing to deduce the phenomena from his principles."[235]

[235] *Ibid.*, 201. As a cardinal example he cites Experiment 16, Book II of the *Opticks*. That Newton's published accounts of how he had reached conclusions is often idealized to the point of misrepresentation is one of the findings of recent Newton scholarship, which has drawn mainly on the detailed scrutiny of his manuscripts. Lohne (1966), for example, has shown that Newton arrived at his theory of differential refrangibility in a manner different from the narrative he gave of it in describing the so-called crucial experiment in 1672. In general, I should say that Marat's suspicions of the published work come closer to an accurate intuition about the kind of window-dressing that Newton sometimes put around his findings than does anything else known to me in the eighteenth-century literature. It is roughly correct to say that the *Opticks* is not really a work of mathematical physics in the sense that the *Principia* is, though it would be more pertinent to say that it is not a work of mechanics. It is entirely correct to say that Newton frequently chose experimental conditions peculiarly suited to exhibiting phenomena in a manner that supported his findings. It must immediately be added, however, that much of the art of the experimental physicist consists in the wit to do precisely that. The question whether it is done artfully or deceitfully is legitimate but not simple, since answering it often turns on judgments of motivation rather than of methodology, and the boundary is correspondingly difficult to define. Modern scholars would certainly concur that in Newton's case the element of art is overwhelmingly predominant and accounts for much of his genius, but they would also agree that he had his disingenuous moments, particularly in dealing with controversy. See Lohne (1961), Bechler (1975a).

As for the points at issue in Marat's mind, the choice between differential refrangibility and inflection as a source of prismatic colors no longer has any meaning, and had less meaning then than he tried to give it. His penchant for inflection was another example of his tendency to leap on a hobby-horse and ride it through a whole science laying waste to everything else. His manipulation of the phenomena of colors in *Mémoires académiques*, as in his earlier writings on heat, light, and electricity, does exhibit manual address and ingenuity. He probably did disclose effects that no one else had produced, though he was never able to develop them into any coherent contribution to science. As for his other main contentions, he was correct about the misleading and inhibiting consequences of Newton's influence in the matter of achromatic refraction, but there he burst through a door opened wide by Euler and Dollond. Finally, Newton's treatment of primary colors was certainly vulnerable, but Marat never gave his criticism any development. There is no hint of the distinction Goethe called for between the study of color as produced in light and as perceived in the eye.

In the Introduction of *Mémoires académiques*, Marat acknowledges that the memoir on the rainbow is entirely negative, and says that in his portfolio he has the manuscript of a companion piece containing positive ideas on the subject, together with other studies treating of colors at sunrise and sunset, of the apparent ellipticity of the moon at the horizon, and of the double image produced by Iceland spar. He has plans to publish them at the end of the year,[236] and meanwhile invites interested readers to write to him, rue du Vieux Colombier (whither he had removed from the Hôtel de Laubespine several years previously). There we may leave Marat the would-be physicist late in 1788, on the eye of his political epiphany, for those plans were overtaken by events and henceforth he wrote of other matters, except when denouncing the Academy of Science and all its works before the bar of revolutionary opinion.

All the foregoing constitutes the minimum background required for taking the thrust of the famous philippic, *Les Charlatans modernes*, that Marat delivered against that body in September 1791. It was written (he said) long before.[237] Comparable hostility for science, or rather what is at bottom the same hostility, has been manifest among the alienated from time to time throughout modern history.[238] Its reappearance was an intrinsic feature of recent political events, and an English-language historian writing of Marat in the 1970s cannot remain unaffected by the experience of the Royal Shakespeare Company's production in 1964 of *The Persecution and Assassination of Jean-Paul Marat as Performed by the Inmates of the Asylum of Charenton under the Direction of the Marquis de Sade*. Indeed, one such historian begs to be forgiven if he digresses from his proper mode for one paragraph to observe that Peter Weiss's play is an instance of art getting at the truth before ever scholarship has done, with its (normally salutary) limitation to the facts. The fact is that Marat never met Sade nor mentioned any of his writings, but it is true that Sade was present at his funeral. The fact is that overtly there is nothing of sexual libertinism or moral diabolism in Marat's campaign against Newtonian science and political authority, nor anything said of Marat's concerns in the writings of Sade, but the truth is that the common enemy was lodged in the same bastion, the structure of respectability, family, rationality, knowledge, and state, which made society what it was. The journalistic hybrid Marat/Sade is no mere shorthand. In the play as in reality both parties scorn reform and cry transfiguration in identical tones, inverting values to the point that the victim in so-called antisocial actions is really to blame through his complicity with society and the perpetrator really innocent through having been victimized. Thus it happens, to look at the matter from the side of the curtain where the actors play their parts, that only

[236] *Ibid.*, x, note. [237] Vellay (1911), 256.
[238] I have explored the comparison a little further in Gillispie (1977).

the mad speak sanely or truly of the facts, while illusion and pretense shelter the spectators in their propriety and wherewithal. It is a further fact that scientists throughout history have considered themselves the benefactors in their work and influence of the whole people whose friend Marat now set up to be, and a final fact that scientists probably have been right, materially at least. But it remains true that science has not always, or perhaps usually, been perceived as benefaction by those subjected to the authorities whose powers it augments. Of course, it is only coincidence that the Marat, whose voice Victor Hugo movingly eternalizes whenever it is a question of speaking for the wretched of the earth, should have traversed most of the misery of his own life vainly and unequally contending with and against science.[239]

The coincidence, however, is singularly appropriate to the truth.

4. DEMACHY

Perhaps, to return now from theater to history, the continuum from Marat to Sade in a counter-Enlightenment could be situated in the career of a lesser in the breed, Jean-François Demachy, apothecary and would-be chemist, pornographer and would-be poet, who boasted of having committed incest with his mother before entering on the lifelong quarrel of his marriage.[240] A onetime student of Rouelle, Demachy kept a shop in the rue du Bac, where he gave well-patronized private courses in chemistry and natural history. On the conversion of the Apothecaries Company to the College of Pharmacy in 1777, he was named demonstrator responsible for teaching those two subjects.[241] Like Marat, he somehow had well-placed friends, and in 1776 Miromesnil, then keeper of the seals,

[239] Massin (1970) quotes (11-12) a paragraph that Victor Hugo had intended for *Quatre-Vingt-Treize*: "Marat n'appartient pas spécialement à la Révolution française; Marat est un type antérieur, profond et terrible. Marat, c'est le vieux spectre immense. Si vous voulez savoir son vrai nom, criez dans l'abîme ce mot: *Marat*; l'echo, du fond de l'infini, vous répondra: *Misère!* . . . On guillotine Charlotte Corday, et l'on dit: *Marat est mort*. Non, Marat n'est pas mort. Mettez-le au Panthéon, ou jetez-le à l'égout, qu'importe, il renaît le lendemain. Il renaît dans l'homme qui n'a pas de travail, dans la femme qui n'a pas de pain, dans la fille qui se prostitue, dans l'enfant qui n'apprend pas à lire; il renaît dans les greniers de Rouen, il renaît dans les caves de Lille; il renaît dans le grenier sans feu, dans le grabat sans couverture, dans le chômage, dans le prolétariat, dans le lupanar, dans le bagne, dans vos codes qui sont sans pitié, dans vos écoles sans horizon, et il se reforme de tout ce qui est ignorance, et il se recompose de tout ce qui est la nuit. Ah! que la société humaine y prenne garde, on ne tuera Marat qu'en tuant la misère; . . . tant qu'il y aura des misérables, il y aura sur l'horizon un nuage qui peut devenir un fantôme, et un fantôme qui peut devenir Marat."

[240] On the life and personality of Demachy, see Toraude's introduction to his edition of the stories and verses, Toraude (1907).

[241] Above, Chapter III, Section 3.

appointed him to the panel of three censors charged with licensing books on chemistry, natural history, and pharmacy. He it was who authorized publication of Marat's *Notions elémentaires d'optique* on 8 July 1784. Demachy too stormed the doors of the Academy of Science with a barrage of memoirs, treatises, prefaces, and apostrophes. Possibly he found some small measure of consolation in the one body that admitted him, where his path crossed Marat's again, for it was the Academy of Rouen. He had failed in Paris because theorists had usurped science from the practitioners who had produced it and whose property it rightfully is. His was a labor theory of scientific value. In 1790 a widely read masonic journal (inevitably Demachy was a Freemason) carried a long series of his articles, a pharmacist's riposte to Lavoisier, his chemical pretensions, and his "algebraic exactness." He, Demachy, has been making experiments for forty years, and "I have always seen that mathematical precision is excellent in calculation, but derisory in fact."[242]

Enough. It is time to bring this chapter to a close. The most sensible way of doing so will be to quote the opinion attributed to a member of the Academy of Science who expressed himself moderately on the matter. He may have been the only one to do so. Certainly he was less conscious than most of his colleagues of some dignity in professional status. The chevalier de Borda was of an older school, and this is what Lacroix attributes to him:

> That probity of mind and infallible tact that nothing could alter led Borda to make a judgment of charlatans in general that seems paradoxical at first. He thought that they are always taken too seriously or not seriously enough. If the credulous give them unlimited confidence, informed people too often reject them without listening. In the former he saw blind ignorance, and in the latter the desire to punish success that comes too easily and angers them. To these considerations, he added that every error that is found out is a sort of progress in the search for truth, if, as the history of the human mind seems to suggest, we are incapable of discovering what is true until we have exhausted all possible absurdities.[243]

[242] Demachy in *Tribut de la Société Nationale des neuf soeurs 1* (August 1790), 116-117, 308; *3*, 252. The series is an attack on Lavoisier's *Traité élémentaire de chimie* (1789), its method and nomenclature, and on all the pneumatic school. He recalls instead the models of Stahl and Boerhaave: "Imitons-les; ils ne raison noient que sur des faits observés et non sur des calculs," *3*, 253.

[243] Sylvestre-François Lacroix, *Eloge historique de Jean-Charles de Borda . . . lu à la Société Philomathique*, n.d. BN [8⁰, Ln27.2403], 38.

PART THREE

APPLICATIONS

CHAPTER V

◇◇

Trades and Agriculture

◇◇

1. SCIENCE AND THE ECONOMY

In 1783 the comte d'Angivillers, director of the Batimens du roi, placed the abbé Tessier in charge of installing and conducting an experimental farm on the royal domain of Rambouillet. Two years later, when composing a preface to the agricultural volumes of the *Encyclopédie méthodique*, which he edited jointly with Thouin and Fougeroux de Bondaroy, Tessier said that he had thereby been enabled to study all the arts even more intensively than before, and trusted that he could "contribute to dissipating certain prejudices among farmers, and impart a healthy distrust of charlatans, from whom agriculture is not exempt."[1]

Tessier formed and expressed those hopes in the context of a set of relations between science and enterprise, agricultural as well as industrial, that had emerged in France by the end of the old regime in a form definite enough to permit specifying what the elements and phases were. So largely have concurrent events in Britain dominated the attention, imagination, and theoretical capacity of historians that it will be prudent to begin with several caveats. Although neither an agrarian nor an industrial revolution occurred in France in the late eighteenth century, great reforms were called for and to an extent effected both in farming and in manufacturing, and it is unhelpful to see in them a series of failures to replicate the British pattern. In France, expertise held proportionally greater importance than in Britain, and invention for private profit proportionally less. In France, greater initiative came from agents of government and less from rising entrepreneurs. Developments in France are better called a rationalization, even an education, of farming and of industry than a new phase of capitalism. (It is, indeed, curious how industry and the economy developed by way of reform in France, reserving the revolutionary mode for political and social change, whereas the reverse was true for Britain.)

[1] *Encyclopédie méthodique. Agriculture*, 6 vols. (1787-1816) *1*, 15. Fougeroux de Bondaroy died in 1789, leaving Tessier and Thouin to edit volumes 2-4 (1791-1796); in 1813 they were joined by Bosc for Volumes 5 and 6. A seventh volume under the same general title appeared in 1821, edited by Bosc and Baudrillard and devoted entirely to arboriculture, the subject that Fougeroux was to have handled. The subtitle is *Dictionnaire de la culture des arbres et de l'aménagement des forêts*.

Clearly these contrasts derive from well-known political and economic factors: differences in geography and natural resources, in social structure and institutions of government, in land tenure and patent law, in mobility of capital and labor, in size and unity of the market and access to it, in fiscal practice and economic theory. We take all that for granted, wishing only to suggest that the French pattern was a variant manner of economic development, not a retarded stage. Important in the variation were the expectations held of science and acted on by officials. We began with that topic at the level of ministers and matter of state in Part One. Having surveyed the institutions of science, let us come back to it at the middle level of lesser persons actually involved—scientists, producers, and administrators. We shall have to do so by means of telling examples. In the present state of knowledge, there is no way of measuring how characteristic they were or how comprehensive a difference the Enlightenment made at the mine face, so to say, or at the third and sweaty level of field and mill. Our subject is not the bread, but the leaven, a history not of the economy, but of science and its possibilities in the productive process.

In the 1770s and 1780s contemporary testimony was widespread to the effect that science was transforming the agricultural and mechanical arts, and there has been discussion recently among historians and historians of science about the nature of the influence.[2] No conclusion has been reached, but there seems to be tolerable agreement about one proposition, which may be stated as a further caveat before pursuing the matter. Whatever the interplay between science and agrarian or industrial production may have been, it did not consist in the application of up-to-date theory to techniques for growing and making things. Again the abbé Tessier makes a valuable witness. Taking up the elements of plant physiology in the same prefatory essay, he observed that

> Messrs. Lavoisier, de la Place, and Meusnier, of the Academy of Science, have just announced experimental results which prove that water is not a homogeneous fluid, as commonly believed, but a compound of inflammable and pure or dephlogisticated air. There would be no point in my discussing that discovery here, or in making any use of it.[3]

It will not do, however, to resort to the other extreme and take Tessier and his collaborators to be literate farmers writing merely of their experience. For Tessier's discussion did draw on science—the science of half a

[2] For a guide to this issue, see Musson (1972), which reprints selections from the journal literature.

[3] *Encyclopédie méthodique. Agriculture,* 1, 15. The reference is to the paper that Lavoisier read at the November 1783 public meeting of the Academy, elaborated in "Mémoire où l'on prouve par la décomposition de l'eau . . . ," MARS (1781/84), 269-283.

century before. Its theoretical categories derive from the *Vegetable Staticks* of Stephen Hales.[4] That classic work, published sixty years previously, had assumed the status of general knowledge among informed persons, much in the way that basic principles of seventeenth-century mechanics had done among Tessier's counterparts concerned with machinery. Indeed, only in the late nineteenth and twentieth centuries have findings of basic science come to be rapidly exploited in the economy. In the eighteenth century, the delay was such that it is difficult if not impossible to follow discrete discoveries into practice. Some of them merged unidentifiably into the increase of awareness; others emerged from it as the findings of scientists studying technology. For it was descriptive, and to a degree experimental, rather than theoretical science that was applied to agriculture and industry. Improvements were proposed where feasible, but essentially the approach was to analyze existing procedures, identify the best of them, and standardize the most effective methods.

The development may be seen occurring in two phases, not to be distinguished too categorically. For convenience they may be called the encyclopedic and the bureaucratic, since they amount to nothing other than aspects of the general process of permeation of culture by science in the French Enlightenment.[5] It is consistent with the over all pattern that the later of these phases should largely fall within the span of fifteen years following the Turgot ministry, when the frequency of transactions between science and government was reaching the point of regularity. The earlier phase corresponds to the literary period of the Enlightenment. We shall have to say something about it for the sake of completeness, and also because the argument will be that agriculture unlike industry could not be moved into the second phase. The pattern, agricultural and industrial, is made fully evident in the technological parts of the *Encyclopédie*,[6] though clearly it had already been formed by practice and was apprehended rather than created by Diderot.

2. DUHAMEL AND THE *DESCRIPTION DES ARTS ET MÉTIERS*

For important instances of what was involved there are no more instructive examples than those afforded by the indefatigability of Henri-Louis Duhamel du Monceau.[7] Born in 1700 and deceased in 1782, Duhamel

[4] London, 1727. Michael Hoskin has edited a reprinting, and given an introduction (London, 1961).

[5] Above, Chapter II, Section 1.

[6] Gillispie (1959a), *1*, xi-xxvi; cf. *Encyclopédie méthodique. Arts et métiers mécaniques*, 8 vols. (1782-1791), *1*, vii-xvi; and for agriculture, note specially Brandenburg (1950).

[7] Interest in Duhamel's writings on agriculture has recently begun to rescue his name from the oblivion in which the reputation of this, the most prolific member of the Academy in the eighteenth century, languished in the memory of later generations. See most

belonged to Buffon's generation, and to Voltaire's. His hallmark was neither style nor wit, however. It was usefulness. Of his writings on shipbuilding, Condorcet remarked in the academic éloge that, like all his works, they were

> immense collections of facts and experiments: everywhere he seeks to establish clearly what is the best practice; to reduce it to definite rules distinguishing it from routine; and even to bottom it on principles of physics, though eschewing any reliance on theories that had no firmer foundation than hypothesis.[8]

Condorcet observed further that Duhamel expected little prior knowledge of his readers and composed his works, not for scientists, but for persons who would put what they had learned to use. Elementary all his writings were, and to confront his pages in their thousands is to despair at first of seizing on the interest that they held, let alone of conveying it. Browsing soon dispels that mood. After a little exposure, they become hard to put down. Like the voluminous correspondence sent in by provincial physicians to the Royal Society of Medicine, they admit the reader into an unwonted intimacy with daily life and manipulations.[9] Another comparison may be made with the records assembled by that body. Its meteorological reports were inspired by the compilations of weather data noted daily at the model farm that Duhamel and his brother created in the latter's seigneury of Denainvilliers near Pithiviers just north of Orléans.[10] From 1741 through 1781 Duhamel published the yearly summary in the *Mémoires* of the Academy with remarks on the agricultural and horticultural effects of weather.

Descended from a prosperous landowning family, Duhamel was drawn by a penchant for natural history and chemistry into the botanical circle centered on the Jardin du roi in the 1720s. He had already completed an unremarkable course of study at the Collége d'Harcourt and followed it with a *licence* in law at Orléans, thus qualifying himself for administration. His hailing from the Gâtinais must have been the principal reason that the

---

notably Bourde (1967), *1*, ch. V, 253-368 and *passim*, whose bibliography locates many of his papers in numerous archives and repositories and also lists the titles of major works. It is worth emphasizing that the Library of the American Philosophical Society in Philadelphia possesses an important collection of the papers and writings of Duhamel and of his nephew, Fougeroux de Bondaroy. There are important biographical indications in the manuscript draft of a preface that Fougeroux prepared for a revision of his uncle's *Traité des arbres et arbustes*, 2 vols. (1775) that he intended to bring out in the late 1780's, but that never appeared. See below, n. 76. See also Plantefol (1969).

[8] HARS (1782/85), 131-155.

[9] Above, Chapter III, Section 6.

[10] The series appeared almost every year in MARS under the recurrent title, "Observations Botanico-Météorologiques faites au Château de Denainvilliers." Pueyo (1968) is a brief notice.

staff of the Jardin proposed his name to the Academy in 1726 to make a study of a saffron blight requested by the government. The herb was an important item in the produce of the region, a source of dye-stuffs as well as culinary and medicinal receipts. Duhamel could find nothing pertinent in the literature. When he interrogated farmers, he got only fatalistic shrugs or attributions of the rot to rottenness. Setting systematically to work, he had patches of the wilted crop turned over with digging fork and hoe, and identified a parasitic fungus, never before described botanically. He named it *tuberoïde*, placing it next to truffles in the Tournefort system. For it lived entirely underground, propagating in the springtime from bulb to bulb and destroying its host in the second season of growth. His experiments showed it to be equally damaging to other tuberous plants, to lilies, tulips, and narcissus. Duhamel prescribed ditching to stop its ravages, and recommended planting it in fields of wheat and other annual crops, which it did not attack, to see whether it might control the spread of bulbous weeds like wild onions. There is no record that anyone adopted the latter suggestion, but Duhamel's preventive measures were practicality itself and his memoir a model of enterprise and clarity.[11] He read the paper before the Academy on 7 April 1728, and it won him election later that year.

Duhamel's good sense and industry soon caught the attention of Maurepas, his exact contemporary, ever on the lookout for scientific talent that could be drawn into the uses of government. Maurepas was then in an early stage of his responsibility for the two Ministries of State, Maritime Affairs and the Royal Household, the charge of which had become virtually hereditary in the Phélypeaux family. (In one of the two main branches its scions held the titles of Maurepas and Pontchartrain, and in the other of La Vrillière and Saint-Florentin.) What with intrigues that need not concern us, Maurepas had become secretary of state for the Maison du roi, including the department of Paris, in 1715 at the age of fifteen. The altogether more important secretaryship for maritime affairs was added in 1723. Under the tutelage at first of his uncle, the duc de La Vrillière, and then of the permanent officials, Maurepas was exercising authority himself by the late 1720s. The Academy of Science came under one of his ministries and the navy under the other, and it was the young Maurepas's intent to make up in modernization and rationalization what the French fleet lacked in ships and firepower throughout the pacific and penny-pinching administration of Cardinal Fleury.[12]

Thus, it was an early instance of policy when in 1731 Maurepas invited

[11] "Explication physique d'une maladie qui fait périr plusieurs plantes dans le Gastinois et particulièrement le safran," MARS (1728/30), 100-112.

[12] On the Maurepas ministry, see Filion (1967), and on Duhamel's service in it, Allard (1968): see also Condorcet, "Eloge de Maurepas," HARS (1781/84), 79-102.

the Academy to name a commission on the provision and preservation of lumber for shipbuilding.[13] Réaumur and Mairan were chosen along with Duhamel, the junior member who was to do the work. He began with the problem of preservatives, having found in his first tour of the ports that techniques varied chaotically from shipyard to shipyard. In 1732 Maurepas appointed him to the ad hoc post of technical counselor. Soon he was making a comprehensive survey, in Dutch and English shipyards as well as French, of methods for heat-soaking timbers so that they might be bent to form the ribs. By the mid-1730s he had expanded his charge into a continuing inquest into the technology of naval architecture, and in 1739 the minister converted his post of personal adviser into the governmental position of inspector general, in which his duties were those of a high official of the state. In 1749 Maurepas had the insouciance to offend Madame de Pompadour and was exiled to his estates, to return to Paris only in 1774, the elder statesman recalled by Louis XVI and looking to science and rationality once again, in the person of Turgot.

Duhamel retained titular status in the ministry after the fall of his patron, and then embarked upon a program of further research and publication. He had already published the first of the treatises that evolved out of his official responsibilities, a manual on the fabrication of ship-rigging.[14] Of that, more in a moment. From there, his interests branched out in three directions. One line, issuing from this treatment of ropemaking, led directly to the realization of the Academy's long overdue *Description des arts et métiers*. A second grafted his investigation of the supply and fashioning of naval lumber onto his landowner's experience in a series on forestry and sylviculture, commercial and ornamental. The third combined the farmer of his own domain with the academic botanist in the writing of memoirs and textbooks of rational agriculture, or agronomy. He also published on naval architecture and on the preservation of health on shipboard and in hospitals,[15] but however well informed, those books were written more at second-hand, and the partition just given will serve to organize a summary of the way in which he thought to bring his proper studies to bear experientially on the improvement of manufacturing and farming.

There is no more instructive example of an eighteenth-century descrip-

[13] Allard (1968), 24 n.16.

[14] *Traité de la fabrique des manoeuvres pour les vaisseaux, ou l'Art de la corderie perfectionné* (1747).

[15] *Eléments d'architecture navale, ou Traité pratique de la construction des vaisseaux* (1752); *Moyens de conserver la santé aux équipages des vaisseaux, avec la manière de purifier l'air des salles des hôpitaux et une courte description de l'hôpital Saint-Louis à Paris* (1759). A translation of the former under the title *A Treatise on Shipbuilding and Navigation* (London, 1754) was reissued in 1765.

tion of an art or trade than *Traité de la fabrique des manoeuvres pour les vaisseaux, ou l'Art de la Corderie perfectionné*, which Duhamel brought out in 1747. Since the writing of books like this ceased with the disappearance of the trades they were meant to reform—the last notable instances being the manuals got out by members of the Academy to serve the revolutionary effort of war production in 1793-1794[16]—it may be well to describe the work on cordage. What is immediately apparent is that it is a very handsome book, itself a credit to the typographer's art. The first edition is a sturdy quarto of some 450 pages, printed on fine paper and beautifully illustrated. The pictures are altogether superior in detail, clarity, and liveliness to most of those in the justly famous set that Diderot commissioned, and sometimes cribbed, in assembling the eleven volumes of plates that accompanied the *Encyclopédie*.[17]

Duhamel's book antedated by fifteen years the appearance in 1762 of the first four of Diderot's volumes, and the arrangement employed by Diderot's engravers was already standard in Duhamel. Tools and machines, and in this case cross-sections of hawsers and cables, are represented in breakdowns and cutaways in one section of the fold-outs and in supporting diagrams, while each successive process is pictured by a vignette of the appropriate shop in operation. In several instances, a by-line attributes the design and engraving to the artist Soubeyran. So natural are the scenes of carding hemp, of making thread, or of combining the cords, that the reader can readily imagine himself stepping right into the rope-walk and lending a hand with comb, spinning-wheel, or top. It is unlikely he would be so deft as the children who keep the lines from fouling. He would then understand the reason for what is puzzling in the headpieces that show façades in the navy yards of Brest, Rochefort, and Marseilles. For the frontal elevations appear to run the whole length of the harbor, more like galleries than workshops. And indeed, the ropemakers' installations did consist of parallel sheds, up to a thousand feet long by twenty to twenty-eight feet wide, one for spinning and one for laying ropes and cables. At both stages, the twisting could be executed evenly only if the lines were fully extended before being wound onto reels.

None of the workers whom Soubeyran depicts could themselves explain in a perspicuous manner the processes in which they were engaged. When Duhamel would question them, they would reply "coldly" that things had long been done that way in their shop because it was the best way.[18] If the shop happened to be one where the crude hemp was cleaned by

[16] There would be no point in citing full titles here. Treatises by leading scientists were published on iron-working, manufacture of small arms, casting and boring of barrels for artillery pieces, production of potash, the soda industry, tanning, the separation of copper from bell metal, etc. Most of these treatises are gathered in AN, AD VIII, 40.

[17] Gillispie (1959a), *1*, xxi-xxii.

[18] Duhamel, *Art de la corderie*, iii.

"scutching" or beating, it was because the fibers were "enervated" by the comb used where carding was the preferred method. If the yarn was then spun coarsely, the purpose was said to be to keep it loose to withstand the strain of twisting; in another establishment, where usage favored fine-spun threads, the reason was that they would pack more closely together. In a word, all Duhamel could get was jargon in lieu of answers, and it was borne in on him that he would need to make his own tests in order to find out what really worked best. He soon perceived that it was more than a question of solving a few problems. The industry was constitutionally incapable of standardizing its own procedures. The thing had "to be taken hold of in its very principles. The ropemaker would need to be followed in every step, and his entire art reconstituted in order to subject it to definite rules."[19]

Clearly, Duhamel would need help, and the most interesting aspect of the story is that, beginning in 1737, he assembled a research group. He started in the navy yard at Marseilles with a young ensign, the comte de Pontis, a mathematically well-trained officer who enjoyed research. In 1739 Duhamel brought Pontis to Paris, and had him elected a correspondent of the Academy of Science, where he attended meetings and pursued his study of the trade under Duhamel's supervision. At Brest, Duhamel enlisted the services of Squadron Commander de Radouay and of another mathematically qualified line officer, a certain Captain Dervaux. Dervaux collaborated in the testing program with a naval engineer, one Ollivier, and with Lieutenant Beaussier of the Port Authority, who was extremely intelligent, energetic, and good with his hands. In 1740 Duhamel moved the seat of operations to Rochefort, where also the ministry authorized facilities, funds, and personnel.[20] In the ten years that the project occupied, he and his associates inspected every step and tested every alternative in the production of cordage, from the planting of the hemp-seed to the warping of the great cables called *grelins* and the tapered lines called rat-tails.

The resulting treatise bespeaks the authority of expertise on every page. Three chapters deal with the planting, cultivation, and supply of hemp, of which the best quality came from Latvia. Standards were prescribed that the raw hemp had to meet before being accepted to be processed in the navy yards. For this is a reforming book. Duhamel set the practitioners straight at frequent intervals—to begin with for having attributed masculinity to the female plant through ignorance of botany.[21] The body of the book is concerned with manufacture—two chapters on scutching and carding, one on spinning, and five on methods for combining threads into cords, cords into strands, strands into hawsers, and hawsers into ca-

[19] *Ibid.*, v.
[20] *Ibid.*, v-ix.
[21] *Ibid.*, 6-10.

bles. Each chapter is conceived on the model of an academic memoir about a particular set of procedures. Every operation is subjected to analysis and testing. Generally speaking, the purpose was to determine in each case the optimal trade-offs between economy, bulk, strength, flexibility, and durability. In chapter VIII on hawsers, for example, Article 14 explains that, although the two-strand bitord and the three-strand merlin are twisted counterclockwise like the threads that compose them in order to acquire elasticity, these same cords should then be twisted in the opposite direction when combined to form hawsers. The reason was that the property now required was tensile strength. To measure the difference, Duhamel tested a hawser composed of twenty-four cords twisted to the point that it was two-thirds the length of the constituent cords. It broke under a weight of 1410 pounds. The cords from the longer end were then unraveled and combined in a hawser "*à maintorse*"—that is, with all the twist in the same direction. It broke under a load of 1190 pounds.[22]

Perhaps that one example of the hundreds of tests reported as "experiments" will suggest that this was no mere manual of handicrafts. It is a book of experimental technology, concerned with determining the strength of materials empirically. Duhamel made a point of conducting the tests as public demonstrations, in the presence of large complements of officers in the important naval bases. For he reckoned with the need to overcome resistance and skepticism embedded in the existing enterprises. Supported by Maurepas, the project moved past research and testing into development. The last chapters describe how the shipyards at Marseilles and Brest produced rigging in conformity with the new standards, to outfit the cargo vessel *Charente* out of Rochefort and the frigates *Mercure* and *Amazone* out of Brest. Under the command of Captain de l'Étanduaire, the *Mercure* ran into a hurricane off Martinique on 9 September 1740 and suffered far less than either of its companion ships, both outfitted with conventional rigging. One misfortune marred the outcome of all these trials in action: the comte de Pontis was killed in an engagement with a British squadron in those same waters in 1745.[23]

The program itself was judged a success in the ensuing decades. It had originated in the complaints about rigging with which naval officers greeted Duhamel on his initial tours of inspection. Ships were top-heavy from weight of tackle; lines were too stiff to run through pulleys; hawsers kinked incessantly, smashing blocks and crippling sailors.[24] The modifications he now introduced made it possible to produce rigging stronger than the old standard by a quarter and lighter by an eighth. A looser twist would have increased strength and lowered weight, but at the price of lessening durability. The new hawsers were so much more flexible than

[22] *Ibid.*, 196-203.

[23] *Ibid.*, 422.

[24] *Ibid.*, i-ii.

the old that one-third fewer hands were required to man the capstans. Kinking was eliminated. Since a ship of 64 guns carried 680 hundred-weight (about thirty-five tons) of rigging above the waterline, the improvement in stability was significant. As workers became habituated to the new methods, these gains were increased.[25]

By that time, late in the 1750s Duhamel was deeply engaged in arranging for the Academy's *Description des arts et métiers*, even while carrying forward programs in sylviculture and agronomy. The publication consists of some seventy-four treatises, printed between 1761 and 1782.[26] The exact number depends on whether certain titles are considered to be independent or subordinate, or in several cases to belong to the set at all. Taken together, they comprise the largest body of technological literature that had ever been produced. Few technical collections can have been more often mentioned by historians and less often, one will not say read, but even leafed through. Miscellany might be a more appropriate term for it than collection. Individual works range in scale from an eleven-page pamphlet on the preparation of starch to a 1356-page masterpiece on the coal industry. There are very few libraries that have the whole run, and no two where they are assembled into volumes in the same sequence or catalogued with anything like uniformity. To attempt to do justice, or even injustice, to the content would obviously be desperate. It is important to be clear about the evolution and certain features of this enormous venture, however, for it was not quite what is commonly supposed, and in certain respects its relation to the Academy is more interesting than the stereotype of forward-looking science modernizing the arts.

The first qualification must be that the Academy's commitment was little more than formal. Of over forty contributors, only eight were regular members. No single leader in the generation of Lavoisier, Laplace, and Condorcet was among them. The most respected of the participants from the earlier generation was Macquer, who is represented by one treatise, on dyeing silk.[27] Drawing on his experience as inspector for the Bureau of Commerce, he carried on when his predecessor and patron, Jean Hellot, who had begun the work, fell ill. Similarly, Pierre-Charles Le Monnier wrote out of his specialty, composing a manual on astronomical instruments.[28] Gabriel Jars supplied a brief account of Dutch methods for

[25] *Ibid.*, xxxii-xxv.

[26] Cole and Watts (1952) give details of bibliography and an inventory of the set, with special attention to its availability in American libraries. J.-E. Bertrand, printer in Neufchâtel, produced a separate edition in 19 volumes (1771-1783). Watts includes the table of contents in his appendix B (36-38). His appendix C on the contributors is to be treated with caution.

[27] *Art de la teinture en soie* (1763).

[28] *Description et usage des principaux instruments d'astronomie* (1774).

tiles and bricks,[29] the abbé Nollet of millinery,[30] and Paul-Jacques Malouin of milling, making noodles, and baking.[31] The mammoth treatment of the coal industry was by Jean-François Morand,[32] librarian of the Academy, whose election in the section of anatomy in 1759 owed more to his father's reputation in surgery than to anything he ever contributed to his titular science.

For the rest, all the academic pieces came from Duhamel himself, from his faithful nephew Fougeroux de Bondaroy, and from Lalande, both of whom he pressed or tempted into service. Duhamel asked no one to try what he was unwilling to do for his own part. Among the twenty trades he described were those producing paste and glue, tobacco pipes, drapery, Turkish carpets, sugar, candles, lamps, charcoal, felt, wire, pins, wax, roofing, earthenware, soap, locks, and ironware.[33] The most astonishing of his productions is a four-volume study of the fishing industry.[34] The first volume deals with boats, tackle, and technique, quite in the manner of his work on shiprigging for the navy and in comparable detail. The remaining three constitute an ichthyology of all the species of commercial value brought into port by French fishing fleets or caught in internal waters. As for Lalande, between 1761 and 1765 he took time from astronomy and atheism to write nine of these manuals, on the trades of Morocco leather, Hungarian tanning, tawing, ordinary tanning, chamois, currying, cardboard, parchment, and papermaking. The work on paper had sufficient success that it was re-issued in 1820.[35] Finally, Fougeroux produced four descriptions for his uncle, on slate-quarrying, gilding and silvering leather, gross cutlery, and barrel-making.[36] The remaining forty-odd entries were farmed out to amateurs, artisans, manufacturers,

[29] *Art de fabriquer la brique et la tuile en Hollande* (1767), a sequel to Duhamel *et al.*, *L'art du tuilier et du briquetier* (1763).

[30] *L'Art de faire des chapeaux* (1763).

[31] *Description et détails des arts du meunier, du vermicilier et du boulenger, avec une histoire abrégée de la boulengerie et un dictionnaire de ces arts* (1767).

[32] *L'Art d'exploiter les mines de charbon de terre* (issued in four fascicles, 1768-1779); on Morand and his father, Sylvain-François Morand, see the Condorcet éloge, HARS (1784/87), 48-53.

[33] See Cole and Watts (1952), appendix A.

[34] *Traité général des pesches, et histoire des poissons qu'elles fournissent, tant pour la subsistance des hommes, que pour plusieurs autres usages qui ont rapport aux arts et au commerce*, 4 vols. (1771-1782).

[35] *L'Art de faire le papier* (1761; new ed., 1820). Oddly enough, an English translation was published in an exceedingly expensive edition in 1976. See the notice by Colin Cohen, *Times Literary Supplement*, 30 June 1978, p. 748. The motivation may have been the plates. The text is less authoritative than the later work of Desmarest (see below, Chapter 6, Section 3, n. 169).

[36] For details of Lalande's other entries and of Fougeroux's, see Cole and Watts (1952), appendix A.

and other enthusiasts. It is true that their books were passed upon in the normal fashion by ad hoc committees of the Academy.

The second somewhat surprising feature emerging from perusal of the *Description des arts et métiers* is that, however beautiful the engraving of the plates, the technical accounts and illustrations were not always up to date, and those stemming from the academic authors tended to be the least so. To understand why that should have been, it will be necessary to reconstruct briefly how the work came to exist at all.[37] Although Colbert had certainly had improvement of the mechanical arts in mind at the founding of the Academy of Science, the earliest formal record of a scheme for systematically surveying trades and crafts occurs in a passage of the *Histoire* for 1699, the year of reorganization under direction of the abbé Bignon.[38] Shortly thereafter, Réaumur began laying the actual groundwork. A member of the Academy from 1708, Réaumur took on responsibility for the enterprise, publishing on the drawing of gold into wire and thread in 1713 and issuing his famous study of the cementation process for steel in three memoirs read from 1720 through 1722 and collected in the latter year.[39]

Réaumur's career was the type, perhaps the archetype, on which Duhamel modeled his own. The older man's studies were both fundamental and extensive in metallurgy, in ceramics, in thermometry, in ornithology, and in entomology.[40] Such an eclectic and celibate pattern (for neither one found time for marriage) was the highest norm to which a man of science might aspire in the first half of the eighteenth century, and to a degree throughout the encyclopedic phase of the Enlightenment. The old Baconian ideal failed to inspire emulation, however, after specialization and differentiation by discipline began to be the order of the day. No doubt that is why the rising lights of the Academy left it to a Duhamel to bring off the *Description des arts et métiers*. When Réaumur died in 1757, his papers contained a vast mass of technological information that he had assembled over the previous forty years and more. There were partially drafted memoirs and treatises, some almost completed; there were reports and details sent in from all over France and all over Europe; there were over 200 copper plates depicting the operation and design of workshops,

[37] There is an account in Lalande, *L'Art du tanneur* (1764), vi-viii, and a prospectus, issued separately in 1759, and printed in the front-matter of Morand (n. 32 above).

[38] HARS (1699/1718), 117-119.

[39] *L'Art de convertir le fer forgé en acier, et l'art d'adoucir le fer fondu . . .* (1722) has been translated by Annelie Grünhaldt Sisco, with notes and introduction by Cyril Stanley Smith, *Réaumur's Memoirs on Steel and Iron* (Chicago, 1956). For the art of gold thread, see "Expériences et Reflexions sur la prodigieuse ductilité de diverses matières," MARS (1713/16), 201-222. Réaumur compares the production, properties, and dimensions of gold thread, spun glass, silk, and spiderwebs.

[40] On Réaumur's career and writings, see Torlais (1958), and J. B. Gough in DSB *11*, 327-335.

tools, and machines in many trades. Here in the engravings that Réaumur commissioned was the main fount of the genre of technical illustration. They date mostly from 1710 to 1730, though several, which he must himself have inherited, go back to the 1690s.

The result was that when Duhamel finally got the designs into print, they were half a century behind the practice of the 1760s and 1770s. The same is true in some measure of the texts. According to Lalande's account, immediately after Réaumur died, the Academy had his papers distributed among twenty of its members, having taken the resolve to complete, revise, or supplement the descriptions as the case might require.[41] It has long been supposed, and reasonably enough, that the Academy felt its hand forced by Diderot's plan for the *Encyclopédie*.[42] Its subtitle is *Dictionnaire raisonné des sciences, des arts, et des lettres*. Publication had begun in 1751, and the first volumes of plates, largely technical, were known to be imminent, a prospect that would have stolen the Academy's lightning along with its thunder. However that may have been, and whatever may have happened to the dozen colleagues whose engagements Duhamel evidently had to fulfill himself, he got his first item, on charcoal burning, off the press in 1761, a year ahead of the first Diderot plates. That memoir is typical of the treatment accorded the more obvious trades. Printed for the most part in the 1760s, the accounts were based on materials bequeathed by Réaumur, which his scientific executors worked up as best they might, in principle by going out to study the processes at first hand, in part by engaging leading craftsmen to be informants, guides, and sometimes collaborators. Most of these descriptions are relatively summary, consisting of twenty to thirty pages of text and six or eight plates. All were printed in a quarto format so that appropriate combinations might be bound together.

Once Duhamel and his colleagues had arranged to salvage all they could of Réaumur's legacy, the emphasis shifted. For the rest, the *Description des arts et métiers* consists of treatises written, not by ordinary members of the Academy, but by persons whose knowledge of their subjects derived from their own experience, either vocational or avocational, and in either case specialized. Thus, certain officers and noblemen supplied essays from their hobbies. Charles-René Fourcroy de Ramecourt, a distant cousin of the chemist, wrote a piece on lime-burning and also contributed to the account of tiles and brick-laying.[43] He was then a colonel of infantry and

[41] Lalande, *L'Art du tanneur*, vi.

[42] Gillispie (1959a), *1*, xxi-xxii. The Duhamel papers in the library of the American Philosophical Society contain an interesting letter from Charles Bonnet in Geneva to Duhamel (18 October 1760) concerning the theft of manuscripts of Réaumur, a loss for which Duhamel apparently laid at least part of the responsibility at the door of Buffon, with whom he was on chronically bad terms. Lalande evidently played the part of talebearer.

[43] *Art du chaufournier* (1766); see also n. 29 above.

chief engineer at Calais, rising later to be commander of the Royal Engineering Corps. Another colonel, Galon, in charge of engineering works at Le Havre, handled coppersmithing.[44] The duc de Chaulnes, amateur of scientific instruments and honorary member of the Academy of Science, composed two treatments, one of the microscope and micrometers and the other of measurement.[45] Similarly, the marquis de Courtrivon participated in the first two sections of the description of the iron trades. The third section consists of Réaumur's technique for softening cast iron, and the fourth is a translation of Emmanuel Swedenborg's *Regnum subterraneum sive minerale de ferro* (1734), rendered into French by Courtrivon's collaborator in the earlier sections, one Bouchu.[46] Like any promoter of a reference work, Duhamel put together what pieces he could find as best he could, availability signifying more than consistency of design or of provenance. Thus also, the comte de Milly's treatise on porcelain originated as a memoir on the industry in Germany read before the Academy of Science in Paris on 13 February 1771.[47] Duhamel thereupon prevailed on Milly to do the whole craft for the *Description des arts et métiers*.[48] The author drew his information from a program of inquiry, observation, and experiment that had occupied the leisure hours of a military career in the previous ten years.[49]

After 1770 or 1771, Duhamel turned increasingly from scientists, officers and gentlemen to artisans, manufacturers, and entrepreneurs. On the artisanal side were Jean-Jacques Perret, master-cutler whose establishment was in the rue de la Tisseranderie in Paris, and our old friend and Lavoisier's future enemy, Jean-François Demachy, master-apothecary and apprentice pornographer. Perret produced a two-volume work, the first volume dealing with domestic cutlery and the second with surgical instruments.[50] Demachy gave two books on the distilling trade, both bristling with polemic, the first on alcohols, acids, and liquid reagents for chemical industry, and the second on brandies and cordials for household consumption.[51] Similar in standing, though altogether more impressive

[44] *L'Art de convertir le cuivre rouge ou cuivre de rosette, en laiton ou cuivre jaune* . . . (1764). This memoir contains an addendum by Duhamel and an extract from Swedenborg.

[45] *Description d'un microscope, et de différents micromètres* . . . (1768); *Nouvelle méthode pour diviser les instruments de mathématique et d'astronomie* (1768).

[46] *Art des forges et des fourneaux à fer* (1762).

[47] "Mémoire sur la porcelaine d'Allemagne, connue sous le nom de porcelaine de Saxe." The report of the commissioners, Lassone, Macquer, and Sage (20 February 1771), recommended that this paper be published in the *Savants étrangers*. Instead, Milly incorporated it in the treatise. See the Foreword, viii-xi.

[48] *L'Art de la porcelaine* (1771).

[49] On porcelain and Sèvres, see below, Chapter VI, Section 2.

[50] *L'Art du coutelier* (2 parts, 1771-1772).

[51] *L'Art du distillateur d'eaux-fortes* . . . (1773); *L'Art du distillateur liquoriste* . . . (1775); on Demachy, see above, Chapter IV, Section 4.

in scale, temper, beauty, and detail, is the four-part treatise on cabinet-making by Roubu fils—André-Jacob Roubu—whose draftsmanship was the equal in point of artistry and geometrical sophistication of the work of the most highly educated architect or engineer.[52] His attainments were an example of the level of technical training available, in this case under the architect Jacques-François Blondel, to the ablest people in the traditional trades of eighteenth-century Paris. Tending more to the industrial is Jean Paulet's three-volume treatise on the manufacture of silk, drawn largely from his own experience in Nîmes.[53]

From among the entire collection, three treatises stand out for permanence of interest, one in the history of industry—Jean-François Morand's *L'Art d'exploiter les mines de charbon de terre*[54]—the other two in the history and practice of fine arts, namely the fabrication of stained glass and the manufacture of pipe organs. It is unfortunate that Morand's massive work on coal mining is so little known. The reason may lie in a tendency to think of Britain as the classic locus of eighteenth-century mining, given the importance of coal for provision of power in the early stages of industrialization. There is no comparable treatment in the British literature, however. Moreover, Morand's description was based in part on observation in the coal-fields of the Newcastle region, as well as in French mines and in those surrounding Liège. Except for this work, the author might be thought to have been a dilettante. Among his other writings were a dissertation submitted to the Faculty of Medicine on the heritability of heroism, a memoir on a well-known set of caves, another on the natural history of certain mineral waters, and an essay on the anatomy of the thymus.[55] It is true that he was a pioneer in the study of population, having assembled vital statistics for the city of Paris covering the years from 1709 through 1770,[56] and true also that Laplace drew on these figures for his first application of probability to demography.[57]

[52] *L'Art du menuisier* (4 parts, 1769-1775). The topics are carpentry, carriage-making, furniture, paneling and inlay, trellises and garden furniture.

[53] *L'Art du fabriquant d'étoffes en soie* (7 fascicles, 1773-1778).

[54] 4 fascicles, paged consecutively, 1768-1779. Lavoisier's very favorable report to the Academy on publication of the second part was printed in Rozier's journal, *Observations sur la physique* 2 (1773), 68-77.

[55] *Quaestio Medica: an ex Heroibus Heroes* (1757); "Recherches anatomiques sur la structure et l'usage du Thymus," MARS (1759/65), 525-537; *Nouvelle description des grottes d'Arcy* (Lyons, 1752). "Mémoire pour servir à l'histoire naturelle et medicale des eaux de Plombières," SE 5 (1768), 128-132.

[56] "Récapitulation des baptêmes, mariages, mortuaires, & enfans-trouvés de la ville et faubourgs de Paris, depuis l'année 1709, jusques & compris l'année 1770," MARS (1771/74), 830-848. According to Condorcet in the éloge (*loc. cit.* n. 32), Morand intended to institute a decennial survey of this type.

[57] Laplace, "Sur les naissances, les mariages, et les morts à Paris . . ." MARS (1783/86), 693-702; see above, Chapter I, Section 5, n. 138.

In addition to population, Morand made the study of coal mines and the coal trade a serious occupation. The treatise issued from the press in four installments dealing, respectively, with the geology and mineralogy of the coal measures; with methods of mining and organization of the commerce in coal throughout Europe; with the state of the industry in France; and, finally, with the practical as well as the theoretical usability of coal in factories, workshops, and households. The reader's most unexpected reward is a digression on steam engines, in effect a parade of industrial dinosaurs about to face extinction through James Watt's invention of the separate condenser in 1776.[58] Morand gave details of the great Newcomen engine pumping water from the Thames through London from a location in York Buildings. The other English machines he had examined were serving mines near Newcastle. He compared their specifications and effectiveness with French counterparts, an installation at Fresnes, and also mining pumps at Montrelay on the border between Anjou and Brittany and in French Hainaut near the present Belgian border. The English phrase "steam engine" seemed to him more apt than "*machine à feu*," and his plates could well have served the needs of contractors called on to duplicate these constructions.

Even if there were no other reason, the gratitude of the historian of science would go out to the editorial largesse of Duhamel and his colleagues for having included in the *Description des arts et métiers* Pierre Le Vieil's charming work on stained glass and the magnificent treatise on building organs by Dom François Bedos de Celles.[59] Persons interested in science might not otherwise come upon these idiosyncratic and beautiful works, which should be known to everyone of whatever specialty who cares about the variety and quality of talents to be met with in the France of the eighteenth century. Published posthumously, Le Vieil's book on stained glass is the slighter and the more personal a legacy. It was not originally intended for the academic series. He composed it in his old age, the last surviving master of a craft fallen into desuetude, recalling its glory, extolling its merit, expounding its techniques, defining its integrity, pleading for its resurrection. His intention was only to dedicate it to the Academy of Science. The commissioners (it is unclear at whose initiative) then accepted the treatise along with the hommage, on condition that a companion memoir that Le Vieil had prepared on glazing be combined with it and that the whole be printed in the double-column format of the *Description*.[60]

Like André-Jacob Roubu, author of the description of cabinet making,

[58] *L'Art d'exploiter les mines de charbon de terre*, 1049-1106.

[59] *L'Art de la peinture sur verre et de la vitrerie, par feu M. Le Vieil* (1774); *L'Art du facteur d'orgues* (4 parts, 1766-1788), paged consecutively.

[60] The commission's report, signed by Duhamel du Monceau, Sage, and Lassone and dated 24 May 1772, is printed in the front-matter.

Le Vieil affords an example that certain of the most accomplished craftsmen needed to fear no comparison in point of culture with contemporaries in other, generally thought more eligible, walks of life. He came of Norman stock, his ancestors having been in the staining of glass for over two centuries.[61] His father had ventured up to Paris from Rouen at the age of nineteen. Related to the famous painter, Jean Jouvenet, Guillaume Le Vieil came to the notice of Jules Mansard, the foremost architect in the favor of Louis XIV. Mansard gave the young man a commission for the glass in the frieze decorating the chapel at Versailles and, even more notably, for windows let in to his masterpiece, the dome of the Invalides. In 1707 Guillaume married the daughter of the leading glazier with whom he worked, one Favier. Of his eleven children, he intended Pierre, the second son, for the church, and was in a position to arrange for the best of formal educations. The lad had his secondary schooling at the Collège de Sainte-Barbe, completed his studies at the Collège de la Marche—where (according to his anonymous biographer) "the élite among young noblemen scintillated in those days"[62]—and entered the Benedictine Abbey of Saint-Vandrille as a postulant. Then his father fell ill, and Pierre renounced the prospect of holy orders so that he might take over the shop and look after his brothers and sisters and their mother.

What with the collapse in demand for stained glass, the business depended on the glazing trade. Le Vieil was saved for higher things in that his education had fitted him for scholarship. He never married. Even in the practice of his craft, his reputation was more scholarly than creative. For it derived from his restoration of the windows in the ossuary of the church of Saint-Etienne-du-Mont, in the Abbey of Saint-Victor, and in Notre Dame. Fortunately, he had a passion for research and was provided with the skills. Part I of *L'Art de la peinture sur verre* consists of the first history of stained glass ever written. How the treatment stands up in the light of later scholarship need not, and indeed cannot, here be judged. It is fascinating eighteenth-century antiquarianism, however. One whose knowledge of stained glass is that of the twentieth-century tourist will learn, to begin with, that the French "*peinture sur verre*" is in fact more descriptive than "staining." Unlike other modes of painting, the art at first consisted in employing refractory pigments to delineate tracery in opaque tones so as to define the figures when the colored segments were re-fired and pieced together into leaded panes. He will also learn, what might have been obvious, that the early windows were adaptations of the ancient art of mosaic, to which Le Vieil had devoted an earlier study.[63]

[61] The main sources of biographical information are Le Vieil's own account of the work of his family in ch. 16 and an "Eloge Historique de Pierre Le Vieil," by "M. S.* * * * * Avocat au Parlement," and a friend. It is printed as a preface to the treatise.

[62] "Eloge," vii, n. 61 above.

[63] *Essai sur la peinture en mosaî que* (1768).

The details have nothing to do with our story, but it is difficult to resist following Le Vieil into the earliest mention he could find of ordinary window glass, in a passage of Lactantius's fourth-century *De Opiscio Dei*: "Verius et manifestius est mentem esse, quae per oculos ea quae sunt opposita transpiciat, quasi per fenestra *Lucente Vitro* aut speculari lapide obductas."[64] Or into the division of labor of such a workshop as produced the glass for the Sainte-Chapelle in five years, between 1242 and 1247. Or into the elaborate privileges accorded the corporation of artists by letters patent of Charles VII in 1430. The temptation is less strong to penetrate into the atelier of Jean de Connet, a sixteenth-century artisan and acquaintance of Bernard Palissy so afflicted with bad breath that his colors would never hold fast. Le Vieil gives a chapter each to the distinctive properties and merits of the glass of the eleventh and twelfth, thirteenth, fourteenth, and fifteenth centuries. His taste is not ours, however. He reserved his greatest enthusiasm for the sixteenth century, when Italian design was joined with the craftsmanship developed by French artisans in the five centuries since stained glass had been the major, and for long the single, genre of French painting. Only then, in the application of color in enameled form to a substratum of clear glass, did the art of the painter entirely predominate over the skill of the glazier, here reduced to provision of a receptacle.

The second, technical half of the treatise amounts to a manual for would-be practitioners. Though never so clear and explicit, it rings a little hollow, for there were none. The affectation imposed by the Academy in the subtitle of treating the art "in its chemical and mechanical aspects" gave an excuse, if any were needed, for it to adopt the treatise. But Le Vieil's heart had been in his history and in his summons to a renaissance. He placed great hopes in the inauguration, under the encouragement of Sartine as lieutenant-general of police, of an Ecole gratuite de dessin[65]—with reason, since it turned out to be a forerunner of the modern Beaux-Arts.

However ignorant, the historian has no hesitation in attributing the highest degree of intrinsic merit to *The Art of the Organ Builder* by Dom Bedos de Celles.[66] For knowledgeable persons have expressed their judgment in the most convincing way. No other items in the *Description des arts et métiers*, and few if any eighteenth-century technical treatises on any topic, have been so much translated and so often reprinted, most recently in 1963.[67] It has also received what may be the surest of accolades denot-

[64] "What is perceived through our eyes appears true and manifest to the mind as if seen through windows furnished with glass or transparent crystal." Quoted in Le Vieil, *L'Art de peinture sur verre*, 10, n. *a*.

[65] Preface, vi, citing the *Mercure de France* (Jan. 1770), II, 160.

[66] *L'Art du facteur d'orgues* (pt. I, 1766; pts. II and III, 1770; pt. IV, 1778).

[67] The earliest foreign recognition was an analysis by Johann Samuel Halle in *Die Kunst*

ing a masterpiece: its authorship was disputed some three decades after Dom Bedos's death in 1779 and attributed to the organist of Saint-Germain-des-Prés, Jean-François Monniot, who died in 1797.[68] Not only was it clearly the first comprehensive work on pipe organs ever written, quite possibly it remains the best. In a way appropriate to the instrument that forms its subject, it combines a classic simplicity in the content with a baroque splendor in the appearance. Not for Dom Bedos the straitjacket of double columns—his format is a roomy folio spreading the print on sheets that folded to twelve by eighteen inches. Though the parts were published at intervals, the pagination and numeration of the illustrations are continuous, running to 676 pages and 137 plates. The most magnificent of the fold-outs portrays the great organ in the abbey of Weingarten in Swabia, with its 6,666 pipes, its 66 stops, and its 32-foot display pipes flanking the console and awing the congregation. Dom Bedos had perfected his knowledge through the study of that instrument, immediately following its completion in 1750 by Joseph Gabler of Ravensburg.

A Benedictine of the congregation of Saint-Maur in Toulouse, where he had taken his vows in 1726 at the age of seventeen, Dom Bedos was of the Languedocian gentry.[69] He built his first organ for the church of Sainte-Croix in Bordeaux in 1748, and is credited with instruments in Clermont-en-Lodève, Le Mans, and Montpellier. For the most part, however, he seems to have been a consultant residing at Saint-Denis after 1762. Ecclesiastical authorities would retain him to oversee the planning, execution, and sometimes the repair of organs, most notably those serving the cathedrals of Autun, Lodève, Narbonne, Sens, and Tours. Sundials were an accessory interest. He designed and mounted a huge vertical timepiece on the basilica of Saint-Denis, after having published *Le Gnomon pratique* (1760). A second, enlarged edition was ready for printing at the time of his death.[70] Although it was addressed to readers who were largely innocent of mathematics, a severe foreword instructs them that they must read the work pen in hand, sine tables on the desk, and the will to work the problems. The publisher's preface distinguishes the manual from an-

des Orgelbaues (Brandenburg, 1779). Johann Cristoph Vollbedig, *Kurzgefasste Geschichte der Orgel aus dem Französich des Dom Bedos de Celles* (Berlin, 1793) is an abbreviated translation; and Jan van Heurn, *De Orgel Maker* (1805-1806), is a Dutch adaptation. The first complete translation was J. S. Topfer, *Lehrbuch der Orgelbaukunst* (Weimar, 1855). The work has been twice reprinted in Paris by the firm of Roret, in 1849 and in 1903. A facsimile edition was brought out in Cassel in 1934-1936, and reprinted with notes by Christhard Mahrenholz in 1963. It forms volumes 24-26 of the series Documenta Musicologica published in Cassel by Bärenreiter for Internationale Gesellschaft für Musikwissenschaft.

[68] For an account of this episode, see the "Begleitwort" by Mahrenholz (n. 67) appended to his edition of vol. I, 6-7.

[69] Raugel (1919) gives a brief notice.

[70] *La Gnomonique pratique, ou l'art de tracer les cadrans solaires avec la plus grande précision, par les méthodes qui y sont les plus propres, et le plus soigneusement choisies en faveur de ceux qui sont peu ou point versés dans les mathématiques* (new ed., 1780).

other with the same title that a Monsieur Garnier had recently brought out in Marseilles, and that would do well enough so long as accuracy was of little moment.

Accuracy reaching to all details was of supreme moment to Dom Bedos. The four parts of his Summa on the organ builder's art deal, first, with the design of large organs; second, with the builder's procedures; third, with the organist's responsibility for care and maintenance; and fourth, with the production of small organs for orchestral and domestic use. The tiniest parts are depicted. The most recondite tools are shown. The most specialized tasks are explained. At the same time, Dom Bedos never lost sight of the work flow. The architect who fails to consult the organ-builder before designing an emplacement is severely chided. The organist is charged not to usurp the artisan's function. Persons, like parts, are to keep their places. Prospective entrepreneurs are supplied with twelve graduated sets of specifications—"*Devis*" in the terminology of French contracting—for organs of as many sizes and purposes. They range from a great affair showing 32-foot pipes to an instrument suited for a village church. All that would have been required to convert these schedules to contracts were the signatures of builders and vestrymen. The concluding part on organs to be played at home was an afterthought, composed for amateurs at the suggestion of his "illustrious" patrons of the Academy. A publisher's note of 1770 promised it for the following year, but Dom Bedos got it out only in 1778. He accompanied it—marvelous to relate about French publishing of any century—with an index, and also with a preface to the whole opus. That was written last, once it was definite what had to be introduced, and it contains a brief history of the organ. Unlike stained glass, Dom Bedos's was a flourishing art, however, and the history is the only perfunctory thing in this fine book.

The last of the descriptions to be mentioned lies on the near side historically of a transition from the practitioner or amateur who loved his art to the government official who would rule it with a rod of freedom, punishing it for its own good with the scourge of competition and regularly consulting with scientists who further prescribed large doses of publicity to cure it of servility to the dictates of routine. The *Description des arts et métiers* was drawing to a close in the late 1770s, and the Academy acceded to pressures emanating from the Bureau of Commerce to issue under that rubric several memoirs by one who, what with the unexpectedness that attended personal destinies in the Revolution, ultimately became the most famous and least fortunate of our authors. The treatment of the woolen industry and of velvet came from the pen of Jean-Marie Roland de la Platière, inspector of manufacturing for the generality of Picardy with his seat in Amiens.[71] Perhaps it is fanciful, and certainly it is anachron-

[71] *L'Art du fabricant d'étoffes en laines, rasées et sèches, unies et croisées* (1780). Roland con-

istic, to sniff the scent of the Girondins in the opening lines of his *L'Art du fabricant d'étoffes en laines*. Clearly, however, the spirit is different from that in which Dom Bedos and Pierre Le Vieil, or the comte de Milly and Jean-François Morand, wrote of their enthusiasms. Roland's "*Avertissement*" challenges the reader with the question, so much tossed about, whether it was advantageous to a nation that industrial processes should be matters of public knowledge, and speaks to it with a sermon on the bracing effects of free trade and a scolding for the workers. Their virtual illiteracy is to be deplored, even in the rare instances when a responsible person like the writer is able to find any of them who are both intelligent enough to understand what they are doing, and sufficiently well disposed to explain it.

This hectoring attitude toward the producers, not only artisans but often also manufacturers, sharply differentiates the bureaucratic and scientific contributions in the *Description des arts et métiers* from the treatment accorded the crafts in its rival and companion collection, the *Encyclopédie* itself. Diderot's instinct was to be critical of the authorities and to think well of the artisans. The instinct of the academic and official experts was to think well of the authorities and to be critical of the artisans. Ideologically, the *Encyclopédie* was of so much greater significance that a comparison on that score would be meaningless. The thrust even of its technical mission had been more democratic than scientific—to teach the artisans "to have a better opinion of themselves," and more generally to redress the prejudice with which exploitation had disarmed self-respect ever since Platonic antiquity. Thus Diderot, in the article "Art," developing the proposition of a moving passage that he interpellated in d'Alembert's *Discours préliminaire*.[72] Technologically, there is a comparison to be made between the two compilations. The *Description des arts et métiers*, however expedient the choice of subjects, was more informative and reliable on the subjects that happened to be chosen, particularly in the later and fuller contributions written by specialists. Even many of its earlier, more dated descriptions were reprinted in the *Encyclopédie méthodique*, whereas it was literary and scholarly articles from the *Encyclopédie* that tended to be chosen for republication.

The immense *Encyclopédie méthodique*, which represents the next chapter in the history of enterprises of this sort, lies mostly beyond our present

---

tributed two companion treatises, *Art de préparer et d'imprimer les étoffes en laines* . . . (1780), and *L'Art du fabricant de velours de coton* . . . (1780). For his early career, see Le Guin (1966), an excellent monograph, in which the circumstances that led to inclusion of these works in the *Description des arts et métiers* are discussed on pp. 48-52. Cf. Claude Perroud, ed., *Lettres de Madame Roland*, 2 vols. (1890-92, Collection de Documents Inédits sur l'Histoire de France) 2, app. G, 625-641.

[72] *Encyclopédie* 1 (1751), xiii, 717.

scope, at any rate in execution.[73] The planning would appear to go back to 1778. The publisher Charles-Joseph Panckoucke had already published four volumes of text and one of plates supplementing Diderot's original *Encyclopédie*, following them with a two-volume index to the entire set.[74] In that year he also launched the project of a new encyclopedia, of which the elements were to be not merely articles arranged alphabetically, but multivolume dictionaries of entire subjects. Gaps would be filled, errors corrected, information brought up to date, plates integrated with text, and facts located where the reader could find them. Some few volumes did reach subscribers before the Revolution compounded the complexities of the original plan—for example, the first volume on *Agriculture* edited by Tessier et al.; six of the eight on *Arts et métiers mécaniques* with six of seven volumes of plates, edited by Jacques Lacombe; three of eight on *Botanique*, edited by Lamarck; one of six on *Chimie, pharmacologie et métallurgie*, edited by Guyton de Morveau; one of thirteen on *Médecine*, edited by Vicq d'Azyr; four volumes of *Histoire naturelle des animaux*, adapted by Daubenton from his collaboration with Buffon; two volumes on *Manufactures, arts et métiers*, edited by Roland in continuation of his memoirs for the *Description*; three volumes on *Mathématiques*, in which Bossut mediated editorially between d'Alembert and readers of a later generation.[75] It will be apparent even from this fragmentary listing that a publisher had better success in enlisting notable contributors than the Academy had enjoyed when it was a question of fulfilling a corporate obligation. For circumstances had changed, and contribution to Panckoucke's venture mingled professional with economic reward. A publishing enterprise it was, nonetheless, completed in over 200 volumes in the 1830s, and no exercise in either ideology or governmental encouragement. We must return to our proper topics, therefore, again by way of Duhamel du Monceau.

3. FORESTRY

In the opinion of Fougeroux de Bondaroy, his uncle's most signal contribution to the public interest was the work Duhamel published on forestry

[73] For a summary account and table of contents, see Watts (1958); Darnton (1979) describes the genesis of the *Encyclopédie* in chs. 8 and 9.

[74] For the career of Panckoucke, and the publication of the *Supplement* (1776-1777) and *Table* (1776-1777) to the *Encyclopédie*, see Watts (1954).

[75] For bibliographical detail, see Watts (1958). Also of technical or commercial interest were one volume of three on *Architecture*, edited by Quatremère de Quincy; three volumes of *Commerce*, edited by the abbé Baudeau; four volumes on *Économie politique et diplomatique*, edited by Démeunier; three volumes on *Marine*, with many authors; three volumes on *Géographie moderne*, edited by Robert; a two-volume *Atlas encyclopédique*, edited by Desmarest and Bonne; and one of two volumes on *Histoire naturelle des vers*, edited by Bruguière and Lamarck.

and lumber.[76] A sequence of five treatises in eight quarto volumes, it appeared between 1755 and 1767, the very years when Duhamel was also organizing and writing for the *Description des arts et métiers*. One of the family properties lay hard by the forest of Orléans, and it was there as a youngster that he began to know the woods. The actual genesis of the project went back to his election to the Academy and appointment to the maritime ministry, and specifically to the charge that Maurepas then laid upon him to concentrate his scientific efforts on objects of importance for the navy. Having composed his treatises on ropemaking and naval architecture, Duhamel had turned aside to produce the commentary on the husbandry of Jethro Tull, publication of which in 1750 is now taken to have inaugurated the critical movement for agricultural reform in France, and followed it in 1752 with a companion work on *Conservation des grains*. It was Louis XV who recalled him from these less exacting concerns to his commitment to sylviculture. Duhamel was accorded an audience to present the *Traité de la culture des terres*, adapting Norfolk practices to French farming, whereupon the king asked him how his work on lumber for shipbuilding was coming along. Except for this expression of the royal attention, he confessed that he might have continued putting off the labor of organizing the report of what his brother and he had been learning all the while.[77]

Though located fairly close together, the three main segments of the Duhamel lands were quite varied in terrain features, as if chosen with a view to experimental and comparative botany. The domain of Denainvilliers had a dry and shallow soil. Through Le Monceau flowed a small tributary of the Essonne, the Œuf, bordered by silt and bottomland. The estate at Vrigny, on the northern fringe of the forest, was sandier and to well suited to supporting groves of long-lived trees to be exploited for their timber. The more retiring of the brothers, Alexandre, who rarely left his manor of Denainvilliers, made acclimatization of foreign trees and plants the special study of his life. His ventures were favored by the friendship of both brothers with a brilliant naval officer, Roland-Michel Barrin, marquis de la Galissonière, governor of Canada from 1745 to 1749. La Galissonière shared in their agrarian and arborial occupations seriously enough to qualify for recognition among agricultural reformers in his own right. He was an associé libre of the Academy of Science, un-

[76] Fougeroux tells of the origin of the series in a draft that he prepared for a revision of the *Traité des arbres et arbustes* (n. 77 below). The work never appeared. The manuscript of the preface is conserved, with others of the papers of Fougeroux and Duhamel, in the Library of the American Philosophical Society in Philadelpha, 580.D881, No. 24. Cf. Ewan (1959).

[77] Duhamel, *Traité des arbres et arbustes qui se cultivent en France en pleine terre*, 2 vols. (1755), ii.

like most members of his rank, who held honorary status. Equally to the point, he had a plan for constructing a chain of forts from New Orleans to Quebec and systematically sent the Duhamels seeds, seedlings, and even saplings selected along the great route from south to north.[78] Duhamel de Denainvilliers would then set out individual plants of the same species in the several locations with a view to determining what the best conditions were for each variety. By the time in the 1780s when Fougeroux (whose property of Bondaroy was also in the region) was telling of their trials, forty years of experience lay behind them, and if his account is to be believed, the stands at Denainvilliers represented the earliest European experiments in acclimatizing North American trees on a significant scale.[79]

Unaware of how much remained unwritten about arboriculture, the modern reader does not at first take the full measure of what Duhamel had in mind. The *Traité des arbres et arbustes qui se cultivent en France en pleine terre* announces his intention of going beyond the provision of naval timber to enlarge on every aspect of planting, management, exploitation, and reforestation, for all commercial and ornamental purposes.[80] Lest he die before the task was done, he had subdivided it into topics, each of which he would treat in a self-contained manner. These first two volumes constitute a natural history, or in the term that the encyclopedic spirit had made fashionable, a dictionary. Accordingly, the arrangement is alphabetical. As always in taxonomy, Duhamel preferred the Tournefort nomenclature for scientific identifications. It was not to scientists that he was addressing the work, however. It was to laymen and mainly to landed proprietors with a view to encouraging them to diversify their plantings while informing them of the characteristics of untried species and the conditions in which they might prosper. Two volumes of *Physique des arbres* followed in 1758, providing the same public with the elements of botany applied to elucidation of the anatomy and physiology of trees.[81]

The next installment, *Des Semis et plantations des arbres*, concerns landscape gardening.[82] In it Duhamel pursued the object of persuading

[78] Duhamel du Monceau and La Galissonière collaborated on a set of instructions for the preparation and handling of specimens to be transplanted, *Avis pour le transport par mer des arbres, des plantes vivaces, des semences, des animaux et d'autres morceaux d'histoire naturelle* (1725). On La Galissonière, see Lamontagne (1961).

[79] Manuscript preface to *Traité des arbres et arbustes* in Library, American Philosophical Society, 580.D881, No. 24.

[80] I, iv-v; on the reception of the work, see Lamontagne (1963). It would unbalance our apparatus to detail the reprintings and translations of the series. They were legion.

[81] *La Physique des arbres; où il est traité de l'anatomie des plantes et de l'économie végétale: pour servir d'introduction au Traité complet des Bois et des Forêts* . . . (2 vols., 1758).

[82] *Des Semis et plantations des arbres, et de leur culture, ou Méthodes pour multiplier et élever les arbres, les planter en massifs & en avenues; former les forêts & les bois; les entretenir, & rétablir ceux qui sont dégradés* . . . (1760).

wealthy landowners that they need not resign themselves to the monotony of oak, elm, and linden in forming grand approaches to their country seats and vistas opening from their formal parks. The availability of some 200 species and 1,500 varieties might well permit them some originality and even profit. Duhamel used the word "*espèce*" as would a gardener rather than a botanist, meaning variety. An interesting digression gives a glimpse into the gardener's trade. Duhamel had found its practitioners to be far more hidebound than their counterparts in manufacturing. In the mechanical trades, a lad left home and set out on the road to roam the country ("*courir le pays*") and learn his craft at the hands of a succession of masters all over France and abroad. Nothing like the institution of *compagnonnage* or wandering journeymen existed to enlarge the horizons of gardening or viniculture. Prisoners of the seasons, those workers had to engage themselves for an entire agricultural cycle and could learn nothing serious in less than three or four years. Local partitioning of the labor force created a greater need than in the case of urban workers for rational supervision by informed proprietors. Not that Duhamel presumed to urge their taking rake or hoe into their own hands. France had evolved too far from Roman ways: "that sort of exercise is not for people of our century." No, no—let the laborers do the work. But oversee it. Make them try things. Show them, and even discuss with them, how their tasks accord with the order of nature. And if, by some happy chance, one were to make a discovery, why the thing to do was publish it so that fellow citizens might benefit. That was the motivation of all of Duhamel's work in agriculture. He had had less success than he could have wished, but at least he had shown landowners how to plant their wastes, their lawns, and their avenues in trees and shrubs.[83]

The remaining two titles on forestry and lumbering are in the genre of the treatise on ropemaking, exercises in experimental technology and civil engineering. Duhamel was fearful lest the reader whose attention had been won with the planting and care of ornamental stock would be fatigued or bored on being led into the forest to penetrate a thicket of detail on the chemical and physical properties of wood; the effect of terrain on the adaptability of various ages and species of tree to the uses of shipbuilding and construction; techniques for cruising and scaling woodlands to estimate their potential yield; methods for felling, cutting, and squaring the logs; procedures for exploiting the underbrush and producing charcoal without brush fires. It is in these volumes, *De l'Exploitation des bois*,[84] that the modern reader can meet with the forest-dwellers in lives led apart from the general structure of rural society in the eighteenth century, and

[83] *Ibid.*, xiii-xiv.

[84] *De L'Exploitation des bois, ou moyen de tirer un parti avantageux des taillis, demi-futaies, et hautes futaies, et d'en faire une juste estimation: avec la description des arts que se pratiquent dans les forêts . . .*, 2 vols. (1764).

largely unchanged since the Middle Ages. Finally, in *Du Transport des bois*, Duhamel twitches his timber out of the woods and moves it to the lumberyard by a choice of cart, wagon, or barge, or by a combination. He reports on hundreds of tests run over many years on methods of seasoning, softening, preserving, and shaping, and on the quality of oak and other kinds of wood from the different provinces of France and from America. In this last installment he permits himself a sigh of relief at having thus unburdened himself of the fardel he had borne for nearly forty years, a work pursued "with persèverance, I might even say with obstinacy," all of which he had now laid before the public.[85]

But no sooner had he caught his breath than out he came with two more magnificent folio volumes on fruit trees.[86]

4. AGRONOMY

Duhamel is one of two heroes (the second being the statesman Bertin) of André Bourde's *Agronomie et agronomes en France au XVIIIe siècle*,[87] a book about books about agriculture, worthy in scale and mode of the first of its protagonists. The treatment occupies 1,740 pages in three volumes of ample format, even without an index. The work is a thesis, a learned one in which argumentation and theory are unobtrusive. These very merits, however, create uncertainty in knowing how to reckon with the information, for Monsieur Bourde never quite comes out and says what importance he attaches to his matter. One course may be to refrain from repeating what is to be learned in so copious a source, while calling its availability to the attention of other readers, in this case those who wish to know what was said in detail about farming in France, mainly from 1750 until 1789 but also back to the time of Henri IV and the *Théâtre d'agriculture* (1600) of Olivier de Serres. Since Bourde gives generous paraphrases of Duhamel's writings on agronomy, it will be consistent with this approach to limit ourselves to identifying these, the most celebrated of our author's contributions to the wished-for welfare of humanity, and to situating them in the overall pattern of his service to utility, before venturing from a more restricted basis to generalize the significance of the

[85] *Du Transport, de la conservation, et de la force des Bois; où l'on trouvera des moyens d'attendrir les bois, de leur donner diverses courbures, sur-tout pour la construction des vaisseaux; et de former des pièces d'assemblage pour suppléer au défaut des pièces simples: faisant la conclusion du traité complet des bois et des forêts* (1767), viii-ix.

[86] *Traité des arbres fruitiers, contenant leur figure, leur description, leur culture* . . . , 2 vols. (1768).

[87] Bourde (1967). The study developed out of an earlier monograph in English on the influence of the example set by English farming on French thought about agriculture (1953). For a full and judicious essay review, see Rappaport (1969).

genre. Even then we shall be alluding to an argument that is latent in Bourde's treatise, whether or not he meant it thus to be extracted from the expenditure of his erudition and the economy of his interpretation.

Jethro Tull published *The Horse-Hoeing Husbandry* in 1731.[88] Duhamel himself tells how he came to adapt the work after several attempts at a translation by others had miscarried. The first had been commissioned from a certain Otter, of the Académie des inscriptions et belles lettres by a marshal of France, the duc de Noailles, one of the enlightened noblemen who were ever urging on their peers the notion of becoming landlords of the improving English type. Translating Tull, a tedious and obscure writer at best, required knowledge of agriculture no less than English, and the floundering Otter turned to Buffon for help. After three months with the manuscript, Buffon threw up his hands, persuaded that Tull was unpublishable in France. His novel ideas were "drowned in much vague reasoning, and such was his prolixity throughout that it would certainly preclude success." Learning that the chancellor of France had referred another translation to Duhamel, this by one Gottford, Buffon sent along the results of his own struggles. Thus, when the first of the six volumes that eventually bore the title *Traité de la culture des terres* appeared in 1750, it represented the work of many hands.[89]

The essential elements of Tull's husbandry were to break up the soil by plowing deeply, to economize on seed by sowing mechanically in drills rather than broadcasting, and to aerate the roots at intervals during the growing period by horse-powered tilling. Like many another system it worked for reasons other than the theory on which the inventor purported to base it. Had that been his most serious claim on attention, Tull would have deserved the strictures of Buffon. He considered that pulverizing the soil permitted the passage of an omni-nutritious principle from its earthen reservoir up through roots to feed the growth of plants in the form of sap. Such was his dogmatism that he contemned other aspects of well-tried husbandry, holding the use of dung for fertilizer to be valueless for plants and noxious to the consumer, and rejecting the practice of crop rotation. Since all vegetation draws on the same infinite fund of nourishment, a field would never be exhausted if cultivated properly, but would support

[88] *The New Horse-Houghing Husbandry: or, an Essay on the Principles of Tillage and Vegetation* (London, 1731). The 2nd edition of 1733 (*Horse-Hoing Husbandry*) bound with a supplement of 1740 became standard. It was re-issued in 1751 and edited with an introduction by William Cobbett in 1829.

[89] Buffon's opinion is reported by Duhamel in the preface to a revision of the first two volumes, *Traité de la culture des terres, suivant les principes de M. TULL, Anglois*, 2 vols. (nouvelle édition, corrigée et augmentée; 1753) *1*, vii. He there recounts the circumstances—how the chancellor, Henri-François d'Aguesseau, had referred the Gottford rendition to him in June 1748, and how it exhibited the same defects as the translation by Otter, who then died in October 1748.

the best of all possible crops in perpetuity. Tull did recommend turnips, though not in whichever third of the old three-field system would otherwise have been lying fallow. Instead, he planted them in alternate rows alongside grain in beds where their roots would assist mechanically in loosening the ground. Duhamel's adaptation de-emphasized Tull's vegetable physiology by combining it eclectically with the views of botanists and others on the functioning of stem and leaves. He reported its tenets, however, not without the appearance of adhering to them, and turned to instructing French readers on the unfamiliar crops and the use of tools and draft animals in the new methods for plowing and sowing. These were the effective reasons for the increased yields and lowered costs of the farming of the future.

Only after publishing the first volume in 1750 did Duhamel begin reporting on the annual results of his brother's and his own program of experimental farming. Together with large excerpts from his correspondence, the accounts occupy the remaining volumes, which appeared at intervals in the next eight years. In effect, the give-and-take of experience became a sort of agricultural forum. He actually used the word "journal," adding to it abridgments of the botanical-meteorological observations at Denainvilliers that he contributed annually to the *Mémoires* of the Academy of Science.[90] For the rest, the nature of the "experiments," as he expressly calls them, can be illustrated with an example from the first set of trials conducted in 1750, some carried out by the Duhamels themselves at Denainvilliers and others at the nearby seigneury of Acou under the eyes of their neighbor and collaborator, a landowner called Saint-Hillaire.

The brothers marked out an experimental area of two arpents (the arpent was about one and one-third acres) in a single field and divided it into halves. One half was prepared, dunged, and planted to winter wheat in the best traditional manner. It took twelve bushels of seed sowed broadcast. The other was prepared in accordance with Tull's system in beds two feet wide running the length of the field and separated by alleys four feet wide. With three rows in each bed, the grain was sowed in drills four to six inches apart. Two bushels sufficed. By winter, the traditional crop showed an expanse of green and the Tull field merely the pattern of the plowing. When shoots appeared in early spring, they were thinned to a separation of four inches, and the crop was given its first hoeing along the alleys. The effect was dramatic. The wheat turned dark green and shot up, bushing out to cover the earth between the rows. The old-method crop was then a yellowish green. Tull's prescription called for a second hoeing when the wheat began to spindle. Comparison at this stage showed that in the traditional crop each grain had produced at most two or three

[90] *Ibid.* *1*, lxix-lxxi, and above n. 10.

stalks capable of bearing ears, and many only one, whereas each grain in the cultivated half had produced a bunch of eight, ten, and even twenty stalks. The former ripened sooner than Tull's arpent, which got its third and final hoeing when the ears had formed. Unfortunately, an unseasonable heat wave penalized the harvest when that wheat was still green. Even so, the new husbandry produced 284 sheaves yielding 70 bushels of large wheat weighing 1,470 pounds, whereas the old method produced 476 sheaves yielding 98 bushels of small wheat weighing 2,058 pounds. In absolute terms, the traditional husbandry had thus given more per arpent. But producing about 25 percent more wheat of inferior quality had required a large and expensive application of fertilizer and twelve bushels of seed as contrasted to no fertilizer and two bushels in the technique under trial. Moreover, in the old sequence oats would have to follow wheat in the next planting, a far less profitable crop, whereas, by the theory of the tilling husbandry, wheat might be planted indefinitely.[91]

The foregoing was the simplest type of measurement that Duhamel adduced. In more elaborate tests, he varied the density of planting, and extended his comparisons to the productivity of a second and a third crop grown in the same place. Obviously, agricultural experimentation was a slow and relatively uncertain art, requiring a full growing season for every trial, and much at the mercy of weather, pests, and the chances that attend the fortunes of any farmer. With all the more reason, therefore, did Duhamel welcome the accounts of innovators elsewhere, many trying other types of produce. The most energetic and prolific of his correspondents was a syndic of the city and republic of Geneva, Lullin de Chateauvieux, whose reports, interspersed with Duhamel's in later volumes, amount to a treatise within a treatise. Among the others—several dozen in all—Diancourt, an officer of the regiment of Grenadiers, tried widening the alleys and planting a row of barley down the middle; Roussell, of Guignes in Brie, began prudently with small patches, but his first cultivated crop was eaten by rabbits before he could quantify it; Delacroix at Verdun realized his hopes despite atrocious weather; Navarre, dean of the Court of Aids near Bordeaux, made experiments with barley, oats, and rye; Bielinski, a grand marshal of Poland, concentrated on barley; Eyma in Bergerac was a forerunner of truck farming and wrote of peas, beans, kidney-beans, carrots, asparagus, artichokes, beets, and strawberries. As always, Duhamel put together whatever information came his way. His publisher, the bookselling firm of Guerin and Delatour, had received a body of drawings from China illustrating the cultivation of rice, and he arranged to have the text translated by a former Jesuit missionary, Father Foureau, so that he might tell how to grow that staple.[92]

[91] *Ibid.* 2, "Expériences et réflexions," Pt. I, Ch. 1, 1-15.

[92] It is rather more convenient to identify the contributions of these collaborators in the

With his experience in the field increasing, Duhamel tended to qualify the application of Tull's system more sharply than he had done while working over its translation at his desk. He early abandoned the purism of the tilling dogma and recommended mingling dung with cultivation. On occasion, he even allowed for alternating crops, for resting fields in fallow, and for growing fodder. As for instruments, the duc d'Orléans had imported one of Tull's drill plows for inspection and trial. Duhamel found it altogether too complicated and also too fragile a device for farming on a large scale in real life.[93] In its place, he designed his own "*semoir*" and also a light plow for the second and third rounds of hoeing.[94] He acknowledged that a different adaptation imagined by his collaborator Chateauvieux might be preferable in many situations.

Into his accounts of all these trials he interpellated another scheme for improving the supply of grain, one on which he had been engaged before d'Aguesseau's initiative had involved him with Tull's system of husbandry. His experiments for the navy on the ventilation of ships and hospitals led him to design and build a model granary in which produce could be stored for years. Lightly roasted in special ovens, and then aerated by a system of donkey-powered bellows, grain was preserved from rot and mildew in a grange secured against vermin. Its vulnerability to these hazards was the main technical impediment to any attempt on the part of government to provide for something like an ever-normal granary on the national scale. Duhamel estimated that in good years French agriculture was capable of producing one-third more staple than the country consumed, which surplus could thus be conserved in selected locations against the lean years that came every three or four.[95]

English abstract, *A Practical Treatise of Husbandry: Wherein Are Contained, Many Useful and Valuable Experiments and Observations . . . by the Celebrated M. Duhamel du Monceau* (London, 1759), ed. and trans. John Mills. See esp. Pt. III, 306ff.

[93] *Traité de la culture des terres* (1753) *1*, lxvii.

[94] *Ibid.* 2, 380-413; cf. *A Practical Treatise of Husbandry* (n. 92, above), 440-453, where the plates are much clearer.

[95] *Traité de la conservation des grains, et en particulier du froment* (1753). Duhamel read a memoir on this subject before the Academy of Science on 13 November 1745, and prints it here as the first chapter. He tells in the *Traité de la culture des terres* (*1*, lxxxi-lxxxii), how he decided to elaborate and print it in the format of that series, where it was to form the third volume. He published the scheme for ventilating ships and hospitals a little later, in *Moyens de conserver la santé aux équipages des vaisseaux avec la manière de purifier l'air des salles des hôpitaux* (1759). Methods for what might almost be called pasteurizing vegetable productions against decay or disease figured in two other works. A treatise on madder, though in the format of the *Traité*, pertained in motivation to the *Description des arts et métiers: Traité de la garance, et de sa culture; avec la description des étuves pour dessécher cette plante, & des moulins pour la pulvériser* (1757). The *Histoire d'un insecte qui dévore les grains de l'Angoumois; avec les moyens que l'on peut employer pour le détruire* (1762), written jointly with Tillet, reports a mission on which the two authors were dispatched by the Academy at the request of the

Such, in brief outline, were Duhamel's agrarian occupations in the 1750s, conducted concurrently with the completion of the multi-treatise work on forestry. The *Eléments d'agriculture*, published in 1762, in which he drew together what he had learned into something like his own school of agriculture, proved to be his most frequently reprinted and translated book.[96] It is indicative of the respective contributions to technology of British practice and French analysis that in 1759, three years before Duhamel produced this textbook, the experimental content of his *Traité de la culture des terres* should have been abstracted and translated back into the language of the country where it all started. Asked for an appreciation of the original of this work, entitled *A Practical Treatise of Husbandry*, the Scottish physician and agricultural writer, Francis Home, observed of Duhamel's first three volumes, "They are distinct, exact, and conclusive so far as they have gone, and stand a model for experiments in Agriculture. What a shame for Great Britain, where agriculture is so much cultivated, to leave its exact value to be determined by foreigners."[97]

The observation is consonant with the burden of André Bourde's massive monograph, to return to that. Agronomy is there said to have been a definable movement finding expression in France in a body of writings and experiments, largely though not exclusively stimulated by the success of the agricultural revolution in Britain. That example was set, not merely by the deep-plowing radicalism of Jethro Tull, which captured attention in Duhamel's rendition of 1750, but more variously by the Norfolk system in general as its other features gradually become known. Bourde insists on a distinction between physiocrats and agronomes. Physiocracy he takes to have been a socio-economic philosophy, a prescriptive analysis of arrangements incumbent on society in consequence of the premise that agriculture is the sole source of wealth. Agronomy he takes to have been a science of agriculture itself, a developing knowledge of techniques by which farming might become progressively more productive.[98] It is a distinction like that between classical political economy and a science or a technology of machine processes based on critical analysis of industrial operations. Agronomists might or might not agree with social and fiscal re-

---

intendant of Limoges, Turgot's predecessor, Pajot de Marcheval. They identified the cause of a devastating blight to be a moth that laid its eggs on the ear of wheat. The larva then bored into the kernel. The reproductive chain could be broken by roasting the grains at about 50° Réaumur.

[96] 2 vols. (1762). Duhamel published a revision in 1779.

[97] Op. cit., in n. 92, above, vi. Home was the author of *Principles of Agriculture and Vegetation* (Edinburgh, 1756).

[98] See, for example, Bourde (1967) *1*, 365-368; 2, 981-985, 1072-1073, 1099. Bourde concomitantly insists on the distinction between his treatment and that of Weulersse (1910), the classic historian of the physiocrats, who in the vein of an older generation of intellectual history made agrarian reform a derivation of the doctrines of the school.

forms advocated in the name of physiocracy. There was no necessary overlap or symmetry. Politically, for example, Duhamel was an arch conservative. The patrons of agronomy, on the other hand, like patrons of science itself, characteristically belonged to the liberal nobility—a Noailles, a Malesherbes, a d'Ormesson, a La Rochefoucauld-Liancourt, a duc d'Orléans—and the ambiance smacks of that ever-so-slightly smarmy attribution of moral value to things rural that attends much rumination about life on the farm by persons—even agronomists—who do not really have to live off the farm.

Of the writers other than Duhamel, Tillet, and Parmentier—of whom more in a moment—the only figure prior to the 1780s who could be called, one will not say a member, but a familiar of the scientific community was the abbé Rozier. (It is true that Bourde assimilates Daubenton and his venture in sheep-raising to the movement, but that does seem to stretch things rather far.)[99] Viniculture and veterinary medicine were two of the enthusiasms with which Rozier accompanied the enterprise that gave him scientific entrée, to wit the editing of his journal.[100] Ever an intelligencer, Rozier was the one to produce the inevitable encyclopedia, the instrumentality with which modernizers instinctively (if illogically) thought to make a body of practice into a body of knowledge, reforming both in relentless strokes of alphabetization. The first volume of his *Cours complet d'agriculture, théorique, pratique, économique . . .* came off the press in 1781, only a year later than promised: the subtitle is *Dictionnaire universel d'agriculture*.[101]

For the rest, Bourde's cast consists of characters whose names will not necessarily be recognized by latter-day habitués of the scientific circles of the eighteenth century. The marquis de Turbilly, more than a patron and landowner, was a farmer and innovator in his own right and author of a work on reclamation of wastes.[102] The baron de Tschudi published on acclimatization.[103] La Salle de l'Etang seconded Duhamel on the introduction of new grasses—lucerne, sainfoin—into productive schemes of rotation.[104] Henri Patullo, a Jacobite exile of Scottish or Irish extraction,

[99] Bourde (1967) 2, 857-859; cf. above, Chapter II, Section 6.

[100] See above, Chapter III, Section 1.

[101] (12 vols; 1781-1805). Rozier was killed in 1793 during the revolutionary bombardment of Lyons. Vol. 10 (1801) was edited by Chaptal, Parmentier, and others, and vols. 11-12 (1805) by Thouin. There is a "Notice sur la vie et les oeuvres de l'abbé Rozier" by A.-J. Dugour in vol. 10, i-xvi; and a discussion of "Économic rurale" by Thouin in vol. 11, i-lvi. The running head calls it an "Essai sur la manière d'étudier l'agriculture par principes."

[102] *Mémoire sur les défrichements* (1760).

[103] *De La Transplantation, de la naturalisation et du perfectionnement des végétaux . . .* (1778).

[104] *Prairies artificielles . . .* (1756).

moderated Tull's extremism and worked out a practical farmer's calculation of cost and profit in the new husbandry.[105] It would be unappreciative to say that Bourde has no trouble in establishing the extent of the literature, for he has gone to enormous trouble. The authors on whom he reports number in the dozens. It is easy to believe that the names appearing in an exhaustive bibliography (his is confined to people actually discussed) would carry the figure up into three digits. There is also abundant evidence that he has identified the main preoccupations of all these persons concerned with reforming French agriculture: clearing and cultivation of waste and woodland, rotation of crops, choice of fertilizers, enclosures of common land and open fields, problems of common rights and especially that of *vaine pâture* or turning livestock into any field to graze between the times of harvest and planting, introduction of new crops, invention of new or modified implements, the role of sheep and cattle and of draft animals, improvement of veterinary medicine, the perpetual battle against pests and blights. As for whether agronomy itself deserved to be called a science, the question is an empty one. At least there can be no doubt that the existence of this literature presupposed a society that set a premium upon bettering things scientifically.

What is left quite undetermined, however, is the difference it all made, if any, beyond the circle of persons who wrote, printed, bought, and read the books. There, Monsieur Bourde leaves his own reader dangling, not to say oscillating between undefined poles. At the outset, he disavows the intention of actually writing a history of agriculture, observing that such an enterprise would be a different work altogether.[106] He lifts his guard periodically by introducing the topics he does handle with paragraphs that further disclaim any pretension to exhaust the several subjects, which he goes on to treat in chapters that often exceed 100 pages.[107] Occasionally he hints that, even if none of these reforms ever got carried into practice in the old regime elsewhere than on model and experimental farms, still the stream of discourse had the significance of raising consciousness of issues and potentialities.[108] He acknowledges that the proposed reforms aroused, if anything, more resistance than support.[109] But then he cannot finally bear to relegate the battle to the books, and a penultimate outburst escapes him. So sustained a commitment, recorded in so large a volume of publication, must have had *some* real influence. It must have *some* importance, also for the history of agriculture, indeed for the very history of France.[110] After all, the published memoirs of the Société d'agriculture

[105] *Essai sur l'amélioration des terres* (1758).
[106] Bourde (1967), *1*, 11.
[107] See, e.g., *ibid.* *1*, 370; *2*, 899, 1077; *3*, 1292, 1355.
[108] *Ibid.* *1*, 20-23; *2*, 952, 1028; *3*, 1487.
[109] *Ibid.* *3*, 1401-1408.
[110] *Ibid.*, 1561-1566; cf. *1*, 22-29.

de Paris contain matter of the highest scientific value (though what value is left unspecified) in the very last years of the old regime.[111] And in 1789 "the rural world began to stir."[112]

Amid this indeterminacy, perhaps it will be well to report the findings of another thesis, of much more modest dimensions, which does purport to study agriculture, no less than thought and policy concerning agriculture, in the same four decades. The author, David Brandenburg, formulates the course of possible developments in a scheme that is both convenient and convincing. It turns on four sets of choices that were in principle open to the whole body of French agriculturalists. First, they might have moved toward large-scale land-extensive farming like that of the United States and Canada in later times. That is only a theoretical and was never a realistic possibility, however. Second, they might have changed nothing and perpetuated a fallow-dominated husbandry, low in productivity and mainly concerned to supply grain for human consumption in the form of bread. That option would have failed to aliment French growth or to sustain the standard of living. Third, they might have followed Britain and Holland in the direction of large-scale land-intensive farming. That also is only a theoretical possibility, for it would have been unfeasible socially and politically if not technically, and was unnecessary economically. Fourth, they had the choice they actually adopted, that of an improved peasant agriculture, with some multiplication and variation of crops and a steadily increasing yield thanks to better informed techniques. "If it is not possible," writes Brandenburg, "to conclude that the *agronomes* of the last four decades of the *ancien régime* determined this choice, at least they accepted it; and to the extent that their recommendations were followed, they were creative."[113] By no means is it self-evident that this is an over-simplified or a vacuous statement.

5. THE SOCIETY OF AGRICULTURE

It is in sects and churches, remarked a great French historian of another generation and quite other subjects, that values and beliefs take a form making them accessible to scientific inquiry.[114] It may follow, methodologically speaking, that the capacity of a body of opinion, thought, or

[111] *Ibid.* 3, 1325.

[112] *Ibid.*, 1563. For the state of French agriculture on the eve of the Revolution, see also Festy (1947, 1950), and the classic in this area of studies, Bloch (1931; new ed., 1955-56). On the question of practical results, see Morineau (1968).

[113] Brandenburg (1954), 220.

[114] Halévy, *England in 1815*, Vol. 1 of *A History of the English People in the 19th Century*, trans. E. I. Watkin and D. A. Barker (New York, 2nd ed., 1949), 383. The remark is à propos of the influence of religion on conduct.

doctrine to issue in the formation of durable and effective institutions offers a useful criterion of their practical significance in the world of facts and events. If so, and if we were to accept unreservedly the evaluation of the most famous and widely read observer of French farming in the eighteenth century, enlightened agronomy would fail the test. On 12 June 1789, Arthur Young attended a meeting in Paris of the Society of Agriculture. The company had been accorded the designation "royal" in the preceding year. Among those on hand—Fourcroy, Desmarest, Tillet, Cadet-de-Vaux, Parmentier, Broussonet, all people known in science—only one person, Cretté de Palluel, was an actual farmer. "I am never present," wrote Young of the occasion, "at any societies of agriculture, either in France or England, but I am much in doubt with myself whether, when best conducted, they do most good or mischief; that is, whether the benefits a national agriculture may by great chance owe to them, are not more than counterbalanced by the harm they effect; by turning the public attention to frivolous objects, instead of important ones, or dressing important ones in such a garb as to make them trifles?"[115]

The Paris society had been founded in 1761 (a year before Duhamel published *Eléments de l'agriculture*) almost simultaneously with a sisterhood of provincial societies of agriculture.[116] Initiative had stemmed, not from the countryside, but from the paternalistic enthusiasm of Henri Bertin, then in the midst of doing his duty by the monarchy in the office of controller-general of finance. A leading agronomist and Duhamel correspondent, the marquis de Turbilly, persuaded Bertin to take advantage of his great responsibility to forward the interests of agriculture, always the administrative sector closest to his heart. The example immediately before their eyes was in Brittany, a Société d'agriculture, de commerce, et des arts, chartered in 1757 by the provincial Estates in backward emulation of the British Society of Arts.[117] Originally, Bertin and Turbilly had no notion of anything like an academy of agrarian science producing research. The idea was rather a network of societies, an extended forum providing for exposure to agricultural enlightenment even as provincial academies facilitated participation in literary culture and philosophy.[118]

[115] *Travels in France and Italy During the Years 1787, 1788, and 1789* (Everyman edition, London and Toronto, 1927), 127-128.

[116] Passy (1912) is the standard work on the Paris Society. The Society in Tours actually antedated that in Paris, however. Between 1761 and 1763, societies were started also in Limoges, Lyon, Clermont-Ferrand, Orléans, Rouen, Soissons, Alençon, Bourges, Auch, La Rochelle, Montauban, Caen, and Valenciennes. Justin (1935) gives an informative if somewhat literal administrative history. Cf. Bourde (1967) 2, 1109, n. 3.

[117] Bourde 2, 1099-1109; Justin (1935), 35-41.

[118] The sociological function of the provincial academies has been studied very carefully by Roche (1978), and the cultural significance of a comparable body in England, the famous Manchester Literary and Philosophical Society, forms the subject of a thought-provoking analysis by Thackray (1974).

It is easy to be wise after the event and to reflect that in farming, unlike discourse, the sharing of progressive views means little if they are not applied. Bertin (it will be recalled) resigned his high office in 1763 to become a kind of fifth-wheel secretary of state in charge of an *ad hoc* ministry defined by and for himself.[119] Breathing life into the societies of agriculture was one of his constant preoccupations until his departure from government in 1780. There, at least, he met with disappointment. Intendants treated his directives to involve the societies in the implementation of policies with a skepticism bordering on non-compliance. The negative attitude of Turgot in Limoges was typical.[120] Even the most enlightened of administrators were used to focusing attention on fiscal measures, on their effect in the economy and on incentives. They had little feel for agricultural technique, little sense that decisions about what to plant where and when, let alone how to cultivate it, fell within the competence of government. Other testimony is overwhelming to the effect that the bodies fathered by agronomists (and perhaps mothered by physiocrats) were premature, and that those which survived their birth had mostly lapsed into apathy and inactivity by the 1770s.[121] If the judgment Young expressed were to turn on the experience of that period, the sole reason to qualify his condescension would be the reflection that he always tended to disdain continental farming in the degree that it departed from the husbandry of Norfolk, while also affecting throughout his writings the stance of the farmer in the field that he never succeeded in becoming.

The same cannot be said of Malesherbes (his lands were in the Gâtinais near Denainvilliers) and perhaps his reminiscence should also be recalled. In a paper read before the same Royal Society of Agriculture early in 1790, he recollected how

> Forty years ago, which is to say in the decade of the 1740s, I was admitted into the Academy of Science, and I remember that it was soon afterward that we had the idea of educating people. There were no Societies of Agriculture then, and it was to have been between the Academy of Science and all the persons cultivating farms throughout the country that we would have liked to establish communications. I have since then been asked why I have never spoken of it. I reply because in those days the people did not trust those who wished to instruct them.
>
> Men working the fields, wine-growers, and other farmers in Europe are usually people without any education and without any ex-

[119] Above, Chapter I, Section 4. Its conduct, insofar as it concerned agriculture, is the subject of Bourde (1967), ch. XIV, 2-3, 1079-1290.

[120] Above, Chapter I, Section 2; cf. Bourde (1967) *2*, 1112-1113.

[121] Bourde (1967) *3*, 1194; Justin (1935), 120-124, 253-261; Passy (1912), 42-43.

perience of theoretical instruction. They find it difficult to express themselves, and scientists cannot communicate with them at all. We have suffered much from the prejudice that, in France as in other countries with a feudal background, seems to ordain that there be two orders of citizenship.

The people, particularly in the country, were defensive about anything that was proposed to them, even if it was for their own benefit, the reason being that the farmer felt obliged to conceal the resources of his enterprise for fear that declaring them would only increase his liability for taxes.[122]

Institutionally at least, the situation began to change in the 1780s. Not that applications of agronomy could yet be observed in the fields, but Bertin was retired in 1780 and Duhamel died in 1782, Baconian Nestors both, one in administration and the other in the Academy of Science. Persons who succeeded to their representation of the agricultural interest in government and in science came from a new mold. The career of François de Neufchateau in the Ministry of Agriculture belonged mainly to the Directory and the early nineteenth century, but already in the 1780s he was the young civil servant rising into expertise out of the Turgot circle.[123] Leading scientists for their part began to treat problems of agriculture analytically as well as descriptively and benevolently. When the Academy of Science was reorganized under Lavoisier's direction in 1785, the old section of botany was redesignated "Botany and Agriculture." Lavoisier's investment of time and attention in agronomy accentuated the shift toward importance.[124] Tessier, Tillet Fougeroux de Bondaroy, Thouin, and the young Broussonet centered their research on agricultural phenomena, and constituted a grouping significant enough to allow the inference that their choice of specialty entailed no self-defeating anxieties about prestige. In 1782 a semi-popular journal began to appear, *économie rustique* having become a well-enough defined subject to be dispensed by possessors and consumed by users of knowledge.[125] It would be a thankless task to try distinguishing between elements of cause and effect in these developments. Certainly, however, no more pertinent evidence needs to be adduced to exhibit the entrance of agriculture into the purview of science than is provided by the early work of Antoine-Augustin Parmentier, whose name has become a household word through his introduction of the potato into the diet of the French people.

In point of career patterns, Parmentier's scientific biography might at

[122] Quoted in Passy (1912), 4.

[123] Above, Chapter I, Section 3.

[124] Below, Section 5.

[125] *Bibliothèque physico-économique, instructive et amusante, contenant des mémoires et observations pratiques sur l'économie rustique* . . . 24 volumes appeared from 1782 through 1797. Cf. Balland (1902), 380-381 n.

first reading seem something of an anomaly. As will appear, however, he had much in common with certain contemporaries working with problems of construction and mechanical processes whom we readily recognize for engineers.[126] Parmentier's technical background was in pharmacy, which craft overlapped more extensively with science than any other, and his vocational background in military service. Although the actual work on which his later reputation depended was chemical, he became known to the public before his fame led to select scientific membership. Like Gaspard Riche de Prony and Lazare Carnot, two of those engineers, Parmentier was never a member of the Academy of Science and was elected to the Institut de France after its foundation in 1795. A little detail is needed to bring out the significance of these facts, and of his place in relation to the preceding encyclopedic phase of agronomy.

Born in 1737—and thus ten to fifteen years older than the mean of the Lavoisier-Laplace generation—Parmentier came from respectable bourgeois stock in the somewhat nondescript city of Montdidier, near Amiens. There he was brought up and educated by a widowed mother. Money was scarce, and he apprenticed himself early to a local apothecary. A leader of the trade in Paris, the apothecary Simonet, was a relation of the family. When Parmentier turned eighteen, he took service in the capital in his well-known cousin's shop. By 1757 he had learned enough to sign on for military service with the army investing Hanover in the Seven Years War. The chief of pharmacy in that command, Pierre Bayen, was an estimable chemist, and Parmentier compounded medicaments under knowledgeable direction and careful scrutiny. He continued his scientific education after he was captured by the Prussians. The hussars stripped him naked (he recalled), helped themselves cheerfully to his accouterments, and did him no other harm. The relaxed confinement of an eighteenth-century prisoner of war allowed him to study in Osnabrück with Johann Friedrich Meyer, a leading German chemist and author of a convincing theory of causticity. Then it was that Parmentier made the acquaintance of potatoes on his own plate, for already they were a staple food-stuff in lower Saxony.[127]

Back in Paris in 1763, he followed courses at the Jardin du roi—Nollet on physics, Jussieu on botany, Rouelle on chemistry—and in 1766 won a contest for the post of qualifying apothecarcy—"*gagnant maîtrise*"—at the veterans' hospital in the Invalides. Reforms were contemplated there, a feature of which was that, on completing the requirements, the success-

[126] Below, Chapter VII, Section 1. On Parmentier, see Balland (1902) for a complete bibliography of his writings and secondary sources, and among the numerous biographical sketches, Mutel (1819) and Blaessinger (1948), 75-118. Kahane (1978) discusses Parmentier's career in the light of problems of food supply today.

[127] Mutel (1819), 9.

ful candidate would be appointed chief apothecary on a permanent basis. So it came about. The position being a newly created one, Parmentier immediately confronted the structure of frustration in the old regime. In 1676, the intendant of the Invalides had agreed with the Grey Sisters that their order would provide nursing care and medicines. A century later its prioress successfully invoked that contract to debar Parmentier from entering on his assignment. The impasse provoked one of the sayings attributed to a hand-tied Louis XV—"If I were a minister, such abuses would not exist"[128]—and was the making of Parmentier scientifically. The crown left him with his stipend, his lodgings, and an injunction to refrain from fulfilling the duties of his station. In this halcyon situation for research, he installed a chemical laboratory in the Invalides and, though he received occasional external assignments in military pharmacy, he struck out along a more original line, investigating the chemistry of nutrition.

A later pharmacologist at the Invalides, Antoine Balland, sharing in the sympathetic tradition through which French scholars often interest themselves in previous incumbents, published a large book of scientific antiquarianism consisting of the chemical analyses extracted from Parmentier's many writings.[129] In contrast to Duhamel and agronomists in the field, whose experiments were exercises in comparative cultivation, Parmentier experimented on agricultural productions in the laboratory. He composed his first paper—in this, too, like certain engineers—for a competition set by a learned society, the Academy of Besançon, the problem being to identify edible produce that might replace ordinary items of consumption, largely wheaten bread, in time of dearth. Parmentier took his point of departure from the work of Jacopo Bartolomeo Beccari, president of the Academy of Bologna in the mid-century, a physician and polymath, who first broke down flour into the factors of gluten and starch.[130]

Unlike Beccari, Parmentier held starch to be the source of nutritive value. He determined that the darker the flour, the larger the proportion of bran and of glutinous matter. If the latter were indeed the source of nourishment, white bread should be less nutritious than black, whereas exactly the contrary was universally acknowledged to be true. The question then arose what other vegetable products might provide starch to the diet. Potatoes being the most obvious candidate, Parmentier started there and to his own satisfaction established the identity chemically of starch extracted from potatoes and from wheat. Preliminary investigations pro-

[128] *Ibid.*, 12.

[129] Balland (1902). On Balland's own career in the third Republic, see Blaessinger (1948), 316-381.

[130] Balland (1902), 21-22.

duced a list of some thirty-odd additional plants with root systems that might be made to yield starch in edible form. Besides potatoes, however, he limited this first research to horse chestnuts and acorns.[131]

Pursuing his analysis in the 1770s and 1780s, Parmentier uncovered the gross chemical properties both of gluten and starch, and frequently noticed the close correlation between the starchiness of substances and the presence of sugar in vegetable tissues.[132] The latter phenomenon led into the main research of his later years, equally famous with that on the potato since it issued in the extraction of sugar from grapes and later beets in consequence of the continental blockade during the Napoleonic wars.[133] The dairy and cheese industry was comparably indebted to his studies of the chemistry and fractionation of milk,[134] and there is general agreement that the refinements he effected in the method of milling flour improved the yield by a third, while the modifications he proposed in the baking of bread increased quality together with quantity.[135] The technicalities are interesting, but it is rather the nature of his research that is relevant here. "When you want to be of service to your fellow-beings," he once remarked, "it is not enough simply to tell them once for all what you have found, what you have done, and what they need to do. You must never tire of repeating it in every way possible and in every possible form—except in that of authority. It is only by means of popularizing science that it can be made useful."[136]

An approach like that of Parmentier could avoid professional odium so long as a specialty was in its nascent stages, since in writing for the public he was going over nobody's head. His was a prospective discipline, moreover, that—again like early engineering—would make sense only in its application. A word about the communication of his research on potatoes will illustrate the pattern. He began his series of chemical analyses in 1771, partly in consequence of an inquiry addressed to the Faculty of Medicine by the controller-general concerning the attribution of various ailments to the consumption of potatoes in Alsace and Lorraine, the only provinces where they were much cultivated. In order to reach a wide au-

[131] "Végétaux pouvant servir en temps de disette à la nourriture de l'homme" (1773) in *ibid.*, 21-32.

[132] *Expériences et réflexions relatives à l'analyse du bled et des farines* (1776).

[133] *Instruction sur les sirops et les conserves de raisin destinés à remplacer le sucre dans les principaux usages de l'économie domestique* (1809). See also Balland (1902), Bibl. nos. 136, 137, 154 and 165.

[134] *Précis d'expériences et observations sur les differentes espèces de lait* . . . (1799), with N. Deyeux. See also Balland (1902), Bibl. nos. 60, 96.

[135] *Le Parfait boulanger, ou Traité complet sur la fabrication et le commerce du pain* (1778). See also Balland (1902), Bibl. nos. 13, 20, 24, 55.

[136] *Traité sur la culture et les usages des pommes de terre, de la patate, et du topinambour* (1789), 9-10.

dience, Parmentier had the later editions of his *Examen chimique des pommes de terre* (1773) redesignated *Ouvrage économique sur les pommes de terre*.[137] The more inviting title was no misnomer, for the accounts of experiments occupy only a small portion of the book. It also contains descriptions of the different types of potato; instructions on planting and cultivation; advice on preparation, cooking, and seasoning; recipes for making bread with potato flour instead of wheaten flour (though he acknowledged potatoes to be better in themselves than in the form of bread); and the argument repeated innumerable times, in this and other works. Since potatoes flourish in precisely the conditions of soil and climate that are most damaging to grain, he was recommending them for a fallback or a supplement and not a replacement. They were to be considered the most valuable boon conferred upon the Old World by the New, which owed the vigor of its native races to their sustenance.[138]

In 1789, some sixteen years later, he gathered up the threads of much intervening research in a second general treatise, equally discursive, in which the main themes from which he digresses pertain to the botany and agronomy rather than the chemistry of the potato.[139] Short sections deal with two inferior, though still valuable, plants with which it often was confounded, the sweet potato and the Jerusalem artichoke. Lest readers feel some want of order in his matter, he admitted that he had not felt quite ready to bring these studies up to date. The bad harvest and appalling winter of the preceding year, followed by devastating hailstorms and consequent food shortage, had persuaded him to hurry this imperfect work into print. There was the greater urgency in that rumors about potatoes exhausting the soil and impeding the growth of cereals had cropped up to compound the perennial false alarms about their effects on health.

In the interval between the two promotional treatises, Parmentier kept the subject alive with continual publication and even with occasional publicity stunts. The government granted him funds and access to certain vacant expanses in Grenelle and in Les Sablons, where the ground was about as inviting to cultivation as in a disused junkyard. There amid the urban rubble he planted experimental patches of potatoes, and astonished the guardians of order by rejoicing instead of taking umbrage at their theft.[140] Looking to educate a more eligible public, he invited persons of fashion and position to dine at the Invalides at repasts in which potatoes were the basis of all the courses. His guests on one occasion were Arthur

[137] Balland (1902), Bibl. no. 3.

[138] His best organized and most comprehensive summary is the article "Pomme de terre" contributed to Rozier's *Cours complete*. . . . 7 (1789), 179-215.

[139] *Traité sur la culture* . . . , note 136 above.

[140] Parmentier, *Mémoire sur la culture des pommes de terre aux plaines des Sablons et de Grenelle* (1787); see also Balland (1902), Bibl. no. 56.

Young (who found Parmentier an epitome of the Gallic conviviality that he met with less often than he had expected), Lavoisier, Broussonet, the abbé Commerel, Mailly (a president of parlement), and others of that ilk. There were two soups, a purée and a bouillon with dumplings. A matelote, or wine and herb stew, came next, followed by two main dishes, one in a white sauce and the other maître d'hôtel. Second servings consisted in five different platters "not less good than the former": a pâté, fritters, a salad, beignets, and "an economic cake," for which Parmentier had published the recipe. The continuation of the feast was not extensive, but "delicate and good"—cheese, preserves, biscuits, tarts, and a brioche, all made of potato flour, after which the company took a starchy infusion imitating coffee. "I could have wished," recorded Parmentier in apparent ignorance of vodka, "that fermentation had enabled me to prepare a liqueur with potatoes to make my company completely happy."[141]

Parmentier was elected to membership in the Society of Agriculture in 1773, when it was in the doldrums, and participated enthusiastically in its reorganization in 1785 and in the activities of the few years that remained before all such societies were undone in the Revolution. There is a certain anomaly here. From having been inadequately institutionalized, the affairs of agriculture quite suddenly became the concern of two organizations that cooperated less than they competed. For it was also in 1785 that the agricultural administration within the Ministry of Finance constituted an advisory Comité d'agriculture, which functioned for all the world like an agency of government during the three years of its existence. Exactly what was at issue here, among administrators, agricultural experts, and scientists, is not easy to determine, and the best course will be to summarize the facts while also venturing a comparison that may explain the contrast, at least in part.

The Society of Agriculture owed its resuscitation to the initiative of the intendant of Paris, François de Bertier de Sauvigny, and to his fortunate choice of a permanent secretary in the person of an energetic and charming young naturalist, Auguste Broussonet. After Bertin's retirement in 1780, the activities of his ministry were distributed among other agencies, agriculture in general reverting to the Bureau des Impositions in the ministry of finance, whereas the relevant institutions in and around the capital, among them the Society of Agriculture and the veterinary school at Alfort along with the Royal Society of Medicine, were detailed to the intendant of Paris.[142] The latter provision was more reasonable than might be thought since the generality included some of the richest farmland in

[141] Balland (1902), 50-52.

[142] For these arrangements, see Passy (1912), 129-135; or Bourde (1967) *3*, 1292-1303.

the country—Brie, the Beauce, the Gâtinais, the Vexin, and the ancient pays de France. As for the intendant himself, Bertier de Sauvigny was a magistrate who combined official ambition with a family tradition of enlightened, patrician agriculture, emphasizing stock-breeding and sylviculture. His father had been intendant of Paris before him and a member of Bertin's ministerial council. Besides the intendancy, Bertier was well placed at court, being superintendant of the queen's household.[143] His was that not-uncommon type of man of the world who would swell his train with academic and scientific associates, who would be a patron of their work, knowingly affecting to share in it, a Maecenas whose tacit price was to be treated like a colleague. It was at his instance that the modest veterinary school at Alfort found itself endowed with a research program and ornamented with professors of the standing of Daubenton and Vicq d'Azyr, whom he had taken up as he also had Parmentier and the cause of the potato, in the mode along with agronomy at large.

Broussonet, on whom Bertier fixed to reanimate the society, was well cut out to be the Vicq d'Azyr of agriculture, bringing to the opportunity the meridional dash of Montpellier.[144] His father, a physician, taught in the Medical University, and his brother became its dean. In 1779, at the age of eighteen, he himself qualified for his medical degree with a thesis on respiration. His examiners were so well impressed with his defense that they thought to secure him the succession to his father's chair then and there. That gambit came to naught, and in any case Broussonet's early scientific pleasure lay in natural history. The Linnaean system prevailed in Montpellier, and the young Broussonet imagined for himself the project of a classification of fishes on Linnaean principles. Off he went to Paris after taking his doctorate. Daubenton readily overlooked his principles, drawn by the lad's personal quality and talent, but Broussonet was disappointed in the materials at the Jardin du roi and moved on to London in 1780. He was delighted with the resources of the British capital, where he cut a swath among the naturalists of the Royal Society and attracted the interest of its president, Sir Joseph Banks, just back from the first South Seas expedition with Captain James Cook. Among the specimens of natural history were many fish from those far parts. Banks made Broussonet free of all he had collected, and description of these creatures formed

[143] On the career of Bertier, see Ardashev (1909); and also the Cuvier éloge of Ollivier, an associate of Broussonet, whom Bertier put on to doing a statistical study of the generality (Cuvier, 1818-1827), *2*, 233-265.

[144] On Broussonet, see the article by Jean Motte, DSB 2, 509-511, with an excellent bibliography. Note specially the éloge by Cuvier (1818-1827), *1*, 311-342; and a manuscript memoir correcting certain details by one Durand, BMHN, MSS 1991, pièces 242-248.

the starting point of an *Ichthyologia*, intended to embrace 1,200 species.[145] Ten sections came out in 1782.

Anglomane no less than agronome, Bertier was also in London from time to time and there made the acquaintance of his rising young compatriot. When Broussonet returned to Paris, late in 1784 or early in 1785, his reputation was one of youthful urbanity and scientific promise. Daubenton, beginning to age and to weary, took him on to supply lecturing both at Alfort and at the Collège de France. Support from such patrons improved upon the reception of his *Ichthyologia* and won him election to the Academy of Science (he was already Fellow of the Royal Society). The drought of 1785 then intensified the inducements of civic service invoked by Bertier, and it was altogether in character that Broussonet should have embraced the seduction of promoting the Society of Agriculture. The permanent secretaryship brought him into immediate prominence in the salons of science and the anterooms of administration, virtually ending his own production at the very outset of his career. For no further installments of the *Ichthyologia* ever appeared.

At the same time, his success in galvanizing the Society into life was an indication that agronomy had in fact developed a constituency. Theretofore, its publications had amounted only to occasional circulars wistfully distributed and never followed up. Now, proceedings and memoirs were printed quarterly and gathered annually. The first volume came from the bindery in December 1785:[146] Bertier used his influence at court to have an audience granted of the king and queen. On 26 February 1786 their majesties were graciously pleased to accept a copy from the entire company assembled, together with a golden jeton, or token of presence, representing a pastoral Louis XVI sceptering cows into the pastures of needy farmers in the Île de France. A public meeting followed in Paris at the Hôtel de l'Intendance on 30 March. Calonne was present on both occasions, having presided at the former. At the latter, papers were read, by Daubenton on the folding of sheep, by Parmentier on potatoes, by the marquis de Turgot (the stateman's nephew) on resinous trees. Prizes were awarded and homage paid to eminent patrons.[147] The doings, in a word, were standard for an eighteenth-century scientific and learned society, and created the visibility which was one of the objects. The quarterly memoirs

[145] *Ichthyologia sistens piscium descriptiones et icones* (London, 1782).

[146] *Mémoires d'agriculture, d'économie rurale & domestique* (1785-1791). Actually only two "Trimestres" appeared in the first year. Thereafter, the numbers appeared quarterly through the first issue of 1790, when publication was interrupted by revolutionary circumstances. For bibliographical detail on the publications of the Society, see Louis Bouchard-Huzard, "Notice bibliographique sur la Société d'Agriculture" (1863), reprinted from *Mémoires de la Société Impériale et Centrale d'Agriculture de France* (1861).

[147] Passy (1912), 144-147.

contain little that will be surprising to anyone who has read in the literature of agronomy. Authors recount the success or failure of agricultural experiments and rehearse the by now familiar themes, urging turnips, artificial grasses, rotation, cultivation, reforestation, and the increase of livestock, and deploring fallow, wasteland, routine, and *vaine pâture*. The content was technical and theoretical for the most part. It would appear that Bertier had the Society avoid juridical, social, and fiscal topics, in politic keeping with its learned and scientific mission.

Retorting upon Arthur Young's gibe about the membership, Monsieur Bourde in his big book on agronomy admits that the Society had enrolled few dirt farmers, if any, and observes that to have reached out beyond the circle of administrators, scientists, rich proprietors, and philanthropic noblemen would have been contrary to its first intention, which had been to enlist the interest of just such persons in the reform of agriculture.[148] For everyone agreed that nothing was possible without the leadership of wealthy, well-disposed landowners who could afford to set an example. It is symptomatic that the Society of Agriculture, in its small way (the number of resident members from 1761 through 1788 was only fifty-eight),[149] should have at last acquired vitality at just the juncture when the government in its fiscal desperation convened the Assembly of Notables in 1787. The persons on the list of the Society were mainly of that sort: among scientists (to name a few) Lavoisier, Vicq d'Azyr, Desmarest, Fourcroy, Fougeroux de Bondaroy, Daubenton; among courtiers and philanthropists, the duc de La Rochefoucauld-Liancourt, the comte d'Angiviller, the inevitable Lamoignon de Malesherbes, the marquis de Gouffier; among magistrates and bureaucrats, Abeille, d'Ailly, d'Ormesson, Dupont de Nemours—clubbable to a man, they were, in the interest of a good cause. Moreover, at Bertier's instance the Society did reach out toward the countryside, forming local *comices agricoles*. With its Catonic overtones of rural Roman virtue, the phrase denoted agricultural associations that would put on farm shows culminating in banquets at the Hôtel de Ville or the great chateau in market towns throughout the Ile de France.[150]

When attention turns to the Committee on Agriculture and to the respects in which it complemented or competed with the Society, only one thing is altogether clear. There was no difference between them over agricultural theory. Institutionally, the Committee was a creation of the Finance Ministry. Politically and ideologically, it would appear to have been the creature of a collaboration between Dupont de Nemours and Lavoisier, its spirit being authoritarian and technocratic rather than pro-

[148] Bourde (1967), 1311-1313.

[149] Passy (1912) gives the list, 212-220.

[150] Bourde (1967) *3*, 1329-1331.

motional and propagandistic like that of the Society. When the several agencies concerned with agriculture under Bertin reverted to the Ministry of Finance in 1780, the functionary in charge of the Bureau des Impositions was one d'Ailly, formerly the chef clerk of Lefèvre d'Ormesson, an inner-circle counsellor of state critical of the departed Bertin. In 1783 d'Ormesson served briefly in the office of controller-general, to be succeeded by Calonne. Calonne then replaced d'Ailly with Charles Gravier de Vergennes, a young magistrate and nephew of the great and famous secretary of state for foreign affairs.[151]

The story of Dupont's part goes right back to the fall of Turgot, when Maurepas had Dupont exiled to his estate in the Gâtinais. Such a precaution was often taken with associates deemed too close to cashiered statesmen. Dupont, it will be recalled, had gone to school to Quesnay and physiocracy before ever attaching himself to Turgot. He now employed his rustication to become a practical farmer. Not being wealthy, he even needed to do that. In the same interval he also became a familiar of the elder Vergennes and his circle. Necker recalled Dupont to Paris and to service in the Finance Ministry in the winter of 1778-1779, and it appears to have been he who put the younger Vergennes onto the notion of promoting the office of taxation into a quasi-ministerial direction of agriculture while seeking out expert advice in the formation and execution of policy.[152]

Lavoisier's interest in agrarian policy and economic theory is well known and needs no further exposition for itself.[153] What does need emphasis here is that in the 1780s the practice of agriculture should have come to seem an inviting and accessible object of analysis, and another sector in which science could properly instruct government, to a leading scientific intelligence and the most influential single member of the Academy of Science. Lavoisier had inherited from his mother a property at Le Bourget, which was farmed by the local postmaster, one Musnier. In 1778, three years after taking responsibility for the Arsenal, he also purchased the more considerable estate of Fréchines in the Loire Valley between Blois and Vendôme. There he had a program of agricultural experiments conducted, coming down from Paris for two to three weeks at seedtime, in the winter and at harvest to estimate the results and to give directions. The purpose was to improve the productivity of a run-down farm in a region of backward husbandry. Livestock needed to be increased for the sake of their manure and fodder in order to feed the livestock.

[151] Bourde (1967) *3*, 1292-1302.
[152] Dujarric de la Rivière (1954), 55-73; Passy (1912), 166-167.
[153] Lenglen (1936); McKie (1952); Dujarric de la Rivière (1949), Sneaton (1956b).

Those related deficiencies compounded each other in what was widely agreed to be one of the critical defects in French agriculture generally.

In 1787 Lavoisier drew up a report on his methods and results after nine years of trial.[154] The contrast with Duhamel's experiments a generation earlier is like that of a qualitative with a quantitative chemical analysis. Lavoisier had had an exact plan surveyed for every field and piece of ground. He maintained a register in duplicate with a section for each field. One copy he kept in Paris, the other in the manorhouse. It gave him a record of the input of seed, manure, and labor, and of the yield, for every crop planted in every field during the entire period. Cross-indexing by the identity of the crop permitted comparison of the results crop by crop, field by field, and year by year. Lavoisier considered that the most distinctive feature of his husbandry was the method of measuring the harvest. When maximum accuracy was important, he had all the sheaves weighed when bound. In ordinary operations, he had the average determined by weighing eight or ten in each wagon. Thus, after every harvest he knew exactly the number and weight of sheaves in the barn. The same precision governed the threshing. The proportions of wheat, straw, and chaff were determined and recorded for every crop grown in every field. Equally significant were the economic data. Lavoisier kept accounts that enabled him to specify the number of sheaves in each harvest that went to pay for tithe, labor, seed, rent, taxes, maintenance of stock and equipment, and the livelihood of the farmer and his family.[155]

Those calculations gave him a basis more definitive than the farmer's perennial lament for asserting that it was impossible under existing burdens for capital to return more than five percent from farming, be the cultivation never so rational. In seven years he had succeeded in increasing his supply of manure from two loads per acre to six or seven; he was keeping a herd of twenty cows and a flock of 300 sheep where there had been a few miserable beasts; his barns and granaries were overflowing; he had potatoes, clover, turnips, and vetches where there had been waste or fallow. He would have done far better economically to invest the money he had ploughed into all this in government paper or other financial opportunities. The conclusions that he generalized from his experience were predictably physiocratic, and other of his agricultural pieces bespeak the fiscal expert more evidently than they do the laboratory analyst. At one time he had thought to write an entire treatise of agriculture, and though he never found time for that, he did draw on what he had learned from

[154] "Résultats de quelques expériences d'agriculture, et réflexions sur leurs relations avec l'économie politique," *Annales de chimie* (Dec. 1792) *15*, 267-285. Printed in *Oeuvres de Lavoisier* 2, 812-813.

[155] McKie (1952), 203-217, gives an extensive précis.

Fréchines in a fragmentary study of the dependence of national wealth on agriculture.[156] The first materials for the analysis went all the way back to his neophyte's mineralogical expedition with Guettard.[157]

In an early meeting of the Committee on Agriculture, Lavoisier harked back to that same enterprise, à propos of a scheme for taking a national inventory of mineralogical and agricultural resources.[158] The committee held its first meeting on 16 June 1785 under the designation "Administration d'Agriculture au Contrôle-Générale des Finances."[159] The drought of that year had hastened its formation. Requests for help and suggestions for alleviating the effects poured in upon the office of Vergennes, whose agents would have been overwhelmed even if they had known how to respond. In the emergency, a small group of experts was self-impaneled. The initial membership consisted of seven persons: Vergennes himself; his chief clerk, Lubert; his éminence grise, Dupont; and from the Academy of Science, Lavoisier, Darcet, Tillet, and Poissonier. It is reasonable to surmise that Darcet was included for his experience in technical administration at Sèvres, where he was inspector of the royal porcelain factory; Tillet for his knowledge of plant pathology and crop parasites; and Poissonier for his influence at court. Lavoisier took upon himself the role of secretary, which meant that he established the agenda, kept the minutes, and ran the committee.

Perhaps the historian may here permit himself a reflection upon the energies that animated people whose acts he is chronicling. In that same year Lavoisier was serving in the office of director of the Academy of Science and effecting its reorganization into eight sections conforming more closely to the actual structure of the scientific disciplines than the original six.[160] It was on 28 June 1785 that he read before the Academy the climactic paper of his research into combustion, the "Réflexions sur le phlogistique."[161] During the ensuing three years of the Agricultural Committee's activity, he was taking the leading part in the development of the modern system of chemical nomenclature, which consummated the chemical revolution, and he was beginning the composition of his *Traité élémentaire de chimie*.[162] All the while his responsibilities in the Tax Farm con-

[156] *Résultats extraits d'un ouvrage intitulé "De la Richesse Territoriale du Royaume de France,"* Lavoisier, *Oeuvres* 6, first published as a pamphlet in 1791, by order of the Assemblée nationale, Reprinted in 1819. Cf. Schelle and Grimaux (1894).

[157] Above, Chapter I, Section 6.

[158] Pigeonneau and Foville (1882), 9e séance, 15 Sept. 1785, 68-69.

[159] *Ibid.*, 2-10.

[160] Above, Chapter II, Section 2.

[161] MARS (1783/86), 505-538; Lavoisier, *Oeuvres*, 2, 623-655; cf. Daumas (1955), 58.

[162] *Méthode de nomenclature chimique, proposée par MM de Morveau, Lavoisier, Bertholet, & de Fourcroy* (1787); cf. Daumas (1955), 61. The *Traité élémentaire de chimie* (1789) was presented to the Academy on 17 January of that year (Daumas [1955], 63).

tinued undiminished, as they also did at the Gunpowder Administration involved in its conflict with the saltpetremen.[163] Such were the other matters on his mind besides the progress of his experiments on the pilot farm at Fréchines, a proving ground for the measures his committee was now urging upon the government and the nation.

The working plan he drew called for executive sessions in an office of the ministry. The group convened weekly at first, and after the first five sessions every two weeks. In accordance with normal administrative practice, the formal record of its deliberations was to be signed by everyone present. "This register," Lavoisier laid down at the outset "will become the repository of the principles of national agriculture, . . . and may serve for the guidance and education of those concerned with the same subject in the future."[164] The tone and touch are unmistakable. Unlike Parmentier, Lavoisier did not believe in imparting novelties by endless repetition; he did believe in the exercise of authority where competent. And he had little sense of how certain persons, who had been occupied with agronomy for thirty years and more, might feel that they had scarcely needed to await the signing of his register, yet another of Lavoisier's registers, to learn the principles of a national agriculture. The work plan further stipulated that if members of the panel felt unqualified to judge of a topic or invention, they would refer the memoir or instrument to the Academy of Science. Nothing was here said of the Society of Agriculture, and when Parmentier and Thouin, both members of the latter, were invited to join the committee, they declined.[165]

Some months later Dupont drew up the committee's view of the distinction between the missions of the two bodies. The Society was essentially an academy, limited in its purview to the science of agriculture. Its competence, moreover, was restricted to the generality of Paris, and the most it could pretend to nationally was the place of *prima inter pares* vis-à-vis the other societies. The "Assembly of the Agricultural Administration," as he now called it, was to be compared rather to the counsel that sat with intendants of commerce in administering the Bureau of Commerce. Potentially, indeed, the new body was even more important since agriculture was the source of subsistence and the richest, most valuable branch of industry. "It is not as scientist that the Government acts through this assembly; it is as Master, as Benefactor, and as Father."[166]

[163] Above, Chapter I, Section 6.

[164] "Rapport sur l'organisation des travaux du comité," Pigeonneau and Foville (1882), 1ère séance, 16 June 1785, p. 6; a slightly different draft appears in *Oeuvres* 6, 186-188.

[165] Pigeonneau and Foville (1882), xvi.

[166] *Ibid.*, "Mémoire sur la différence qui existe et qui doit exister entre l'Assemblée d'administration de l'agriculture et la Société d'agriculture de Paris," séance du 24 mars 1786, 199-202.

More can be known about the committee's would-be exercise of this benevolent and paternal function than about the actions of the Society, for Lavoisier's register survived and has been published.[167] Broadly speaking, the matters discussed were of two sorts, technical and administrative. Early on, Lavoisier offered the use of his farm at Le Bourget for purposes of experiment. His colleagues preferred a location nearer Paris and took a lease on the Clos de Verneuil in Asnières.[168] No record of any trials remains, and for the most part the technical aspect of the committee's work consisted in its response to initiatives from others. Memoirs rained in upon it outlining every imaginable improvement in cultivation, crops, and implements, most of them contributed by persons of some consequence. Clearly, whatever else ailed French agriculture, it was not a shortage of ingenuity on the part of thousands of local worthies. When dealing with these proposals, the committee's procedures and tone were those of ad hoc commissions of the Academy of Science passing upon the works of aspirants to scientific favor and upon the designs of inventors seeking subsidy or other advantage. In the very first session, the "sieur Hébert," a tax collector, was told that his ideas for enriching the soil contained nothing either new or useful, and on the contrary reflected the fruit neither of judgment nor experiment. The next week the mayor of Confolens was rebuffed for planting a species of weed among the wheat to discourage weevils. The details of his process, far from arousing confidence, served only to create doubts.[169]

The governmental aegis comes out in the minutes of the other, administrative aspect of the proceedings. By what means might the committee impart its information to the people who must put it to work? How might it reach from the meeting rooms of the Finance Ministry down into the village? The obvious instrument would be a journal. The committee was clear that such a vehicle should emanate from its authority, rather than from the Society (unfortunately, Bertier had the same idea) in order that it might become a force for public education.[170] A more economical approach would be to persuade the abbé Mongez, who had taken over the *Journal de physique* from Rozier, to enlarge the coverage of agriculture, and then to contract for a certain number of copies to be distributed gratis by the ministry. Printing up circulars and manuals and sending them out would also be expensive, and the committee explored the practicality of substituting for letter press a system of polytyping by which plates would

[167] Pigeonneau & Foville (1882).

[168] *Ibid.*, 2nd, 8th, & 9th séances, 23 June, 1 Sept., 15 Sept., 1785, pp. 7, 52-53, 58-59.

[169] *Ibid.*, 1st and 2nd séances, 15 and 23 June 1785, pp. 2, 3.

[170] Pigeonneau and Foville (1882), séance du 7 juillet 1785, p. 11.

be cast for each page and distributed to provincial centers where the sheets could be run off.[171]

More critical than merely multiplying publications, however, would be establishing a serious administrative connection with the localities. As a start, the committee put in hand a correspondence with provincial societies of agriculture, academies, and agencies of government. Needed was a combination of two elements; first, an intercourse like that between the Society of Medicine and its correspondents; and second, authority over a set of subordinate officials like the inspectors of manufacturing in the various generalities, who received directives from the Bureau of Commerce in Paris.[172] The problem with adopting the former model, if anyone even thought of it consciously, was the absence in agriculture of anything comparable to the structure of the medical profession with the country doctor or surgeon there on the scene. The problem with replicating the procedures for commercial regulation was that the notion presupposed the existence of what the administration of agriculture was trying to become. In this institutional vacuum, Lavoisier and his associates thought to enlist the services of the one person in the village who might be made accessible to instruction and who also had the farmer's ear and could speak his language. It happened that the abbé Lefebvre was an agronomist by avocation and by profession procurator-general of the Order of Sainte-Geneviève, which supplied clergy to an enormous number of country livings. It is unclear whether the idea of making a parish agricultural agent of the village curate originated with Lefebvre or with the committee. In either case, both parties entered into it enthusiastically. Lefebvre became a member of the committee, and Lavoisier put his colleagues in touch also with the general of the second principal order of secular clergy, the Prémontrés, urging him to have agriculture included in the subjects taught in a training school for practical subjects open to its priests at Soissons.[173]

Lacking in all this, together with the most rudimentary sense of political feasibility, was money—which, indeed, came to the same thing. Calonne, unmoved by an apostrophe on the part of Lavoisier summoning him to become the Trudaine of agriculture,[174] allowed the committee exactly 3,000 livres. Most of that went for beet seeds from Bavaria. Thus, the correspondence could never graduate into anything beyond the exchange of views, fine views of betterment. Naturally enough, in the cir-

[171] *Ibid.*, 23 Juin 1785, 8; "Réflexions sur les moyens de faire parvenir aux habitants de la campagne les instructions publiées par le gouvernement," Lavoisier, *Oeuvres* 6, 227-229.

[172] Below, Chapter VI, Section 3. For the procedures of the Bureau of Commerce, see Bonassieux and Lelong (1900), introduction.

[173] Pigeonneau and Foville (1882), xviii-xix; séance du 9 Novembre 1785, 82-83.

[174] *Ibid.*, 15 February 1786, 175-176.

cumstances, the discussions at the fortnightly sessions grew more and more physiocratic, more and more concerned with prevailing by argument at the seat of government and persuading the powerful and privileged to modify the fiscal impositions and legal impediments obstructing the progress of agriculture. Equally naturally, ministers began to weary of the committee's recommendations, which increasingly took on the character of importunities. Feeling its influence slipping, the committee began to enlarge its membership, adding persons who might strengthen its credit in the lobbies and corridors of power, most notably the duc de La Rochefoucauld-Liancourt. He attended relatively seldom in person and was represented by one of those highly articulate retainers of Polish extraction who are not infrequently to be observed gravitating around French statesmen of high lineage. Maximilien Lazowski's father had come to France in the retinue of Stanislas Lesczinski, and he had himself arrived in Paris by way of military service and an English exile. His discourse tended to dominate the later sessions, marked more by anglophilia than by agronomy.[175]

Calonne was dismissed from the Finance Ministry in April 1787, to be replaced in interim by Fourqueux and then by the scarcely less obscure Laurent de Villedeuil, the over-all fiscal responsibility (if that is the word) being confided to Loménie de Brienne, archbishop of Toulouse, at the conciliar level. Gravier de Vergennes survived in office until 17 June, and the committee staggered on through the summer, holding its sixty-ninth and final session on 18 September.[176] There is no indication in the register that the members knew they would not convene again. Indeed, it was at this session that Lavoisier read his report on the nine years of experience he had acquired in farming Fréchines. On 28 December 1788 he repeated the reading before a ceremonial assembly held in the presence of Necker at the Hôtel de Ville by the Society of Agriculture, which had been crowned with the designation "royal" by letters patent of the previous 30 May.[177] For the government could not finally ignore the institutional claims of agriculture, much though its own nature and limitations might lead it to prefer the approach of Broussonet and public relations to that of Lavoisier and executive action armed by science.

Lavoisier's futile defense of the administrative committee of agriculture in the summer of 1787 gave him the first taste of political failure, his earliest experience of justifying an enterprise, worthy of the nation in its science and its prospects, before a group of politicians too beset to heed or even hear a mandarin. Then it was that he observed, in a memoir often quoted though seldom situated in the background of his preoccupations,

[175] *Ibid.*, xxiii-xxv.
[176] *Ibid.*, 440-442.
[177] Passy (1912), 193-201, 286-287.

that deficiencies in knowledge and education were not the only factors working against the progress of agriculture in France: "in our institutions and our laws it meets with obstacles that are more real."[178] Lavoisier was even then serving as a delegate in the assembly of the Orléanais, where ownership of Fréchines made him a landed proprietor, and composing a statement on the needs of agriculture for the guidance of elective bodies representing other provinces in this year of portent.[179]

There were eight counts in his indictment of the regulations and practices weighing on the countryside: (1) the arbitrary imposition of taille; (2) the obligation of the corvée; (3) feudal and ecclesiastical tithes and dues, which often consumed half the real income of a locality; (4) excise taxes on salt, tobacco, and other articles of consumption; (5) the banality or monopoly over the milling of grain exercised by local lords; (6) the right of free pasture; (7) the interference of millers with watercourses and the consequent flooding of much bottomland; and (8) prohibitions forbidding the exportation of grain or other agricultural products. It was not merely on economic grounds that Lavoisier deplored this structure of prescriptive abuses. He was equally vehement on the moral damage they inflicted through the systematic humiliation of the productive classes in the name of law.

It is clear from the events that soon followed, no less than from the wealth of detail and refinement of analysis in the writings of social and economic historians, that Lavoisier was largely right.[180] What is interesting is that it should have been a scientist who came to these conclusions. For what could science do about any of these, the real problems? From a technical standpoint, the enormous literature of agronomy leaves a sense of overkill. If institutions and attitudes could have been changed, a mere fraction of it would have offered guidance enough. And until institutions and attitudes could be changed, reform would be limited largely to a few model farms, or at best to change by permeation at an imperceptible rate, and science would continue to be invoked in the literature rather than applied in the countryside. It was otherwise with the mechanical arts: in industrial matters, officials could give leads here and there; in engineering, new ways could come about without an immediate change in the basic rhythm of life for the whole society.

[178] "Mémoire sur le département de l'agriculture," Pigeonneau and Foville (1882), séance du 31 juillet 1787, 400-417, p. 408-409. A revised version is printed in Lavoisier, *Oeuvres* 6, 189-190.

[179] "Instruction sur l'agriculture pour les assemblées provinciales," *Oeuvres* 6, 203-215.

[180] For a review article on the state of the question of agricultural growth in eighteenth-century France, see Forster (1970).

CHAPTER VI

◇◇◇

Industry and Invention

◇◇◇

1. THE STATE AND INDUSTRY

In the manufacturing sector it is even more notable than in the agricultural that two closely linked factors differentiate the French pattern of technical modernization from the British in the later eighteen century. On the one hand, French entrepreneurs habitually looked first to the state rather than to the financial markets or to private savings to provide the capital they clearly needed together with the protection and privileges they felt they needed.[1] On the other hand, government officials charged with economic responsibilities systematically resorted to scientific information and personnel in judging of projects and proposals and sometimes in implanting them. The mercantile aspect of those tendencies dated back to the reign of Henry IV and the ministry of Sully, while the scientific aspect derived from the reign of Louis XIV and the ministry of Colbert. The evolution of these relations in the later eighteenth century, following the elder Trudaine's tenure in the Bureau of Commerce,[2] was marked by a continuing tension in governmental agencies between the novelty of liberalism and the persistence of paternalism, combined with rapid increase in the influence of scientific expertise across the whole gamut of these relationships. In the latter respect, moreover, there can be no doubt about the reality of the effects in industry, if not in agriculture.

After Daniel Trudaine, the most effective administrator of the Bureau of Commerce was Jean-François Tolozan. Trudaine (it will be recalled) was succeeded on his death in 1769 by his son, called Trudaine de Montigny. Less energetic in his nature, leading a life more ornamental than instrumental, the younger Trudaine allowed the clerks to run the agency. Turgot, for his part, preferred to handle economic matters himself and made little use of its personnel for anything beyond routine. It fell to Tolozan to reanimate the bureau and make it once again the nerve center of commercial policy. He was already fifty-five years old in 1776, when, in

[1] After writing the above, I began to realize that the comparison is one that must in part have been recollected from the classic, uncompleted work of Ballot (1923), 3, 10, and *passim*, whose concern is with mechanization of industrial production. See also Martin (1900).

[2] Above, Chapter I, Section 1.

the reorganization of the government following the dismissal of Turgot, Louis XVI named him to be one of the four intendants of commerce. Tolozan had made his entire career in the bureaucracy, where his reputation for probity matched that of the elder Trudaine, or indeed of Turgot. When Necker came to office in the Finance Ministry in 1777, he moved Tolozan into Trudaine's old suite of rooms, where he quickly became the leading spirit of the bureau, charged specifically with responsibility for the program—practice would be a more accurate word—of premiums and subsidies paid out to manufacturers and inventors, with oversight also of the entire hosiery industry, and finally with supervision of commerce in general in a region comprising seven generalities in central and western France.[3]

For all that he represented the civil service at its best and most conscientious, Tolozan never carried the weight in government that Trudaine had done. He is to be identified, not with some brave new departure, but with adapting the bureau's mode of encouraging commerce to changes in the condition of the economy and also in the opinion of informed people. The middle of the century, when Trudaine was in the prime of his statesmanship, had been a period of economic growth. Vincent de Gournay's liberalism then made sense, and what we have called the encyclopedic phase of the application of science to agriculture and manufacturing was in keeping with it, not to say a feature of it.[4] Trudaine could favor planting or transplanting industries and expect that, after a little nurturing, entrepreneurs would run their undertakings progressively in a climate of laissez faire. In fact that never came to pass, and in the final decade of the old regime circumstances had changed. Financial stringency and recurrent fiscal crisis called for regulation and for selective response to the perpetual demand for government involvement emanating from industry itself. Tolozan thought to reform where possible, and to rationalize, not to initiate or liberate. His essay on the commerce of France and of her colonies, composed in the early months of 1789, is a lucid overview of the whole economy at the very end of the old regime.[5] It is also the best single statement of the case for a moderate degree of regulation. To call the point of view neo-mercantilist would exaggerate the reservations he entered against the proponents of unlimited commercial liberty and the advocates of British

[3] On Tolozan and the Bureau of Commerce, see Bonnassieux and Lelong (1900), lxi; Parker (1965), 87-88; Biollay (1885), 369-384; and the very rough draft of Tolozan's own memorandum on the service he directed, AN, F^{12}.657: "Compte rendu par M. Tolozan des différents objets qui concernent son département," 13 May 1787, to be addressed to the controller-general, Laurent de Villedeuil.

[4] Above, Chapter V, Section 1.

[5] *Mémoire sur le commerce de la France et de ses colonies* (1789). It also affords a reminder of the importance of the Caribbean islands in the French economy, so that the decision of the government to retain Guadeloupe and Martinique rather than Canada in the peace negotiations of 1763 seems less irrational than it is sometimes said to have been.

solutions to French problems. He himself used terms like "middle way" and "common sense."

The difference of opinion divided his own service.[6] We have already noticed how one of the most dogmatic, even punitive, enthusiasts for forcing French industry to be free was Roland de la Platière, inspector of manufactures at Amiens.[7] All parties agreed on the importance of scientific rationalization in technology, however, and that consensus was more significant than these divergences in philosophy. In this, the bureaucratic phase of scientific penetration into the economy, government would supply consultation in technology, guiding management and drawing on members of the scientific community. The relation was both more sophisticated and more indirect than that out of which it had developed in Trudaine's time, the immediate subsidy by the Treasury of whole enterprises and industries in the belief that what was being primed would prove to be a pump.

The notion of an encyclopedic followed by a bureaucratic mode of applying science is a matter of nuance, more descriptive than definitive in value. In both phases, and indeed throughout the eighteenth century, the technological undertakings of the French government appear to have fallen naturally into four classes that admit of rather more categorical distinctions: they consisted, first, of the conduct of certain enterprises under the immediate ownership of the crown; second, of measures involving direct governmental stimulus to innovation and industrial development; third, of a set of responses and encouragements to initiatives coming from artisans, inventors, and entrepreneurs; and fourth, of new educational departures, the less important at the lower level of certain artisanal schools, the more important at the higher level of serious technical schools looking to the professionalization of engineering, civil as well as military. Not that these categories correspond to any settled plan of policy, nor even to the administrative purview of departments and ministries—they simply allow a canvass of things that agencies of the state did in fact do amid the interplay of civic needs with technological opportunities.

2. STATE-OWNED INDUSTRY: SÈVRES AND THE GOBELINS

In addition to the age-old responsibility of the crown for military and naval construction and production of munitions, four manufacturing enter-

[6] The issue was discussed in responses made by the Deputies of Commerce, representatives of the important provincial cities; by local chambers of commerce; and by the corps of inspectors-general of commerce, to a memoir circulated by Tolozan asking for their opinion of a plan for a "Système d'administration intermédiaire entre la stricte exécution des anciens Réglemens et la liberté générale et indéfinie." The exchange will be found in AN, F[12].654.

[7] Above, Chapter V, Section 2.

prises were owned by the state and administered by the Batimens du roi in the eighteenth century: the tapestry looms at the Gobelins, the porcelain manufactory at Sèvres, the oriental and Turkish rug shop called the Savonnerie on the site of an ancient soap works at Chaillot, and the establishment for upholstery and hangings in Beauvais.[8] Terminology is confusing, for the phrase *Manufacture royale* was also accorded to other concerns. Sometimes it referred to an initial impetus and an elaborate structure of privileges, monopoly, and close regulation. The most famous example was the Manufacture Royale des Glaces producing mirrors, mainly at Saint-Gobain.[9] In other cases it connoted little more than the recognition of quality signified nowadays by the "appointment" of certain products to the Queen of England. The grant of the usage in 1784 to the Montgolfier paper factory in Annonay will afford a characteristic instance.[10] Of the state-owned industries, the Savonnerie, which dated from 1626, was a relatively minor affair. In the mid-eighteenth century, it employed only twenty workers and nine or ten apprentices.[11] The installation at Beauvais, on the other hand, was in a flourishing state, with a labor force of 120 specializing in the finest fabrics both for hangings and for covering the canapés and fauteuils that made the reigns of Louis XV and Louis XVI great periods in the history of furniture.[12]

For the present purpose, however, Sèvres and the Gobelins were the significant establishments. The very names evoke productions of art rather than industry, of course, and it is a surprise to find that the number of workers who earned their living at Sèvres in the 1770s and 1780s was more than three hundred.[13] The organization of the Gobelins was more complex, and also more entrenched. The crown had taken title in 1667, three years after instituting the manufactory for mirrors later moved to

[8] Havard and Vachon (1889) celebrate the history of all four, the occasion being the Exposition of 1889.

[9] On Saint-Gobain, see Cochin (1865) and Frémy (1909), the older work being the more informative. For an overview of the later history of Saint-Gobain, which moved in the nineteenth and twentieth centuries from mirrors by way of chemical industry into atomic energy to become one of the giant conglomerates of our own times, see Choffel (1960) and Hartemann and Ducousset (1969).

[10] Below, Chapter VI, Section 3.

[11] Harvard and Vachon (1889), 317.

[12] *Ibid.*, 590.

[13] "Memoire en faveur du Sr Boileau," 23 January 1772. BMS, carton A-1. The account in the present chapter is based almost entirely on the very rich documentation in the library at Sèvres. Personnel records with evaluations of each worker occupy five registers, classified V^e^, with further documents in cartons D-1, D-2, D-3. There are also very full accounts, minute-books of administration, and descriptions of early processes. The present organization dates from 1903, when Emile Bourgeois brought order out of the chaos into which the archives had been allowed to slide and prepared an inventory, *Les Archives d'art de la manufacture de Sèvres. . . . 1741-1905* (1905). There are also important materials in AN, F^{12}.1493-1498; O^1 2061-2063.

Saint-Gobain. Originally, the mission included provision of furniture together with tapestries. Five galleries of high-loom weaving and three of low-loom, each with its own entrepreneur in the earlier part of the eighteenth century, were served by a studio employing an entire team of painters and by extensive spinning and dyeing shops.[14] Both at Sèvres and at the Gobelins, moreover (and this rather than the scale is the reason for noticing them here), the government began assigning chemists to the staff for the purpose of developing products and procedures and controlling the composition of materials in the laboratory. The sources are fuller, or at least more accessible, at Sèvres, however, and afford a rare and welcome entry into the technical detail of artisanal "secrets" and into the actual course of developments that brought them under the scrutiny of science.[15]

The story begins on the side of Paris opposite from Sèvres, in the Chateau of Vincennes. An edict of 24 July 1745 accorded a privilege to a company newly installed there, which undertook "to manufacture in France porcelains of the same quality as those made in Saxony, in order to spare consumers in this kingdom the need to disburse their funds in foreign countries in order to obtain curiosities of this kind."[16] It was at Meissen in 1709 that a true or "hard" porcelain, like that imported from China and Japan since the sixteenth century, was first produced in Europe. Friedrich Böttger, a chemist and mineralogist, then learned that the property of translucence that distinguishes porcelain from other forms of pottery depended upon using a paste of kaolin mixed with feldspar. Kaolin occurs in Saxony.[17] None was yet known in France, however, and thereby hangs an irony in the tale. For policy precluded using foreign materials, and the reputation of Sèvres was made, not by fulfilling the charge to produce real porcelain, but by surpassing it artistically through the development of a soft (*tendre*) porcelain that owed its translucence to a glassy base.

The *fritte* was composed of a melange of Fontainebleau sand, crystallized nitre, sea salt, Alicante soda, alum, and gypsum. Vitrification required fifty hours. One-third its weight of clay was then mixed in, this "body" consisting of about two-thirds chalk and one-third Argenteuil marl. A little black soap made the paste malleable. Unglazed forms emerging from the kiln were called "*biscuit*." Sculptured groups and figurines were often left in that state. Normally, however, a lead-based glaze in liquid form was poured over the objects to be "covered." They then went back into the kiln. Colors were applied over or under the glaze ac-

[14] Fenaille (1903) *1*, 73-78, for the founding, and *4*, 10-15, for outline of eighteenth-century arrangements, which tended toward centralization.

[15] The most important printed source is Brongniart (1844), though see also Garnier (1892).

[16] BMS, carton A-2.

[17] Brongniart (1844) gives a clear and authoritative account of the early techniques.

cording to whether they could stand the firing. The famous Sèvres blue, mainly cobalt oxide, was a color of "*grand feu*" to be baked on under the glaze. Gilding had to be super-imposed afterward. Writing down the ingredients is easy. Their identity and proportion were a closely guarded secret, naturally, and here as in cookery the art was in the mixing and the handling. The detail of the proportions and manipulations occupies nine closely written pages of a manuscript memoir of 1781, a summary that Régnier, the director, ordered the chemists to prepare for the eyes of the responsible minister, d'Angiviller.[18]

Developments may be followed at three levels: on the high plane of policy, patronage, and official expectation (to which much supposed history of technology is restricted for lack of information beyond the drafts of regulations and decrees and surviving examples of the finished product); below that, in the intermediate range of administration, management, and accountability; and, down, finally, on the ground floor of technique, artistic design, execution, and labor in the shop. At the outset, the corporate model was the Manufactory of Mirrors at Saint-Gobain, which after early vicissitudes had become a commercial no less than a technical success by the mid-eighteenth century.[19] In porcelain, the entrepreneur was a former director of the Compagnie des Indes, Jean-Henri Orry de Fulvy. In the 1740s Orry de Fulvy was intendant of finance, a member of the Council of State, and a close friend of the marquis du Châtelet, governor of the Chateau of Vincennes (and complaisant husband of Voltaire's learned mistress, the translator of Newton). An elder half-brother, Philibert Orry de Vignory, was controller-general in 1745. Drawing on loans from the Treasury, the initial subscribers invested 90,000 livres. In 1747 the Council of State accorded the new company a monopoly over the manufacture of porcelain in France. The privilege was to be exploited for thirty years and to bear the name of a straw man. "Privilège Charles Adam," the letterhead reads. Such a person in fact existed, being the father-in-law of the first master-potter, François Gravant (of whom more in a moment). The real investors, as usual, preferred anonymity.[20]

Engaging sculptors, molders, painters, gilders, turners, masons, and laborers; constructing kilns, wheels, troughs, mortars, mills, racks, and work-tables in the old riding school of the chateau—all that soaked up capital. Orry de Fulvy kept going to his backers for further funds. In 1751 he died, having extracted some 250,000 livres from his increasingly reluctant associates. Machault was now controller-general. Madame de

[18] Memoir of 21 May 1781, signed by Mignot de Montigny, Macquer, Millot, and Dufour, BMS, register Y-60.

[19] Cochin (1865), 41-55.

[20] BMS, register Y-1 is a record of administrative documents; carton C-1 contains memoirs on technical processes.

Pompadour liked porcelain. In 1753 the company was reorganized under the financial wing of the Tax Farm, borrowing the name now of one Eloi Brichard and offering new shares, one-third of which were taken by Louis XV. It was then that the designation "manufacture royale" was authorized and that the interlaced double *L* began serving as trademark.[21] For artistically Vincennes was already a success. A public showing of the dinner service commissioned by the Empress Maria Theresa in 1754 created a sensation and established the superiority of French porcelains in the world of fashion and luxury. In 1977 an exhibition in Sèvres reminded the connoisseur of the beauty of the objects produced even before 1756 and the move there, where Madame de Pompadour owned the land that has been occupied by factory and museum ever since.[22] Financially, however, Vincennes proved a disaster. In 1759 the investors petitioned the king for relief from their engagements. Yielding to the Pompadour, Louis XV bought them out and took title to all the shares, stock, inventory, buildings, and terrain at a cost to the crown of 1,937,509 livres.[23]

Thus did Sèvres become an enterprise of state. Regulations multiplied. An ordinance of 21 April 1779 had to rehearse the provisions of conciliar decrees of 24 July 1745, 19 August 1747, 8 October 1752, 19 August 1753, 7 December 1753, 17 February 1760, and 15 February 1766.[24] In sum, that legislation required private manufacturers of earthenware, faïence or china to register their trademarks. Any objects they made in imitation of porcelain were to be white or colored only in blue. Cameo decoration was also to be monotone. No gold might be applied, either in leaf or in boss, and no figurines or sculptures of any sort produced. Other employers must not seek to hire away artists or other persons who had learned their trade at Sèvres. As for the workers there, unauthorized absence even for a single day was a criminal offense. They might not quit their jobs without six months notice and a binding oath never to take other employment in ceramics and never to impart anything they had learned. Penalties were severe: fines and confiscation of all stock and equipment in the case of manufacturers; 1,000 livres fine or three years in prison for a faithless laborer on the first offense, and corporal punishment for a repetition.

Such penalties were actually inflicted. Enforcement in the region of the capital fell to the lieutenant-general of police of Paris, whose service now

[21] BMS, register Y-2 contains administrative records; carton B-1 contains many supporting documents.

[22] An issue of *Le Petit Journal* (14 October 1977), new series, no. 52, "Porcelaines de Vincennes," prepared by Antoinette Faÿ-Hallé and Tamara Préaud, was devoted to the exhibition.

[23] An admirable summary of the accounts was drawn up in 1780, probably by J.-B. Montucla, covering the two decades 1760-1780, BMS, carton A-1.

[24] BMS, carton A-1.

had to oversee the ceramic industry among its many other duties. The detail that had in consequence to pass under the eyes of a high magistrate is astonishing to contemplate. To cite only one instance, in 1772 a respectable banker in Paris, one Freinet, is dismayed that a renegade potter called Brolliet has somehow been given a permit to install a shop in the property next to his own, once-quiet country house in Vaugirard. Freinet's denunciation begins:

> For some years Paris had been infested with a swarm of men so dangerous to Society that the good faith of the Citizen has no shelter from the assaults of their importunity and infidelity, despite the precautions that the Ministry in its wisdom takes for the maintenance of public order: the details that follow offer a striking example. . . .

Those nuisances included noise, dirt, bad debts, shoddy pots, and the flaunting of sexual revels that might have served the marquis de Sade for source material.[25]

When the king assumed title to Sèvres in 1760, the director of the factory was Jacques-René Boileau, who was continued in the post and made accountable to the Finance Ministry. Bertin was then controller-general, and took the government's oversight with him to the special (or "little") ministry he headed after 1763. There was some feeling, notably on the part of Jean-Jacques Bachelier, the artistic foreman, that Boileau exercised an authority overreaching his competence in the several specialties. Nevertheless, the director evidently combined effectiveness in the training and management of artists and craftsmen with modest courtliness in keeping Sèvres in favor with Versailles. He it was who established a school where apprentices were treated as "pupils and children of the manufactory."[26]

Boileau was not, perhaps, a man of business. In no year from 1760 through 1769 did the receipts equal the expenses. By the end of the first decade the crown had had to make up a cumulative deficit of almost 750,000 livres, and had come to accept the proposition that artistic emulation, accrual of prestige, and a favorable balance of foreign trade in porcelain were worth the cost.[27] The administration was honest, however.

[25] Mémoire, August 1772, BMS, carton A-2. For various prosecutions, see cartons A-3, D-1.

[26] "Mémoire en faveur du Sr Boileau," 23 January 1772, BMS, carton A-1; Jean-Jacques Bachelier, "Mémoire historique sur l'origine et le régime de la Manufacture Royale de Porcelaine de Sèvres," fol. 7, BMS, register Y-37.

[27] Memoir cited n. 18, where it is observed that Louis XV had always considered Sèvres as an "établissement de luxe en partie, en égard à l'élégance, le perfection, et les ornements," with which it was to be conducted.

On his death in 1772, Boileau left creditors and workers paid and a balance of 300,000 livres in the working funds. His notions of a succession were those of his century. In arranging for a pension that he never lived to draw, he requested that one Leviston be named vice-director and successor, on condition that this Leviston marry Boileau's maiden sister-in-law. Such had been his own probity and devotion to duty that he had been unable to provide the dowry that would secure her future by other means. The démarche succeeded only partially. Leviston got the post of cashier, together with Mademoiselle Briais and 40,000 livres.[28]

Possibly Bertin would have done better to accept the original proposition. Instead, he appointed another subordinate, Parent, who turned out to be neither competent nor honest. Parent replaced Leviston with his own man Roger, whose accounts showed a shortage of 247,000 livres in 1778. Thereupon Bertin had Parent packed off to the Bastille and installed Régnier, a good technician and sound administrator. In 1780, it was Bertin's turn to be retired, and Necker detailed the manufactory to d'Angiviller in the Batimens du roi, while also giving some thought to selling it into the hands of private industry.[29] D'Angiviller immediately tightened up the accountability by assigning J.-B. Montucla to oversee it for the ministry. In 1784, after careful consideration, it was decided to reaffirm the privilege, thus recognizing the propriety of governmental ownership of what had evolved into a conservatory of ceramics, a center at once of fabrication, training, and fashion—as, indeed, it still remains.[30]

So much for high policy and management. Memoirs written in 1781 for the information, and perhaps the instruction, of d'Angiviller are among the documents that enable the historian to peer beneath the letter of rules and regulations into the actual operation of the manufactory.[31] Bachelier, foreman in the painting shop, and Millot, foreman of kilns, had been present soon after the creation, and both composed reminiscences that exhibit the artisan's side of the transactions between craftsman and patron, all the more saltily in Millot's case for his uncertainties in

[28] "Mémoire en faveur du Sr Boileau," cited note 26; Bachelier, Mémoire historique, cited note 26, fol. 5.

[29] "Mémoire [1780] suivi d'un tableau, sur les differentes manières dont la Manufacture de Porcelaines de Sèvres pourrait être vendue . . . ," AN, F^{12}.1493.

[30] "Arrêt qui confirme le privilège de Porcelaine de France établie à Sèvres . . . ," 14 May 1784. AN, F^{12}.1493. For a later confirmation of the conception of the enterprise, see Alexandre Brongniart, *Du Caractère et de l'état actuel de la Manufacture Royale de Porcelaines de Sèvres . . .* (1830).

[31] Millot, "Origine de la Manufacture des Porcelaines du Roi en 1740," BMS, Memoir conserved with registers Y-53—Y-59, fol. 5; Bachelier, memoir cited n. 26. The latter has been published twice, during the Revolution in an undated printing, and again edited by Gustave Gouellain (1879). Both versions are modified, and the second bowdlerized. It is preferable to refer to the original text.

orthography and choice of auxiliary verbs. The notion that there might be money in porcelain had not been new with Orry de Fulvy. He had before him the example of Louis-Henri de Bourbon, prince de Condé, who had been principal investor in a workshop producing porcelain-like ware in Chantilly since 1725. The process or "secret" (of the fritte, that is) belonged to one Cirou, sometimes spelled Siront, who may have had it from the first such establishment in France at Saint-Cloud.

Around about 1740 two faithless, and apparently feckless, workers, the brothers Dubois, one a turner and the other a modeler, either left Chantilly or were dismissed. Somehow they set themselves up in the Tour du Diable at Vincennes, attracted Orry's attention through du Châtelet, the governor, and offered to exploit Cirou's process under protection within the purlieu of the chateau. For four years they consumed Orry's advances, ever failing to achieve the promised quality of translucence in the mediocre ware that came out of the kilns, and also refusing to communicate to their patron the secret of the fritte. To Orry's rescue came another refugee from Chantilly, François Gravant, a grocer whose business there had failed. The Dubois took him on. Learning that Orry in despair and anger was about to send them packing, Gravant had to choose between a livelihood and his old friends. He chose the livelihood, and one night when they were dead drunk, as they often were, he copied out the detail of the fritte from papers where it was written down. Then he let Orry understand that he knew the secret too.[32]

An arrangement followed. The terms on which Gravant now sold Orry's company what he had stolen were characteristic of the acquisition of trade secrets by entrepreneurs. Those terms included his own services, for unlike the Dubois he was generally credited with being skilled and intelligent, though difficult personally. Also party to the affair was the original kiln foreman, Claude Gérin, or Gérain. Their specifications for a soft paste scrawled in a barely legible hand and in barely decipherable spelling contain the earliest surviving account of the process clearly and elegantly set forth by a team of chemists for Regnier and d'Angiviller thirty-five years later. Only when an experimental firing produced a whiter fritte, and then a clearer porcelain than ever before, did Orry de Fulvy have a contract drawn. It is subscribed by Gravant and Gerin and their wives, who sign shakily "Anne fleur faime de gerin" and "Marie henriette mille famme de gravat." For Gravant the provision, in addition to wages, called for a bonus of 10,000 livres at the end of ten years on condition that he still be working for the company. In case he should die before that, his wife and children would receive the 10,000. They would also have a pension of 600 annually whenever he died, if he had continued

[32] Bachelier, memoir cited note 26, fols. 1-2.

in service and if the company was still making porcelain. Three years later, this arrangement was transmuted into an annuity of 1,200 livres on the head of himself and his wife, that being the two-life income on 24,000 livres, which was the value set on his secret.[33]

In this second agreement, no condition was placed on Gravant's continuing in the factory, where (according to Millot), Gravant and "le nommé" Gerin obstructed improvements in the construction of kilns that he had been brought in to effect. In his reminiscence Millot tells how he himself finally succeeded in designing a reverberatory chamber five feet high inside the ten-foot kiln, which otherwise could never be got to heat evenly. The first test firing lasted forty-eight hours. At noon of the second day all the workers gathered in the painting studio, where the new furnace was being tried. One by one the porcelains were removed and ranged on a table, each greeted by gasps of admiration. Orry de Fulvy was delighted and sent his valet off with two louis to buy wine for the staff of the paste shop. There was no more work that day. Violins played, and the afternoon passed joyfully amid singing and drinking.[34]

It remained to perfect the decoration. Word spread, and the company was inundated with proposals from artisans and inventors, many of whose techniques existed only in their imagination. The second important purchase was the process for gilding exploited until the 1770s. In this instance, the supplier, Brother Hippolyte, a Benedictine of the Convent of Saint-Martin des Champs, had conceived and developed the method himself. He would deal only with Orry de Fulvy, from whom he demanded and received a price of 9,000 livres—3,000 down and 600 a year representing the life interest on the balance.[35] Colors were an even more delicate matter. In 1754 the company paid one Taunay, an artist, 6,000 livres for shades of carmine, purple, and violet. It was in connection with these pigments that there is an early, and perhaps the first, record of formal verification by a chemist. Taunay was paid his money on the strength of a certificate by Jean Hellot stating that the painter had dictated the instructions to him, that he had tried the ingredients and operations and found that they worked exactly as claimed, Taunay "having kept nothing back."[36]

It is seldom possible to fix so precisely the moment at which science entered into an industry or to specify so fully the circumstances that brought it about. Orry de Fulvy, "convinced"—the account is Bachelier's now—"of the advantageous influence that science and talent might have

[33] The first contract is dated 1 January 1746; the second 27 December 1748, BMS, carton C-1. For further detail, see registers Y-40 and Y-41 (the record of tests by Gérin, 1745-1750) and Y-42 (Gravant's process for the paste).

[34] Millot, Memoir cited note 31, fol. 2.

[35] BMS, carton C-1.

[36] 10 July 1754. BMS, carton C-1.

upon the progress of the Manufactory," engaged Hellot to consult on chemical aspects.[37] He began that employment in 1751, just before Orry's death and the reorganization of the company under the purview of the Ministry of Finance. A directive defining his responsibility is worth giving at length. It bears the signature of Machault in his own hand:

> The Sieur Helot of the Academy of Science, having been named on 25 June to take cognizance of the various secrets concerning the operation of the Manufactury of Porcelain established in the Chateau of Vincennes; and His Majesty, having recognized by virtue of the tests made by the Sieur Helot, by the Memoir that he has submitted to Us on 7 October 1751, and by the improvements that he has made in the said compositions, that these latter are capable of development to a high degree of perfection, His Majesty by decree of 8 October 1752 has revoked the Privilege accorded to Charles Adam . . . and has accorded it to Eloy Brichard, by decree of 18 August 1753. . . . And His Majesty, having by Article 9 of the said decree, reserved for Himself the secret of the said compositions, in order that they may be implemented by the person named for that purpose, has named the said Sieur Helot to continue to oversee them, to perfect the operations, to have all the experiments made that may be needed to simplify the compositions, and to procure a larger number of colors and enamels; to that end, the said Sieur Helot will give whatever orders he judges appropriate to the workers and artists charged with the work and with the preparation of pastes, glazes, and enamels; which orders the said workers and artists will be required to execute, each in the sector that concerns him, without being allowed to communicate to anyone, no matter whom, the secrets of the said compositions.
>
> Every month the said Sieur Helot will draw up a statement of the funds needed, both for procuring the raw materials that go into the compositions, and for the different tests, without being obliged to break down the detail of the expenses carried on the said statement. . . .
>
> Since it is important to take cognizance both of the compositions already found, and of the different discoveries that may be made in the future, the Sieur Helot will continue to maintain in the form of regular minutes the Memoir of all the knowledge acquired to date concerning paste, glazing, colors, and enamels, and the use of gold, of which the original is entered in a register bound in blue Morocco,

[37] Bachelier, memoir cited note 26, fol. 2. On the career of Hellot and its significance in industrial chemistry, see Todériciu (1977).

locked by a key to remain in his hands, in order that it may be consulted according to need: which Register shall be returned to Us, by his widow or heirs, locked and sealed, immediately after his death; in order that We may make such disposition of it as seems to Us proper.

The Sieur Helot will make a copy of the said Register, in greater detail than is contained in the memoirs that he has already submitted to Us, which copy will be deposited in a safe which shall be installed for that purpose in the interior of the Manufactory, and sealed into the main wall of the location intended for it: to which Memoir he shall add new discoveries as they occur, so that in case of a fire at the Manufactory or in the residence of the said Sieur Helot, it will be assured that the said secrets cannot be lost. The key to that safe shall remain in the hands of Sieur Helot, and will be transmitted to Us, along with his Register, after his death.

Done at Fontainebleau the first of November one thousand seven-hundred and fifty three.

MACHAULT[38]

It is largely owing to Hellot's fidelity to these instructions that we owe the survival of much of this early technical record. Nothing if not practical, Hellot was closer in temperament and mode of knowledge to the foreman he had to oversee than were his successors later in the century. The early rosters classify his place "*artiste-chimiste*," whereas by the 1770s it had become "*académicien-chimiste*." The evolution of the role of chemistry at Sèvres, indeed, exhibits a steady increase in distance from the shop, and simultaneously in status and authority, through the incumbencies of Macquer and Darcet until (to overreach the boundaries of this book) Napoleon had Alexandre Brongniart appointed director in 1800. Brongniart's administration, reputed the most effective until the present century, lasted until 1840. In these early days, Hellot worked directly on the product and even sold the management several "secrets" of his own.[39] Similarly, in 1760 Louis-Claude Cadet set himself up in the shop and saved the day with a method for eliminating a rash of brown spots that for years had been breaking out during the glazing of white porcelain, and that had reduced Gravant to despair and even to tears in his declining

[38] BMS, carton A-1. The Hellot registers are classified Y-49 and Y-50. When Macquer succeeded Hellot, he made a copy of the latter's register, Y-52. In 1772, Louis XV personally signed the directive to turn the locked register and safe over to Parent, the new director.

[39] BMS, carton C-1, which together with carton C-2 contains the records of tests made on various processes that were purchased. Carton C-3 contains records of others that were rejected on advice of the chemists.

days. Cadet was then a young military apothecary, seeking reputation and entry to the Academy of Science. On the strength of his success, he petitioned Courteille, the intendant of finance who was then controlling Sèvres for the ministry, to be named Hellot's successor in advance.[40]

Courteille preferred Macquer, whose reputation was already made, and who gradually took over Hellot's work before the latter's death in 1766, first as assistant and increasingly as successor. Macquer's register of porcelain research begins in 1757. It is evident from the first folio that he was conducting the work in a different spirit from that of the empirical approach of Hellot, and equally of Réaumur and Guettard, whom he cites. Guettard had collected samples of many clays in his mineralogical excursions. Réaumur had discovered, surprisingly, that ordinary bottle glass was better suited than fine, clear flint glass or crown glass for mixing with gypsum to produce ware with the appearance of porcelain. In the language of the potteries, the manifesting of the desired properties when the clay was dissolved in a batch of molten glass was called "*sucement*." This sucking out, or perhaps secreting, the porcelain was the critical step. Macquer disliked the term, for its inacurracy no less than its vulgarity, considering that what the process really involved was a chemical combination rather than an extraction of some sort.

Not content simply to identify the clays and types of glass that together would yield the phenomenon, Macquer planned a campaign of experiments with the object of forming a theory to explain it. Cheap glass, he observed at the outset, contained "*terre absorbante de cendre*"—impure potash—which was absent from fine grades. Gypsum consisted of a compound of vitriolic acid and "*terre calcaire*," that is to say calcium sulphate. His hypothesis was that the action of vitriol on potash was responsible for the properties of porcelain, the "violence" or intensity of the firing having vaporized the reagents and effected a true cementation. It would be tedious to follow the detail through which Macquer confirmed his hypothesis by direct experiments and elevated it to the status of theory by its power to resolve certain anomalies—among them the refractory properties of a chalky gypsum from Virofley used to line the kiln. This early register contains the record of over a thousand tests, carefully numbered for reference, that Macquer ran in order to determine by qualitative chemical analysis of laboratory samples whether given clays were worth the expense of trying on the factory scale.[41]

The management at the manufactory, however, appeared to be more interested in artistic triumphs than in technical improvements. Boileau was quite content to fabricate objects for display to be adorned with the

[40] Courteille to Cadet, 25 September 1760, BMS, carton C-1; Cadet Mémoire, undated but in 1761, carton D-1.

[41] Macquer's registers are Y-57, Y-58, Y-59.

designs of a Boucher, a Pigalle, a Van Loo, or in their style, after having been modeled by a Blondeau, a Van der Voorst, a Fernex, or a La Rue, sculptors who, as is often true of their art, are less remembered than the painters. When Macquer reported eventual success in producing hard porcelain to the Academy of Science in June, 1769, he was very scornful of the "fake porcelain," the mere mixture of clay and glass, which was all that Vincennes and Sèvres had yet produced. It was easily scratched, chipped and broken. It could not stand even moderate changes of temperature. Coffee or hot chocolate would often crack the cup.[42] There was a story that Louis XV in his carriage stopped Boileau one day on the road between Versailles and Sèvres to tell him of an experiment that Madame de Pompadour had just performed in her salon. Someone had sent her a saucer of hard porcelain produced by one Hannon in Frankenthal, and she had tried having an egg shirred in it by means of a burning mirror. The dish from Germany showed no mark, whereas one from Sèvres cracked straight off. The king understood that this Hannon had set up in the Palatinate after agents of his own Manufacture Royale had harried him out of Strasbourg.[43]

Ensuing negotiations between Sèvres and Hannon came to naught, despite a hastily arranged journey that Boileau made to Frankenthal in company with the faithful Millot. According to the latter, Hannon wanted too high a price for his secret.[44] At about the same time, and unbeknownst at first to Boileau and the staff at Sèvres, Courteille on behalf of the ministry quietly charged Macquer to resume investigations into the possibility of producing a hard porcelain from clays occurring in France. Those experiments occupied Macquer from 1763 to 1765, when he succeeded—to his own satisfaction at least—in producing samples the equal of Hannon's in quality, though no better. In his letter to Courteille, Macquer does not say where the clay had been found. He says only that he is not sure whether the supply is adequate, and asks permission to impart the process to Boileau and Millot so that it might have a full-scale trial.[45] There the matter dropped for the time being. Macquer never said why, but a probable explanation emerges from Millot's reminiscence. Courteille had also supplied the factory with a ten- or twelve-pound sample of kaolin smuggled back from Saxony by some agent of the government, and someone had handed part of that along to Macquer without remembering where or how it had been obtained. He, Millot, had produced small

[42] Macquer, "Mémoire . . . sur une nouvelle porcelaine," *Mercure de France* (July 1769), p. 191.

[43] Millot, Memoir cited note 31, fol. 5.

[44] *Ibid.*, fol. 5. Negotiations, and eventually litigation, with Hannon dragged on into the 1790s. BMS, carton C-1.

[45] Macquer to Courteille, 28 April 1763, BMS, carton C-2.

plaques of hard porcelain from his trials with that kaolin. He and Boileau kept the best of them—identified as Number 7—as a touchstone of hardness and clarity or whiteness.[46]

Thereafter, Macquer collaborated in the research with a member of the Academy of Science whose instrumentality in its liaison with the Bureau of Commerce and the ministry has failed to catch the attention of historians. Etienne Mignot de Montigny—not to be confused with the younger Trudaine, also "de Montigny," where they were neighbors—was a nephew of Voltaire, born in 1714. After publishing a single memoir in rational mechanics, a clever piece on the motion of rigid bodies, Mignot de Montigny inherited a post in the Treasury. Whatever his mathematical talent, he evidently preferred practical matters and served under Daniel Trudaine (the father) in the Bureau of Commerce, his main interests being the street pavings of Paris, the mechanization of the textile trades, and the applications of chemistry. The tone of Macquer's correspondence bespeaks real respect for a colleague seen to be superior in standing and insight and also affection for one who was evidently a shrewd judge of men in their handling of materials and machines. It is not too much to say that the presence of Mignot de Montigny in officialdom will be felt in this correspondence as that of a kind of éminence grise, albeit a kindly one, a counterpart in technology to Vincent de Gournay in economics.[47]

Not until 1767 did the quest issue in the discovery of a true kaolin in France. The first such sample was sent along by the archbishop of Bordeaux, who had it from an apothecary called Vilaris. Macquer reports the discovery in a memoir on hard porcelain prepared for the Academy of Science and printed also in the *Mercure de France*:

> M. Vilaris having agreed to identify the location where that earth occurs, I repaired there in accordance with the orders of the King transmitted to me on the 8 and 20 August last year by M. Bertin, Minister and Secretary of State; I took with me M. Vilaris and the Sieur Millot, one of the principal workers in the Royal Manufactory, in order to make excavations and send to the Manufactory a sufficient quantity of the same earth to continue testing on a larger scale than has been possible hitherto.[48]

Though earthier, Millot's account is less dry. He tells how Boileau had provided Perronet, director of the Ponts et chaussées, with samples of the

[46] Millot, Memoir cited note 31, fol. 7.

[47] See especially, Macquer to Mignot de Montigny, 8 October 1765, BMS, carton C-2; on Mignot de Montigny, see the éloges by Vicq d'Azyr, HSRM (1780-1781/85), 85-101, and by Condorcet HARS (1782/85), 108-121. Mignot's unique memoir was "Problèmes de dynamique," MARS (1741/44), 280-291.

[48] Millot, Memoir cited note 31, fols. 8-12.

Saxon kaolin, in hopes that the highway engineers might come on something similar. The archbishop of Bordeaux had also taken a lump after a visit in 1765, saying that he knew a clever apothecary who traveled a lot with his eyes fixed on the ground. This was Vilaris, who took into his confidence a sometime companion, a surgeon called Darnet from Saint-Yrieix. He it was who came upon what looked to be an identical deposit. A three-pound sample passed from him to Vilaris, from Vilaris to the archbishop, and from the archbishop to Sèvres. Millot washed it clean and tried a paste in the kiln. The first piece out was a small Bacchus, followed by several goblets and saucers, all as lucent and hard as the finest porcelain of old Japan.

Now began a game of treasure hunt. Vilaris would not tell the archbishop the location of the kaolin, lest the government withhold a reward. Boileau failed to persuade the ministry to satisfy Vilaris's greed. Exasperated, Millot said to Boileau "You'll be surprised not to find me in the manufactory one of these mornings." "Why?" asked Boileau. Because, said Millot, he was about to take matters into his own hands and go find the kaolin. At that threat, Boileau went to Bertin, who hastily arranged to dispatch Macquer with Millot at this side. Furnished with a portable laboratory and assorted samples, they departed from Paris on 28 August 1768. On their arrival in Bordeaux, the archbishop welcomed them with enthusiastic courtesy. Vilaris came round to call at their inn, all coyness and cupidity, still unwilling to lead them to the kaolin until he should have his recompense. For nine days Macquer and Millot dangled, awaiting a reply from Bertin. When it came, it forbade further dealings with Vilaris and ordered them to prospect on their own. So they did, canvassing the country from Bordeaux down to Bayonne and along the Pyrenees. They turned up several decent clays. Millot would try them in a kiln he improvised in a forge belonging to a locksmith and show them to the archbishop. None was really satisfactory, but the archbishop decided to bluff. He called in Vilaris, showed him the old No. 7 plaque and said that the commissioners had evidently found kaolin.

No longer "so proud," Vilaris again called at the inn, offering his services. Meanwhile, Macquer and Millot had been joined in their researches by an interested medical man, a Doctor Camouty, physician to the court of Parma. Now Vilaris conducted them all three in the direction of Limoges, to Saint-Yrieix, where they were lodged on the outskirts of the town, so as to be as remote as possible from the house of Darnet, supposedly Vilaris's old friend, the surgeon who had recognized the deposit in the first place. The party was roused before dawn and led through back yards and alleys to a bank of clay opposite the cemetery of a neighboring village. It bordered on a lane almost hidden between its banks. Out came pick and shovel. Unluckily, two villagers heard the digging and hastened to report the intrusion to the owner of the property, a Madame du Montois. She immediately sent her son to demand by what authority they were thus trespassing. The young man threatened to ring the church bell, call out the village, and have these strangers arrested. With the production of passports and explanations, work resumed, and the party dug out about 400 pounds and filled a barrel with kaolin the first day.[49]

Turning back and forth between Millot's recollection and the further correspondence of Macquer with Mignot de Montigny gives an effect of counterpoint between the voices of artisan and scientist. The documentary echo is probably faithful to what passed between them in the course of developing the capacity to produce hard porcelain at Sèvres. Macquer designed and Millot constructed a high-temperature kiln. Laboratory-scale tests succeeded in late 1768 and early 1769. The problem from Macquer's point of view then became one of persuading Boileau in the management and Millot in the shop to adapt their procedures to the requirements of the new materials. "You read Boileau like a book," acknowledges Macquer in a letter to Mignot, à propos of the director's passive resistance to making the workers test more than one type of model. "You know the man by heart, . . . but you understand that M. Boileau will always do only what he wants."[50]

Millot for his part kept blaming the glaze for accidents that really resulted from the buckling of racks incapable of withstanding the higher temperatures. Macquer seized on the occasion of a particularly bad batch to impress on him the necessity of following Mignot's advice and washing the clay for the racks with a meticulous care that was uncalled for when a kiln was charged with soft paste. Confronted with lopsided saucers, bowls with the lid baked on, and misshapen figurines, Millot looked beaten and discouraged and began complaining that this new porcelain was too much work. Macquer tried to revive his enthusiasm with the carrot of a bonus—to be paid when they had achieved a complete success. At last in November, they brought it off thanks to a modification in the draft that achieved an exceedingly intense and smokeless heat. Two complete batches, one in a square and the other in a beehive kiln, succeeded well, the latter almost perfectly. Thirty-six of the forty pieces were without a blemish. Macquer was present when Millot opened the door to unload the charge. Boileau (he wrote Mignot) now suddenly seemed to like the new porcelain better. Indeed, he kissed Millot on both cheeks in a transport of joy, and an-

[49] "Memoire . . . sur une nouvelle porcelaine," *Mercure de France* (July 1769) 193-194. The new hard porcelain that resulted was verified and approved by Duhamel du Monceau and Bernard de Jussieu, commissioners for the Academy of Science. The draft of the above memoir is in BMS, carton C-2. Other items in the Macquer correspondence concerning this mission are in BN, MSS Fr. 9134, fols. 86-96; 9135, fols. 65-96.

[50] Macquer to Mignot de Montigny, 10 October 1769, BMS, carton C-2.

nounced that he would write to Bertin forthwith. What sealed his pleasure in the next few days was the further discovery that the new porcelain absorbed only half as much gold as the old for the same extent of gilding, and that, since pigments held fast after one coat, he could save the labor that went into the three or four retouchings and refirings needed for coloring soft porcelain.[51]

Boileau was even then nearing his retirement. The evil times on which the direction fell between 1772 and d'Angiviller's re-ordering of the management after 1780 may explain the failure to move aggressively into production of hard porcelains. It may also have been that taste under Louis XVI turned still more effeminate. At all events, not until the direction of Brongniart after 1800 did the manufactory at Sèvres shift to exploiting kaolin and turning out ware like that of ancient China and Japan in its durability, though not perhaps in beauty of design or execution.

One area of lacuna is to be regretted in the documentation, which in other respects affords the intimate record just reported of the imposition of scientific controls over manufacturing processes under governmental supervision. Virtually no evidence survives of how d'Angiviller came to choose Darcet to succeed Macquer after the latter's death in 1784. There is evidence of earlier consultation. Darcet joined with Macquer and the younger Rouelle in a certificate of 1772 reporting that various pieces of pottery submitted for a privilege by the egregious Brolliet were nothing but low-grade faïence.[52] It is sometimes stated that Darcet, like Brongniart after 1880, was himself director. That is wrong. Like Macquer, he was a consultant with authority in the chemical aspects. Darcet maintained his laboratory at the Collège de France, where he was professor, and traveled out to Sèvres several times a week, also as Macquer had done. There are indications that the minister's first preference was Cadet, he who had been disappointed of the succession to Hellot.[53] Evidently, the government continued the practice of attaching two chemists to the manufactory, one representing the Academy and the other the ministry. Desmarest, with long experience in the corps of inspectors of manufacturing, followed Mignot de Montigny in the latter responsibility. Assorted démarches and proposals from the early 1790s indicate retrospectively that Darcet and Desmarest were active in their respective sets of duty, but there is little from the 1780s to show what they actually did.[54]

A much earlier memoir is of special interest, however, for it exhibits Darcet making the transition from medical man to chemisty by way of a

[51] Macquer to Mignot de Montigny, 25 October 1769, BMS, carton C-2.

[52] 2 August 1772, BMS, carton A-2.

[53] D'Angiviller to De Mauroy, Inspecteur des Manufactures, 30 June 1784, BMS, carton B-3.

[54] BMS, carton A-5.

flyer in porcelain that never quite got off the ground. After taking his medical degree in Paris, Darcet became personal physician to the comte de Lauragais, a nobleman versed in chemistry and not averse to improving his fortune. In addition to the prince de Condé, the duc d'Orléans—first Prince of the Blood—was among the great interested in porcelain in the early days of the Manufacture Royale, and even before. After his death in 1752, his maître d'hôtel, a certain Montamy, claimed that his laboratory had found a method for making true porcelain from raw materials found in France. The new duke, his son, showing no special interest, Lauragais resolved to pursue the matter for his own account. He secured the remaining materials from Montamy together with several finished pieces and the name of the artisan responsible, one Leguay. To Leguay he joined Darcet, his physician, who had been following Rouelle's lectures at the Jardin du roi, and another medical man, a longtime, older friend of Darcet, a Doctor Le Roux. All three were to collaborate in Lauragais's laboratory. Apparently Lauragais, who was of an impatient disposition, decided to drop the affair just at the moment when the chevalier Turgot, brother of the statesman, sent in from his travels a clay that may well have been kaolin. Thereupon, to get the enterprise back into his own hands, or partly so, Lauragais had to agree to provide the capital in return for fifty percent of the prospective profit, the other fifty percent to go to Darcet, Le Roux, and Leguay. All of this was predicated on the grant of a royal privilege, which was never forthcoming, what with Bertin's intention of liberalizing access to the industry. The interest lies in the interplay between entrepreneurship and technical proficiency.[55]

To return to the decade of the 1780s, all the indications are that the demands put upon technically proficient people were increasing. In 1784, when Darcet was named academic chemist at Sèvres, the government chose Berthollet to succeed to Macquer's other post, that of director of dyeing at the Gobelins, reporting to the Bureau of Commerce.[56] Fourcroy then said jealously that Berthollet had secured the most eligible position open to a scientist.[57] Official enlistment of chemistry went back further at the Gobelins than at Sèvres, the establishment itself being older. Dufay, Buffon's predecessor in the intendancy of the Jardin du roi, had been

[55] "Mémoire sur la porcelaine," BMS, carton A-2, together with a dossier concerning the proposals and démarches of Lauragais, minuted by Bertin, BMS, carton B-2.

[56] The letter of appointment was signed by Calonne and dated 24 February 1784. (See Berthollet to Tolozan, 2 February 1790, AN, F^{12}.1329.) The post was also sometimes called "administrateur du commerce au département des teintures." Sadoun-Goupil (1977), 21, a meticulous biography, with complete bibliography and calendar of correspondence.

[57] Lanthenas to Madame Roland, 1 May 1784, *Lettres de Madame Roland*, ed. Claude Perroud, *1* (1900), 374-375, note.

the first member of the Academy to be government administrator of the department of dyes (as the post was sometimes called). On his death in 1739 he was followed by Hellot, who in 1751 added his duties concerning porcelain to his prior responsibilities. The dye trades were far more minutely regulated than was pottery by codes instituted in Colbert's administration.[58] They were also much closer internally to the tradition that identified crafts with mysteries. Professional chemistry was correspondingly slower to rationalize the processes, while also sounding greater depths of practice. As compared to the ceramic trades, where Macquer could introduce a theoretical approach into the laboratory analysis of clays and pastes in the late 1750s, the dyeing industry had to await Berthollet's *Eléments de l'art de la teinture* in 1791 for the earliest work to apply principles of chemical theory to the comprehension of its procedures. The lack was felt in the Bureau of Commerce. The major charge laid upon Berthollet in his letter of appointment was precisely that he should produce such a work, in addition to his normal responsibility for conducting inspections of the Gobelins and for controlling quality.[59]

Unfortunately, there are no sources at the Gobelins preserving a record of the work of chemists, or else there is no access to them. In either case, it may reasonably be supposed that Hellot's and Macquer's modes of discharging their duties would have been much the same there as at Sèvres, the more so since there is evidence that the two installations were comparable in point of the artistic direction. In 1755, when the comte de Marigny, then superintendant of the Batimens du roi, appointed François Boucher to be chief inspector for designs, the great artist was instructed by the minister to visit the Gobelins one day a week in order to examine the work of the looms, there to criticize and to advise. The artisans were thus to be kept up to that level of merit and urbanity requisite for artistic mastery.[60] Thirty years later, the demand on Berthollet's energy was several times more intense than that. Luckily enough, he left something like a scientific autobiography in the form of a report written early in 1790 at the request of Tolozan in the Bureau of Commerce.[61] Drawing on the testimony of that document, together with knowledge of the work Berthollet published then and later, the historian may be categorical about the influence of his position upon the configuration of a career that assumed gathering importance in French scientific life, commandingly so in the revolutionary and Napoleonic periods.[62]

[58] Berthollet, *L'Art de la teinture* (1791), xxvii-xxix.

[59] Berthollet to Tolozan, 2 February 1790, AN, F^{12}.1329.

[60] Marigny to Boucher, 6 June 1755, Fenaille (1903) *4*, 226.

[61] The document is that cited in notes 56 and 59. On the career and importance of Tolozan, see above, Section 1 and n. 3.

[62] Crosland (1967), esp. 56-146.

It was definitely opportunity in applied chemistry that drew Berthollet, even like Darcet before him, from the service of the duc d'Orléans into that of government, and thereby allowed him to make the transition from a livelihood in medicine to one in science. Berthollet recalled for Tolozan how Calonne's letter of appointment precluded his continuing to practice medicine. He was allowed to keep his hand in for the purpose of attending the duke and a few chronic patients, but nothing else. Otherwise, he was to spend all his time on experiments that would cast light upon the chemical arts, and on the preparation of his treatise of dyeing. His stipend was 6,000 livres, of which he diverted 600 to pay for the services of an assistant. Given another pair of hands, he could arrange the laboratory so that several investigations were always going on concurrently. He learned German and collaborated on a translation of Pörner.[63] He assembled a library of treatises and memoirs on dyeing. He would have liked to follow the procedures in the workshops themselves, but "the mystery that the dyers make of their operations has prevented me from benefiting from the observations I could have made."[64] Thwarted at the vat-side, he tried the next best recourse and prepared a questionnaire to submit to manufacturers known for their enlightened and informed attitudes. Chaptal and Puymaurin in Languedoc, Maier in Lyons, Haussmann in Colmar, Décroisille in Rouen kindly obliged.

The yield of these inquiries, and of his research and reading, was an immense array of procedures and recipes that were different for the application of every color to every fabric in practically every shop. Swatches of a few of the fabrics that he inspected, dyed crimson, green, and ochre, remain in a fragmentary dossier preserved in the Academy of Science in Paris.[65] Lacking was any set of principles, any scheme either classificatory or theoretical, by which this congeries of processes might be converted from a miscellany into a technology (the word had yet to be coined). It was true that Hellot had prepared a manual on dyeing woolens that remained standard after forty years, having also been translated into English and German. But its gesture at theory was physical, a relic of mid-century Newtonian corpuscular mechanism. Particles of color were made to penetrate pores in the fiber and were held in place by the throttling effect of mordants. What was needed was a chemical theory, the more urgently

[63] Karl Wilhelm Pörner, *Chymische Versuche und Bemerkungen zum Nutzen der Farbekunst*, 3 vols. (Leipzig, 1772). Berthollet and Desmarest edited a French version, *Instruction sur l'art de la teinture* (1791), in which they abridged "les longueurs que les allemands se permettent." AN, F[12].1329.

[64] AN, F[12].1329.

[65] "Manuscrits de C.-L. comte Berthollet legués à son élève et ami, J.-E. Bérard." These papers came into the possession of the family of André Paillot, by whom they were given to the Academy of Science in 1928.

since chemistry itself was just emerging from a revolutionary transformation of its theoretical structure.[66]

In accordance with the preference for theoretical or "pure" over applied—or impure?—science conventionally imputed to leading figures, it has sometimes been inferred that Berthollet was constrained against his deeper wishes and better self thus to study dyeing.[67] It may have been so. Equally, however, or rather more probably, it may not, for nowhere did he himself leave a record of harboring such sentiments. Quite the contrary, he felt his duties to be a liberation from the increasingly uncongenial demands of practicing medicine. Favorite studies—*études chéries*—is the phrase he chose for his investigations in mentioning to Tolozan his appreciation of the opportunity.[68] The expression might be thought simply politic, perhaps, except that it is consonant with the substance of the book he wrote. *L'Art de la teinture* appeared in 1791 and is both less and more than a manual for the trades: less, in that it does not contain detailed specifications and designs, like those of treatises in the *Description des arts et métiers* or of a book such as Bélidor's *Architecture hydraulique*;[69] more, in that the execution is faithful to Berthollet's intention of producing a theoretical work, one that explicated chemically the procedures used in dyeing. The first volume deals with the physical theory of color, with the properties of wool, silk, cotton, and linen, with the nature of the operations involved in dyeing, and with the reagents employed—acids, alkalis, and solvents—their preparation and characteristics. The second volume concerns the dyestuffs themselves, color by color, and the methods of applying them to the several fabrics. The theoretical standpoint throughout is that of elective affinity, and the underlying problem is to comprehend the forces that lead the dyestuffs to combine more or less stably with the materials to be colored.

In order to judge of what was beginning to happen here, it will be necessary to anticipate for a moment. Berthollet achieved prominence in the years after 1791, when he brought out his book on dyeing, and before 1803, when he published the celebrated *Essai de statique chimique*. In that interval, the nature of chemical bonding received much discussion throughout the entire discipline.[70] The two books have a lot more in common, however, than might be supposed from any contrast that would set

[66] Berthollet, *L'Art de la teinture* (1791), xl. The Hellot treatise was *L'Art de la teinture des laines*, . . . (1750).

[67] Sadoun-Goupil (1974), (1977), 137-138.

[68] Letter cited n. 56, AN, F^{12}.1329.

[69] Above, Chapter V, Section 2; on Bélidor, 2 vols. (1737-1739), see Gillispie (1971a), 102-103, and DSB *1*, 581-582.

[70] Holmes (1962) gives an excellent discussion of the transition from elective affinities to ideas of mass action.

a handbook of industrial chemistry over against the theoretical wellspring for laws of mass action and chemical equilibria. Already in *L'Art de la teinture* Berthollet is concerned with the effects on chemical combination of the heat and rate of reactions, and of the amount and concentration of reagents. To estimate the force with which they will combine, it is not enough merely to know their identity. In short, both books belong in the early history of physical chemistry, which specialty like others in this period may thus be seen issuing from theoretical analysis of industrial procedures.

A coincidence, or rather a conjuncture, shaped the theoretical model. The forces of chemical combination that Berthollet imagined in *Essai de statique chimique* were micro-gravitational in nature. It had happened almost twenty years earlier, in 1784, that when he was starting his project for rationalizing the action of dyes in the light of chemical theory, Laplace was concluding a collaboration with Lavoisier on the theory of heat. The prospect for quantifying the action of force in chemical combination, which was to say at molecular dimensions, had been the focus of Laplace's interest in the studies that produced their joint *Théorie de la chaleur* (1783).[71] Thereafter, Laplace drew away from Lavoisier and toward Berthollet in an affinity that ultimately became a program of research occupying their respective sets of protégés. The formation of related schools of chemistry and of physics, in the guise of the Société d'Arcueil of Napoleonic fame, followed on completion of their two treatises, *Essai de statique chimique* in 1803 and Volume IV of *Mécanique céleste* in 1805.[72]

A good deal less deep in conceptual consequence, and incomparably more important in economic consequence, was the most celebrated of the contributions to come out of Berthollet's service to the Gobelins, the procedure he developed in the 1780s for bleaching by means of chlorine. In the nineteenth century, chemical bleaching became one of the two most important industrial processes to have emerged from French chemistry of the late eighteenth century (the other being the Leblanc soda process). The tale of its displacing what might be called the open field system—dependent on buttermilk, lye, and pegging out the cloth for long exposure to sunlight—has been often told.[73] It is a story that involves exploitation by private enterprise of project research initiated by government, and entails ironies typical of the early stages of many a technology. The most obvious is that the government should have been the French and the enterprise largely British. Watt and Boulton moved immediately to adopt

[71] On this collaboration, see Guerlac (1976) and Gillispie, "Laplace," DSB *15*, 312-316.

[72] Crosland (1967); Fox, "Laplace," DSB *15*, 356-363; Sadoun-Goupil (1977), 61-84.

[73] Ballot (1923), 529-533; Lemay (1932); Musson and Robinson (1969), 251-337; Sadoun-Goupil (1974).

Berthollet's process, and other British firms took it up aggressively. It is true that certain French manufacturers, notably Décroisille in Rouen, were also trying it on a factory scale by 1789, but it was mainly the British industry that benefited during the revolutionary and Napoleonic wars. A second irony is that Berthollet, for all his ambition to enlighten the arts with theory, should have failed to recognize chlorine for an elementary substance. That, too, was left for the British, in the person of Humphry Davy. Berthollet followed Scheele, who had first isolated the gas, in taking it to be a combination of hydrochloric (marine or muriatic) acid with oxygen (or dephlogisticated air, in the older language). He then attributed its bleaching action to oxidation of the dye. Faithful like many inventors to his original idea, Berthollet always preferred the use of chlorine itself in dilute solution to that of the hypochlorite (eau de Javelle, in French commercial usage), which came to be generally employed and which does furnish oxygen.

Those matters are well known, however, and need no further rehearsal here. A companion point does deserve emphasis. Berthollet began to work with chlorine in 1785, the year after accepting responsibility from the government for the Gobelins and dyes. It is indicative of the stage to which the aspiring scientists' role in such research had evolved that he should have published his discovery instead of retaining it for exploitation (as—to anticipate again—did Leblanc, the obscure surgeon simultaneously struggling up out of artisanal status).[74] Not for Berthollet the padlocked register or the walled safe, and not for him the scrounging for subsidy characteristic of French inventors or the litigation over patents characteristic of the British. Naturally, he was interested in the application of his discovery. He corresponded with James Watt, following its development, and continued his research into the 1790s.[75] What he emphasized to Tolozan, however, was the technical triumph of the thing, the management of a gas that suffocates its handlers and devours their apparatus, and equally the victory over a routine entrenched in centuries of

[74] Gillispie (1957a), and below, Chapter VI, Section 4.

[75] Berthollet announced his discovery in a paper read before the Academy on 8 April 1785 and published in Rozier's journal, *Observations sur la physique* under the title "Mémoire sur l'acide marin déphlogistiqué," 26 (1785), 321-330. A fuller version is printed in MARS (1785/88), 276-295. He gave the first complete account, "Description du blanchiment des toiles et des fils par l'acide muriatique oxigéné . . . ," in *Annales de chimie* 2 (1789), 151-190, with an "Additions à la description du blanchiment . . . ," in *Annales de chimie* 6 (1790), 197-203. Section 3 of part I of the second edition of *L'Art de la teinture* (1804) amounts to a treatise on bleaching. There is an excellent English version of that edition, *The Art of Dyeing* (London, 1841), translated, annotated, and illustrated by Andrew Ure. J. G. Smith (1979) gives the fullest and most scholarly account of the early history of chlorine bleaching. Unfortunately, it came into my hands too late for me to take advantage of the findings.

ignorant practice.[76] It was in the course of those experiments (he also recalled) that he had discovered the detonating properties of potassium chlorate and conceived the idea of substituting it for saltpetre in the fabrication of gunpowder of unprecedented power.

We have already met with Berthollet at the explosive intersection at Essonnes in 1788 of his research with Lavoisier's responsibility for the provision of munitions,[77] and the remaining assignment of importance, his collaboration with Monge and Vandermonde on the conversion of iron into steel, took him far from the Gobelins and the dyeing trades. That will be better treated in the next section, dealing with various governmental stimuli to technical innovation in the private sector of industry.

3. STATE-ENCOURAGED INDUSTRY: TEXTILES, MINING AND METALS, PAPER

An account of the favor lavished on Jacques Vaucanson and John Holker by the Bureau of Commerce and other agencies will serve to introduce the topic of government, science, and private industry. The stories of those two, respectively the archetypal inventor and the archetypal industrialist, begin in the 1740s and 1750s, coincidentally with Hellot's introduction of chemical controls into the fabrication of dyes and porcelains in the enterprises of state. Vaucanson was then the most ingenious and systematic of a legion of innovative mechanics, native sons zealous to profit from the creations of talent and wit, while Holker was the most successful among a swarm of English, Scottish, and Irish textile workers all too eager (from the British point of view) to sell their skills in France. In neither instance are their motivations to be identified with those of scientists. Government rather than inclination brought them into connection with the scientific community.

In 1746 Maurepas, at the zenith of his influence in the Maison du roi, forced election of Vaucanson upon a reluctant Academy of Science, probably at the instance of the Bureau of Commerce. The occasion was an announcement in the *Mercure de France*—not in any academic publication—of his success in constructing an automatic silk loom.[78] True to type, Vaucanson refrained from communicating details of the design. One of those stories that is apocryphal in fact, and true in principle, has it that he parried the snobbery of mathematically minded colleagues who increas-

[76] AN, F^{12}.1329.

[77] Above, Chapter I, Section 6.

[78] T. Doyon and Liagre (1966), 214-220. This excellent biography is the only serious study of the career of Vaucanson. After his election to the Academy, he served on many commissions evaluating the work of fellow inventors until the year of his death, 1782. The entire list has been abstracted from the Minute-books of the Academy by Doyon and Liagre, 443-449.

ingly set the tone at the Academy by observing, à propos of their disdain for his ignorance of geometry, "I could have built them a geometer."[79]

The allusion was to stunts that had made Vaucanson famous, a trio of automatons that were the rage of Paris in the season 1738-1739. The first of these creatures, an Apollo later transmuted into a faun, played the flute, fingering the stops and all, with a repertory of a dozen airs. He was soon joined by a drummer-boy rattling out martial rhythms on a tambourine. The third and most sensational was a duck, more true to life than any decoy in that it quacked, switched its tail, and consumed and digested quantities of corn, excreting the remains. Vaucanson was the one who pocketed the proceeds from charges for admission to the hall and from the grand tour his robots subsequently took of the British Isles and the continent. Backers who had enabled him to live, to equip his shop, and to lease the public rooms of the Hôtel de Longueville, in what is now the place du Carrousel, were constantly worsted in the strife of successive partnerships.[80] A country boy from Dauphiny and anything but innocent, Vaucanson had the wit to join the Renaissance tradition of clock-work figures onto the eighteenth-century fascination with the animal-machine, the man-machine, and the hypothetical statue endowed one-by-one with discrete sensations to be analyzed by the associationist psychology. The authors of a recent, and admirable, biography of Vaucanson are themselves technical men professionally and are naturally tempted to make him a precursor of automation and cybernetics.[81] An historian may feel equally tempted to bring him on stage in the closing act in some theater of machines. The temptations are complementary. Yielding to both and combining the results would produce a balanced point of view, in that Vaucanson clearly had a highly important gift for thinking about systems no less than parts.

There was a time after the success of his automatons, when his penchant for system almost undid him. In 1741 the controller-general, Orry de Vignory (the same whose brother patronized the manufactory of porcelain at Vincennes), thought to capitalize on Vaucanson's ingenuity for the sake of the interest, as well as the amusement, of the public and appointed him inspector-general of manufactures. The commission was twofold. Vaucan-

[79] "Eh, que ne le disaient-ils, je leur aurais fabriqué un géomètre." A real letter twenty years later contained the following reflection: "Celui qui a inventé le rouet à filer la laine ou le lin ne serait regardé par les Académiciens de nos jours que comme un artiste et serait méprisé comme un faiseur de machines. Il y aurait cependant de quoi humilier ces messieurs s'il faisait réflexion que ce seul mécanicien a procuré plus de bien aux hommes que n'en ont procuré tous les géomètres et tous les physiciens qui ont existé dans leur compagnie." Vaucanson to Trudaine, 10 May 1765, AN, F^{12}.1449. Quoted in Doyon et Liaigre (1966), 423.

[80] Doyon and Liaigre (1966), ch. 3, with an illustration facing 50.

[81] *Ibid.*, 1-2.

son was to investigate and analyze the silk industry in order to explain its superiority in Piedmont and to propose measures for perfecting it in France. The prévot des marchands, or municipal administrator, in the silk capital of Lyons picked a progressive manufacturer, one Jean-Claude Montessuy, to accompany this enfant-terrible of an inventor sent to inspect an industry of which he had no prior experience. They spent the better part of two years in northern Italy and southern France, entering shops on the strength of a royal passport, informing themselves about every step in the procedures, taking back to Paris samples of raw and finished silk for testing, trying the tools in the Hôtel de Longueville, disassembling and reassembling pieces of machinery—reels, bobbins, looms, and mangles.

The report that issued from this inquest attributed French inferiority, not to any natural advantages across the Alps, but to the ignorance, indiscipline, and disorder of the industry. Vaucanson despaired of reforming it piecemeal and considered that only a radical solution would suffice. His recommendation called for seven pilot plants for throwing silk, two to be installed in Dauphiny, Provence, and Languedoc, and one in the Vivarais. Each would employ 100 women in reeling silk, another 100 workers in making yarn, and 80 people in preparing warp and weft and in weaving, together with supervisory personnel and certain specialists. The investment for buildings and machinery would come to 600,000 livres, the annual payroll and cost of maintenance and fuel to 243,000, and the consumption of raw materials to 980,000 per year. Sales should amount to 1,655,000, which allowing for interest charges of six percent on the investment would leave a yearly profit of 336,000 livres. No document could make a more startling contrast with the vagueness and hand-wringing characteristic of much of the written record of eighteenth-century entrepreneurship. That a kind of mechanistic Saint-Just should think in terms of a clean sweep is less surprising than that seasoned officials—themselves swept off their feet, evidently—should have attempted to act on so technocratic a recommendation. The controller-general did have the prudence to pare the scope to that of a single installation for a start, and since it was an affair of silk, the place to begin had to be Lyons.

In Lyons as in other great cities, the corporations of arts and trades constituted the municipality. Work was regulated by an elaborate code governing wages, prices, and conditions of employment, while quality was verified by the agents—*gardes-jurés*—reporting to the Bureau of Commerce. In the silk industry the community consisted of some 250 master merchant-manufacturers supplied by over three thousand shops kept by master-workers. A small, diminishing class of independent masters fabricated and sold the finished product in their own shops. In 1737 the workers had succeeded in improving their terms and situation vis-à-vis the merchant-manufacturers, and had secured a more democratic code and a stronger representation in the counsels of the guild. Into this situation came Vaucanson, inspector-general of silk for the kingdom, with authority from Paris to found a Manufacture Royale that would set the pace for the entire industry and show it how to rationalize production. Local investment would be tempted from its conservatism by the provision of guarantees from the Treasury against loss. Montessuy would be director, responsible to the investors. Vaucanson would be inspector, representing the state and overseeing the technical side. The plan exhibited, indeed, that mixture of private capital with public backing and governmental oversight that has come to characterize the modern French economy.

Such a scheme could never have been accommodated to the regressive regulations of the existing industry. For months Vaucanson and his associates in the ministry in Paris worked on a new set of provisions to supersede the municipal code. Their draft was adopted by the Council of State on 19 June 1744. On 22 July Vaucanson and Montessuy took the coach for Lyons with 1,500 printed copies in their baggage. The document was clear and categorical. In 181 articles under fourteen headings it called for converting merchant-manufacturers into employers and master-workers into employees. All parties would learn their jobs by emulation of the model Manufacture royale now to be created. Rumor had naturally preceded posting of the proclamation. On the night of 6 August the silk workers of Lyons rose and ran our reformers right out of town. In the opinion of Pallu, the intendant, Vaucanson and Montessuy owed their very lives to having lodged in his residence, whence Vaucanson escaped in the disguise of a monk.[82] In less than three weeks after their departure from Paris, they were back in the capital, having succeeded only in provoking the most serious strike in eighteenth-century France. As often happens, the people knew their enemy, and he did not know them.

Thereafter Vaucanson concentrated on machines, evolving for himself the role of professional inventor and supplier of technology. "Whenever he steps out of practical mechanics, he is more a machine than those he makes," observed an acquaintance.[83] The years from 1744 to 1751 were the time of his great inventions. The silk loom already mentioned was automatic in that all the assemblies were operated from a single input of power, which might be supplied by man, mule, donkey, or mill-wheel. A draw-loom for brocade and figured silk followed. In it the pattern was replicated on the principle of a player piano. Needles carrying threads of different colors were sent through the fabric by the action of a coded pattern punched into perforations in a sheet of paper passing over a roller. Like the later sequence in the British woollen and cotton industries, Vaucanson succeeded in mechanizing the weaving of the fabric before he did

[82] *Ibid.*, 198-20.

[83] *Ibid.*, 421-422.

the production of thread or yarn. A throwing mill for converting fibers reeled from the cocoon into raw silk was the third of his important machines. It was an adaptation of devices he had observed in his tour of Piedmont. More original and more sophisticated was his solution to the problem of producing organzine, the silk cords for warping. Vaucanson's biographers find the principle of automatic regulation embodied in the technique he imagined for synchronizing the great squirrel-cage reels on which strands were stretched. If so, feedback antedates James Watt's governor on the steam engine by half a century.[84] Finally, Vaucanson designed a mangle to achieve the effect of watering or moiré, traditionally accomplished by crushing folds of silk under a roller bearing an enormous load of masonry. Curiously enough, this seemingly simple problem gave him more trouble than the intricate machines he devised for throwing, weaving, and figuring silk.[85]

Vaucanson's inventions clearly belong to the same generation technically as the famous devices that opened the revolution in the British textile industry slightly later: the flying shuttle, the spinning jenny, the water-frame, Crompton's mule, and ultimately the power-loom. The relation of any of these machines to the science of mechanics is enigmatic at best. As for their relation to the organization of industry, it would have been entirely feasible to attach Vaucanson's silk looms to a source of power external to the workshop and thereby convert it into a factory. Strategically, that was the significance of Richard Arkwright's application of the water-frame in the English midlands. What made that impossible in France, or at least largely delayed it until the time of the Revolution, was the considerable immobility of capital and labor. The riots in Lyons were evidence enough.

For, comparable though Vaucanson was to his British counterparts in point of the technical and scientific import of his inventions, his situation relative to the structure of industry and government was quite different. Inventors in Britain obtained from government only the protection of a patent. The success or failure of an invention was determined by its profitability. Whether they themselves profited was a matter of shrewdness tempered by luck. In any case, their lives were lived and fates decided within the early and expanding industrial system. Not so Vaucanson, whose success was conditional on his ability to keep the flow of subsidy coming through the Bureau of Commerce from the Ministry of Finance. Despite recurrent friction, he managed to retain Trudaine's confidence

[84] For a general account, see *ibid.*, ch. 9, and for technical detail Borgnis (1819-1820) and Razy (1913). The best place to study the machines themselves is the Musée des tissus in Lyons. The collection has been much improved since Razy published the guide and analysis just cited.

[85] Doyon and Liaigre (1966), 270-277.

and that of officialdom, though without ever gaining it among proprietors of silk mills. On the contrary, they had to be enjoined or granted incentives to adopt his machines, and often they resisted altogether. That is understandable, since the only person who consistently benefited financially from his inventive activity was Vaucanson himself.

In contrast to the stereotype of the inventor who dies in the poorhouse while some exploiter makes a fortune, Vaucanson became a wealthy man while loss accrued to the proprietors of businesses he supplied, even as it had to investors in his automatons. The most durable of his connections with a manufacturer, that with Henri Deydier, was a very different matter from the partnership of James Watt with Matthew Boulton, to take for contrast the most intimate of British arrangements between inventor and manager. Vaucanson kept his operation in Paris, where he took a lifetime lease on the Hôtel de Mortagne, rue de Charonne. There he installed a machine shop with the purpose not merely of developing but of producing looms. In one exchange of correspondence with the ministry, he invited the controller-general together with other officials to visit his *laboratoire* in order to see for themselves what measure of technological value society was reaping in return for its money.[86] After his death in 1782, Lefèvre d'Ormesson, briefly the controller-general, resolved to maintain the laboratory of machines and tools Vaucanson had made and assembled, and to open the collection for the edification and emulation of qualified members of the public.[87] In the revolution, this *Cabinet des mécaniques du roi* became a nucleus of the modern Conservatoire des arts et métiers, where models and relics of his inventions continue to be on display.

Following the fiasco in Lyons, officials in the Bureau of Commerce thought to encourage modernization of the silk trades in a more modest and decentralized fashion. Later in the 1740s and early in the 1750s the government accorded certain mills the appellation "royal," making small grants in aid of specific improvements, and paying small premiums when the organzine met a certain standard. There were half a dozen of these enterprises in Dauphiny and another two or three in Provence. Schoolmasterish as ever in instinct, however, the administration kept coming back to the notion of a model plant that would do for silk what Saint-Gobain had done for mirrors and Sèvres was about to do for porcelain. Vaucanson still had the confidence of Trudaine in all that pertained to silk, and they found their man in Henri Deydier.

Deydier was scion of a dynasty that had been respected in silk manu-

[86] Vaucanson to Fourqueux, 24 March 1776, AN, F[12].654. There is extensive correspondence between Vaucanson and officials in the ministry in this carton, much of it regarding his troubled relations with two important silk manufacturers, Enfantin in Romans and Jubié in La Sône.

[87] See the preface by Bertrand Gille to Doyon and Liaigre (1966), xii.

facturing and related commercial activities in the Vivarais (now the Ardèche for the most part) for three generations. Patriarchal, philoprogenitive—he was one of fourteen children of his father's first marriage—the family was of a type that set the tone in scores of lesser urban centers throughout France, their industriousness having been a major element responsible for the real wealth generated in the commercial sector of the economy. Other examples are to be met with in the Montgolfiers of the paper industry in Annonay.[88] In the middle of the eighteenth century, fathers in these milieux began sending sons away for formal education rather than simply training them up in the business. Henri Deydier took a degree at the University of Paris and returned home to Aubenas an innovator, not a rebel, to be sure, but still not content with the way things had always been done. He was a pioneer, for example, in conversion from wood to coal for industrial fuel. Like many a milltown of the old industrial regions, whether in New England, the English midlands, or the tributary valleys of the Rhône, Aubenas is a small city on a small river, the Ardèche. The improvements Deydier made in the family mill brought him to the attention of Vaucanson, who in 1751 or 1752 nominated the young silkmaster to Trudaine to be their representative within the industry, their Judas-goat the die-hards might have said.

Formally the proposal to the state came from Deydier, but like his counterparts in some program of development today, he had been given to understand exactly what the government wanted to support. He was to buy land and build a mill at his own expense. He engaged an architect, Guillot Aubry, who designed buildings in accordance with Vaucanson's specifications, the two being close associates. The structures were to accommodate twenty-five mills for reeling silk and twenty-five for organzine, which machinery was to be furnished by Vaucanson from the Hôtel de Mortagne at the expense of the state. For a period of ten years Deydier would be paid a premium on his production, the amount varying from ten to forty sous per pound of finished silk according to its fineness. At the end of that time, the machinery would become Deydier's property. Thereafter, the plant would continue to enjoy the privilege of a "Manufacture Royale," and the porter would wear the royal livery. Deydier himself would be exempted from billeting troops and his foremen from serving in the militia and from certain other civic liabilities.[89]

The Council of State ratified the terms on 5 September 1752. The buildings were to be ready to house the first consignment of mills that Vaucanson would construct by July 1754. The machinery was shipped dismounted from Paris and assembled on location. Vaucanson had to ap-

[88] See below.

[89] On the model factory at Aubenas, see Doyon and Liaigre (1966), ch. 11.

prove every detail of the installation. The arrangement was analogous to the way in which makers of very sophisticated equipment nowadays will lease or sell not only their products but their servicing of them; whatever the legal ownership, there is a sense in which machines that are too complicated for their users remain the offspring of their makers. So it was with Vaucanson in his laboratory of invention. The hope that other proprietors would emulate Deydier, and replicate the factory at Aubenas out of enlightened self-interest, proved illusory, not to say visionary. The Bureau of Commerce kept holding out inducements to other manufacturers to adopt the mills and looms perfected there. There were negotiations, protracted and often acrimonious, with Enfantin in Romans, with Jubié in La Sône, with the estates of Languedoc over abortive enterprises in Montpellier.[90] None succeeded. Aubenas itself, after returning Deydier modest profits in the early years, slid from technical eminence into financial failure and went under in 1775.[91]

After Vaucanson died, in 1782, his inventions did have the effect, in other hands and other times, of transforming the industry he had tried to revolutionize *a priori*. That they should have had to await the evolution of the economy may seem ironical at first thought, for Vaucanson prided himself, and justly, on his perception of the economic component of invention. In this respect he was like Thomas A. Edison (as also in the philistinism of his attitude to mathematics and scientific theory). His strategy may be compared with Edison's creation of a laboratory and pioneering of a profession of invention more justly than with the gadget-by-gadget approach of contemporaries. But however tempting schematically, the comparison is bound to be anachronistic, and not only because of the intervening century. In the attempt to disseminate the inventions of Vaucanson, a limiting factor was always the unavailability of trained personnel to operate his machines, let alone to maintain them.[92] He had his team of assistants in the Hôtel de Mortagne, to be sure. The formation of cadres in the field, however, was one feature of his program that never succeeded. His sense of system was abstracted from the human term in the equation; his economics of invention was predicated on the substitution of skill, which was scarce, for labor, which was plentiful.

Through the leadership of John Holker, Trudaine and his associates in the Bureau of Commerce had the satisfaction of seeing the cotton industry achieve the kind of mechanization that the genius of Vaucanson and the model of the factory at Aubenas failed to bring about in silk. At Saint-Sever, a suburb of Rouen, Holker mounted a Manufacture royale in 1752 for the operations of carding and spinning cotton, of weaving cloth, and of finishing corduroy and velveteen. His pilot plant succeeded commer-

[90] *Ibid.* [91] *Ibid.*, 351, ch. 14. [92] *Ibid.*, 288.

cially from the outset, as did an offshoot started at Sens in 1760. In these establishments Holker trained workers who then moved out into the industry, qualified both to fabricate and operate the flying shuttle and other devices of the English midlands in the 1750s and 1760s, the spinning jenny in the 1770s, and the water-frame in the 1780s.[93]

Some weight should be given to personality, and more to circumstance, in explaining the difference between the results in silk and cotton. Personality was certainly a factor. Vaucanson's disposition was difficult and contentious, and Holker thought him unrealistic about the technical sophistication that could be expected of workers. By contrast, Holker was adept in the management of men no less than machines. He was also a good judge of both although—or perhaps because—he was nothing of an inventor on his own account. There was never any question of intruding him into the company of the Academy of Science. Whenever Holker had to do with that body, the intermediary was Etienne Mignot de Montigny, Trudaine's staff scientist, whom we met in the affair of hard porcelain at Sèvres.[94] In the matter of inventions, Holker's part was to "surprise their secrets," as the phrase went, by importing the new techniques and enticing—the word here was "*debaucher*"—skilled people to emigrate from Britain to France. The contrast between Vaucanson and Holker was that between creator and impresario.

At the same time, circumstances are bound to count more tellingly than temperament in an explanation of the very possibility of mechanizing the cotton industry in France, and first of all in Normandy. Other textile trades had to await the Revolution and the striking down of the many regulations deriving from government, guild, and *compagnonnage*, or system of apprenticeship. Those impediments to change and progressive entrepreneurship did not exist in cotton, a new industry that was itself a change in the eighteenth century. Modernization there represented no threat to old techniques embedded in a spirit of routine, as it did in silk, and there was no corporation of cotton masters ensconced in the structure of municipalities. Importation of cotton cloth from India and the Levant did indeed suffer, but merchant manufacturers in Rouen and Caen needed to be concerned only to the extent that cotton competed in the market for woollens and linens. For cotton was a rural industry at the outset. In villages throughout Normandy, peasant workers, men and women, occupied the winter months with spinning wheel and hand-loom in their own cottages. The development was highly welcome to the government, ever so-

[93] Rémond (1946) is a highly condensed monograph on the career of Holker, based on extensive and detailed knowledge of the archival materials.

[94] The Condorcet éloge of Mignot de Montigny contains an interesting appreciation of his relations with Holker and the textile industry, HARS (1782/85), 108-121.

licitous over matters of subsistence. The Council of State legitimized and regularized the practices in an edict of 7 November 1762.[95]

This cottage industry was already widely dispersed in the Norman countryside when, early in 1750, Marc Morel, inspector-general of manufactures in Rouen and a confidant of Trudaine, chanced to fall in with Holker. A Jacobite and lieutenant in the expatriate regiment commanded by David Ogilvie, Earl of Airlie, Holker was then thirty-one years of age and casting about for a more eligible lot than that of junior officer in a forlorn cause. He was descended from Catholic gentry in Stretford, near Manchester; had learned the cotton business; and had started his own mill and a family in the early 1740s. In 1745, when the young pretender and his army penetrated to Manchester, Holker volunteered to join the Stuart forces, only to be captured and jailed after the Battle of Culloden. In 1746 he escaped from Newgate Prison and made his way through Holland to France. His overtures for a pardon meeting with no response from the British government, he was ready to hear Morel's proposal that he resume the cotton trade in Normandy and become the protagonist of modernization and concentration of production.

To train a local cadre single-handed in a language not his own and without equipment would have been a desperate prospect, and in the course of these discussions, Holker conceived the notion, tinged with the romanticism of his jailbreak, of returning to England incognito, there to recruit a team of workers large enough to mount a factory on the scale of enterprises in Manchester. Meanwhile, he joined Morel in tours of inspection. Morel translated Holker's highly critical memoirs on the state of manufacturing in Normandy and forwarded them to the Bureau of Commerce. Impressed, Trudaine had Morel bring his English associate to Paris for an interview with the controller-general, Machault. The conversations went well, and Machault authorized creation of a secret fund to defray the expense of a clandestine talent hunt and purchasing mission.[96]

The plan succeeded. Holker spent three months in the Manchester region over the winter of 1751-1752. At the same time, the French government openly dispatched Mignot de Montigny to London, apparently for scientific discussions with colleagues of the Royal Society. Thus, Holker could have access to counsel and to a legal presence in the country. He bought machinery under dummy names and sent the dismounted parts through various shippers to different destinations on the continent. He put himself in touch with workers he had known, or known about, and held out the prospect of rising higher in another world. Those who responded were slipped out through London and Dunkirk. Holker himself returned through Ostend. The summer of 1752 sufficed to assemble men

[95] Ballot (1923), 42.

[96] Rémond (1946), 53-54.

and machines in Saint-Sever and to take on local hands. An edict creating a "Manufacture royale de velours et draps de coton" issued from the Council of State on 19 September.

The published terms were nothing strange. Ostensibly, an association of four investors active in the textile business furnished capital and was granted the exemptions and privileges normally accorded a new enterprise. Really governing the affair, however, was a set of secret provisions containing features that were quite extraordinary. Behind the sleeping partnership of the capitalist quartet, Holker was to receive 20 percent of the profits without making any investment at all. The direction, moreover, was to be entirely in his hands.

Combining proficiency in British technology with experience in French commerce, he formed a managerial committee consisting of two English foremen, Guy Hall and James Leatherbarrow Anson, to oversee production and two Rouennais merchants, Guillebaut and Leclerc, in charge, respectively, of accounts and of sales. In 1754, the second full year of operation, Holker was employing a work-force of 92 people. The British cadre consisted of 20 skilled workers. Under the two foremen were three finishers, called Morris, Molloy, and Michael; three joiners and machinists, a father and son combination named Wills and an Irishman, Richard Smith; seven weavers (one from India); four calandarers; and a dyer. Of the 72 French laborers, the largest number, 15, were weavers supposed to emulate the operations of their British counterparts. By 1760 the 58 looms had increased to over a hundred, despite the down-turn of exportation in the Seven Years War. In the 1770s Saint-Sever kept 180 looms busy, and in the 1780s over two hundred.[97]

More significant than this local success was Holker's influence in the industry at large. The effect of mounting a Manchester factory in Rouen surpassed expectations from the outset, and Trudaine had Holker appointed to a post in the administration. On 15 April 1755, the controller-general commissioned him an inspector of manufacturing at an annual stipend of 8,000 livres. Holker thereby joined his friend and sponsor Morel in the service of the Bureau of Commerce with responsibilities of a special sort. He was to oversee factories of the foreign type and to look after the well-being of foreign workers.[98] The incongruity of thus installing the exemplar of free enterprise in the government was more apparent than real. In his annual tours of inspection, Holker visited manufacturers in the more conservative regions, in Lyons for example, and held out incentives to change their ways. To requests for subsidy or other forms of government assistance, he would respond with proposals for hiring a foreman

[97] *Ibid.*, 58-60.

[98] Bonnassieux and Lelong (1900), li; Rémond (1946), 81-82.

experienced with up-to-date machinery, and he would name a candidate.

Operatives who had learned to construct looms and jennies from the two Wills and Richard Smith at Saint-Sever were moved on to other establishments, at first in Normandy and Picardy, and increasingly in the later 1750s and 1760s elsewhere. The policy makes a diametric contrast with the restrictions imposed on artisans at Sèvres, in the Gobelins, or in the normal master-apprentice relation, where the object was to guard the secret of production in each particular establishment. Saint-Sever supplied tools and machinery as well as manufactured goods and trained people. Other mill-owners, not always content to buy equipment there or to hire away its well-paid people, took on their own British artisans, also with the encouragement of the Bureau of Commerce and often through Holker's good offices. Traces frequently turn up in the archives of inducements held out to a certain Jeanne Law, to la demoiselle Hayes, to Mademoiselle Offlanegan, spinsters adept with the jennies of the 1770s, and to others. Like many pensioners of the French state, they sometimes needed the services of lawyers in collecting the promised premiums, dowries, and annuities.[99]

Holker himself was exempted from the normal regulations that precluded government officials from engaging in the commerce that they regulated. Concurrently with his bureaucratic responsibilities, he enlarged his own interests beyond the original factory at Saint-Sever. First he moved into the woollen business in Picardy, where again he employed techniques from Manchester for scouring and fulling and also for puckering and glazing finished cloth. In the 1760s he expanded his operations to Amiens, to Rheims, to Sens, to Carcassonne, to Nîmes, and was constantly on the move between his various factories and the nerve center of industrial policy in Paris. For Holker saw beyond the profits of the moment or the season. His was the spirit of adventure in industrialization. A sorcerer's apprentice, he carried Trudaine's views further, it may be, than his patron had imagined, or perhaps wished, and thought to force the pace and revolutionize the entire French economy. From cotton Holker moved into textiles in general, and from textiles into hardware, leather, carpentry and joining, ceramics, and even chemical industry.

He had little sense of measure, and jealousy and resentment inevitably stirred in the breasts of less favored manufacturers, finding an excuse in the all too glaring conflict between his public responsibilities and his private interest. Throughout the 1770s Trudaine de Montigny continued the friendship shown Holker by his father. It was, perhaps, more in keeping with traditions of French officialdom than with his own career that in 1768 Holker's son, also called John, should have been appointed his as-

[99] Ballot (1923), 44; Bacquié (1927), 29.

sociate with succession to the inspectorship.[100] The younger Holker compromised himself by shady doings in Philadelphia, where he was French consul and navel agent in 1777 and 1778, in the early days of the Franco-American alliance. The magic of the Holker name failed to rub off on Tolozan, and Roland detested him, constantly girding against his dwindling influence in the antechambers of the Bureau of Commerce, and in the Corps of Manufacturing Inspectors.[101]

That Corps, in which Roland served and in which the controller-general commissioned both Vaucanson and Holker, had been founded by Colbert in 1669 and was abolished by the constituent assembly in 1791. In the interval over four hundred inspectors filled its ranks. Formally they were responsible to the Council of Commerce, the panel of the Council of State which set industrial and commercial policy and which is not to be confused with the administrative Bureau of Commerce within the ministry of finance. Initially, the mission of manufacturing inspectors was to superimpose the authority and sanction of the state upon the responsibility already exercised by gardes-jurés within the guilds for overseeing quality. By 1730 the detail of the regulations that the Colbertist regime had laid down, most minutely for textiles and related trades, filled seven quarto volumes. Between 1730 and 1785 over a thousand additional edicts and amendments were issued.[102] Liberalization in the mid-century affected mainly fiscal policy, and the Council of Commerce never seriously contemplated abandoning the defense of standards to the forces of the market. Carrying out the routine operations of the corps at any one time was a staff of 42 inspectors, assisted by a slightly larger number of under-inspectors.[103] The chief inspector in each generality reported directly to the controller-general rather than to the intendant. In principle, every manufacturing plant was to be visited four times a year. That was seldom possible, even though certain industries exploiting natural resources—mining, for example, and fisheries—were exempt from provincial jurisdictions and subject to the surveillance of specialists traveling out of Paris. In the course of their rounds, inspectors had the duty, when satisfied, of stamping bales of merchandise with the seal of approval, in the absence of which commodities could not legally be placed in trade. If they turned up in-

[100] Rémond (1946), 119-120.

[101] Roland brought his hostility into the open in the prefatory remarks to *L'Art du fabricant de velours de coton* (1780). The attack was answered by Holker, or perhaps by a partisan, in a pamphlet, *Lettre d'un citoyen de Villefranche à Monsieur Roland de la Platière* (1781). Roland resumed his offensive in *Réponse à la lettre d'un soi-disant citoyen de Villefranche* and together with an associate, one Baillière, *Lettres imprimées à Rouen en octobre (1781)*.

[102] Roland de la Platière, "Manufactures," *Encyclopédie méthodique: Manufactures, arts et métiers 1* (1785), 116.

[103] Roland, "Inspecteur," *ibid.*, 69.

fractions of the code in their visits to shops and warehouses, they initiated corrective action ranging from warnings, through impoundment of defective merchandise, to fines and imprisonment. Theirs was not a vocation making for popularity.[104]

Increasingly the service took on a more active role in the development of specific industries and even in research. The tendency is identifiable during the tenure of Philibert Orry as controller-general (1730-1745), and it became a feature of commercial administration with Trudaine's entry into the Council of State in 1744. The senior officials of the corps, the five inspectors-general, were influential in making policy and might accompany Trudaine in a consultative capacity at sessions of the Council of Commerce. Besides Vaucanson and Holker (the latter was appointed to the highest rank), others actively concerned with the development of technologies related to their knowledge were Jean Hellot for mining (as well as dyeing), Gabriel Jars for mining and metallurgy, and Nicolas Desmarest for papermaking (as well as wool-growing), all three being members of the Academy of Science. Among progressive administrators, the most famous who held appointments in the Corps of Manufacturing Inspectors later in the century were Dupont de Nemours and Roland de la Platière. The encyclopedic concept of inspection that had evolved from the merely regulatory aspect appears in the definition of an inspector's function that Roland included in the article he prepared on the service for the *Encyclopédie méthodique*:

> The Inspector is an agent of the Council [of Commerce]. He is sent out into the provinces in order to investigate the state of the arts and of commerce; to identify the factors that make their progress so slow; to seek out and specify measures appropriate for encouraging their growth and guiding them in the improvement of which they are capable. Detailed steps conducing to these great objectives constitute his occupation, the result of which is to expand education and to bring prosperity to birth. Knowing what is done and how it is done, and rationalizing the practise of the arts by analyzing the products and calculating the benefits—such are the elements of his job.[105]

In the three volumes on "Manufactures, arts et métiers" that Roland edited for Panckoucke, the articles treat only textile and related industries, even like the first Colbertist code, on which many of the *obiter dicta* heap much scorn for other reasons.[106] The word still connoted the fabri-

[104] Bacquié (1927) is a very informative work of historical piety by an inspector of manufactures in the Third Republic.

[105] Roland, "Manufactures," *Encyclopédie méthodique* *1*, 62.

[106] Those volumes of *Encyclopédie méthodique* came out in slightly garbled order: *1* (1785); *2* (1784); *3* (1790). A fourth volume, ed. G.-T. Doin, appeared in 1808, treating

cation of things by hand, although now by many hands gathered into a factory. The commissioning of Gabriel Jars as inspector of metallurgical factories in 1768 was thus a recognition at once of the prospects for heavy industry and of French backwardness in mining, the exploitation of metals, and the use of coal.[107]

Coal deposits in France are less readily exploitable than in Britain, Germany, and central Europe, and since forests were correspondingly more abundant, there was no incentive to turn to coal until depletion of woodlands began threatening the supply of fuel and charcoal. That menace began to loom early in the eighteenth century. In part, the problem of developing a coal-mining industry was juridical and in part technological. French law vested the right of exploiting mineral deposits in the proprietors under whose land they lay. Noblemen as a rule, landowners were normally deficient both in technical capacity and motivation. Trudaine set about creating a rational code of mining regulations early in his service on the Council of Commerce. A decree asserting the prior rights of the crown and providing for systematic grants of concessions to qualified entrepreneurs issued from the Council on 14 January 1744. Technological incapacity was compounded by obstructive litigation, however, and the effect disappointed expectations.[108] Trudaine and his associates were well aware of the superiority of British, German, and Scandinavian practice in the extractive industries generally. Needing ammunition about the organization of mining no less than information about the best methods, the Council of Commerce, drawing also on the advice of Jean Hellot, resolved to send emissaries on reconnaissance abroad. Trudaine's first choice fell upon Gabriel Jars, one among several budding specialists whom he had already brought up to Paris with a view to forming a cadre of persons qualified in mineralogy and metallurgy.[109]

Jars was the youngest of three sons of an entrepreneur and landowner of Lyons, among whose properties were important copper mines at nearby Chessy and Saint-Bel. The family had connections with higher social circles, and the marquis de Vallière recommended the boy's technical promise to Trudaine. Trudaine thereupon arranged for him to enter the École des ponts et chaussées in order to complete an education begun in the Col-

---

dyes, oils, and soap. Roland's *Discours préliminaire* to volume 3 has a certain value for its history of the several arts and of the interest taken in them by government.

[107] Bacquié (1927), 43-54.

[108] The matter is well summarized in Arthur Birembaut, "L'Enseignement de la minéralogie et des techniques minières," in Taton (1964), 372-376. For the French coal industry, see Rouff (1922).

[109] Taton (1964), 376-385. On the genesis of the Jars missions, see J. Chevalier (1947) and the éloge of Jars by Grandjean de Fouchy, HARS (1769/72), 173-179, printed also in Gabriel Jars, *Voyages métallurgiques 1* (1774), xxi-xxviii.

lège de Lyons and continued in his father's enterprises. Not that formal instruction in mining and metallurgy was offered in Paris, but at least students at Ponts et chaussées were taught how to learn something technically, and in the early 1750s the Bureau of Commerce sent eight or ten young men there to prepare for careers in mining. The director, Perronet, worked closely with Trudaine in matters of development.[110] Jars was nineteen in 1751 when he arrived in Paris. After only two months, Perronet sent him off to visit, inspect, and report on the lead mines of Poullauen in lower Brittany, and a little later to the mines at Sainte-Marie and Giromagny in Alsace. In 1754, after an interval at home, where Jars designed and constructed a new type of furnace for refining copper,[111] the government again enlisted his services, this time for an investigation of mines in Forez, the Pyrenees, and the Vosges. On this expedition, Jars had a companion, also trained briefly at Ponts et chaussées, a young highway engineer François Guillot-Duhamel (not to be confused with the famous agronomist).

So much was preparation, and testing, for the main work of a life lived much abroad and arrested early. In 1756 Jars and Duhamel were dispatched to make a study of the mining industry of central Europe. Combining diplomacy with technology, that mission took them to Saxony, Bohemia, Hungary, Styria, and Carinthia, and occupied them for two and a half years.[112] In 1764 Jars departed alone for Britain. Prior to setting off, he consulted closely and at length with Holker, whose success with transplanting textile technology the Council of Commerce now thought to emulate in mining.[113] Jars spent fifteen months touring the mines of Yorkshire, Cumberland, Staffordshire, Derbyshire, the west country, and the region around Edinburgh. In 1766 a third expedition took him to Liège and Limburg, the Harz mountains of central Germany, and (in company with his brother) to Sweden and Norway. Clearly, his reports gave the officials of the Bureau of Commerce just the sort of information they most valued. Jars had been a correspondent of the Academy of Science since 1761. On his return from Scandinavia in 1768, a new post of inspector of manufacturing for metallurgy was created for him. At the same time, the ministry exercised its prerogative and preferred him over Lavoisier for election to the Academy of Science despite his placing second—a close second—in the vote to fill a vacancy in the section of chemistry.[114] (Lavoisier, it may be recalled, had also made his first reputation scientifically on mineralogical expeditions with Guettard.)

[110] Below, Chapter VII, Section 1.

[111] "Description d'un grand fourneau à raffiner le cuivre . . . ," MARS (1769/72) 589-606.

[112] "Description d'un nouvelle machine exécutée aux mines de Schemnitz en Hongrie, au mois de Mars 1755," SE 5 (1768), 128-132.

[113] Bacquié (1927), 44.

[114] Above, Chapter I, Section 6.

Thereupon, Jars might look forward to a career combining the kind of leadership that Holker was giving in the textile industry with continued development of the family copper interests in the Lyonnais. He completed one swing through the foundries and forges of northern and eastern France in the autumn of 1768, to which we shall return. In the summer of 1769, he was ordered on another into Auvergne, Forez, and the Massif Central. On an August day in the mountains above Clermont-Ferrand he suffered sunstroke and died soon afterward. Jars had then published only an incidental pamphlet on tiles and bricks for the *Description des arts et métiers* and one piece on a Hungarian force pump in the *Savants étrangers*.[115] The ministry had preferred to keep the reports of his journeys in manuscript and in quasi-confidence. That cannot have been intended as a permanent policy, for several memoirs were accepted by the Academy and printed soon after his death.[116] His brother reprinted them and made a selection from the remainder to publish in the three-volume *Voyages métallurgiques*, his monument more durable than brass.[117]

The *Voyages métallurgiques* constitutes a cardinal document in the history of the technology and organization of mining, and is at the same time a seductive travelogue, guiding the twentieth-century tourist of eighteenth-century Europe into regions where he would not otherwise penetrate. On the instructions of Jean Hellot, who drew up the initial directive for Jars and Duhamel, the young men began their first mission with three or four months of language lessons in the mining center in Freiberg. It was not enough to speak German and a smattering of Hungarian, and later English and Swedish. They had to be able to ask the miners questions in all their own dialects and to understand the answers.[118] There are etymological nuggets to be picked up in *Voyages métallurgiques*. The reason that the term for coal, *charbon de terre* in most of France, becomes *houille* in Alsace, Lorraine and French-speaking Belgium is that the oldest mines to be exploited, those in the vicinity of Liège, go back to Germanic times in the thirteenth century, and the word *houille* has a Saxon root, as indeed does coal.[119] To one reader it was also news that a colloquial word applied to unexpected inconveniences goes back to the coal mines of Cumberland, where minor faults often required climb-

[115] See n. 112, above; *Art de fabriquer la brique et la tuile en Hollande* . . . (1767).

[116] "Procédé des Anglois pour convertir le plomb en minium," MARS (1770/73), 68-72; "Observations métallurgiques sur la séparation des métaux," *ibid.*, 423-436, 514-525; "Observations sur les mines en général, & particulièrement sur celles de la province de Cornwall, en Angleterre," *ibid.*, 540-557.

[117] 3 vols. (1774-1781).

[118] Birembaut in Taton (1964), p. 380; Grandjean de Fouchy, "Éloge de Jars," in *Voyages métallurgiques 1*, xxiii.

[119] "Quatorzième mémoire: Sur plusieurs mines de charbon & quelques forges de fer, d'Allemagne & des Pays-Bas," *Voyages métallurgiques 1*, 283.

ing over or around obstacles in order to follow a seam: "On nomme ces dérangements *hitch* ou *smal-trouble*."[120] The temptation to recite anecdotes of travel will be resisted, however. It is more important to notice that the British sections of *Voyages métallurgiques* afford another instance, along with Duhamel on Norfolk husbandry and Morand on coal mining and steam engines, in the literature of French technology reporting on procedures that were simply practiced across the channel and not written up.[121] The same is not true of the continental sections: Jars had studied Swedenborg religiously, and carried the French translation as a bible to illuminate his own observations in Germany and Sweden.[122]

The fullness of the descriptions both of mining and metallurgy may conceal the pointedness of Jars's purposes, which in accordance with the motivation of the government were twofold. In the first place, he wished to analyze coal and iron, not merely for their own sake, but primarily for their relation in production. The instructions he carried to Britain in 1764 directed him "to ascertain the various uses made of the different types of coal, of their prices at the pit-head; if it were true that large coal was employed in blast furnaces to smelt iron and also copper ore; if it were necessary to treat it for this purpose and reduce it to the material called 'coucke' in England." In the second place, the political economy of the extractive industries was as important for his mission as were matters of mineralogy and technology. After further injunctions to find out the reason for the superior polish of English brass and the incomparable hardness of English files, he was told that he must, "above all, ascertain the reason why industry is pushed much further in England than it is in France, and whether this difference, as there is every reason to suppose, is due to the fact that the English are not hindered by regulations and inspections, and that they have few means of getting wealth other than by trading and manufacturing."[123]

To pursue this question of industrial regimes briefly before turning to coke, Jars's accounts make a qualitative comparison of British and continental practices hard to resist. Although he composed two papers on a technique he had devised for improving the circulation of air,[124] Jars offered no judgment about the conditions under which miners lived, worked, and died. A reader in an age of social conscience, one who follows

[120] "Douzième mémoire: Sur quelques mines de charbon, des forges de fer, & plusieurs autres établissements utiles d'Angleterre," *ibid.*, 239.

[121] Above, Chapter V, Section 4, n. 89; Section 2, n. 54.

[122] In the edition translated by Bouchu for inclusion in the *Description des arts et métiers*, above, Chapter V, Section 2, n. 46.

[123] Passages from the directive are translated in Chevalier (1953), 57-58.

[124] "Observations sur la circulation de l'air dans les mines . . . ," MARS (1768/70), 218-228, 229-235.

Jars in imagination down the shafts and along the galleries of the English collieries, and who also accompanies him vicariously into the Carinthian and Styrian mines, not to mention those of Dannemora, will surely decide that it would have been better to be an Austrian or a Swedish than an English miner. The 530 employees of the two companies exploiting the Eisenartz in Styria worked from seven to eleven in the morning and from noon to four in the afternoon five days a week.[125] If Jars is to be believed, the six- to eight-hour day and the forty-hour week were normal in central European mining, and miners had their allotments and gardens around the pithead.[126] There were mines in England where the hours were no longer, except for children doing light work, but the kind of thing that stays in the mind from the memoir on Cumberland is exemplified by an encounter he records from a coal mine near Workington. Operations in that pit were much plagued with explosions of fire-damp, "*faul-air*" as our reporter has it. One man's face was marked by the scars of seven or eight such accidents. He told Jars how the surest way to escape burns and survive concussion was to cast yourself face down in the deep mud that always floored the gallery until the shock wave passed over.[127]

Jars never answered the concluding question in his directive about the comparative advantages of freedom from regulation, and the reader is left dangling with the impression that, tacitly at least, his actual experience belied the liberal presuppositions of his superiors. His memoirs generally begin or end with a summary of the jurisprudence of mining in the region discussed, and unfettered exploitation by private enterprise is not the model he commends to those who sent him. In 1759, early in their collaboration, Jars and Duhamel devoted an entire memoir to the regulations in Saxony and in certain Hapsburg provinces.[128] They concluded with a draft of legislation for adapting the provisions to operations in France. The proposed edict would suppress the patchwork of private ownership and piecemeal concession that was vitiating the exploitation of French mineral resources. A corps of mining engineers would be created holding commissions from the controller-general. Their function would combine the kind of oversight exercised by inspectors of manufacturing with the responsibility for designing actual constructions vested in the highway en-

[125] "Second mémoire: Description des mines et des forges de fer et d'acier de la Styrie," *Voyages métallurgiques* *1*, 32.

[126] "Treizième mémoire: Sur la jurisprudence des mines de Saxe, & des différens états de l'impératrice-reine de Hongrie," *ibid.*, *3*, 433.

[127] "Douzième mémoire: Sur quelques mines de charbon, des forges de fer, & plusieurs autres établissements utiles d'Angleterre; forges et mines du duché de Cumberland," *ibid.*, *1*, 247.

[128] "Treizième mémoire: Sur la jurisprudence des mines de Saxe, & des différens états de l'impératrice-reine de Hongrie," *ibid.*, *3*, 385-458.

gineers of the Corps des ponts et chaussées. Their chief would hold general rank and would advise the controller-general on the granting of concessions. For the state itself would not undertake mining. It would simply assert its eminent domain and retain one-tenth of the value of the minerals exploited. That claim might be lessened or forgiven for appropriate periods in order to encourage new enterprise, but never alienated. Specifications, safety, drainage, access to woodlands, provision for creating new communities around the mines, relations with landowners, with the tax farm, with local communities—all that would occupy the new administration, along with prospecting for additional resources.[129] The document represents an early adumbration of what ultimately became one of the respected engineering services of the state, the modern Corps des mines.

That blueprint, modeled on centralized Germanic administration, was drawn before Jars had traveled to England or to Scandinavia. After he had been there, he included the mining codes of Cornwall, Devon, and Derbyshire among documents that he translated and that his brother included in the appendix to his book.[130] Jars rarely referred to them in actual recommendations, however. Instead, the Swedish regime, which he studied in his last expedition abroad, seemed to him the most responsible form of organization for mining and metallurgy. The quality of Swedish iron was the initial attraction, naturally enough. Discriminating steel makers and tool makers, most notably in Sheffield, used only Swedish, never English ingots. So pure was the ore in the mines of Dannemora that the foundries needed only charcoal for smelting and dispensed with a limestone flux. Responsible parties in the Swedish government had taken the fullest advantage of this good fortune. Already in the eighteenth century, the Swedes appear to have known how to socialize their resources and to mingle public ownership with private enterprise.

Since early in the seventeenth century, administration of mining, the principal branch of Swedish commerce, had been vested in a distinct department of state. A council consisting of a president and ten colleagues was served by a permanent staff of secretaries, clerks, notaries, and lawyers. Also of the council were an assayer and an engineer of mines, each with a corps of technically trained people. Both nationally and locally administrative and technical responsibilities were parallel functions. Sweden was divided into twelve mining districts, each under the charge of a bergmeister qualified in law and in mining, and all installations were visited by a set of inspectors trained in geometry, mechanics, mineralogy, chemistry, and natural history. At the same time, iron founders and even working miners—bergsmen—had a financial interest in the product of the mines, and the administrative and technical direction centered in Stock-

[129] *Ibid.*, 448-458.

[130] *Ibid.*, 522-548.

holm had to mesh with the rights of the communities. As for foundries and forges, the basis was mixed. Some belonged to mine operators, others to peasant proprietors, and still a third class to noblemen and large landowners. Forests like minerals remained state property, and each iron-master was regulated as to the district from which he could draw fuel and charcoal and also in the price. There was a hierarchy in the order of the metals, moreover—gold, silver, copper, lead, tin, and iron—and the higher the rank, the greater the priority in access to timber. Such was the system that Jars described with greater enthusiasm than any of the organizational arrangements he reported from Britain.[131]

From Britain he learned, among other things, the uses of coke. Apparently, he never visited Coalbrookdale in Shropshire, where in 1709 Abraham Darby had succeeded in smelting iron with coke.[132] Jars initiated his discussion of the substance almost casually, in connection not with the saving of charcoal, but with abatement of pollution produced by the extremely bituminous high-sulfur coals of Newcastle. An installation of nine furnaces along the Tyne fractionated crude coal by distillation into tar and a greyish brick-like residue sold for fuel. These "cinders" burned cleanly enough to roast the malt for breweries without spoiling the beer, and were also employed for jewelers' furnaces, domestic heating, and in powdered form for fertilizer.[133] In Sheffield, conversion of iron to steel by cementation, and also the tempering of files and cutlery, depended on coke, both as a source of carbon and as an even-heating fuel. The substance was denser and darker than the cinders of Newcastle, and was produced in ovens operated like charcoal burners.[134] In his reports for the ministry, Jars wrote only fleetingly of how blast furnaces were charged with "coaks" in Cumberland at Clifton Furnace and in Scotland at the Carron Iron Works.[135]

Evidently, however, Jars seized on the importance of the technology, for he took extensive notes, which he discussed in detail with his older brother and with others on his return to France. Furnished with the authority of inspector, and armed with all he had learned abroad, he set forth on his tour of the domestic iron industry in September 1768.[136] His

[131] "Huitième mémoire: Sur les principales mines et forges de fer de la Suède," *ibid.*, *1*, 95-103.

[132] On the English iron industry, see H. R. Schubert, "Extraction and Production of Metals: Iron and Steel," in *A History of Technology*, ed., C. Singer *et al.* *4* (1958), 99-117.

[133] "Dixième mémoire: Sur les mines de charbon de Newcastle en Angleterre," *Voyages métallurgiques* *1*, 209-212.

[134] "Douzième mémoire: Sur diverses mines de charbon . . . ," *ibid.*, *1*, 255-260.

[135] *Ibid.*, 235-237; "Treizième mémoire: Sur les mines de charbon et les forges de fer de l'Ecosse," *ibid.*, *1*, 272-279.

[136] Bacquié (1927), 47-51, reproduces lengthy passages from his journal. The original is in the AN, F^{12}.1300.

journal records a visit to Montbard to consult with Buffon on the exploitation of his iron mines, foundry, and forges. Though nothing is said of coke, Jars made two suggestions. Buffon's furnaces were square in cross-section rather than round, and it would repay the expense to rebuild them in the latter form, generally recognized to give a better draft and more complete combustion. Jars also advocated the Swedish practice of lining them with bricks made from their own slag instead of from local limestone. He found the lordly old naturalist interested, cooperative, and receptive to criticism, and left instructions for other experiments that Buffon promised to have his people try.

From Montbard it was a good day's journey to Montcenis, a locality that has been altogether overshadowed in the industrial age by the neighboring hamlet of Le Creusot, where Jars repaired to inspect and advise on old coal mines. There, a pair of progressively minded entrepreneurs called Jullien and de La Chaize had taken over shafts that opened out of the floor of a deep valley and had been mined in a desultory way for centuries. Before beginning serious operations, La Chaize and Jullien were awaiting a *privilège* or permit from the Council of Commerce, and perhaps also a subsidy, as well as certain local exemptions. Jars thought their prospects good. Le Creusot was only four leagues from two highways, one leading to Châlons on the Saône and the other to Toulon on a navigable tributary of the Loire. Access to two important shipping arteries would thus be easy. Iron ore occurs in the region, which was dotted with small foundries and forges. Development of metal industries to work up their product into nails, small-arms and hardware could also be expected to create a steady local market for coal.

More exciting to Jars than these, the conventional concerns of an inspector, was a further chance for fortune. Were the opportunity to be grasped by the forelock, he could imagine a success that would be emulated throughout the country and transform its entire metallurgy. The coal he had seen was bituminous, of good quality and plentiful. Old miners told of seams and surfaces barely scratched. If that coal were to be "prepared the way the English do it," that is to say converted into coke, it could be used to smelt the nearby iron ore.[137] Jars once remarked that in all his visits he took a hand in the operations he was shown, since it was impossible to learn a process merely by watching someone else.[138] That practice now stood him in good stead. He constructed a small coke oven and tried the coal of Le Creusot. The experiment succeeded better than he had dared hope. He showed La Chaize the sample of coke he had

[137] *Ibid.*, Bacquié (1927), 49. On Jars and Le Creusot, see Gueneau (1924).

[138] "Observations métallurgiques sur la séparation des métaux," MARS (1770/73), 423-425.

made and explained the process in detail. The two partners determined to continue trials on a large scale, and Jars urged them to send a couple of barrels to his family enterprises in Chessy. His brother would then try the coke with copper ore, a thing (according to his journal) never yet attempted in any country.[139]

Thereafter, Jars's route took him to Lorraine, where extractive industry was on a larger scale than elsewhere in the country. The ironworks of the Wendel family at Hayange near Metz was the most considerable in northeastern France. Its two blast furnaces had an annual capacity of almost 700 tons of pig iron. The head of the family, Charles de Wendel, like many provincial manufacturers of his generation, had given his son, Ignace, an education qualifying him for a technical career, which the young man was beginning with a commission in the artillery. In common with other branches of the service, the artillery allowed junior officers much freedom from military duty. Jars arrived in Hayange in January 1769, full of his vision for converting French metallurgy to coke. Jars's tale of his experiment at Le Creusot fired the imagination of Ignace de Wendel, who persuaded his father to try a full-scale test. Jars rigged an oven to prepare two hundred-weight of coke. The newer of the furnaces was blown out and charged with coke in lieu of charcoal together with iron ore and limestone. The mass was ignited and blown by as strong a blast as water power could produce. It sufficed, and the first pig iron smelted by coke in France flowed from the hearth on tapping twelve hours later.[140]

Meanwhile, the elder Jars brother had constructed coke ovens on the family premises at Saint-Bel. Whether or not he ever received the promised sample from La Chaise and Jullien at Le Creusot, he was ready to experiment with coke in refining copper by the time that Jars reached home with word of the triumph at Hayange. That was in February or early March, 1769. The brothers arranged two reverberatory furnaces side by side, one to be fired with charcoal and the other with coke. They ran the trials with the precision of a laboratory experiment on the industrial scale and kept careful account of cost, yield, and quality. Except for the effect on the lining of the furnace, the advantages all lay with coke. Costs were reduced by 25 percent and the time of operation by a slightly larger fraction. The younger Jars tried similar tests on iron in a foundry at Saint-Etienne, also in March, with even more encouraging results. In the summer he set forth on his second official tour of inspection, into regions of the Massif Central. In August he was dead. The surviving brother, who (confusingly enough) also used the name Gabriel, wrote up the experi-

[139] Bacquié (1927), 49-50.
[140] Ballot (1923), 440-441; Chevalier (1953), 64.

ments on copper for the Academy of Science and published them in 1774.[141]

Throughout the 1770s all attempts to convert iron foundries to coke proved ephemeral. One sequence of efforts involved Rhenish and Jacobite investors who had access to blast furnaces at Hayange and to Buffon's iron-works near Montbard. At the same time, La Chaize and Jullien continued at Le Creusot with attempts to put Jars's process onto a profitable bottom, associating themselves with other sources of capital from time to time. These two ill-defined groups alternated between competition and rickety combinations and never developed the momentum that the ministry—in the person of Bertin during this period—hoped to impart through doling out privileges and governmental favors. By 1776 La Chaize and Jullien lapsed back into coal mining and disappeared from the metallurgical scene. The initiative that ultimately succeeded, albeit very indirectly, came from a different quarter altogether, the collaboration of another mineralogist, Antoine-François de Gensanne, with an army officer, Marchant de La Houlière, a brigadier of infantry.[142]

With a background in Montpellier, Gensanne (like Jars) had been sent abroad to gather mineralogical information.[143] His partner, La Houlière (like the young Ignace de Wendel) was an enterprising, technically imaginative officer with interests transcending the military. In 1773 Gensanne and La Houlière started an ironworks in the diocese of Alais, encouraged by the bishop and by the estates of Languedoc, economically the most active of the provincial authorities of the old regime. By now, word of coke and English methods was everywhere. Guyton de Morveau, Lavoisier's future associate, read a paper on the subject before the Academy of Dijon and published an abstract in Rozier's journal.[144] Jars had written too little to provide a manual for practice, however, and La Houlière resolved to go to England and to see for himself. Off he went in 1774, to return many months later with English mastery in the shape of an English iron-master, William Wilkinson, quarrelsome younger brother of the famous John Wilkinson of Broseley in Staffordshire.

The French government was fully party to this invitation, motivated by the imminence of intervention in the American war and deficiencies of French ordnance. The initial project was restricted in scope, a foundry utilizing scrap iron for casting cannon. It was constructed on the isle of

[141] "Quinzième mémoire: Manière de préparer le charbon minéral, autrement appelé HOUILLE, pour le substituer au charbon de bois dans les travaux métallurgiques," *Voyages métallurgiques* 1, 285-338.

[142] Ballot (1923), 440-469.

[143] Gensanne's major work was *Traité de la fonte des mines* (1770).

[144] "Observations sur la réduction de la mine de fer par le charbon de pierre de Mont-Cenis," *Observations sur la physique* 2 (1772), 450-452.

Indret in the Loire near Nantes. Production began in 1778 and disappointed expectations in 1779. Alarmed, the responsible minister, Sartine at this juncture, resolved to thrust a French associate upon the prickly Wilkinson. His choice fell naturally upon the scion of the foremost metallurgical dynasty in France, none other than Ignace de Wendel, who ten years before had accommodated Jars's first large-scale trial at Hayange and whose background in the artillery was a further asset. By 1780 two things had changed. Wendel's father had died and his mother, the formidable "dame de Hayange," was running the ironworks. More important, after 1776, James Watt's steam engine, fitted with precision cylinders bored by John Wilkinson, could furnish adequate and reliable power for blowing the blast in a coke-charged furnace. Apart from everything else, dependence on water power had been a severely limiting factor in that operation. Such was the origin of the collaboration that brought Wendel and Wilkinson together and led them beyond the rather routine ordnance plant at Indret to the creation of an up-to-date iron works with reduction by coke, power from steam, and division of labor.

The collaboration was never an easy one. In prospecting for a site, Wendel preferred Saint-Etienne, but Wilkinson saw what Jars had seen and insisted on Le Creusot. There a royal manufactory was chartered. Financing was complicated. The company was combined (for fiscal rather than technological reasons) with a new glass concern licensed in the name of the queen, and shares to the value of 10,000,000 livres were floated, a one-twelfth interest being taken by the king.[145] Molten iron flowed from the first hearth tapped on 11 December 1785. The four blast furnaces were thirty-nine feet high. Each was furnished with a blower delivering 9,000 cubic feet of air per minute in contrast to a blast of 2,000, which was all that could be obtained from bellows run by water-power. Twelve to fifteen miles of railways (the first in France) carried coal from the collieries and ore from the iron mines. Four reciprocating steam engines provided power for the blast furnaces and for pumping water from the coal shafts. A fifth, rotatory engine drove mechanical hammers in the forge. Soon after the opening, a further machine of that type was installed to revolve the bit in a boring mill for cannon and cylinders. A specialized workforce of over 1,300 somehow occupied a hamlet where in 1781 the population had consisted of seven or eight families.

The mills that rose up in Le Creusot were thus as dark and satanic as any in England. "A wonder of the world," the old Daubenton called the installation on an early visit from his sheepfolds near Montbard.[146] For it

[145] The text of the "*acte*" of the Society is given in Ballot (1923), 458-460.

[146] Quoted *ibid.*, 463. See the description in Rozier's journal by its third editor, Jean-Claude de Lamétherie, "Mémoire sur la fonderie et les forges royales établies au Creusot près Mont-Cenis en Bourgogne . . . ," *Observations sur la physique* 30 (1787), 60-67.

was half a century before French metallurgy in general began to resemble the model established at Le Creusot.[147] Even there much disappointment and vicissitude, economic and technological, ensued until the Schneider interests entered the scene, and the hyphenation of Schneider with Creusot in 1836 makes a more significant milestone than the first tapping of a furnace to mark the passage of the French economy into the age of heavy industry.

Among the many visitors to the foundry in its early years were a trio from the Academy of Science—Vandermonde, Berthollet, and Monge—come to observe the operation of blast furnaces using coke—"fired by the English method," in their phrase.[148] In May 1786 Vandermonde presented their "Memoir on iron considered in its different metallic forms" before the Academy. Any reader wishing to sample the clarity, the urbanity, the cogency, and the rationality that distinguished the style of French science at the height of its powers could do no better than turn to this memoir, along with Lavoisier on phlogiston, Lavoisier and Laplace on theory of heat, and Laplace on attraction of a spheroid and on secular inequalities of the planets.[149] Indeed, the incongruousness of the subject matter may make the metallurgical paper an even more effective witness to the importance of those qualities. At all events, Vandermonde, Berthollet, and Monge had a momentous discovery to announce: the properties of steel depend primarily upon its content of carbon in a "true solution."[150]

Their inspection of Le Creusot was the culmination of a thorough survey of the literature and of the industry. Despite all the experience of the trade, the characteristics of iron varied so widely from one foundry to another, and even from one batch to another in the same foundry or forge, that traditional lore held the metal to be "constant" like gold, that is to say capable merely of physical modification.[151] Only in recent decades had the chemical basis of a few effects at last been recognized. For example, arsenic was known to increase the brittleness of red-hot iron, and phosphorus that of wrought iron. The problem that Vandermonde, Berthollet, and Monge now set themselves was to generalize attributions of that kind, and to explain both physically and chemically, but specially chemically,

[147] Bourgin and Bourgin (1920) give a census of the entire industry on the eve of the Revolution, when it was still largely an affair of domestic foundries and family forges scattered widely wherever there were veins of ore.

[148] "Mémoire sur le fer, considéré dans ses différens états métalliques," MARS (1786/88), 132-200; see esp. 137-138 (note).

[149] For discussions of these memoirs, and bibliographical references, see Guerlac, "Lavoisier," DSB *8*, 66-91; C. C. Gillispie, "Laplace," DSB *15*, 273-403.

[150] "Mémoire sur le fer," MARS (1786/88), 191.

[151] *Ibid.*, 132-133.

what happens with iron in each of the stages of extraction, refining, and conversion. Why is cast iron brittle, melting at a low temperature? Why is wrought iron malleable, melting at a high temperature? Why is steel very hard after tempering, also melting at high temperature? Why does steel overexposed to cementation fracture uselessly on the anvil and melt at a low temperature again? What, in short, are the substances to which iron owes these properties in its three principal forms?

Two investigators had preceded our authors in an equally comprehensive way, Réaumur and Torbern Bergman, both of whom they treated with respect. Réaumur in his *L'Art de convertir le fer . . .* (1722) had been concerned in principle to uncover and publicize the best procedures actually employed in the trade.[152] His practical purposes were to find empirical methods for converting iron to steel by cementation, and for mollifying cast iron and giving it a malleability like that of wrought iron. His explanation of the former of these modifications was that cementation suffuses the metal with "sulfur" and salts, about which discovery Vandermonde, Monge, and Berthollet observed mildly that it might have been valuable in its day but was too vague for theirs.[153]

Bergman was another matter. His *De praecipitatis metallicis* was a purely theoretical work.[154] Part of its purpose was the same as their own, to explain the chemical basis of the different forms of iron. That object the great Swedish mineralogist thought to have accomplished by specifying the ways in which the metal was combined with phlogiston through the processes practiced in foundries. He had been put on to his theory by noticing that metals dissolve in acids only after losing a portion of their phlogiston. Thence he concluded that metallic substances contain phlogiston in two different modes or degrees. In the first, the combination is relatively weak. Metals may be deprived of phlogiston by calcination ("oxidation" in the parlance of combustion theory). In the second mode, the principle is evidently more tenacious, since calxes (metallic oxides) retain a portion of their phlogiston. Its presence, indeed, is all that distinguishes them from acids.

The two types of phlogiston were called "reducing" and "coagulating" by Bergman, and he measured the proportions as a function of the volume of inflammable air (hydrogen) released when he immersed various metals in acid solutions. After phlogiston, the second important variable for his theory was the material basis of heat. Bergman estimated its quantity in the three forms of iron by dissolving samples in a fixed amount of nitric acid and measuring the increases in temperature. The third variable was

[152] See above, Chapter V, Section 2, n. 39.

[153] *Ibid.*, 146.

[154] Uppsala, 1780. A complete bibliography of Bergman is Birgitta Moström, *Torbern Bergman, A Bibliography of His Works* (Stockholm, 1957).

graphite, which he took to be a combination of fixed air (carbon dioxide) with phlogiston. From analysis it followed that, of the three principal forms of the metal, wrought iron is virtually free of graphite and contains more matter-of-heat and phlogiston than the other two; steel contains more graphite but less of the other principles; and cast iron contains more graphite but less matter-of-heat than steel. The refining of cast iron, therefore, consisted in eliminating or decomposing its graphite while imparting phlogiston, and the cementation process for converting iron to steel consisted in the formation of graphite by the combination of "reducing phlogiston" from the metal with carbon from fixed air.[155]

Although Bergman and Réaumur differed in many respects, their systems were alike in making steel a middle form between cast and wrought iron. Vandermonde, Berthollet, and Monge thought that unconvincing from the outset.[156] Their memoir also makes much of an anomaly in the Bergman analysis. He had found less hydrogen evolved by the action of sulfuric acid on cast iron than on wrought iron, and also less by the action of the acid on steel than on wrought iron. It was natural to explain the former difference on the grounds of incomplete reduction of the mineral. If incomplete reduction is also to be the explanation of the (rather smaller) disparity with steel, the conclusion had to be that cementation restores some portion of the oxygen. But reduction and cementation both involved exposure of the metal to carbon, and the notion that cementation would partially reverse reduction appeared untenable.

So much for criticism: in the main, the memoir for the Academy was less a refutation of Bergman than a new explanation of the same phenomena on the basis of the new chemistry. The authors mentioned neither oxygen nor Lavoisier. There is no reason to suspect a slight, however. The modern nomenclature was not published until the year after their paper was written,[157] and their terms for oxygen, oxide, and oxidation were dephlogisticated air, calx, and calcination. That made no difference to the argument. The point of view was that of Lavoisier's theory of combustion, and the approach was gravimetric. To call their method quantitative and that of Bergman qualitative would be too pat: the latter did depend on volumetric analysis, after all. His discourse also required property-bearing fluids, however. In this respect, the contrast is sharper than might appear on a casual reading. Although our French authors, too, retained the matter-of-heat, the use they made of it was the same as Lavoisier's with caloric. It was a physical agent responsible for change of state from solid to

[155] "Mémoire sur le fer" devotes an entire section to a summary of Bergman's experiments and theory, MARS (1786/88), 149-154.

[156] *Ibid.*, 154-155.

[157] *Méthode de nomenclature chimique, proposée par MM. de Morveau, Lavoisier, Bertholet, & de Fourcroy* (1787).

liquid and liquid to gas and not a weightless, or at least an unweighed, chemical agent. Equally telling of Lavoisier's influence was the method. At every step, reagents and products were weighed, and every conclusion established by analysis was confirmed by synthesis.

The findings are worth summarizing in modern terminology.[158] In the first place, the properties of cast iron were explained by incomplete reduction of the ore in the blast furnace. That is why the crude metal evolves less hydrogen and takes less oxygen from water than does wrought iron in reacting with sulfuric or hydrochloric acid. Further evidence of incomplete reduction was the whitening that the black or grey form of crude iron undergoes on heating in the absence of air. The two impurities of oxide and carbon combine. The supposition that carbon is responsible for the dark hue of black and grey cast iron is proved in two ways. Wrought iron may be converted into steel by cementation in the presence of grey or black crude iron; and when samples of the latter are dissolved in sulfuric acid, the residue forms carbon dioxide on combustion.

Steel, in the second place, is characterized by complete reduction of the metal and its combination with carbon. That these precisely are the effects accomplished by cementation is evident since (1) the gain of weight corresponds to the amount of carbon added, (2) the residue left by the action of acids is carbon according to the same tests as those employed on cast iron, and (3) the bubbles in blister steel consist of carbon dioxide formed of carbon and residual oxide eliminated from the metal.

The properties of wrought iron, finally, are explained by the relative purity of its state chemically. Further, and conclusive, evidence that carbon is the substance mainly responsible for varying the states of iron comes from graphite. When the molten metal cooled, whether in the form of crude iron or of steel, a residue floated up consisting simply of carbon saturated with iron—"nothing else" (observed our authors) but the material for "English pencils."[159] (The misnomer "*plombagine*" for graphite survives in the phrase "lead pencils.")

By no means is it true of most scientific memoirs, even the most important, that they seem simply right. That is the kind of satisfaction to be found here, however, and one or two further comments on the historical significance may be in order. It is in keeping with a host of indications about the place of French science in the late eighteenth century that the nature of steel should have been determined, not where it was made best, in Damascus, in Sweden, or in Sheffield, but in Paris. An additional observation about the development of French science itself will narrow the chronological reference to the last ten or fifteen years of the old regime. If

[158] "Mémoire sur le fer," MARS (1786/88), 198-201 for the authors' summary.
[159] *Ibid.*, 192.

the memoirs by Jars, written between 1756 and 1769, are compared to this treatise by Vandermonde, Berthollet, and Monge, the difference in elegance and sophistication is marked. Much might be said about the contrast, but its central feature is that the mode of the former is descriptive and of the latter analytic. Comparable instances abounded in other sciences so that, in thinking of astronomy, mathematics, and botany, no less than of chemistry and mineralogy (though not quite yet, perhaps, of physics and zoology), the change will seem intrinsic to a process of maturing and not merely incidental.

More to the present point, investigation of iron manufacturing was entirely characteristic of that stage in the relations between science and technology when science supplied, not so much theory that would change or create technology, as analysis that would explain it.[160] To exaggerate will not do, however. Even this early on, theory must not be overly discounted. What guided Vandermonde, Berthollet and Monge, after all, was precisely the atmospheric theory of combustion, applied to metallurgy in the very year of its formal enunciation by Lavoisier, though of course it had been in gestation and under discussion for some time. That reflection may serve to temper the recurrent fashion in historiography for diminishing the difference that progress makes in science and vice versa. For the new chemistry was more than the chemistry of principles, qualities, and affinities with other names. True, Bergman had not entirely failed to explain the properties of the form of iron, but coming close amid some confusion was not the same thing as succeeding.

Of the three authors, Berthollet came to the collaboration a recent convert to the new chemistry. His involvement with metallurgy grew out of his wider responsibilities as commissioner for the Bureau of Commerce.[161] It was also the beginning of a lifelong association with Monge, personal as well as professional, a relationship that became of cardinal importance to French science in revolutionary and especially Napoleonic times. It will be more natural to discuss the early career of Monge in connection with military engineering and the school of Mézières, where he was professor of mathematics and physics until December 1784.[162] Suffice it to note here that before moving to Paris, as he then did, Monge had already assisted Lavoisier during visits to the Arsenal. His own experiments created a claim for him, along with Cavendish and Lavoisier, in the discovery of the composition of water,[163] and critical passages in the metallurgical memoir bespeak his familiarity with that aspect of chemistry. Perhaps more important, Monge also brought to the research the experi-

[160] Above, Chapter V, Section 1. [161] Above, Section 1.

[162] The standard work on Monge is Taton (1951); see also Taton, "Monge," DSB 9, 469-478, with more recent references.

[163] Perrin (1973).

ence and economic interest of a practicing iron master, for he had taken responsibility for oversight of a forge in Lorraine inherited by his wife from her first husband.

The member of the team whom the Academy named first in attributing authorship, Alexandre Vandermonde, was the moving spirit of the research and its principal organizer. Much less remembered than Berthollet and Monge, Vandermonde, born in 1735, was twelve to fifteen years older than his colleagues.[164] He was one among several members of the Academy for whom mathematics had been a point of entry into science without becoming their vocation, and who, once elected, were free to follow their real interests in technology. Other examples were Etienne Mignot de Montigny, whom we have met at Sèvres, and Monge's most original pupil at Mézières, a brilliant engineering officer, Jean-Baptiste Meusnier, whom we shall meet shortly formalizing laws of flight. Monge himself, after his translation to Paris, put more thought and energy into problems of physics, chemistry, and metallurgy than into mathematical investigations of the kind that had made and that sustained his reputation. The pattern, therefore, is not the obvious one of intellects uncertain of their powers seeking refuge from exacting fields in matters respectable for public utility. Early in the 1770s, Vandermonde contributed four papers that were important enough in the theory of equations to have attracted the respectful interest in the 1930s of so demanding a master as Henri Lebesgue, and he has been credited with having started the theory of determinants.[165]

On the strength of this work, Vandermonde was elected to the Academy in May 1771. The opportunity that shaped the remainder of his career arose in the following August, when he was named to serve with Vaucanson on a commission to report on a writing automaton. The son of a merchant who had amassed a considerable fortune trading into China from Macao, Vandermonde had an independent competence. The attraction that machinery and technology held for him was intellectual and civic rather than self-serving, and he was one of the few who could get on with Vaucanson. We have already seen how, after the latter's death in 1782, the controller-general resolved to institute a Cabinet des Mécaniques du roi in the Hôtel de Mortagne, where Vaucanson had maintained his laboratory of invention. By then Vandermonde had made his bent clear, and on 15 October 1783, the government named him curator of this museum and laboratory of technology.

Earlier in the same year, the commanding officer of the forges at Amboise, which produced ordnance for the artillery, assigned a subordinate, a Captain Dulubre, to yet another French mission to investigate the use of

[164] Birembaut (1953) gives a biographical summary.
[165] Lebesgue (1937-1939).

coke in England. On his return, Dulubre enlisted the cooperation of J.-F. Clouet, professor of chemistry and Monge's colleague at Mézières, where the laboratory at the school was available for analysis of ores. Needing funds, the military command approached the Ministry of Finance, which in turn assigned Vandermonde and Berthollet to inspect the work of the laboratory at Mézières and of the ordnance forges at Amboise.[166] Vandermonde was now installed at the Hôtel de Mortagne, and Berthollet had just been appointed director of dyes at the Gobelins and was being drawn into wider consulting responsibilities. At the behest of the Ministry of Finance, for example, they were called on as a team to evaluate the drawn-out claims of one Jean-Baptiste Delaplace, an artisan who insisted that he had discovered a secret for smelting native ores and converting French iron to a steel the equal of Sheffield in fineness and temper.[167] All concerned with metallurgy were of a mind to get to the bottom of the questions posed by the state of the iron industry in France, and that determination was the genesis of the memoir on the metal in its different forms.

An altogether less predictable dénouement was the emergence of aviation from modernization of the paper industry, which story also begins with initiatives by an inspector of manufacturing. Nicolas Desmarest began his service in the Corps of Manufacturing Inspectors at Limoges in 1762, during the time of Turgot's intendancy. The bureaucracy frequently coupled generalized responsibilities for a region with specialized responsibilities throughout the country, and Desmarest concerned himself with industry as a whole in the Limousin together with wool-growing (we have already noticed his collaboration with Daubenton in the latter's experimental sheep ranch at Montbard), and most notably with the manufacture of paper. Desmarest was elected to the Academy in 1771, the same year as Vandermonde, in recognition of the geological observations demonstrating the volcanic origin of the basaltic peaks, the puys, of Auvergne.[168] He thus could read before that body a pair of papers, the first in 1771 and the second in 1774, in which he summoned the paper industry to mend its ways in the light of reason and experience.[169] The pro-

[166] On the state of the forges at Amboise, see the reports of Berthollet and Vandermonde, AN, F[12].656.

[167] Below, Section 4.

[168] "Mémoire sur l'origine & la nature du basalte . . . ," MARS (1771/74), 705-775. Cf. Kenneth L. Taylor, "Desmarest," DSB, *4*, 70-73. For the collaboration with Daubenton, see above, Chapter II, Section 5.

[169] "Premier mémoire sur les principales manipulations, qui sont en usage dans les papéteries de Hollande, avec l'explication physique des résultats de ses manipulations," MARS (1771/74), 335-364, read on 20 February 1771; "Second mémoire sur la papéterie, dans lequel on traite de la nature et des qualités des pâtes hollandoises, ainsi que des usages auxquelles les produits de ces pâtes peuvent être propres," MARS (1774/78), 599-687, read in December 1774. Both mémoirs were separately printed, and are cited in that pagination.

gram he developed, and its execution in the Montgolfier mills at Annonay, will serve as a final example of governmental reliance on scientific expertise in its encouragement of rationalized technology. Here are the concluding sentences of his second paper, published with the usual delay in 1778:

> The government is convinced that the best method for the administration of manufacturing is to invest it with the spirit of research and instruction, and that the most effective means for fitting instruction to the needs of manufacturers is to put it to work in their shops. Government officials know full well that workers read very little but observe narrowly and imitate readily, and that while in their prejudice they resist mere reasoning and idle speculation, by the same token they may be convinced of the value of a new process if they can relate it to one they know.
>
> It is thus that the paper industry is to be transformed by a revolution that is called for in the interests of commerce and that will be helpfully forwarded by the zeal and enlightenment of certain manufacturers. I shall count myself fortunate if, now that I have prepared the way, I may follow its course and progress.[170]

The problem, in the eyes of these same officials and manufacturers, was both technological and organizational. Technically, French paper compared unfavorably with Dutch, though only in certain respects. Organizationally, rationalization was impeded by inertia among many mill owners who acquiesced in the conservatism of the labor force. To take the relation of technique to quality first, Dutch paper was universally acknowledged to be superior for stationery and for drawing. The surface was smoother, the grain finer, the finish harder, and the tensile strength greater. Scholars handling eighteenth-century paper may still recognize the Dutch product by its light-blue tint, compared to the ivory tone characteristic of French paper. The latter, on the other hand, was preferred for printing and for packaging. Its more porous surface absorbed printers' ink without smearing, and it could more readily be processed into cardboard and wrapping paper. The industry was relatively new in Holland. Huguenots from the region of Angoumois had transplanted it after the revocation of the edict of Nantes in the late seventeenth century. When it was sufficiently established for the differences in the product to become apparent, they were attributed to the effects of more efficient machinery. In pulping the raw material, the Dutch used horizontal cylinders studded with spikes to shred the rags of flax or linen, whereas French papermakers pounded it in a stamping mill consisting of great mallets or pestles, four to a trough, worked by a single axle-tree.

[170] "Second mémoire," MARS (1774/78), 88-89.

Several enterprising manufacturers in France experimented unsuccessfully with these cylinders or "hollanders" in the 1750s. The failure was attributed to resistance among workers and to the lack of skilled mechanics to maintain the strange machines. Those obstacles were certainly important, but Desmarest found that there was more to the Dutch technique than the mode of maceration, and that the other aspects were quite unknown in France. He had begun his study of the trade in 1763, when Turgot requested a report on the paper makers remaining in Angoumois. Thereafter, Desmarest visited installations in Auvergne and Burgundy, and inspected the famous L'Anglée plant in Montargis, chosen by Diderot to illustrate the industry in *L'Encyclopédie*.[171] Puzzled by the lack of success with hollanders even in mills run by progressive entrepreneurs, Desmarest proposed to Trudaine, then in his last years at the Bureau of Commerce, that he be authorized to make an inspection of the industry in Holland. Trudaine had confidence in Desmarest, after his success with sheep and wool, and sent him off on this new mission in the spring of 1768. What Desmarest learned in several months in Holland is very interesting for the interrelatedness of the various aspects of a technology and their dependence on the availability of power.

The first difference between the Dutch and French systems, and that from which the others all derived, was in the preparation of the rags. The French followed the ancient practice of easing their disintegration by rotting them and allowing the sour linen to ferment in bins for six weeks to two months before maceration. The Dutch fed clean, fresh rags directly to their cylinders. The use and design of those machines was the result rather than the cause of that difference, however. Dependent on windmills, the first generation of Huguenot paper makers had too often found the power failing just when their rags had fermented to the point requiring maceration, after which the whole batch would spoil while they waited for the wind. Accordingly, they had to devise a technique for pulping sound rags with no more preparation than a good laundering. Other virtues followed out of this necessity. The critical operation in paper making was the formation of the sheet. The vatman dipped it out of the suspension of the pulp with a flat sieve or mold, matted the fibers by giving the frame two sharp shakes, right to left and front to back, and then passed the form to the coucher, who tipped the contents onto a pile, alternating wet sheets with layers of felt. Two hundred and fifty sheets (an additional margin of ten was allowed for defective seconds) made a post—*porse* in French usage—or half a ream. After the post went under the press, the felts were removed and the sheets compressed a second time against each other.

[171] The plates illustrating papermaking in the Montargis plant are reproduced and explained in Gillispie (1959a) 2, 359-368.

Thereupon, the Dutch and French processes diverged again. The grainier quality of the Dutch sheets required an additional operation that, in Desmarest's view, was more significant than any other step in accounting for the differences in the product. In order to compact the fibers and smooth the surfaces, the posts were disassembled and, like the shuffling of a deck of cards, reassembled in random order to be pressed again. That *échange* was repeated several times, in order that the contact with a succession of surfaces might flatten the unevennesses in each. The lesser porosity of sheets made from unrotted rags also forced the Dutch to dry their paper more gradually and to size it while still moist. After sizing, sheets intended for the finest grades would be topped off by still another round or two of échange and pressing.[172]

There is no evidence that Desmarest in 1768 had yet visited Annonay, seat of the provincial estates of Vivarais, a small city built into deep clefts at the confluence of two small rivers, the Cance and the Déume. Along the banks of the latter, four paper mills were situated in the eighteenth century, as indeed they still are, and with the same names: Marmaty in the parish of Boulieu; Vidalon-le-haut and Vidalon-le-bas in the parish of Davézieux; and Faya in the heart of Annonay itself. The Vidalon mills (which are now one) belonged to the Montgolfier family, and the two that bracketed them upstream and downstream to the Johannots. The Déume was specially adaptable to the needs of paper making. A short river with a steep pitch and a fast current, it turned the mill-wheels strongly except during a few dry months in the summer when the level might be low. Since the hills surrounding the pass of Tracol, where it rises, are granite, its waters are clear and soft. Rags came through the laundry good and clean. Annonay was well situated to supply paper to Lyons, and had ready access to the markets of Orléans and Paris by the waterway of the Loire and equally so to Marseilles and abroad by the Rhône. The Saône and Rhône made easy the supply of rags, of which the best quality came from Burgundy. The climate was excellent for sizing and drying paper the year around.[173]

[172] The foregoing account and comparison of Dutch and French methods, and of Desmarest's 1768 mission, is drawn from his "Premier Mémoire" (MARS [1771/74]), supplemented in certain details by his article "Papier (Art de fabriquer le)," *Encyclopédie méthodique: Arts et métiers mécaniques* 5 (1788), 463-592, which is the most authoritative contemporary source for the entire industry. Also useful, particularly for commercial aspects, is the Lalande treatise in the *Description des arts et métiers: Art de faire le papier* (1761): a revised and enlarged edition appeared in 1820. There is an admirable and detailed account of the technology in Rosenband (1980), section I-C.

[173] This account is based on an important report drawn up by Desmarest in 1779 for the information of the Estates of Languedoc: "Mémoire sur l'état actuel des papéteries d'Annonay et sur les améliorations dont elles sont susceptibles," Archives départementales de l'Ardèche (Privas), Series C 960.

The two leading families, the Montgolfiers and the Johannots, were respectively Catholic and Protestant in a city much divided since the religious wars, where even at the turn of the twentieth century, well brought-up Catholic children were never supposed to consort with the offspring of Protestants.[174] In the interests of business and factory discipline, these differences could be bridged, however, and the Montgolfier and Johannot mills had much in common besides a supply of water. Mathieu Johannot and Pierre Montgolfier had collaborated in experimenting with hollanders in 1751, and had jointly abandoned the attempt after a few years of frustration. In the 1770s, when Desmarest had come to be familiar with their operations, he reported that in their four mills they kept going twelve vats each of which was good for about 50,000 pounds of paper annually, about twice the average capacity per vat in the country as a whole. He attributed their productivity to the "liberty" they allowed the workers to labor during all the hours of the day not consumed by sleep.[175]

Between them, the Montgolfiers and the Johannots had thus made Annonay, with a production of 600,000 pounds annually, one of the leading centers for paper in France. Of the two, Pierre Montgolfier had the greater reputation. It was to him—"the senior and most important manufacturer in the province"—that Bacalan, intendant of commerce in Paris, addressed a questionnaire in 1769 when the Bureau of Commerce was requesting information from factory owners concerning the needs of the industry.[176] The mills in Annonay had come into the hands of the family—whose tradition traces their connection with papermaking back to an ancestor in Bavaria in the twelfth century—in 1693 on the marriage of Raymond Montgolfier to Marguerite Chelles, the daughter of the proprietor of Vidalon, and of Raymond's ne'er-do-well brother, Michel, to her sister, Françoise. The management descended to Raymond's son, Pierre, the fifth of nineteen children, who was born in 1700 and died in 1793. After a brief turn at the seminary in Lyons, he returned home to learn the business from his father. It had over one hundred employees in 1735, when Raymond Montgolfier petitioned the archbishop of Vienne to authorize consecration of a chapel in the factory, and about 140 in 1769, when Pierre replied to the inquiry from Paris.

"Nothing," he then answered the ministry, "is more revolting than the tyrannical power that the worker wields with respect to his master, noth-

[174] So the present writer was informed by an elderly resident in 1978. For the history and genealogy of the Montgolfier family, see Rostaing (1910), and for the Johannots a recent, though undated and unpublished, thesis presented to the Sorbonne, Jean-Pierre Le Moine, "Les Johannot—famille protestante et papétière d'Annonay aux XVIIe et XVIIIe siècles." A copy may be consulted in the Bibliothèque municipale d'Annonay.

[175] Desmarest, Mémoire cited note 173, above, fol. 2.

[176] Rostaing (1910), 108-112.

ing more degenerate or more insolent than this miserable bunch of rascals, and at the same time nothing so urgently demands the attention of the Council [of Commerce] as these seditious upstarts."[177] The frustration he took the occasion to vent concerned the practices called *modes*, by which working paper makers, like other itinerant journeymen joined in institutions of *compagnonnage* (or trade guilds), had acquired the right to set working conditions, to collect dues on hiring and promotion, and to celebrate traditional rites and festivals. Demands for higher wages were not the problem at this stage in labor history. Obstruction of all attempts at innovation especially incited Pierre Montgolfier's ire. Beyond that, he took the central government to task for the impediments placed in the way of trade by the taxes, tolls, and customs duties of which provincial manufacturers in all trades ritually and rightly complained. For the rest, he acknowledged the high reputation of Dutch paper and asked for a subsidy to send one of his sons to Holland to learn about procedures there. The intendant Bacalan replied, temporizing as befitted a bureaucrat, and alluded to the mission Desmarest had just completed for that very purpose. He promised that the latter would put himself in touch with Montgolfier on his next tour through the province, and it appears that through this exchange Desmarest learned of the Montgolfiers and they of him.[178]

Like other industrial patriarchs of his generation in the provinces—like Deydier, for example, whose son was qualified by technical education to operate Vaucanson's machinery—Pierre Montgolfier sent his boys to school instead of simply putting them in the mill to learn as he had done under his father's eye. Of the two who mainly concern us and history, the inventors of ballooning, Joseph, the twelfth child, born in 1740, studied in the college at Annonay and later in the nearby Rhône town of Tournon. Etienne, or Saint-Etienne as he was often called by his familiars, the fifteenth child and youngest son, born in 1745, was sent to the college of Sainte-Barbe in Paris. Both were good at mathematics, mechanics, and exact science in general, but they were of very different temperaments. Joseph was a rebel and a dreamer, the stereotypical inventor, absent-minded, shy, fantastical, a failure in his attempts at business, impulsive, and the father in his turn of a spendthrift who in the early nineteenth century wasted his widowed mother's substance and left her a charge upon the conscience of the family. Etienne was a pillar of rectitude, a disciplinarian, never one for nonsense, totally reliable in responses to the calls of duty, friends, and family, a man of order and of system, who excelled in all aspects of the business, technical, commercial, and financial, and who left only daughters as issue of a stable, peaceful marriage.

[177] *Ibid.*, 88-89.

[178] Bacalan to Pierre Montgolfier, February 1769, Rostaing (1910), 88.

As a teen-ager, Joseph ran away from the college in Tournon, with all its rules and all its priestly keepers, and headed down river for freedom and the Mediterranean. There along the coast he wandered for a time, picking up the odd jobs by which a gifted mechanic could live casually, and learning chemistry on his own somehow. By 1760 the prodigal was home again, and his father thought to set him up in company with his brother Augustin and his sister Marianne to run the mill at Vidalon-le-bas, then being mismanaged by their cousin Antoine. It did not work to be so near home. For a time Joseph thought of the Antilles, and then went off with the same brother to Dauphiny, to try the only business he knew far from the paternal eye. The two mills in Voiron and in Rives were a fine success technologically—they experimented with hollanders in the 1770s—and a failure commercially. Rives resembled a factory less than it did a laboratory of mechanical and chemical engineering.[179]

Etienne was very young when he was packed off to Paris. Since his older brother, Raymond, was being groomed to succeed their father in the management of the mills, and since in addition to Augustin and Joseph there were three other older brothers (one of them defective, however, and one in orders) it was not thought that the benjamin of the family would be needed in the trade. Accordingly, he was educated to be an architect. Etienne succeeded brilliantly in school, as he did in everything, and took service for professional experience with the famous J.-G. Soufflot. He was assigned several of the great man's commissions to design on his own, among them a factory for the manufacture of wallpaper in the faubourg Saint-Antoine. Its proprietor, Jean-Baptiste Réveillon, not a great deal older than Etienne, became his close, staunch friend. The relationship graduated into so close a business association that the Réveillon factory became in effect the Paris office of the Montgolfier paper mills.[180]

Etienne was thus well started into a career when in 1772 his brother Raymond died, the hope of the family. By then Pierre was 72 years of age. A good judge of the qualities of his sons, the old man sent for Etienne to come home and assume the heir apparent's place. Hence the irony that the youngest son became the head of a family concern, in whose counsels he was never more than first among brothers, however. As Pierre grew older, his children found their father harder to manage than was the business, wherein there was a tacit division of responsibility. Etienne did the

[179] The account of Joseph depends on Rostaing (1910), 187-221, and many items of family correspondence in private hands, to which the writer has kindly been given access by Monsieur Charles de Montgolfier.

[180] The account of Etienne depends on Rostaing (1910), 229-252, and on the materials cited in note 179. The registers of the factory are still extant at Vidalon-le-haut. Much of the record was microfilmed by the Archives nationales, where the microfilm is designated Archives Canson-Montgolfier, 53 AQ, 131-MI.

planning and exercised general over-sight; Jean-Pierre, "Montgolfier l'aîne" in the usage of the country, managed the plant and kept the books; Marianne oversaw the work of the women and children in the rag-picker's shop; Alexandre (the abbé), though less continuously involved, represented them in commercial negotiations in Lyons and to some extent in public relations; Marguerite ran the commissary and oversaw the housekeeping, for the mill at Vidalon was an enormous pension, almost a barracks, where the workers lived on the job under a domestic no less than an industrial discipline.

Clearly, Etienne Montgolfier and Desmarest were cut out to understand each other very well indeed. In 1771 Desmarest was transferred from Limoges to be inspector of manufacturing at Chalons-sur-Saône. The journey from there to Annonay was relatively easy. He and Etienne began collaborating soon after the latter's recall from Paris in 1772, which is to say in the interval between Desmarest's presentation of the first memoir on paper to the Academy of Science and preparation of the second for delivery in 1774.[181] The first memoir (it will be recalled) emphasized the value to the Dutch of the practice of échange, the shuffling of the sheets between repeated pressings, and Desmarest reported later how delighted he was on a subsequent visit to Annonay to find that his suggestions had been adopted in all four paper mills, Johannot as well as Montgolfier. Their doing so had required courage and sacrifice, he acknowledged. Inevitably workers unused to the new procedures had spoiled a large proportion of the sheets, and the loss was initially considerable. Of all the mill owners who tried échange after publication of his memoir, only the Montgolfiers and the Johannots had persevered to make it work and to reap the benefit in the greatly improved quality of their finer grades of stationery.[182]

In the second memoir, Desmarest took up the preparation of pulp and the deleterious effects of using rotted fabric. Apparently, Etienne made available to him one of the stamping mills and vats at Vidalon and collaborated in a lengthy series of experiments. Together they determined that it was materially possible to beat fresh cloths to a proper pulp by means of mallets, but only when they restricted the choice of rags to the very finest and softest linen. Commercially the cost would have been prohibitive, and the trials established that substitution of hollanders was economically a precondition to using fresh and healthy rags.[183] Such a transformation would have had the subsidiary advantage, less important than in Holland but still real, of freeing the mills from dependence on a steady flow of waterpower. For it was not, in the final analysis, a question of simply comparing the respective advantages of natural versus fermented

[181] Cited in note 169.

[182] "Mémoire sur l'état actuel des papéteries . . . ," note 173 above.

[183] *Ibid.*

pulp (Desmarest admitted that there might be uses for a little rotting); of hollanders versus stampers; or of slow drying, and compression before and after sizing, to the speedier French methods of finishing the sheets.

No, the question was nothing less than transforming an entire system of production, in which each step entailed the others. That was to be accomplished by a process of rationalizing the industry and of educating its practitioners in new and better ways. Desmarest's favorite word was "*Instruction*." The superiority of the Dutch derived, not simply from their cylinders or from their natural pulp, but from what lay behind all that, the tact and single-mindedness with which their entrepreneurs had learned how to get the most out of their machinery, to substitute it wherever possible for labor, and to run the machines rather than to let the machines run them. Ignorance was the problem in France, the lack of overview of an entire process, the substitution of labor for operations performed by machinery when the latter failed or wore out, improvisation rather than steadiness. He, Desmarest, had attempted to show the way in his two memoirs, and he planned a third on the construction of the machinery itself. But he had also in mind a more effective instrument of instruction than all the treatises that he could write, though they might be its prerequisite:

> That would be the creation of a workshop where all the processes and machines would be in operation and that would be open for the observation and research of those who would like to inform themselves more or less thoroughly. Such a workshop, designed according to a rational plan, would show the order and connection of the operations, their sequence and progression. It would exhibit, in short, a general system for manufacturing every type of paper, in accordance with which manufacturers could adopt the features best suited to their ideas and their business. They would find there only those processes tested and verified by results.[184]

For lack of a plan of development based on "infallible principles," all the enormous funds poured into French industry in the past thirty years had produced no observable improvements. Instead, ignorance and uncertainty had prevailed among manufacturers, and only when those capable of emulating the appropriate models invested resources in sound projects instead of in ill-conceived schemes would matters change for the better.

Evidently Vidalon-le-haut was Desmarest's candidate to become for paper what Le Creusot was for iron, Aubenas for silk, and Saint-Sever for cotton—the pilot plant that was ever the favored recourse of the Bureau of Commerce; and Etienne Montgolfier was to be the philosopher-manu-

[184] "Second mémoire. . . ."MARS (1774/78), p. 88.

facturer. By good fortune Annonay was the northern tip of the provinces represented in the Estates of Languedoc, the most energetic economically of provincial bodies in the eighteenth century. Desmarest might hope to elicit a subvention from Montpellier that would have been obtainable with greater difficulty in Paris. The memoir he addressed to the estates in 1779 reviews the progress already achieved in the mills of Annonay. It would be unreasonable to expect private manufacturers to undertake the entire expense of a demonstration intended to transform an entire industry. For it was not simply a question of installing hollanders ready made. They would have to be designed for local conditions. They would then need to be run by experienced personnel. Desmarest held little hope that laborers used to the old routines could be trained to operate the new technology. A manufacturer who meant to succeed had better be prepared to break in a new work force. As it happened, he knew of a highly accomplished foreman called (improbably enough) Ecrevisse—"Crayfish"—a Dutchman who had just finished converting a paper mill in Lille to the methods used in neighboring Flanders.

Desmarest estimated that it would cost 18,000 to 20,000 livres to mount a brace of hollanders, the first for shredding rags into threads and the second for pulping the threads into paste. Entrepreneurs who wished to submit a proposal should undertake (1) to install these machines within the term of eighteen months, (2) to furnish the shop with other items of machinery needed for the Dutch process, (3) to follow every step of those procedures in the fabrication and finishing of paper, (4) to produce by the new techniques paper of every grade from the finest stationery to the sturdiest material for packaging, (5) to open the shop to visitors so that other manufacturers could freely observe all the operations, copy the design of the machines, and bring along their own workers to learn the entire series of manipulations, (6) to hire and pay the Flemish foreman (Ecrevisse) whom Desmarest would recommend to oversee the construction of the machines, the training of workers, and the execution of the Dutch system as a whole.[185]

The text of the recommendation mentions neither names nor specific mills—Desmarest's compliments over progressiveness and enterprise in adopting the échange refer to all four and to the manufacturers of Annonay as a group. They now were Matthieu Johannot at Faya, his son Pierre-Louis at Marmaty, Pierre Montgolfier's nephew Antoine-François Montgolfier at Vidalon-le-bas, and of course Etienne in Pierre's name at Vidalon-le-haut. Patently, however, the stipulations were such that the last would be the one to meet them. The estates at Montpellier adopted Desmarest's recommendation on 3 January 1780. Etienne submitted their

[185] "Mémoire sur l'état actuel des papéteries . . .," note 173 above.

proposal over his brother Jean-Pierre's signature on 7 January, revising it on 3 February when it was objected that he had not met the fourth of the above conditions.[186] Matthieu Johannot did not then make a bid. He later said that he was never informed of the resolution of the estates.[187] Evidently Desmarest assured Etienne that approval would be forthcoming, for Ecrevisse was installed at Vidalon-le-haut before the summer of 1780 although only in the next session of the estates, on 4 January 1781, was the subsidy actually voted, and three years later only half of it had in fact been paid out of the provincial treasury.[188] The money—it may be surmised—was not the important thing. On the scale of the Montgolfier operations, the amount was a small one. What mattered was official backing and recognition. Johannot and Antoine Montgolfier acknowledged as much in their complaints over the favoritism shown their competitor after the new machinery was in place and operating successfully.[189]

Anticipation of trouble with the workers was almost certainly one of Etienne's motivations in seeking sanction from the provincial authorities, though not only and not necessarily the major one. Etienne may even have provoked the trouble on purpose, for he and his brothers were if anything more determined than their father to rid the plant of the inefficiencies and obstructions deriving from the corporative *modes*, the traditional rights of craftsmen in the industry, and to become masters of production in their own house. They had before them the examples of Augustin and Joseph in the branch factory at Rives, near Voiron in Dauphiny. Theirs being a new and progressive establishment, Augustin had resolved to break the stranglehold of the old ways and to turn his labor force from journeymen and apprentices into employees with no extramural allegiances or recourse. He had prevailed, surviving threats of arson and of violence and many acts of vandalism, and he counseled resolution to his relatives in Annonay. Let them not suppose that old workers could be reformed by contact with a better-trained and educated cadre. If any were to be kept on, they must be segregated and confined to their own tasks.

Installation of the hollanders and operation of the model process in one of the shops at Vidalon precipitated matters there. Grumbling and incidents of insubordination multiplied while Ecrevisse and his team con-

[186] Montgolfier l'aîne (Jean-Pierre) to Rome (Syndic-général . . . des trois états du païs de Languedoc) 7 January 1780, Archives de l'Ardèche, C-143; Rome to de la Chadenède (subdelegate at Vienne), 19 January 1780, *loc. cit.*; Pierre Montgolfier, certificate 3 February 1780, *loc. cit.*

[187] "Matthieu et Pierre-Louis Johannot à Nosseigneurs des Etats de Languedoc," printed remonstrance, undated but 1781 or 1782, *loc. cit.*

[188] "Etienne Montgolfier à Nosseigneurs des Etats de Languedoc," printed petition, undated but 1783 or 1784, *loc. cit.*

[189] Remonstrance note 187 above, and "Antoine-François Montgolfier à Nosseigneurs des États de Languedoc," assemblés à Montpellier en 1780, *loc. cit.*

structed and broke in the new equipment in the summer of 1780 and into the following year. In October 1781 the leaders of discontent persuaded their fellows to down tools and stop work altogether. They left the plant to establish solidarity with their counterparts in the other mills in Annonay. Matthieu Johannot, a syndic of the community, mediated this first phase of what had become a strike, and the laborers returned to Vidalon, threatening to walk out en masse at the end of six weeks unless their *modes* of employment were restored and respected. Écrevisse's trainees now refused to pay the dues exacted from apprentices, however, and Pierre and other members of the family were abused when they upheld the new men. Finally, at the end of November, the whole body of veteran workers did abandon Vidalon. That may well have been Etienne's intention from the start, for although he gradually took back some who renounced the *modes*, for the most part the new Vidalon was staffed by new employees. In this course, he said in a report to the estates, he was further following the advice of Desmarest, who had insisted that the surest way to break loose from the spirit of routine among the workers was to hire new ones.[190] But Etienne Montgolfier hardly needed Desmarest to tell him that or to steel the resolution that converted a walk-out into a lock-out.

The course Etienne took could not be expected to make for popularity among the laboring elements in the region, while his pre-emption of patronage from the province for the installation of hollanders had for the time being estranged his peers. Success, in other words, had left the Montgolfiers with a problem in point of reputation, or at least of sympathy. No document survives to prove that it was with an eye to public relations that Etienne now began showing interest in a visionary project of his brother Joseph, but certainly he need have acted no differently if that had been in his mind. We do not know when Joseph first thought of flight. He had probably read a science fantasy, *L'Art de naviger dans les airs, amusement physique et géométrique*, published in Avignon in 1755 by a clerical predecessor of Jules Verne, Father Joseph Galien, a Dominican

[190] His language may be quoted: "La difficulté qu'il (Montgolfier) a éprouvé à plier les anciens Ouvriers à partie de ces nouvelles manipulations, à cause de la routine vicieuse qu'entretient parmi eux un esprit de corps, que le Gouvernement a, jusqu'à présent, inutilement tâché à détruire, lui a fait prendre le parti, d'après les avis de M. Desmarest, d'en former des nouveaux. Convaincu par expérience qu'il ne pouvoit espérer de plier à ces nouvelles manipulations des gens qui lui alléguoient toujours les anciens usages pour s'y refuser, il n'a pas craint la dépense qu'entraînoit son plan, d'instruire des jeunes gens dans les nouveaux procédés, & il a sacrifié son intérêt momentané à l'espoir d'acquérir la perfection à laquelle il aspire." Etienne Montgolfier à Nosseigneurs des États de Languedoc, undated but 1783 or 1784, a printed petition, Archives de l'Ardèche, C-143. I am indebted to Mr. Leonard Rosenband for sharing his copies of these documents. His thesis (Rosenband, 1980) gives an account of work and labor in the Montgolfier plant. For the strike, see section 2-B.

who was professor of philosophy and theology at the University of Avignon.[191] Joseph Montgolfier was then fifteen years old, and ran away from school soon afterward. Family tradition has it that further suggestion came to him from the sight of a pretty peasant girl blowing soap-bubbles by the sea-shore. According to the éloge after his death by J.-M. Degérando, the siege of Gibraltar during the Seven Years War led him to reflect that, if the fortress was impregnable to attack by sea and by land, there remained the air.[192] Henry Cavendish first isolated hydrogen in 1766, and the lightness of "inflammable air" was a property much discussed among everyone knowledgeable about chemistry in ensuing years. Whatever the influence of these and other factors, Joseph during his peregrinations in the late 1770s began calculating the relations between the capacities of globes filled with fluids lighter than air, the relative lightness of such agents, and the weight they could lift by displacement of an equal volume of air. He also attempted experimental measurements of the variation of the specific gravity of air with temperature, working with toy balloons in his quarters in Avignon, where for a time he lived. His imagination extended to designs for parachutes, which he tested for their effectiveness in slowing the fall of heavy objects dropped from the towers of the palace of the popes.[193]

Etienne was skeptical of his brother's enthusiasms, and their aging and irascible father even more so. At best they indulged Joseph when he was at home; at worst he irritated them, when they had all the problems of the plant, the hollanders, and the workers on their minds. One day in the autumn of 1782 (so it is said) Etienne settled down in his study for refreshment and a brief respite from these cares. Called to the far end of the room, he put a paper sack over the coffee-maker to keep it hot—and turning to look back saw that the bag had floated right up to the ceiling! Whether the tale is apocryphal or no, certainly Etienne changed his mind and began working seriously with Joseph from at least November 1782. His correspondence with Desmarest (inspector at Sèvres after 1781) tells of their hopes and mounting excitement. They refined the calculations. They tried small models of balloons made of paper reinforced with taffeta, in the garden just north of the factory. So deep and narrow is the valley there that their doings passed quite unobserved. In the spring, they were ready with a large-scale model. It flew admirably, rose above the high banks of the Déume, and settled harmlessly onto the property of a vintner less than a kilometre distant. Close friends knew of the project, and Jo-

[191] *Rapport fait à l'Académie des Sciences sur la machine aérostatique, inventée par MM de Montgolfier* (1794), 8-9. See below, Chapter VII, Section 3, n. 148.

[192] J.-M. Degérando, "Notice sur Monsieur Joseph Montgolfier," *Bulletin de la Société d'Encouragement pour l'Industrie Nationale* 13 (1814), 91-108.

[193] Rostaing (1910), 187-221.

seph and Etienne told of their success in the course of a convivial dinner at the house of an intimate of the family, Boullioud de Brogieux, a considerable personnage in the town and province. He it was who imagined what a lark it would be to launch the "machine" (as the Montgolfier always referred to their invention) before the eyes of the États particuliers, the local estates of Vivarais, scheduled to convene in June 1783.

On Wednesday 4 June the deputies of the province were accordingly invited to attend a demonstration in the place des Cordeliers. A small plaque marks the spot where aviation began. Hanging limply inside a sixteen-foot scaffold was a bag of canvas lined with paper. The two layers were fashioned in four segments buttoned together along the seams, and ribbed with string like a huge fishnet. The day was fine and still. Under the open mouth of the machine the brothers Montgolfier kindled a small blaze of dry straw and woollen shreds in a brazier. Gingerly they blew up the fire with a small bellows. The bag filled and swelled into a sphere 100 feet in circumference. It rose against the ropes, tugged for a few minutes, and was off when released, rising above the astonished assemblage to a height of 6,000 feet in about ten minutes. As the air cooled and escaped through the buttonholes, the balloon slowly descended, drifting about a mile and a half the while, to settle into a vineyard so gently that the grapes were not bruised nor the vines broken.[194]

The sensation when the word—in the inevitable form of a procès-verbal—reached Paris, the appointment of a commission on flight in the Academy of Science, the exhibitions before the court and the public, the launching of manned aircraft, the choice between hot air and hydrogen—all that pertains to engineering rather than to industry, and will be deferred to the concluding chapter. Almost immediately, it was a question of who would respond to the invitation that the Academy, on instructions from the controller-general, issued to bring the balloon to Paris for demonstration and study. The choice fell on Etienne with his feet on the ground, and not on Joseph with his head in the clouds. With all his travels, Joseph had no sense of the capital, and he could be awkward in company. Etienne was a man of urbanity and he knew Paris. Now, however, he was rubbing elbows in a world of science and fashion he had never entered as a student from Vivarais. His letters home tell of meeting Lavoisier, La Rochefoucauld, d'Ormesson, Malesherbes, the maréchal de

[194] *Ibid.*, 253-262; Barthélemy Faujas de Saint-Fond, *Description des expériences de la machine aérostatique de MM de Montgolfier* . . . (1783), 1-28. Faujas gives the date as Thursday 5 June, thus starting an error that has been perpetrated in most later accounts. In fact, that was the date on which the procès-verbal of the Etats particuliers certified to having witnessed the experiment with a "machine diöstatique" performed before them "hier 4 juin," in the words of a petition submitted by Etienne and Joseph to establish their priority. A photocopy is in the possession of the author.

Castries, the comte d'Antraigues, Monge, and Faujas de Saint-Fond, among many others. In all this, he had several purposes besides staging the two flights he launched for the public. Of the time and trouble that the balloons took, and the difficulties encountered, he complained on more than one occasion. "People talk only the machine to me," he wrote his wife irritably, "I get to talk only of the machine; I do nothing but the machine."[195] Weeks in the late summer of 1783 turned into months in the autumn and into the winter of 1784, and he worried about the business, and about the machinations of the Johannots. Still, the business was a large part of the reason he was there. The shame was that waiting on scientists, courtiers, and ministers in offices and drawing rooms was always at their pleasure.

Etienne persisted, as was his way, and as was equally his way, he succeeded, at least in two of the three objectives he had set himself. In December 1783 the king accorded the title of nobility to his father, henceforth to be styled Pierre de Montgolfier. A coat of arms was accorded on 7 January 1784. The letters patent recognized the feat of his sons in aviation equally with his achievement in paper making. Specially emphasized was the development of vellum at Vidalon in the early 1770s, together with the pioneering of hollanders and Dutch methods and the grant from the Estates of Languedoc. Next came distinction for the factory itself. On 19 March Calonne authorized Etienne the designation "Manufacture Royale" for his establishment at Vidalon. It had taken months of negotiation to secure that accolade, and also to turn back the pretentions of the Johannots to the same distinction. Only in one respect was Etienne disappointed, and Desmarest with him. It had formed part of their plan to have Vidalon further recognized as an educational establishment no less than a model factory, an École nationale de papéterie. They never did achieve that ultimate goal of the enlightened manufacturer and bureaucratic scientist. Vidalon prospered nonetheless under Etienne's direction until the revolution and became the leading supplier of the industry. Desmarest for his part was appointed in 1788 to a post newly created by Tolozan, that of director-general of manufacturing, in which he would give to all French industry the kind of leadership and guidance that he had developed in his responsibilities for paper.

At lower levels Etienne de Montgolfier (as he, too, now was) provoked resistance on occasion. A memoir of 9 August 1784 minuted "Mr de Montgolfier papéterie erigée en Manuftre Royale" in the hand of Rome, the provincial syndic in Montpellier, reads:

> Another thing that you will see in the enclosed copies of letters of Monsieur the intendant and of Monsieur de Montgolfier . . . is that

[195] Letter undated, but September 1783; copy in possession of the author.

Monsieur de Montgolfier, capitalizing on the good luck that has brought him his so-called (and so far, at least, altogether useless) discovery, wanted to try to augment the marks of favor that he has already received, whether with respect to that or the papermill at Annonay. I could not, however, refrain from observing to him that, as to the latter object, the Estates had already amply carried the views of the minister into effect.[196]

4. INVENTION

The phrase "*brevet d'invention*" for a deed vesting in an inventor his right to private property in a discovery was unknown to French law before the revolution. Legislation adopted by the Constituent Assembly on 5 January 1791 provided the first comprehensive patent code in France.[197] Nevertheless, from the extent of testimony and the intensity of feeling attending passage of that law (to which we shall return briefly), it is clear that the notion of what a patent should accomplish was already well developed. Since codification was rudimentary (merely a royal declaration of principle of 1726), the consensus among inventors must have had its basis in common experience of actual operations. A discerning historian of French law has, indeed, argued that only the designation was lacking in the old regime and that main elements of a working patent system did then exist in the practice of awarding inventors exclusive rights to the exploitation of their ideas for a limited time.[198] The elements in question Isoré takes to be the following: an invention had to be new; it had to be capable of industrial application; it had actually to be exploited within a certain period; the benefits of a monopoly accrued to the author for a number of years in anticipation of its eventual entry into the public domain; complete specifications and if possible models had to be furnished and made public; the merit and novelty of the process or device had to be verified by qualified persons before this tacit contract between an inventor and society could take effect.

Examples of letters patent conferring such rights may be adduced from the sixteenth century onward, the earliest being a monopoly accorded

[196] Archives départementales de l'Ardèche, C-143.

[197] S.-J. de Boufflers, *Rapport . . . sur la propriété des auteurs de nouvelles découvertes* (1791) (BN,Le[29].1206), the committee report with the text of the law. See also C.-A. Costaz, "Notice sur les brevets d'invention," *Bulletin de la Société d'Encouragement* (1802) *1*, 81-85. In 1966 the Institut national de la propriété industrielle mounted an exhibition on the occasion of the 175th anniversary of the law of 1791. The brochure, "Le brevet de 1791" contains many valuable indications, as do the preface and foreword contributed by Marcel Plaisant and Guillaume Finniss, respectively, to the catalogue, *Brevets d'invention français* (1966), published by the Institut to accompany the exhibition.

[198] Isoré (1937).

Abel Foullon in 1551 for fashioning typographical characters of a new design. Similarly, in 1559 Naufredo Bathany was allowed a monopoly for a cistern of his "invention." All told, Monsieur Isoré uncovered some sixty-odd grants of this kind between 1551 and the onset of the Colbert ministry in 1667. Juridically and commercially, the writs were comparable to those entitling an entrepreneur to the exclusive exploitation of a foreign process in return for importing it to France. A famous case in point was the concession offered in 1551 to Thesco Mutio of Bologna for installing the manufacture of Venetian glass at Saint-Germain-en-Laye.[199] Technologically, however, the significance of brevets for invention was clearly different and greater. In the reign of Louis XIV and throughout the eighteenth century, the grant of privileges of the latter sort became increasingly common. An exhaustive search would be an excellent enterprise for quantitative history, and would probably yield instances in the thousands.

Such a search would need to be conducted among the records of the Parlement of Paris and the other parlements, which had to register acts of the royal government before they could take effect.[200] For whatever the functional precedents for a modern patent system, the juridical standing of letters patent conveying monopolistic rights to inventors was very different. Legally, such concessions partook of the nature of commercial privileges accorded to guilds, corporations of craftsmen, and manufacturing establishments. Economically, moreover, inducements to invention would often be perceived as threats to existing enterprises. Consequently, the parlements, ever defensive of corporative rights menaced by governmental initiatives, would scrutinize these measures, which like other exertions of royal authority tended to innovation. The sovereign courts would often shorten the duration of a monopoly granted an inventor by the crown, and would at the least require to be convinced of the value and novelty of a device or process before registering the privileges accorded its author by royal dispensation.[201]

To these ends, expert testimony often had to be invoked. In the modern world, patent systems vary between two extremes in regard to prior verification of the claim.[202] Some national offices require demonstrations of

[199] For Foullon, see AN, X^1.A 8617, fol. 178; cit. catalogue ***Brevets d'invention*** (note 197 above), no. 437; for Thesco Mutio, see AN, X^1.A 8649, cit. *ibid.*, no. 438.

[200] The manuscript registers of the Parlement of Paris are preserved in the AN, X^1.A, where they are being inventoried by a research group under the direction of Professor P.-C. Timbal. The work has reached the sixteenth century. Acts of the Council of State are preserved in manuscript registers in the Bibliothèque de la Chambre des Députés, where they are known under the name of the compiler, Recueil de Le Nain. A table of contents, the "Tables de Le Nain," may be found at the AN. These were the main sources available to Isoré.

[201] Isoré (1937), 115.

[202] Boehm (1967), 3. The first chapter gives a brief comparative survey of the provi-

its validity. Such is the German practice. Others simply register the design, leaving the questions of newness and merit to be determined by litigation, if need be, and by the market. The latter is the French practice, stemming from the law of 1791. That measure precluded technical evaluation of the subject of a patent, mainly in reaction against the exposure of inventors to scientific scrutiny by the authorities of the old regime. The practice of refereeing had its origin long before there was an Academy of Science, whose procedures in this respect were so similar to those of the Parlement of Paris that they may have derived from it. When letters patent granting economic privilege were delivered to the court, the procurator appointed a "*rapporteur*" to examine the substance of the matter. If the affair surpassed ordinary competence, the parlement would also name a commission composed of bourgeois notables and leading members of the trade or craft concerned. Theirs was the responsibility of pronouncing on the technical and economic aspects of an invention. On occasion the parlement refused registration, in which case the crown might force it to proceed, but only by exercise of a peremptory authority that ministers were reluctant to invoke.

In the three decades immediately following the foundation of the Academy of Science in 1666, its technological duties consisted largely in reporting on the devices invented by its own members. The most prolific in that respect was Claude Perrault, architect of the Louvre and of the Observatory of Paris, who imagined improvements for firearms, telescopes, and drawbridges, and addressed his ingenuity to the abatement of friction in general. Only as a feature of the reorganization of the Academy in 1699 was it vested with the responsibility for determining, on the request of the king, whether inventions for which the crown issued letters patent were both "new and useful." Agents of the government thereafter had the option, though not the obligation, of consulting the Academy. The parlement also might still refer proposals to the Academy. Even in the eighteenth century, however, the Academy was simply the most visible tribunal for rendering judgment on claims by inventors. It was never the sole recourse, especially in the provinces. Often, moreover, its members were asked to advise rather in relation to their employments than through the medium of an academic commission—Berthollet at the Gobelins, for example, and Vandermonde at the Hôtel de Mortagne.

Precedents accumulated without benefit of legislation until 1762, when the crown made a first attempt at enunciating comprehensive principles. Commercial privileges were then declared to serve the complementary purposes of rewarding the efforts of inventors and stimulating the in-

---

sions of the main national patent systems and of the history of British provisions. For the French practice of granting exclusive privileges, see Bondois (1933).

genuity of those subject to competition but precluded from innovation by existing regulations. The normal duration of a monopoly was fifteen years. Such exclusive privileges were held to be personal rights and neither salable nor heritable (a step backward in that many precedents made them patrimonial like other forms of property). If they were not exploited, they lapsed after a year.[203]

The duties of the Academy in respect to these proto-patents were closely related, obviously, to its charge to conduct a *Description des arts et métiers*, and also to its power of giving its approbation to writings and technical designs and devices, some of which were submitted for approval and publicity and some for subsidy and sale to the government, rather than as candidates for the right to monopolistic exploitation. The Academy early established a repository for the machines and ideas it had approved in the public rooms of the Observatory. Starting in 1699, brief notices appeared in the annual *Histoire*, and in 1729 an engineering officer called Galon, the same who contributed a treatise on copper working to the *Description des arts et métiers*, proposed to publish adequate accounts of these resources.[204] The Academy welcomed his initiative, and the work appeared in six volumes in 1735. It contains descriptions of some 377 inventions—pumps, cranks, mills, drawbridges, fire-arms, time-pieces, calculating devices, scientific instruments—and makes an impressive display of ingenuity. The plates are well drawn and the instructions specific enough so that a competent mechanic could construct and operate the devices illustrated. With that one set, however, the publication lapsed until 1777, forty-two years later, when a seventh and last volume appeared.[205] Edited by Meusnier on the initiative of Vandermonde, it contained a selection of inventions submitted between 1735 and 1755. In the later 1770s Vandermonde was collaborating with the aging Vaucanson in his laboratory of invention at the Hôtel de Mortagne, and during the Revolution the collection there was joined with the machines and instruments at the Observatory to become the basis of the Conservatoire des arts et métiers.[206]

All the evidence is that most of the scientists called on to appraise the ideas submitted by eager and often avaricious inventors found the duty

[203] Isoré (1937), 115-117.

[204] *L'Art de convertir le cuivre rouge* . . . (1764).

[205] *Machines et inventions approuvées par l'Académie Royale des Sciences, depuis son établissement jusqu' à présent, avec leurs descriptions*. The volumes are of equal size, and it seems probable that the years covered by each give a measure of the acceleration of inventive activity: I (1666-1701), II (1702-1712), III (1713-1719), IV (1720-1726), V (1727-1731), VI (1732-1734). They contain tables with the names of the inventors. For the seventh volume, see below, Chapter VII, Section 3.

[206] Above, Section 3.

irritating and time-consuming. A large fraction of the so-called inventions were chimerical, and an even larger number so vaguely set forth as to be incomprehensible. In the 1770s and 1780s, a member of the Academy with mechanical qualifications would annually serve on ten or a dozen committees charged with reporting on technological proposals of one sort or another. Though but a small proportion of the whole, the total number of designs or models that won approval in the course of the century was probably well over a thousand, and more likely close to two thousand.[207] It remains unclear, however, whether any of these grass-roots inventions ever did enter into industrial production. Things that did—Vaucanson's looms, Holker's cotton machinery, coke-fired smelting at Le Creusot, the Montgolfier hollanders—began, not with the initiative of some obscure mechanic or artisan, no matter how ingenious, but with initiatives from on high. Sometimes the impetus came, as we have seen, directly from the government, the source of will-power and often of capital. Sometimes it came from the Academy of Science, in the form of prizes set for developments desired by agencies of government. Artificial saltpetre for gunpowder is an obvious example.[208] Artificial alkali by the conversion of salt from seawater into soda is another, equally famous and far more important technologically. The story of the Leblanc process belongs for the most part to the 1790s, but it began in 1783 with a prize instituted by the Bureau of Commerce to be administered by the Academy.[209] Initiatives of that sort, and also referrals of artisanal approaches from below, came before the scientific community from several ministries, from war, navy, the Maison du roi, and Bertin's Petit Ministère. During Tolozan's administration, however, the Bureau of Commerce was far and away the most active technologically of all the agencies. The sort of experience an ordinary inventor might have of it will appear if we give some detail of the failure of one of them, Jean-Baptiste Delaplace (no relation to his great homonym), to win recognition and reward.

Delaplace (as he signed his name on the dozens of documents that his affair generated) called himself a chemist and was a manufacturer of rouge, whether jewelers's or cosmetic or both is left unsaid.[210] He lived and kept a shop in Saint-Germain-des-Près, at the corner of the rue des Petits Au-

[207] These figures are estimates based on impressions derived from many readings of the Procès-verbaux of the Academy of Science for other purposes and topics. It would be possible to get an exact figure, but that has not been done as yet. It does seem clear that the acceleration indicated in note 205 above slowed appreciably.

[208] Above, Chapter I, Section 6.

[209] Gillispie (1957a) is to be supplemented, and perhaps superseded, by J. G. Smith (1979), which reached me too late to affect the present writing.

[210] The ensuing account of the Delaplace affair is based on documents all of which are preserved in AN, F^{12}.1305A. The notes in Parker (1965) give a useful entrée into the location of records of inventive activity.

gustins (now the rue Bonaparte) and the rue des Marais (now the rue des Beaux-Arts). The coloring matter that he fabricated had won approbation from the Academy of Science at the time he discovered it.[211] Apparently Delaplace had once been a servant in Necker's household.[212] On 5 August 1778 he wrote to his former employer, now the director-general (or minister) of finance, with a pair of related claims, this time in metallurgy. He had found, first, a secret for refining (*mollifier*) iron smelted from native French ores so that pigs were as pure as Swedish or Biscayan iron. In a second step, he could convert his iron by cementation into steel as fine as the best English grade. His condition for communicating the technique to the government was anything but modest. He asked to be named an inspector of mines in charge of a foundry that the king would mount as an enterprise of state to exploit his process.[213] Necker referred the proposal to Tolozan, who in turn consulted the inspector of manufacturing mainly responsible for advising the Bureau of Commerce on the iron industry. This subordinate, Pierre-Clément de Grignon, took a set against Delaplace and his pretensions from the beginning. The initial reply was short: "The secret that you propose is already known in France."[214]

If Grignon had had his way, the matter would have ended there. Tolozan was more guarded. A minute in his hand instructs an underling to draft a letter leaving Delaplace the option of requesting the appointment of a commission from the Academy of Science.[215] To a further intimation that more detail of the process would be welcome, Delaplace replied by expatiating on its beauty and simplicity. He had tried a large number of experiments both in the laboratory and on an industrial scale, and was ready to demonstrate the latter. He could do so on a charge of 2,500 to 3,000 pounds in a furnace at a cost of only four livres for the material. His operation would prolong the smelting by forty-five minutes at most and inconvenience the workers not a whit. The effect would appear in forging. The crude iron would be far more homogeneous in texture than the normal product, and hence more malleable. He could draw it into threads as fine as hair. He could also convert it into steel, producing sixty-four hundred-weight a week from a single furnace at a cost of six and a half sous per livre. The same quality of English steel sold in Paris for a price over four times that amount.[216]

[211] Delaplace to Necker, 5 August 1778.

[212] Delaplace to Necker, 5 December 1780.

[213] Delaplace to Necker, 5 August 1778.

[214] Tolozan to Delaplace, 18 December 1778. Grignon was author of works on metallurgy, notably *Mémoire sur l'art de fabriquer le fer, d'en fondre et forger des canons d'artillerie* (1775). On his career, see Bacquié (1927), 375.

[215] 18 December 1778.

[216] Delaplace to Tolozan, 11 January 1779.

A further minute in Tolozan's hand requests Grignon to arrange these tests.[217] Obedient to his instructions, Grignon had an interview with Delaplace and remained unimpressed. A report to Tolozan of 21 January 1779 says that, after talking with this artisan, he doubted that Delaplace knew the first thing either about mining or metallurgy. He had no qualifications whatever to be an inspector of mines, and his secret was in all probability illusory. He had no samples to show and was reluctant to get into the expense of a demonstration for fear of losing his *propriété* in his discovery. The draft of a brief dismissal denying Delaplace a privilege accompanied this memorandum. Again, however, Tolozan minuted his instruction that the applicant must be given a chance to show what he could do. Instead of naming an ordinary academic commission, it was decided to let him have his day in court in an ironworks, and specifically at Buffon.[218] None other in France (wrote Tolozan to the lordly old proprietor, bespeaking his cooperation) was so well suited to testing a new technique that might have important consequences.[219]

The conditions were drawn by Grignon, and appear severe enough to accord with his skepticism. The principal commissioners were to be the forge-master of the county of Buffon, Jacques-Alexandre Chesnau de Lauberdière, and his counterpart at the foundry of Aisy, Edmé Rigoley. Tests were to be run under their eyes and in the presence of Buffon. Four pigs were to be cast, the first and fourth in the normal manner, the second and third under the direction of Delaplace. All four would be fractured at one end and examined physically. The crude iron would then be refined, and the yield in wrought iron weighed. Three samples of refined metal, two inches long by an inch square, would be forged from each batch. Time of smelting, consumption of charcoal, facility of forging, peculiarities encountered in refining, texture and appearance in cross and linear section, tensile strength, ductility, and hardness—all those characteristics would be recorded in a procès-verbal accompanying each chip. One sample from each batch would be wrapped and sealed with the signet of Buffon and his two commissioners and also of Delaplace. The package would also contain one ounce of the secret preparation. If (and only if) the report of the commissioners was favorable to Delaplace, was the container to be opened. Its contents would be examined in his presence after the government had agreed to grant a privilege and an award proportional to its value. The method would then be published. Should Delaplace wish to proceed to the second stage of his discovery, conversion of iron to steel, he was to have a cementation furnace constructed according to his specifications. A procès-verbal and samples would be sealed and forwarded along

[217] *Ibid*.

[218] Grignon to Tolozan, 21 January 1779.

[219] Tolozan to Buffon, 22 July 1779.

with the small iron bars to the minister of finance, in case the government wanted further trials made. All expenses were to be born by Delaplace, "except that in the case of a full and entire success," he would be "reimbursed by the government, and only in that case."[220]

In late February or early March 1779, Tolozan summoned Delaplace to his office and read out these stipulations. The applicant made no doubt of his prospective success. He could ask for no more enlightened a judge than Monsieur le comte de Bufon (sic). He agreed—with one reservation. Under no conditions would he part with his secret, as Monsieur de Grignon was requiring him to do. He would, however, throw in a hint of how the additive worked:

> This ingredient . . . allows no foreign materials to remain interspersed between the metallic parts, not even ferruginous earth. The effect is readily observable during the forging of the pig iron. Since its parts have no other matter interspersed among them, they unite readily with each other and under the hammer yield an iron that is pure and for that reason soft, malleable, and ductile. . . .[221]

Again Tolozan ceded, adding to the agreement a codicil in his own hand: "The Sieur Delaplace will consent to turn over his secret only when the government deals with him concerning its value and merit."[222]

Thus emended, the contract went all the way to Necker for approval. Thereupon Delaplace was again haled into Tolozan's office to sign the accord and receive a letter of authorization to Buffon. Buffon left Paris for Montbard late that year, in June, and needed three weeks before receiving Delaplace at the foundry.[223] The latter meanwhile had somehow hurt his leg and spent two months abed while the wound healed.[224] The historian who frequents the archives that preserve the record of these exchanges comes to recognize the contrasting styles: impatient and condescending on the part of the authorities, who, however, do not quite dare simply to dismiss the suit, lest something of potential value be lost; dogged, a touch meaching, uncertain in syntax and spelling on the part of the artisan, who is momentarily ill-used but appeals nonetheless to a putative paternalism. His posture is an unstable balance between respect, obsequious resentment, and ambition. His wife and children are pinched of the necessities of a decent life and are hanging on his hope of a breakthrough and a just reward for the fruits of talent and devotion. At the outset, it is to be ex-

[220] Delaplace printed his own account of the affair, *Mémoire sur l'art de mollifier et purifier les fers; découverte du sieur de la Place*, undated. A copy is contained in the dossier, AN, F^{12}.1305A. The conditions are set forth, pp. 5-8

[221] Delaplace to Tolozan, 4 March 1779.

[222] *Mémoire* (note 220), 8.

[223] Tolozan to Delaplace, 8 June 1779.

[224] Delaplace to Tolozan, 11 June 1779.

pected that the Delaplace initiative, like numberless others, will never be put to the test even if capable of it, and will simply evaporate, a wish invention, the technological counterpart of a thought experiment.

On the contrary, the thing worked. By 27 July all was ready. The procès-verbal, executed on 31 July, is one of those categorical, exact, and explicit documents, accompanied by a detailed journal or log, that bespeak the centuries of notarial discipline through which French bureaucratic procedures have evolved.[225] The stipulated tests were executed to the letter. In smelting the second and third pigs, which bore serial numbers 50 and 51 since the most recent relining of the blast furnace, Delaplace was given a free hand. Half an hour before the tapping he had the slag skimmed off. His secret preparation was contained in a wooden box eight inches long by three inches square. He affixed it to the end of a ten-foot iron bar, plunged it through the hearth into the middle of the molten metal, and stirred until the rod melted. The flame appeared suddenly whiter. The metal on tapping appeared slightly more liquid than usual. Pig 50 weighed 1,025 pounds and pig 51 weighed 975. After cooling, their properties were compared with those of pigs 49 and 52, smelted according to the normal procedures at the blast furnaces of Buffon, which produced the best quality of iron that could be extracted from French ores. The grain and texture in the Delaplace castings appeared to be superior and the surface brighter. They exhibited a number of gray-white flakes, a sign of good quality, in the surfaces that had been in contact with the sand of the form. The Delaplace iron filed more easily, took a brighter polish, and was less brittle. A strong laborer wielding a large sledge hammer failed to break it. In all these respects, it appeared superior to the normal product in the crude stage.

On refining, number 50 gave the smith a little trouble: the metal melted more readily and the worker was taken by surprise. Consequently, the consumption of charcoal was greater and the yield of wrought iron lower than might have been the case. With number 51, the forger had got onto it, and worked the metal into wrought iron giving all the signs of high quality. The Delaplace samples were brighter and of finer grain than their competitors though slightly less sinewy. But would the differences prove advantageous in practice as well as in appearance? That remained to be determined by the outcome of the different operations of forge and foundry, the polishing, the fashioning into nails, the rolling into sheets: the detail is reported in the procès-verbal, from which the conclusion will suffice:

> The result of all the above experiments is that the process of the sieur de la Place, far from having had a deleterious effect on the iron of

[225] A printed copy of the complete procès-verbal is in the dossier, F[12].1305A.

Buffon, which however it has visibly modified, has made it acquire body and ductility and has enabled it to go through more heats without becoming stringy or brittle. Of course, these effects are less noticeable in iron of naturally good quality, like that of Buffon, than in the case of other iron (where we are bound to conclude that it would be very considerable). That is why we think that the secret of the sieur de la Place merits the *attention of the government*, the more so in that no manipulations that could interfere with the ordinary work of the forges are required, and the operation is as simple as it is easy.

Done and delivered at the Buffon foundry, 31 July 1779 [signed] Rigoley and de Lauberdière.

When the two forge-masters communicated the procès-verbal to Buffon, he ordered further tests and experiments in the interest of confirming so marked a set of results. Accordingly, Lauberdière and Rigoley consulted also the foremost *méchanicien* of the region, the well-known Régnier of Semur-en-Auxois. To him they sent pieces of wrought iron, coded by letter, from all four batches. He made two bolts, one from the Delaplace iron and one from the ordinary, and found the latter easier to forge and the former superior for polishing. A trace of scale on the Delaplace product might have impeded the forging, however, and comparative samples were also tried on a locksmith (Bréan) and a knife maker (Urse), both of Montbard. The verdict was unreservedly for Delaplace. Asked to make a second trial, Régnier was even more emphatic about the polish of the experimental iron and far more tentative about the problem of forging it. He now thought that it contained less cinder. In a second certificate of 25 August, Rigoley and Lauberdière reaffirmed their earlier conclusion: they now urge ("*nous supplions même*") the government to pay attention. The matter is very promising. They will gladly accommodate Delaplace's further operations in their furnaces, "persuaded as we are that great advantage may result for the improvement of our iron."[226]

Still to be tried was the conversion into steel. For that purpose, Delaplace had a special furnace constructed in a courtyard of the ironworks. Unfortunately, the bricklayer was incompetent. The bricks had not dried properly, and the shape was irregular. Never mind—time was wasting, and Delaplace decided to take a chance. In the center of the furnace he placed a metal cube eighteen inches on a side. It contained substances unknown to the observers together with 125 pounds of his iron in the form of small ingots about an inch in cross-section and half again as long. An equal amount of the Buffon iron was fashioned in the same way for normal cementation. It took several days to dry out the furnace sufficiently to try a heat, which would be of forty-eight hours duration. On 12 August at

[226] *Ibid.*, 12.

7:00 p.m. Delaplace fired the charcoal. Alas, thirty hours into the conversion, the ill-baked bricks in one of the supporting arches crumbled and the casket tilted over onto the bottom of the furnace. Delaplace feared last the casing might be broken. Even if it was intact, the flames could no longer bathe it on all sides. There was nothing for it but to continue, however. Later it appeared that several cracks had indeed flawed the procedure. Nevertheless, Delaplace thought his steel good enough to be evaluated. Again, the commissioners addressed samples to Régnier, who tried them along with a cutler called Pernet, a specialist in razors. Pernet failed to follow the procedures appropriate to good English steel, and his results were inconclusive. For himself, Régnier found it a little recalcitrant on the anvil, like the iron from which it came, but concluded unequivocally that it was "*réelement bon*" and the equal of steel from England.[227]

Delaplace rejoiced, thinking his fortune made and his future assured.[228] Rigoley and Régnier were of his sort, and Lauberdière was an iron master even though a cut above the others socially. Having worked with Delaplace daily for a period of four weeks in the heat and sweat of the foundry, they could not have concealed their excitement and enthusiasm over the results, and fragments of correspondence show that they gave him every encouragement. So also did Buffon, who had been an eyewitness. True, it would have been improper for Delaplace to see the formal procès-verbal, or report, before it reached the authorities in the Ministry of Finance. He was furnished merely the abstract, containing the favorable judgments just quoted, but no detail. Back in Paris he thought to reinforce his claim by furnishing samples of his iron and steel to four or five prominent jewelers, metal workers, and cutlers, all of whom also responded with very positive testimonials. Armed with these, Delaplace got in to see Tolozan, who meanwhile had received a copy of the procès-verbal, which Buffon accompanied with a highly favorable covering letter.[229]

Tolozan congratulated Delaplace on the success of the trials—whereupon nothing further happened. Delaplace never was shown the procès-verbal. In one of their subsequent interviews, Tolozan observed that it was contrary to protocol for artisans and inventors seeking privileges to see the reports, since referees would thereafter feel inhibited from serving as confidential consultants.[230] At one point Delaplace was told that he had given offense by accepting an invitation to dine with Buffon, who after all was lord of the manor on which his foundry stood.[231] More systematically

[227] Régnier's judgment is conveyed in a memorandum of 22 August 1779. See also the "Journal de travail de la forge de Buffon pour les expériences de M. de la Place," p. 12. This is a printed log, accompanying the procès-verbal.

[228] *Mémoire* (note 220), 15-18.

[229] Buffon to Tolozan, 1 September 1779.

[230] Tolozan to Joly de Fleury, 12 July 1782.

[231] His response to this reproach in a desperate appeal to Necker is worth preserving:

he attributed the inaction of the authorities' reports to the enmity and jealousy of Grignon, who had not attended the tests and whose hostility he had felt from the beginning. In the course of months of awaiting official replies that never came and besieging bureaucrats in their offices, he did manage to obtain an interview with Grignon, who told him that his so-called secret was probably nothing more than refined saltpetre, and that there was nothing new about his cementation furnace, the design of which had been published by a certain Tara.[232]

Grignon was, in fact, the obstacle. The entire dossier from the foundry at Buffon, together with the sample strips of iron and steel, was routed to his office.[233] Sitting at his desk, he did a cost analysis and concluded that smelting the two Delaplace pigs had consumed significantly greater quantities of charcoal than the normal process, and that refining them had also been more expensive in time and material. To that would need to be added the cost of the secret ingredient itself, which could not be estimated because Delaplace obstinately refused to reveal its nature unless he was first assured of an award. In Grignon's eyes, the reports even failed to establish conclusively that his iron and steel were of better grade than Buffon's. The covering memoir by which Tolozan forwarded these evaluations to Necker bears the dismissal "*Néant*" in the minister's own hand, dated 21 January 1780.[234]

"il suffit qu'un homme soit pauvre et malheureux pour faire croire qu'il n'a aucuns facultés. M[r]. de Tolozan, au mois de juillet dernier en m'annonçant qu'il alloit me faire délivrer un ordre pour toucher 800# me diste que je m'étois mal conduit envers Mr de Buffon, de m'être mis à sa table, ayant été laquais de madame Necker. Ce compliment m'a d'autant plus surpris que je ne m'étois mis à la table de Mr de Buffon qu'après bien des invitations réiteré de sa part et même d'un ton à me faire craindre sa disgrace si je le refusois; et son maître de forge est venu par son ordre m'arracher de mon ouvrage, me tenant par le bras, me conduisit dans la salle à manger & me fit assoire à table à côté de lui? auroit-il été naturele que j'allasse dire à Mr de Buffon et à trois ou quatres maîtres de forge et leurs filles ou femmes qui etoient là; "dominé non sum dignus." Je n'ai certainement brigué aucun état, et je me suis conduit comme tout honête homme doivent faire. Mr de Buffon me croyoit donc un gentilhomme, et c'étoit donc pour cela qu'il m'avoit accordé sa protection après avoir vu l'effet de mon secret et m'avoir dit lui-même qu'il étoit très satisfait de moi, ainsi que MM les Commissaires. Voilà son terme et qu'il alloit écrire à Mr de Tolozan. . . . Personne ne sait mieux que vous que les arts sont à tous ceux qui savent se conduire par la connoissance de causes à celle des effets; et que tous les hommes se ressemblent, par l'organisation, par la manière de sentir, et par les besoins de première necessité; et c'est souvent à ces derniers à qui nous devons le développement de nos facultés. Vous savez Monsieur que j'ai eu l'honneur d'être votre très humble serviteur et que je le suis toujour. Paris ce 5 X[bre] 1780. J. Bte. Delaplace."

[232] Delaplace to Necker, 30 April 1780; also an undated "Réponse aux objections faites verbalement par Monsieur la chevalier Grignon au Sr de la Place."

[233] Tolozan to Buffon, 13 September 1779.

[234] Memoir: "Le S[r] de LaPlace a présenté . . ." Grignon's criticism occupies a memoir of eight long columns.

Nothing like so categorical a decision was ever imparted to Delaplace, who was left dangling in a state of diminishing expectations. Gradually giving up hope for a privilege, let alone a subsidized installation, he was reduced to haunting the anterooms of these newly unresponsive officials and importuning them for the reimbursement of his expenses. The only thing more degrading than the letters he composed must have been the gratitude he felt bound to express out of the necessity of salvaging a little something more.[235] One of the astonishing features of these and similar records is the doggedness of these inventive artisans, their refusal to take a non-answer for the "no" it really meant. In July 1782, during the brief tenure of Joly de Fleury as controller-general, Delaplace was at the Ministry of Finance again with resumés and renewed demands to demonstrate the effect of his secret. Again the file was passed among persons whose surnames appear and reappear in later French science and administration. From Tolozan it went to Douet de Laboulaye (an inspector of mines) and from him to Liouville, inspector-general and administrator of the foundry operated for the crown at La Chaussade in Lorraine.[236]

During the late spring and summer of 1783, Liouville put the Delaplace processes for iron and steel through another series of trials and comparisons with the normal production of his foundry. The procès-verbal has not survived, but the findings were evidently more favorable than not. The Delaplace samples absorbed heat more readily on refining, were firmer under the hammer, and shed impurities more easily than standard iron. They were slightly more economical in material and slightly less so in labor. Comparison of nails fabricated from the two sorts of iron at the nearby forge of Cosne was a stand-off. Both were excellent. Conversion into steel again gave ambiguous results. Of the two batches that were tried, one gave excellent and the other mediocre results.[237] Liouville sent both sets to Paris for more intensive examination in the Hôtel de Mortagne, where (after the death of Vaucanson in 1782) Vandermonde was curator of the laboratory of machines. He and Berthollet, together with Monge, were even then starting upon the research into the nature of steel

[235] Delaplace to Tolozan, 23 April 1780; Delaplace to Necker, 24 May and 20 June 1780. Only after reiterated appeals did the Treasury grudgingly pay Delaplace 500 livres toward the expenses of Rigoley and Lauberdière (whom Delaplace had had to reimburse), and then dole out another 300 for the cost of the conversion that half-miscarried. Delaplace never did get back his own costs, which he put at 1,427 livres—"un objet de conséquence," he allowed himself to observe, "pour un homme chargé de famille, peu riche, et à qui ses essais avoient déjà coûté trés-cher avant de parvenir à un résultat heureux." *Mémoire* (note 220), 19-20.

[236] Tolozan to Joly de Fleury, 12 July 1782; Tolozan to Laboulaye, 3 May 1783; Laboulaye to Tolozan, 8 May 1785.

[237] Berthollet gives a resumé in his "Rapport sur les fers et aciers préparés par le procédé de Mr de la Place," 30 June 1785.

published in the memoir of 1788.[238] This episode may quite possibly have been its point of departure.

However that may be, samples of the Delaplace steel and normal steel were tried on leading artisans of Paris: the locksmith Dumier, the cutler Daumy, the razor-maker Le Petit (by appointment to the cardinal-prince de Rohan), the specialist in springs Raulin. All of them found both types to be of excellent quality, the slight differences tending to favor the better of the Delaplace pieces. Vandermonde himself found the latter "steelier."[239] A surprising episode in his laboratory led him to send Berthollet a special message. The iron ladle he had used in re-founding Delaplace's steel had itself been converted to a good grade of steel. Astonished, he repeated the trial with another ladle, and the effect proved constant.[240] Inconsistent results in conversion between one batch of Delaplace iron and another remained troublesome, but such unpredictability afflicted the entire industry. That was one reason that systematic research was needed. When Berthollet went over the record, he much regretted that the inventor refused obdurately to allow his cementation process to be tried on samples of ordinary La Chaussade iron, as Liouville had suggested in the interest of further controlling the comparison. Nevertheless, Berthollet too was persuaded that Delaplace had something very well worth pursuing, if not for ordinary steel castings, then for the finest grade of steel produced by cementation. If for no other reason, the matter was important because the quality of English steel was deteriorating. English steel makers knew they had no competitors and were no longer as scrupulous as formerly. Berthollet made his report on 30 June 1785. Here is the conclusion:

> It thus seems to me that Mr Delaplace ought to be furnished with the means of repeating his operations carefully in a suitable forge; of trying his cementation on different kinds of iron which have not undergone his preparation; and of perfecting his process sufficiently to produce a uniform and constant effect. If he succeeds, he should expect from the wisdom and justice of the administration the recompense due for important services.[241]

And surely now, it might be thought, Delaplace would have his chance? Nothing of the sort: the better part of a year later, on 17 March 1786, he was again importuning the controller-general—this time Calonne—at least to be reimbursed for his expenses, which had climbed to 7,254 livres:

238 "Mémoire sur le fer . . ." MARS (1786/88), 132-200. See Section 3 above.

239 Vandermonde, "Épreuves faites à l'Hôtel de Mortagne des différens échantillons de fer et acier remis par Mr Berthollet" (undated).

240 Vandermonde to Berthollet, 24 June 1785.

241 Rapport cited note 237.

> If Monseigneur does not condescend to come to his aid, he (the petitioner) will lose the fruits of twenty years of work and the small competence belonging to himself, his wife, and his children. In this extremity, he takes the respectful liberty of begging Monseigneur, if he judges it improper for the government to buy his secret, at least to grant him a recompense that will enable him to go into business with the proprietor of a forge. That would be a way to preserve an important secret for the state, and would be the last recourse of a large family ruined by the discovery that its head had every reason to regard as a source of good fortune for himself and his dependents.[242]

The Revolution found Delaplace trailing these claims before committees of the National Assembly and technological agencies of the emerging republic. The results were the same. He never did get satisfaction, unless his steadfast refusal to reveal his secret without regard could be called that. The consequence is that we do not know what it was.

It died, or disappeared, with him.

Among numerous similar attempts at innovation, potentially real or really illusory, many were too episodic to have left more than echoes of aspiration and chagrin in the archives. Let us, however, choose one other example among those that produced a record full enough to permit reconstituting the story. Berthollet figured in this episode too, though now he was the obstacle. The affair of Dino Stephanopoli is interesting because not only did it divide bureaucrat from expert, and artisan from scientist, but it reached higher and divided the Academy of Science against itself. "Le sieur Dino," as he is often called by the officials, though obviously of Greek extraction, had been a military surgeon in his native Corsica. From a series of "Votre Grandeur" petitions casting himself on the benevolence, first, of d'Ormesson and, next, of Calonne in the Ministry of Finance, we learn that he had been born in 1726, and that he was a man near sixty with an aged mother, a wife, and two children.[243] The well-being of his family had been sacrificed for the canonical twenty years during which a "*travail pénible*" had unearthed the secrets of certain dyestuffs. Among the costliest items in the trade was the jet-black dye prepared from "*noix de galles*," the core of galls that bulge out on certain species of white oaks in sub-tropical climates, and that France imported at great expense from the Levant. Dino had found a substitute that he could prepare from a substance occurring very widely throughout the country. It would save 95 percent of the current cost, all poured out in "tribute" abroad. In June

[242] "Monseigneur le Controleur Général," Mémoire.

[243] The ensuing account of the Dino Stephanopoli affair is derived from documents to be consulted in AN, F[12].1330.

1782 Dino offered his secret to the government in return for a pension proportionate to its value to the state. An exclusive privilege would not have been appropriate; it was the sort of thing that should be adopted throughout an entire trade to the benefit of all.[244]

Preliminary reports on the decoction by Macquer were favorable.[245] In his official capacity as director of dyes, he consulted a well-known hat-dyer called Guenault, who was equally well impressed. Tolozan's colleague, Maurille Michau de Montaran, one of the four intendants in the Bureau of Commerce, handled the Dino application. On d'Ormesson's instructions, he informed Dino that nothing could be decided until he stated his conditions and imparted the composition of his dye to Macquer. That Dino refused to do. Through "a mistrustfulness natural to everyone of his nationality," noted Montaran, "he fears lest the profit he hopes to make from his secret will be lost to him if it gets out of his hands before his future is assured."[246] Dino did show his hand, or make his bid, on the score of money. His views were not modest. He wanted 200,000 livres, or the equivalent, a pension of 12,000 livres a year, to be divided between his wife and one child after his death and continued for their lifetimes. His secret (he went on to say) was applicable both to wool and silk. There the matter stopped until Calonne succeeded d'Ormesson early in November 1783.

Without agreeing to a specific sum, Calonne promised Dino that, provided he would confide his composition confidentially to the government, and provided that further tests were positive, he would be rewarded in proportion to the value of his discovery for the arts.[247] Macquer died in February 1784, having been unable to work for some months, and the affair was thus among the unresolved matters awaiting Berthollet on his appointment to the direction of dyes of the Gobelins. Meanwhile Dino had enlisted the interest of leading hatters and dyers in the capital and supplied them with batches of his decoction for their use commercially. Their experience of the dye was excellent although its basis remained unknown to them.[248] New in his post, Berthollet felt it to be his duty, not merely to collect empirical testimony from practitioners, but to verify the efficacy of the composition chemically. Accordingly, he tried a standard

[244] Montaran to Macquer, 13 July 1782.

[245] Macquer to Montaran, 5 September 1782.

[246] Memoir by Montaran (undated, but late 1782 or early 1783).

[247] Memoir by Nicolas Desmarest, 28 December 1789, giving a resumé of the affair, fol. 2.

[248] Evidently Dino had tried his dye in Marseilles before coming on to Paris. There is a certificate of attestation signed by four dyers and validated by a subdelegate of the generality, 20 December 1781. The Guenault certificate is dated 15 June 1782 in Paris. Dino obtained a similar document signed by three additional Paris dyers, Chol, Jannin, and Beaujolin, 14 June 1784.

laboratory test for the strength of dyestuffs, and precipitated the particles of coloring matter by adding vitriol (sulfuric acid) to the solution. He was now privy to the information that Dino's secret was a decoction of ordinary oak bark. From it he obtained only one-eighth the quantity of black precipitate that he did from an equal volume of a bath of gall at commercial strength. Further tests in his laboratory convinced Berthollet that, in order to achieve the desired effect, a solution of oak bark would have to be so much stronger than a solution of gall that any saving was deceptive. Moreover, only the first few samples dipped into each bath gave the lustrous appearance that pleased the artisans, and even in these the black was far from fast. For all these reasons, he returned a negative verdict: "It is beyond doubt that oak bark cannot replace core of galls, and I do not know what can have been the source of the illusion among the hatters whose testimony was so favorable to it."[249]

Dino immediately demanded to be tried by a jury of artisans, appealing to Calonne over—or perhaps under—Berthollet's head:

> Although persons may possess the most complete knowledge of chemistry, it is a fact, Monseigneur, that they will frequently fail in attempts to employ the procedures of a craft when they are not accustomed to its practice. The most limited of artisans succeeds better than chemists in that respect.[250]

According to Desmarest, whose resumé of the affair in late 1789 is sympathetic to Dino, Calonne had no "illusions" about Berthollet's findings. He directed that tests be made in the shop by hatters and dyers who were to determine the value of the oak bark that "M. Berthollet has contested."[251] Thus placed on the defensive by the controller-general himself, Berthollet arranged with Dino for a series of trials by leaders of the trade. Selected for this commission, with the endorsement of Montaran in the Bureau of Commerce, were the dyers Henriet, Pinel, Digeon, and Duvivier, and the hatters Devilly, Trubert, Jeannin, and Chol. Instead of Berthollet's laboratory, they used the premises of Henriet in the rue de la Bûcherie. In four all-day sessions in October and December 1785, 220 hats and a large number of swatches of felt were passed through the competing decoctions. The samples were coded so that no one judging the effect would know which dye had produced it. Each procés-verbal is

[249] "Rapport sur la propriété que le Sieur Stephanopoli prétend avoir découverte dans l'écorce de chêne. . . . ," 22 March 1785.

[250] Dino Stephanopoli to Calonne, 1 June 1785. On the same day he wrote to Lenoir, lieutenant-general of police, recounting his version of Berthollet's laboratory test. See also note 258 below.

[251] Desmarest memoir (note 247), fol. 2; Montaran to De Crosne (lieutenant-general of police), 4 September 1785.

signed by all participants, Berthollet in his educated hand, Dino and the tradesmen in their uncertain scrawls and inconsistent spelling of their own names. What was consistent was their repeated vote to place in the pile of superior quality the objects that had in fact been dyed by Dino's preparation.[252]

Despite their choices, Berthollet remained convinced that they were wrong, and that for chemical reasons the oak bark could not be preferable. His second and fuller report gives further experimental detail from the laboratory. It renews the finding that the Dino dye faded rapidly, and cites the literature to the effect that there was nothing new about it anyway. Oak bark had long been employed by unscrupulous dyers (though even then fortified by a little gall), and when cheaper dyes were wanted for products of inferior quality, the trade exploited domestic oak in the form of sawdust and shavings from the sawmills. He concludes:

> Although oak bark can give a good black under certain conditions, it cannot be compared to core of gall: because much less black material precipitates out of it; because the material precipitates much less speedily; and, finally, because the material that does precipitate has much less body than that due to core of gall.[253]

Whatever Calonne's man-of-the-world skepticism about scientists, the ministry could scarcely overrule its own director of dyes in deference to the artisans, and Dino was left with one recourse: to appeal Berthollet's second report to a higher rather than a lower tribunal, to the Academy of Science itself. That he did, and the Academy named two commissioners, Baumé and Vandermonde. Baumé being the chemist, the work fell mainly to him, and, after an investigation occupying eight months, he also found for the Dino preparation.[254] Normally the Academy would have adopted the recommendation of its commissioners and given the product its approbation. Berthollet still objected, however. In deference to his position, the Academy held off and referred the whole matter to the entire section of chemistry to whom they added Darcet, still in supernumerary status.[255] That made a committee of the whole of eight members, which proceeded to split right down the middle.

The nature of the division is interesting. Lavoisier and his two closest associates, Guyton and Fourcroy, rallied to Berthollet on what may be called analytical grounds. The four members with a pharmaceutical back-

[252] The four procés-verbaux are dated 28 October, 31 October, 7 November, and 6 December 1785.

[253] "Second rapport sur la propiété que Mr Dino Stephanopoli prétend avoir découverte dans l'écorce de chêne de pouvoir remplacer la noix de galles dans la teinture en noir," 15 December 1785.

[254] Desmarest memoir (note 247), fol. 5.

[255] *Ibid.*, fol. 6.

ground—Baumé himself, Cornette, Cadet, and Darcet—refused to sign their report. They drew up a dissenting report supporting the favorable judgment of the tradesmen on empirical grounds.[256] There can have been no clearer example of the reality of a division in the ranks of chemistry between a theoretical and a practical faction. The outcome makes an equally telling example of the influence of the former. Condorcet in his report to Montaran regrets the impasse. Under the circumstances, the Academy can make no recommendation at all, and both sides have agreed to keep the respective reports confidential. Accordingly Condorcet cannot leave copies of the texts in the ministry. Officials will understand that circumspection is necessary "because of all the ridiculous things that Stephanopoli says."[257]

Thus stalemated, Dino turned in something like desperation toward a remaining recourse, the College of Pharmacy, where again he might expect to find an outlook like his own. That body (it will be recalled) was under the patronage of the lieutenant of police of the city of Paris. His administration traditionally looked to the interests of artisans and craftsmen whose companies composed the municipality. Already after Berthollet's first rejection of the favorable reaction of hatters and dyers, four of them rallied to Dino in appealing to that magistrate.[258] The commission that the College now appointed found in favor of his process unreservedly once again. Alas, in the course of their doing so, his worst and first fears were realized: the identity of his dye became known to all, and his secret was one no longer.[259] Like Delaplace, he pursued his claims into the more favorable political ambiance of the Revolution, laying his sad case before legislative committees and bodies constituted by the new order to encourage arts and trades. There is no evidence that any official of the state ever made good on Calonne's promise.

Whatever the merits of oak bark, and however well founded Berthollet's chemical reservation, it would be difficult to find a more transparent illustration than the Stephanopoli episode of why inventive artisans sometimes had occasion to perceive science less as patron and guide than as overlord and enemy. Given the nature of the subject, which generated little overt comment before 1789, perhaps in this one instance the historian may be forgiven a moment of revolutionary hindsight. The patent law of 1791 was drafted by the committee of argiculture and commerce of the

[256] Lavoisier to Montaran, 31 December 1786; Berthollet to Montaran, 26 March 1787; Montaran to Condorcet, 29 March 1787; Lavoisier to Montaran, 20 May 1787. The two reports are summarized in Desmarest memoir (note 247), fols. 6-8.

[257] Condorcet to Montaran, 21 April 1787.

[258] Dino Stephanopoli to Lenoir, 1 June 1785; also Jannin, Beaujolin, Duport fils, Morel to Lenoir, 14 June 1785.

[259] Desmarest memoir (note 247), fol. 9.

Constituent Assembly. Its reporter was the amorist and *littérateur*, Stanislas de Boufflers. He rejected the view of the Academy that novel techniques should undergo scientific scrutiny, and adopted the proposition that patents should be had for the asking. How can any panel judge fairly of an invention since, by definition, the thing does not yet exist?

> As for the scientists themselves, have they not sometimes been charged with conflict of interest? Have they always been fair to inventors? Let's face it: the studious can scarcely believe in inspiration, and persons who are used to carving out the paths that lead to knowledge find it difficult to believe that it can be reached in a single bound.[260]

[260] S.-J. de Boufflers, *Rapport . . . sur la propriété des auteurs des nouvelles découvertes* (1791), p. 12. The report was read before the Constituent Assembly on 20 December 1790. BN (Le^{29}.1206).

CHAPTER VII

Engineering, Civil and Military

1. THE ECOLE AND CORPS DES PONTS ET CHAUSSEES

In the eighteenth century, engineers, hitherto concerned mainly with siegecraft, became in the literal sense civilized. Among the administrative reforms of Turgot that lasted into the Revolution were regularization of the École des ponts et chaussées and enlargement of the corps of civil engineers trained there. An edict of 19 February 1775 accorded the school the designation "royal." The recognition might be thought overdue. The corps had long since been discharging its responsibilities in a manner to justify the hopes of Daniel Trudaine, who had made it his favorite service among many charges. Beginning in 1730, Trudaine had striven to improve roads in the almost trackless reaches of Auvergne, where he was intendant at the beginning of his career. Promoted to Paris and an intendancy of finance in 1734, and named to the Council of State in 1744, Trudaine welcomed assignment of the detail of roads and bridges to his office. Throughout the entire period of his administration of the Bureau of Commerce, he kept the significance of transport prominently in view in governmental circles concerned with trade, having delegated direct oversight of the system of highways and waterways to Jean-Rodolphe Perronet.[1]

A happier choice would be difficult to imagine. Perronet devoted himself to the corps and to its school from the moment of his appointment in

[1] The regulations approved by Turgot are printed in F. de Dartein (1906), 107-123, under the title "Instruction concernant la direction . . . des Ponts et Chaussées." For details on the administrative history of the Ponts et chaussées, see Vignon (1862) and also Petot (1958), the latter with a good bibliography. An evocative popular highway history is Héron de de Villefosse (1975). Imberdis (1967) is a monograph on the roads of Auvergne. The archives of the Ecole nationale des Ponts et Chaussées are maintained in its library, rue des Saints-Pères, and are rich in documents exhibiting the mode of operation of the school and of the corps in the eighteenth century. The *Catalogue des manuscrits de la Bibliothèque des Ponts et Chaussées* (1886) is supplemented by a recent, though undated inventory to be consulted in typescript. This collection will be referred to as BPC. A report of 27 brumaire an IV (18 November 1795) is particularly illuminating (carton 2629 bis, #657 [9]). It consists of a memoir on the history and regime of the school prepared at the request of the minister of the interior.

1747 until his death in 1794 at the age of eighty-six. His title at court, "premier ingénieur du roi," was symmetrical with those of the heads of other skilled services, first physician, for example, and first surgeon. Having begun a career in military engineering, Perronet had been forced by lack of means into architecture and civil construction. He was an accomplished bridge builder and the originator of a design for arches wherein the curvature was composed of the segments of several circles. In the Pont de Neuilly, which he finished in 1774, the effect was so pleasing that Hubert Robert, the architectural painter, chose the opening ceremony for the subject of a canvas. One of Perronet's bridges is still standing in Paris, the pont de la Concorde, originally the pont Louis XVI.[2] In raising the standards and enhancing the standing of civil engineering, Perronet was sustained by a succession of intendants in the Ministry of Finance: Trudaine de Montigny after the death of his father in 1769, Jules-François de Cotte from 1778 until 1780, and finally Antoine Chaumont de La Millière until the Revolution. In 1777 Necker pressed Perronet to take the post himself, but the old engineer demurred on the grounds that the technical director must not also be the magistrate who judged of the corps and, if satisfied, defended its interests politically.[3]

The modern approach to educating engineers may be traced right back to the opening of collaboration between Perronet and Trudaine. An edict of 11 December 1747, issued with the authority of Machault in the office of controller-general, provided that young men entering the Ponts et chaussées were to pass through a period of training in the drafting and cartographic section of the Finance Ministry, the Bureau des géographes et des dessinateurs.[4] There, rough and unready roadbuilders—already called "ingénieurs" though not yet "civil"—would be furnished with competence to draw maps, plan roads, and draft architectural elevations. Perronet expanded this bureaucratic scheme for technical instruction into an institution commonly called a school by the late 1750s, long before Turgot legitimized the usage. The respect it won carried the establishment through a wartime administrative and budgetary crisis in the early

[2] Perronet published a treatise of bridge building in the form of an account of his own major constructions, *Description des projets et de la construction des ponts de Neuilly, de Mantes, d'Orléans* . . . , 3 vols. (1782-1789). Otto Mayr wrote a notice for the DSB *10*, 527-528. There is a longer notice in Dartein (1906). The principal éloge is Gaspard Riche de Prony, *Notice historique sur Jean-Rodolphe Perronet* (1829). There are lists and drafts of various writings on administrative, professional, and technical matters in BPC, carton 1021, no. 408; 1026, no. 397; 2074, no. 394; and 2081, no. 395. [3] Petot (1958), 158.

[4] *Ibid.*, 139-144. See also Gaston Serbos, "L'Ecole Royale des Ponts et Chaussées," in Taton (1964), 345-363. An original draft of the scheme for training engineers is preserved in BPC: "Mémoire sur les moyens de former des sujets propres à occuper différens emplois des Ponts et Chaussées." It is undated, but evidently of 1747. Carton 2629 bis, no. 657 (1).

1760s, when enrollment fell from thirty-odd to a low of sixteen. After the Peace of Paris in 1763, the number climbed steeply to a high of 110. Success created an over-supply of engineers, which it was one of the purposes of the Turgot reform to correct, partly by absorbing a portion into an expanded corps and partly by limiting to twenty the size of each of the three classes within the school, while also raising the requirements.[5]

Candidates were subject to the discipline of the school even while awaiting vacancies in the third and lowest class, and were then called "aspirants" or "supernumeraries." These hopefuls came from bourgeois families more likely to be provincial than Parisian. In principle, they were not to be drawn from the ranks of craftsmen and artisans "among whom education and feeling are rarely to be encountered, qualities absolutely essential in the personnel of the Ponts et chaussées."[6] Low rather than middling background was never an insuperable handicap, however, since genealogical certificates were not required, as they were for admission to the military schools. Instead, candidates had to present a letter of recommendation from an appropriate sponsor who might be a local worthy, a high nobleman, an ecclesiastic, a member of the Academy of Science, or a member of the corps itself. The last source proved to be the most reliable. Engineers in the field, animated by a zeal natural to men proud of their service, kept an eye out for talented youths to recruit into the corps. Surest of admission were sons of their own colleagues. In all other cases, fathers had to furnish evidence of ability to provide the prospective student with 600 livres a year for his maintenance in the capital.[7] Many parents also incurred the prior expense of sending their boys to a boarding school kept by a well-known architect, Jean-François Blondel, whose establishment was one of several preparing teen-age boys for entrance to Ponts et chaussées, Mézières, and sundry military and naval schools with technical requirements.[8]

Organization of the Ecole des ponts et chaussées into three classes derived from gradations in the old Bureau des dessinateurs. Distinctions of skill among the draftsmen employed there had simply been carried over into criteria for marking stages of education. The beginners' class, the third, was concerned with geometry, trigonometry, elementary surveying, and mapmaking. Subjects to be mastered in the second class were mechanics, hydraulics, advanced surveying, conic sections, theory of curvilinear surfaces, stereotomy, and strength of materials. Finally, students in the top class were occupied with design, architecture, and construc-

[5] "Instruction . . . ," Articles 1 & 2, Dartein (1906), 107.

[6] Petot (1958), 149, quoting a pamphlet of 1777, *De l'Importance et de la nécessité des chemins publics . . . avec un précis historique de l'état actuel des ingénieurs des ponts et chaussées.*

[7] BPC, carton 2154, no. 683, contains a folder of parental certificates of support.

[8] Serbos, in Taton (1964), 356.

tion, and spent much of their time in the field on practical projects. The arrangement was different from that of a modern school, in which a new class would have entered annually and progressed to graduation in three years time. Although standards were high, people might take as long to meet them as their courage lasted. The majority never did. Of 387 students admitted between 1769 and 1788, only 141 qualified for commissions. Many of the remaining 246 failed of promotion to the first class or even to the second. Those who flagged often found employment with private architects or builders, and sometimes on a hired basis with the corps itself, or else with the military engineers or with local authorities in the pays d'états, the provinces that conserved their regional assemblies.[9]

A quality of naturalness about the evolution of the Ecole des ponts et chausées makes its history a revealing instance of the transition from a system of training by example and emulation toward provision of professional schooling. The process was not quite completed at the end of the old regime, when certain features of a modern educational institution were still missing or rudimentary. The most startling of these lacunae might be taken as evidence that an organized body can function well in the absence of structures later deemed essential, for the Ecole des ponts et chaussées had no faculty. After 1763, Perronet turned over day-to-day direction to an associate, Antoine Chézy. Two additional commissioned engineers served on the staff, one as inspector of discipline and the other of the learning process. The latter did have a small teaching responsibility for engineering drawing and mapmaking. Otherwise, the facilities for instruction that he coordinated were provided both inside the school and outside of it.

Internally, the subjects central to the curriculum, namely mathematics, elementary mechanics, stereotomy, and strength of materials, were taught by the three leading students in each class. Like the tripartite division into classes, that practice of monitoring represented simply a continuation, albeit at a rising level of technical sophistication, of the training of less advanced draftsmen by more accomplished geographers in the Bureau des dessinateurs of the 1740s and earlier. The student-professors were paid a small stipend, and given academic credits toward promotion, while also improving their prospects for desirable assignments on being commissioned. For subjects of broader scientific and educational import, all students had to seek external instruction. They followed courses approved by the school in higher mathematics, rational mechanics, hydrodynamics, chemistry, mineralogy, natural history, architecture, and drawing. Among them were the public offerings at the Jardin du roi, the Collège de France, and the Academy of Architecture. Other, private lec-

[9] Petot (1958), 152-153.

ture series were open by subscription, and small scholarships were available to help the needier students pay the fees for a course such as Fourcroy's on chemistry.[10]

The method of evaluating the students' progress may be taken as an index to the progress of the school itself in its evolution toward professional status. In the early years, success in developing the several skills was rewarded by prizes. In the modern technical schools, which emerged from this background during the Revolution, merit was measured and talent defined by competitive examinations. An early form of the latter system was superimposed on the former just at the time of Turgot's codification of the curriculum, which thereupon exhibited elements of both until the significance of the prizes dwindled to the vestigial. That reform may, therefore, be taken to mark the transition. Every year in April a competition was set. School then recessed for six months, and students went off to the provinces for practical work in the field. Awards were announced in the following March. Students thus had the better part of a year to prepare their entries. Subjects were as follows (the meaning of the numbers in the first column will be explained below):

	Degrees or Points	*Value in Livres*	
MATHEMATICS			
Mechanics, hydraulics, differential and integral calculus			
1st Prize	40	200	
2nd Prize	35	100	
Algebra and conic sections			
1st Prize	30	180	
2nd Prize	29	90	
Elements of geometry			
1st Prize	28	150	
2nd Prize	27	75	
			795

[10] These arrangements are outlined in the "Instruction . . ." of 1775, Articles 1-25, Dartein (1906), 107-111. See also BPC, carton 2629 bis, no. 657 (9), "Rapport relatif . . . à l'Organisation de l'École des Ponts et Chaussées . . . le 27 brumaire an 4e [18 November 1795], l'ère époque." The opening section is a contemporary history of the school from its foundation through 1788.

	Degrees or Points	*Value in Livres*	
ARCHITECTURE			
Architecture of bridges			
Complete project for a bridge in stone, or in wood, with arches, drawn to scale on the profile of a river, together with specifications and estimates in detail on the basis of given costs for materials			
1st Prize	26	200	
2nd Prize	25	100	
Architecture of docks, jetties, locks, dikes, or canals—as above			
1st Prize	24	180	
2nd Prize	23	90	
			570
Architecture of a civil building			
1st Prize	22	150	
2nd Prize	21	75	
			225
Stonecutting (Stereotomy)			
1st Prize	20	100	
2nd Prize	19	50	
			150
STYLE AND METHOD			
Essay on a subject to be announced			
1st Prize	18	100	
2nd Prize	17	50	
			150
Drafting geographical and topographical charts			
1st Prize	16	70	
2nd Prize	15	35	
			105

	Degrees or Points	*Value in Livres*	
Theory and practice of leveling and volumetric calculations applied to earthworks			
1st Prize	14	50	
2nd Prize	13	25	
			75
Estimating work-tasks in construction of buildings, in accordance with usage in Paris			
1st Prize	12	40	
2nd Prize	11	20	
			60
DRAWING			
Geographical and topographical maps			
1st Prize	10	70	
2nd Prize	9	30	
Ornamentation and figures			
1st Prize	8	60	
2nd Prize	7	30	
Landscapes			
1st Prize	6	40	
2nd Prize	5	20	
			250
WRITING			
Block lettering			
1st Prize	4	40	
2nd Prize	3	20	
Penmanship			
1st Prize	2	40	
2nd Prize	1	20	
			120

Total annually 2,500.

Clearly, the divisions are arranged in descending order of importance and the subjects within each division in descending order of difficulty. Students might present entries in these competitions in any year in which they felt prepared. They were strongly encouraged to proceed as rapidly as possible through the entire gamut. The governing council of the corps, the Assemblée des ponts et chaussées (of which more in a moment) judged all the projects, coopting members of the Academies of Science, of Architecture, and of Painting and Sculpture to sit on the appropriate juries. The controller-general himself often presided at the formal session in which awards were announced or, in his absence, the intendant of finance. Prizes were never given in cash, but in credits toward the purchase of books or instruments. Before 1775 and the Turgot regulations, the efforts of entrants who failed to win as much as an honorable mention were noted with verbal evaluations of the type still familiar in French educational practice—"*sait bien*," "*assez bien*," "*entend peu*," and so on down. Promotion then depended partly on performance, partly on the time spent in school, and partly on summer field work and occasional special missions.[11]

A principal object of the new regulations was to stiffen these easygoing, personal procedures. Alongside the prizes, which became ornamental, a numerical system for evaluating progress was instituted in order to afford a criterion for promotion entirely on the basis of attainments and without regard to seniority or other personal considerations. That is the purpose of the points or degrees to be assigned to the relative merit of a student's performance of every exercise. Points were credited for the two honorable mentions accorded in each contest. Below that, every student whose project was accepted for the competition received one fourth the number of points assigned to the first prize. Students also earned credits by successful completion of both internal and external courses of instruction, and in more generous measure for their field work from April until October.

At the opening, mid-point, and conclusion of each term of residence, the grades amassed by each student would be tallied and the ranking established. The three most accomplished in each class, those chosen to be professors, took precedence in the order of the leaders, respectively, of the first, second, and third classes, followed by the second and third of the first, of the second, and of the third. Only these nine students had the right to wear the uniform of the corps while still in school. The top trio in the first class also had first call on vacancies in the corps of commis-

[11] "Instruction . . . ," Article 27, Dartein (1906), 112-113; cf. Serbos in Taton (1964), 361-362, and the Report to the minister of the interior, BPC, carton 2629 bis, no. 675 (9).

sioned engineers, to be followed by their classmates in the order of their ranking. The size of each class being restricted to twenty, promotion also awaited the creation of vacancies, again in the strict order of grades.[12] One effect was to increase the disparity between hares and tortoises. The ablest and most energetic completed the requirements in two and a half years. A span of seven or eight was the mean. In 1777 the oldest student was forty-one and the youngest seventeen.[13] Certainly this reform offers one of the more striking examples of the tendency in the Turgot entourage toward basing responsibility on merit where merit meant technical competence. All that remained to convert the system into the competitive recruitment of elites was to eliminate the prizes and to call the contest an examination, while all that remained to convert the Ecole des ponts et chaussées into a fully professional school was to provide it with a faculty.

Educational institutions priding themselves on school spirit have often fostered esprit-de-corps by instigating the neophytes to police each other, in regard both to academic probity and genteel behavior. Students at Ponts et chaussées were expected to exercise influence on any of their number whose conduct tended to violate standards of decency and respectability, and if offending parties proved obdurate or vicious, to report their derelictions to the director.[14] Evidently, these expectations were fulfilled—at least in the case of unpopular comrades. In 1758, a petition subscribed by a right-thinking group requested Perronet to expel seven of the company whose amorous adventures and taste for low company had disgraced the school: "the Sieur H . . . (they reported of one of these rotters), is the son of a well-known butcher in Paris. If his talent and his education compensated for the defect of his birth, the corps would feel respect for the confidence with which his patrons honored him. We are convinced there is nothing along these lines to redeem him."[15] By the regulation of 1775, a student's preparation of projects for the prize competition, or examination as it was becoming, must be carried out in the presence of a number of his fellows, who were to be in a position to certify that the work was really his.[16] The term "honor system" later adopted for similar procedures in character-forming institutions in Britain and America may seem an Anglo-Saxon euphemism, since there too enforcement depends on substituting delation for supervision.

After a series of temporary locations, all in the Marais, the École des

[12] For the system of grading and promotion, "Instruction . . . ," Articles 31-50, Dartein (1906), 113-117. A notebook survives containing a record of mathematical problems set for students between 1778 and 1790, BPC, carton 115, no. 4.

[13] Petot (1958), 152.

[14] "Instruction . . . ," Article 57, Dartein (1906), 118.

[15] Petot (1958), 154, quoting a letter of 23 March 1758.

[16] "Instruction . . . ," Article 26, Dartein (1906), 111-112.

ponts et chaussées occupied a town house in the rue de la Perle at the corner of the rue Thorigny from 1771 until 1788, when it moved to larger quarters in the rue Saint-Lazare. Its prestige increased steadily in the 1780s, so much so that foreign governments began to send fledgling engineers to Paris. Indeed, letters of foreign students and their patrons are among the more informative documents, conveying detail that the French took for granted. In 1787 four young Neapolitan officers were enrolled, Messrs. Dillone, Tironi, Pichichelli, and Constanzio, together with a Swede, one Liungberg, who had special permission to observe the construction of Perronet's Pont Louis XVI.[17] In 1790 the electoral prince-bishop of Trier was supporting the studies of an officer of his artillery called Kirn.[18] Somewhat earlier, there was even a project for planting a military offshot of the École des ponts et chaussées in Cadiz, the genesis of which is worth recalling.

In 1778 a graduate and engineer of fortune called Groüillier took service with the Spanish government as a mining consultant in Guadalcanal. The local authorities failed to understand that an engineer could not be expected to correct the "caprices" of nature, and matters went badly.[19] Perronet thereupon interceded with the duc de Crillon, who was much in the service of the Spanish crown, in order to arrange a commission for his protégé in the military engineers during the American war.[20] Groüillier justified his patron's confidence, and brought credit on his training, by engineering the construction of a highway through the difficult Sierra Morena north of the Guadalquivir, whereupon the funds ran out. Posted to Mahón in the island of Minorca, he occupied himself in idle moments with a scale drawing of the port in perspective, and sent it along to the responsible minister in Madrid, who thought it might please Charles III. In fact the king was delighted, sent for Groüillier, and charged him with the task of rendering all the ports of Spain in comparable exactness and detail. His draftsmanship in the course of executing that commission brought him the esteem of the chief minister, Floridablanca, the architect of Spain's enlightened despotism, the same who had thought of founding an Academy of Science, and to whom Marat was recommended to be its permanent secretary.

Others among the great wished to see Groüillier's attainments emulated. Drawings of the port of Cadiz were to be the chef-d'oeuvre of his series, and there in 1783 he came under the eye of the Count O'Reilly,

[17] BPC, carton 2636, no. 659 (18), marquis de Circello to Perronet, 5 October 1787; Baron de Staël de Holstein to Perronet, 14 July 1787.

[18] BPC, carton 2636, no. 659 (18), Baron de Dominique to Perronet, 17 April 1790.

[19] Groüillier to Perronet, undated but written in August or September 1783, BPC, carton 2636, no. 659 (18).

[20] Duc de Crillon to Perronet, 9 October 1778, BPC, *loc. cit.*

governor of the province and inspector-general of infantry. Observing what could be accomplished with compass, ruler, and transit, O'Reilly conceived the notion of starting an academy at Cadiz, or to be more exact in its Port Santa Maria, where mathematics, fortification, and design might be imparted to military officers, and he bespoke Groüillier's good offices to send back to Perronet in Paris requesting that a professor be named from the Corps des ponts et chaussées. With Spanish largesse, the governor left the salary and conditions to Perronet's discretion.[21] Even while assuring O'Reilly that he could place full confidence in Perronet's judgment and pride in the quality, both technical and human, of the corps, Groüillier took leave to caution his former mentor that the nominee must be of "gentle character and not abrasive (non rampant), but fit to gain the confidence and friendship of all those with whom he would be dealing."[22] O'Reilly need not have worried, for Perronet's choice had fallen on one well launched in a career, Henry Bouchon de Bournial, thirty-four-year-old inspector stationed in Paris.[23]

The exchange of correspondence brings out the elements of paternalism amounting almost to domesticity, and of courtesy tempered by realism, that governed in the life of the Ponts et chaussées. Bouchon is gratified that Perronet should think of him. It is a further mark of the benevolence that has been shown him since his student days, beginning in 1767. He will do his utmost not to disappoint his patron or to bring anything but credit to the corps. The opportunity is a fine one, and of course he will accept. At the same time (and he apologizes for intruding personal considerations) the prospect of leaving Paris is anything but welcome to his family and himself—better to enjoy an obscure life in the capital than to shine in exile. There is, moreover, the worry whether the Spanish authorities can be counted on to make good their commitments. He relies on Perronet to impress on Count O'Reilly the obligation to assure the security of his employment and the payment of a pension when the time to retire should arrive. For then he would return to France.[24]

Mindful of the risk and sacrifice, Perronet informed O'Reilly that an annual salary of 3,600 livres would be equitable (half again what Bouchon was receiving in France, though Perronet did not say so), and that an additional 2,400 would be required for travel expenses and the purchase of books and instruments.[25] With the courtliness of men in high places, O'Reilly agreed and was delighted to learn in January 1784 that Bouchon

[21] Groüillier to Perronet, [August or September 1783], BPC, carton 2636, no. 659 (18).

[22] Groüillier to Perronet, 17 November 1783, BPC, *loc. cit*.

[23] Perronet to Groüillier, 14 October 1783, BPC, *loc. cit*.

[24] Bouchon to Perronet, 5 October 1783, BPC, *loc. cit*.

[25] Perronet to Groüillier, 14 October 1783, BPC, *loc. cit*.

had duly equipped himself and was on the point of sailing from le Havre.[26] We do not know how he fared, but he must have settled in to life in Cadiz, at least for a time, since his name does not appear on a roster of the corps for 1786.

That roster gives ages, dates of commission, assignments, and places of residence of the personnel.[27] Fortunately the archives in which it is preserved contain sufficient documentation to permit an account of the scale and mode of operation of the Corps des ponts et chaussées in the last decade of the old regime.[28] Its ranks consisted of the three grades of engineer, inspector, and second-engineer (or "*sous-ingénieur*"). Since their commissions were issued by the controller-general, the responsibility of the corps extended to the regions administered directly by the Ministry of Finance.[29] Throughout those provinces, the pays d'élection, amounting together with certain special jurisdictions to some seventy percent of the area of France, Perronet was assisted in his direction of the Ponts et chaussées by a staff of four inspectors-general. Each was set over a group of generalities comprising about a quarter of the total territory beyond the generality of Paris, wherein Perronet retained immediate responsibility.[30]

In principle, an engineer of the first grade—sometimes called "ingénieur-en-chef"—was assigned to a single generality and served on the staff of the intendant, over whose signature proposals were originated, contracts let, and invoices paid. The legislation codified in 1775 contemplated 25 of these places, a provision that by 1786 had been increased to 32 by the addition of supernumeraries for special purposes, four of them serving in the generality of Paris. The second grade of inspector numbered 67 persons in 1786, the official quota of 50 having also been stretched by ad hoc arrangements. In age these field engineers ranged from thirty-two to sixty-nine. Two were in their sixties, 18 in their fifties, 24 in their

[26] Comte d'Oreilly to Perronet, 23 January 1784, BPC, *loc. cit.*

[27] "Etat général des ingénieurs, inspecteurs, et sous-ingénieurs au 1er juillet 1786," BPC, carton 1342, no. 660.

[28] The archives are divided into three main chronological divisions for the period down into the nineteenth century: 1. the period of Perronet's direction, 1747-1794; 2. from 1749 until 1815, when Lamblardie, Sganzin, and others headed the schoool; 3. the period of Prony's direction from the end of the Napoleonic regime until 1839.

[29] In Alsace and Corsica, highways like other facilities depended on the War Department, while in the more important pays d'état, the resurgent estates preferred autonomy to provision of services, however skilled, from Paris. Only in the wealthiest of the great provinces, Burgundy and Languedoc, did the quality of the roads come up to the national standard, for the authorities in Dijon and Montpellier followed the practice, even like foreign princes, of subsidizing the training of young engineers at the Ecole des ponts et chaussées and bringing them home to work for local government. On these provincial Ponts et chaussées, see Petot (1958), 269-318.

[30] A minute headed "Tournées de Mrs les Inspecteurs-Généraux" gives the distribution of territories in 1775. BPC, carton 2636, no. 659 (19).

forties, and 23 in their thirties. As for the second-engineers, there were 112 in the service in 1786, the youngest being twenty-four and the Nestor, Laguette père of La Réole in the generality of Bordeaux, being sixty-four. This was a young group for the rest, with only one in his fifties, nine in their forties, 53 in their thirties, and 48 in their twenties. The duties of the two lower ranks appear to have been more or less interchangeable. Thus, each chief engineer in a generality disposed of the services of six or seven subordinates, who would be aided by students from the school for half the year. In addition, Chézy (the associate director of the school) headed a group of five responsible to the Treasury for maintenance of the streets in Paris, and a further detachment of eight was charged with oversight of dikes and levées in the Loire Valley, four at Nevers, two at Tours, and one each at Orléans and Châteauroux.

All told, the corps consisted of 230 commissioned engineers. Before 1777 only three or four a year had been taken in; after that the average was nine, except for 1784 when for some reason 18 were added. Prospects for promotion from second-engineer to inspector were also improving. From 1781 through 1786, it was normal to have spent about 12 years in the lowest rank; before that, the expectation had been 16 years. After 1775 the merit system introduced into the school pursued its graduates into their careers. Alongside the formal ranking, members of the corps were graded on their performance of particular tasks by their superiors, and distributed independently of seniority into three classes according to evidence of ability. A second-engineer, therefore, might have a higher score than an inspector, and an inspector might be above a chief engineer. The ordering in class soon surpassed the formal ranking in real importance, if not in dignity, for it became the basis of choosing people for the more desirable assignments, those more interesting in themselves and also in the opportunities they afforded for extra-curricular employment.[31]

As was true of most governmental offices, the stated salaries were inadequate despite a system of supplements to meet the cost of lodging and expenses. Soon after becoming controller-general, Calonne observed that the second-engineers did not earn enough to keep a horse for riding around their districts.[32] Their stipends began at 1,200 livres and rose with length of service to 1,500 and then to 1,800. Inspectors drew 2,400, chief engineers 3,000, the inspectors-general 6,000, and Perronet 8,000. Members of the corps augmented these incomes by accepting chances to oversee constructions other than those paid for out of the budget of the Ponts et chaussées—barracks, public buildings, parks, canals, locks, and dikes. It was understood that they should not ask fees equivalent to what

[31] "Instruction . . . ," Articles 63-67, Dartein (1906), 119-120.

[32] "Calonne à MM les Intendants," 27 January 1784. BPC, carton 2636 no. 659 (18).

could be earned in the private sector. Custom in the building trades allowed architects a commission equal to 10 percent of the total expense for small jobs and 5 percent when the cost exceeded 100,000 livres, although leading members of the profession could command 10 percent even for major contracts. The practice was much what it is nowadays: a retainer of one-third payable on completion of the drawings, another third for oversight of the construction, and the final third when the structure was accepted by the client. Chief engineers of the Ponts et chaussées, being already in the public service, normally received fees at half that rate, retaining two-thirds and passing along one-third to the subordinate who actually ran the job.[33]

Formally, directives on projects to be undertaken at public expense issued from the authority of Trudaine and his successors in the Ministry of Finance, who transmitted them to the intendant of the generality affected. The chief engineer then saw to their execution. Practically, however, most proposals originated out in the provinces on the initiative of officials on the scene, who had to submit their ideas to the scrutiny of a central committee meeting weekly in Paris. Trudaine had begun convoking the chief persons of the corps at the time of its reorganization in 1747. They would foregather at his residence on Sundays and stay on to dine after finishing their deliberations. As precedents built up, these consultations received the official-sounding designation of "Assemblées des ponts et chaussées," although their legal basis continued to be simply the intendant's pleasure. Perronet kept a record of acts and attendance in a register and presided on occasions when Trudaine was absent. Others normally present were the four inspectors-general, the head of the section of dikes and levées, and the chief engineer of the generality to be affected by proposed improvements. Trudaine also invited appropriate members of the Academy of Science to attend when it was a question of deciding on the adoption of some invention or technique involving principles of hydrodynamics, cartography, or calculation. Mignot de Montigny was a regular participant, since he exercised the responsibility in the Treasury for the funds expended by Chézy in maintenance of the streets in Paris. Two or three advanced students from the school were also admitted each week, in order that they might observe how the proposals they would one day be drafting were evaluated.

Meetings of the Assemblée des ponts et chaussées were working sessions with none of the formality of academic proceedings. The chief engineers would submit detailed plans and specifications for all new constructions. Only rarely did a project get through without modification. The assembly

[33] "Mémoire sur les honoraires qui sont dûs aux Ingénieurs des Ponts et Chaussées pour les travaux étrangères à leurs services, remis à Mr de Cotte," 11 June 1780; BPC, carton 2636 no. 659 (15).

would normally require plans to be revised, drawings to be redrafted, and estimates to be reviewed. Occasionally it would reject a proposed route and substitute another. It frequently had to override the resistance of property-owners to condemnation proceedings or to withstand the pressure that a great landowner was bringing in favor of constructing a road or bridge that would serve his private interests rather than the public. Nothing was more likely to arouse local passions than the routing of highways and canals. Although Trudaine made clear that he was not bound to follow the advice of the assembly, in fact he always did so. In practice, therefore, its recommendations had the force of governmental directives.

In addition to this quasi-political and technical control over the location and design of work in the field, the assembly was also the examining body for the school. It judged the entries in the competitions for prizes and graded the performance of the students. Its authority over persons followed the engineers into their careers, for it had an important advisory voice in decisions about assignment and promotion. In appearance, those invited to sit in the assembly were no more than the inner circle of the corps. In actuality they constituted its governing council.[34] Indeed, the Assemblée des ponts et chaussées affords a significant instance of a proto-professional occupation evolving its own mode of self-regulation, under the aegis of government and with the justification of the public interest. As a consequence, complaints cropped up in the 1780s to the effect that the corps, like other services, suffered from excessive centralization, and that Perronet would do well to delegate a measure of decision-making power to local subordinates.[35]

Although he never heeded these sentiments, morale and discipline were excellent. Other suggestions and proposals that came in from the field bespeak a healthy interest in the good of the service and a freedom in approaching Perronet. Once in a while people had to be reminded of what was expected. In 1773 Trudaine de Montigny ordered the dismissal of a second-engineer who had scandalized the audience in a respectable theater by rowdy behavior while in uniform. He instructed Perronet to address a circular to the corps enjoining decent behavior while on leave, and restricting the grounds for furlough.[36] In August 1778 Cotte (at Perronet's

[34] The procedures are described in a memoir "L'Administration des ponts et chaussées," of 1 July 1777 drafted (probably by Perronet) for the information of Necker on the latter's appointment to direct the Ministry of Finance (BPC, carton 2636, no. 659 [22]). See also "Observations de l'Assemblée des Ponts et Chaussées . . . ," 26 May 1791, BPC, carton 1342, no. 660; "Instruction . . . ," Article 59, Dartein (1906), 119; Petot (1958), 158-162.

[35] "Examen fait par le Sr Perronet du mémoire de M. Lamandé . . . ," 19 April 1785. BPC, carton 2636, no. 659 (16). Perronet's rejoinder is an admirably clear summary of the administrative functioning of the corps.

[36] "Circulaire aux ingénieurs," 21 May 1773, BPC, carton 2636, no. 659 (12).

request) chided the chief-engineers for tardiness in their reports, and directed them to assemble their subordinates into conclave every November without fail. Even more seriously, Perronet observed that inspectors and second-engineers had been absenting themselves from the job and delegating traverses and the determination of graded slopes to underlings. Lest these abuses continue, younger personnel must be required to reside in the localities where they worked, whatever their predilection for life in a city.[37] Later, in 1785, La Millière in the ministry was shocked to be told that certain inspectors, second-engineers, and students on their field training were taking private commissions and even borrowing money from contractors whose constructions they were supervising. Perronet was to investigate whether such dereliction really had occurred, and to remind the corps that its tradition was unbending in regard to avoidance of conflicting interests.[38]

Such were the admonitions provoked by occasional slippage in a service much respected by the public and by itself. An apologia of 1791, drafted to educate civil officials come to power in the Revolution, conveys a typical engineer's own sense of the inwardness of his duties. It came from the pen of a seasoned member of the corps in Caen called Didier.[39] In the course of a normal enterprise, the engineer in charge would incur responsibilities in four domains: administration, construction, accountability, and arbitration. Before enlarging on construction, the heart of the matter, let us indicate briefly the nature of the paperwork and negotiation that came his way under the subsidiary headings. In deciding whether to propose a project at all, the bureaucrats who came and went needed the advice of the engineer settled in the country, not only for technical reasons but for his knowledge of the region, of its web of local influence and interest, its pattern of communication, and its contractors with their special skills and reputations. When authorization to proceed then came down from Paris, the engineer had to verify and approve the specifications and estimates—the *devis*—before contracts could be let. As the work progressed, invoices were paid only over his signature. Complications might naturally be encountered no matter how carefully the costs had been calculated and the plans drawn. No one could foresee what obstacle to a foundation or a pier might be encountered thirty feet under ground or at the bottom of a river. In such cases, only the trained engineer was in a position to pass on the legitimacy of the contractor's claim for adjustment in the estimates. When disputes arose between local authorities and contractors, between

37 "Circulaire écrite par Mr de Cotte aux Ingénieurs des Ponts et Chaussées," 22 August 1778, BPC, carton 2636, no. 659 (11).

38 La Millière to Perronet, 9 August 1785, BPC, carton 2636, no. 659 (17).

39 Didier to Perronet, 16 February 1791, BPC, carton 2630, no. 658 (8). The memoir is a 16-folio memoir, very carefully thought out and expressed.

contractors and property owners, or between property-owners and the administration, and litigation ensued, the court would call on the engineer for expert testimony.

Didier speaks of the exercise of engineering skills in actual construction as "the art," consisting, first, of the creation of roads and bridges and their maintenance and, second, of all other types of building (*travaux d'art proprement dites*. In his view, the felicitous routing of a highway, involving as it did, topographical, political, and commercial considerations, presupposed a faculty almost more intuitive than analytic, to be met with only in persons whose technical expertise had been tempered by civic experience. Once roads were projected on the landscape and traced out upon the map, it was for the engineers in the field to run surveys, draft plans in detail, determine altitudes, measure for embankments, grade slopes, balance volumes of earth between cuts and fills, assess property for the indemnification of owners, keep track of costs, and oversee the progress of construction. Main roads were built employing a technique perfected by Pierre Trésaguet, who was chief engineer in the Limousin during Turgot's intendancy and an early graduate of the Ponts et chaussées, and whom Turgot brought to Paris to become inspector-general. The construction was much the same as the type named after MacAdam in Britain. A road bed was hollowed out and roughly lined with large stones. A second layer consisted of broken stones the size of walnuts, packed and fitted into place by hand to make a slightly convex surface. The depth was twelve to fifteen inches along the center line. In villages and stretches liable to flooding, paving was sometimes superimposed and the stones set on a six-inch bed of sand over the gravel.[40]

The good new roads of France were the wonder of foreign travelers. En route to Narbonne, on 23 July 1787 Arthur Young writes:

> The roads here are stupendous works. I passed a hill, cut through to ease a descent, that was all in the solid rock, and cost 90,000 livres (£3937), yet it extends but a few hundred yards. Three leagues and a half from St Jean to Narbonne cost 1,800,000 livres (£78,750). These ways are superb even to a folly. Enormous sums have been spent to level even gentle slopes. The causeways are raised and walled on each side, forming one solid mass of artificial road, carried across the valleys to the height of six, seven, or eight feet, and never less than fifty wide. There is a bridge of a single arch, and a causeway to it, truly magnificent. We have not an idea of what such a road is in England.[41]

[40] Petot (1958), 321-322; Héron de Villefosse (1975), 161.

[41] *Travels in France and Italy during the years 1787, 1788, and 1789* (Everyman ed., London, 1927), 39.

An engineer naturally took pride in such a feat, and liked to place obelisks or other markers at crossroads and special terrain features, along with crucifixes erected by the pious and pieces of sculpture representing some legend or personnage of the region. However human such touches, Didier sternly wished it remembered that, while imagination and fancy were all very well, maintenance was the touchstone of an engineer's professional reliability. His sense of order and devotion to the service would appear in the regularity of his regime of inspection and repair: in his keeping an eye on bridges, culverts, aqueducts, and streams; in his recommending corrective action in timely fashion; in his exercising vigilance over contractors and their charges; in his reporting local business enterprises to the authorities when their operations caused excessive wear and tear or otherwise abused the road system.[42] The engineer's work thus consisted of designing, drafting, and verifying. We are not to picture him standing over the bent backs of peasant laborers hacking at the ground with pick and shovel, nor herding the women with their pack-baskets of crushed stone or sand from gravel pit to wagon. In some regions the local authorities supplied the contractor with forced labor raised by the corvée. In others the roadside parishes had commuted that ancient servitude into a form of taxation, and the contract stipulated that the builder hire his own labor. Turgot had suspended the corvée under the six edicts, and its restoration some eighteen months later involved complexities onerous to all concerned. Such was the confusion that in 1777 Perronet requested Necker to appoint a fifth inspector-general in the corps whose main duties would be coordinating the supply of labor by local authorities.[43] At all events, that was never the responsibility of the engineer charged with overseeing the job. Like the architect of a building, he would visit the site no more than two or three times a week.

Even so, he and his colleagues were stretched thin. If evenly distributed (and they were not), two or three inspectors or second-engineers would have been allocated to an area equal to that of a modern department, a region the size, say, of Rhode Island or Delaware. Since the architecture and oversight of public works "*proprement dites*"—bridges, canals, civil buildings, and commercial harbor works—also fell to them, their days were fully occupied. Thus pressed, the engineers had to take on the services of subordinates, of whom the two categories were called *conducteurs* and *piqueurs*. Gradually, these employees acquired the collective identity of a class, although to specify the juncture at which that occurred is probably impossible. Their functions were the product of necessity rather than of a settled act of policy, like Trudaine's decree of 1747 providing formal

[42] Didier to Perronet, 16 February 1791, BPC, carton 2630, no. 658 (8).

[43] Perronet to Necker, 22 July 1777, BPC, carton 2636, no. 659 (19).

training for the commissioned members of the corps. The authorities kept regretting the necessity, moreover, and what can be known of the work of conducteurs and piqueurs must be gathered from fugitive passages in administrative documents deploring excessive recourse to their services. Their place in the corps (it may be surmised) was comparable to that of sergeants and corporals, respectively, in the armed forces, and their relation to the engineers that of non-commissioned to commissioned officers. They were paid by the corps and appointed by the local administration on the recommendation of the district engineer, who might also call for their dismissal.

Conducteurs—"operator" conveys the sense—might substitute for the engineer within their range of competence. They needed to be skilled in arithmetical computations and capable of running traverses, drafting topographical plans and profiles, determining grades, and calculating areas. These operations were to be executed under the eye of the engineer, who was himself supposed to measure the principal bases for a survey and to verify the calculations and closings of triangles. The conducteur should also possess the elements of stereotomy and cartography and write well enough to clerk for the engineer. All this presupposed schooling, obviously. Many among the conducteurs were onetime pupils at the Ponts et chaussées who had failed to complete the requirements. Piqueurs, on the other hand, tended to be former employees of builders and contractors, older workers who had learned their skills on the job. They had to be able to write a little, and to figure and measure well enough to read maps to scale and to place stakes and markers properly. In this and other occupations, the word has the sense of foreman or overseer. They would remain on the job in between inspections by the engineer, seeing to it that the contractor observed instructions.[44]

Perronet sometimes worried about a certain incongruity between the title of inspector and the actual duties of his people in the field. In 1766 he suggested changing the designation to "*Ingénieur avec commission*."[45] The idea was in keeping with realities in other sectors involving the application of knowledge under the aegis of government. We have already

[44] Duties of the conducteurs and piqueurs are outlined in the Didier memoir. Evidently their functions antedated the formation of the school, for the prospectus of 1747 (carton 2629 bis, no. 657 [1]) alludes to their existence in its last paragraph. The author, probably Perronet, says "On ne parle point icy des conducteurs, apparailleurs, chefs d'atteliers, et piqueurs qui sont des emplois trop subalternes mais cependant nécessaires; les employés supérieurs doivent s'occuper à les former. L'émulation dans ces états pourroit donner de bons sujets pour remplacer les entrepreneurs qui doivent avoir su conduire les travaux avant d'estre chargé de leurs entreprises."

[45] Legendre to Perronet, 17 March 1766, BPC, carton 2636, no. 659 (9). Perronet's correspondent was an engineer and is not to be confused with the mathematician, Adrien Legendre.

seen how the gradual assumption of technological leadership by the Corps of Manufacturing Inspectors bespeaks a more positive attitude toward power than the Colbertian assumption that authority consists in verification of regulations and standards.[46] A constant hovering of inspectors buzzing and peering into everything continued to permeate French administration, however, and Perronet's notion of a more spirited and operational job description may have been premature. The École des ponts et chaussées itself had budded out on the stem of the bureaucracy, after all, and the small staff directly responsible continued to inspect the studies instead of conducting them.

Others found nothing wanting in the standing accorded civil engineers, then or later. On the contrary, in 1785 when Roland elaborated a scheme for reform and professionalization of the Corps of Manufacturing Inspectors (which he represented in Amiens), he proposed modeling the school he wished to see created on the École des ponts et chaussées. In his opinion the corps to which he belonged was staffed by amateurs and mediocrities even as the Ponts et chaussées had been earlier in the century. Factory inspectors like civil engineers needed to be educated amid the cultural and scientific resources of the capital although their work, too, would lie out in the provinces. There in Paris they would find the elements they needed—drawing, mathematics, natural history, chemistry, physics, and the arts of reasoning and persuasion.[47] That suggestion came to nothing, even though the Ministry of Finance had moved to institute an École des Mines along virtually identical lines only two years previously. Indeed, opening of the School of Mines in November 1783 must certainly have prompted Roland's initiative.

2. THE ECOLE AND CORPS DES MINES

In mining, too, inspectors were enforcing compliance with regulations well before there was provision for education of engineers who might shape and change things as well as scrutinize them. The practice in the mines administration of calling the officials "inspecteurs" and their juniors "élèves" would seem to cry out for formal schooling, except for the reflection that Ponts et chaussées was the very first school outside the educational structure of the church to achieve more than temporary existence. Even in the Academy of Science the designation for the lowest rank had originally been "élève," becoming "adjoint" in 1719. After the middle of the century, a handful of early mining inspectors got what training

[46] Above, Chapter VI, Section 3.

[47] The Roland proposal, "Projet d'une École des Inspecteurs des Manufactures et de Commerce," is contained in his article "Inspecteur," *Encyclopédie méthodique: Manufactures, arts et métiers, 1* (1785), 68-70.

they received in science and mathematics as supplementary students at the École des ponts et chaussées. Such (it will be recalled) was the case with Jars, who had to be commissioned in the Corps of Manufacturing Inspectors, there being as yet no Corps des Mines but only a miscellany of ad hoc appointments.

The notion of starting a school of mines in France was certainly stimulated if not actually planted by the organization in 1765 of the Bergakademie in Freiberg.[48] High reputation rapidly accrued to this, probably the most famous institution devoted to technical education in the eighteenth century. Apart from promotional ventures, like the Jars missions, originating in the Ministries of Finance and War, the normal administration of mines fell to Bertin and his Petit Ministère during the time of its activity from 1763 until 1780. Bertin was a careful man, sparing of funds. When it was a question of new-fangled training and book-learning about mines, he had to reckon with the skepticism of Antoine Monnet, the best ensconced mineralogist in the service of the government in the 1770s.[49]

Monnet was named inspector-general of mines and mining in 1776. He owed his preferment to the patronage of Malesherbes, and he knew a lot—there is no question about that. He was also an extremely difficult associate. Indeed, it is probably more than mere chance that a number of persons drawn to mining and mineralogy should have presented the scientific community and the government with severe problems of personality. Guettard, Lavoisier's mentor, was notorious for his gruffness and his stiff, out-dated Jansenism. His abrasiveness was collegiality itself, however, compared to the experience with Monnet and with Balthaser Sage. From the point of view of the Academy, their behavior, like Lamarck's, was only marginally less objectionable than that of a Mesmer or a Marat. Theirs was typically the conduct of apostles of a subject on the borderline between lore and science. Not that anyone doubted where the future lay for mineralogy, and the prospect of seeing it escape its self-taught adepts only intensified their difficulties with security and self-esteem.

An Auvergnat, born in 1734, Monnet came to his knowledge of minerals by way of the apothecary's trade and Rouelle's lectures on chemistry at the Jardin du roi in the early 1750s. Analysis of mineral waters brought

[48] For the foundation and early history of the School of Mines, see Aguillon (1889), corrected in many details and supplemented in others by Arthur Birembaut, "L'Enseignement de la minéralogie et des techniques minières," in Taton (1964), 365-418. The bicentennial celebration volume of the Bergakademie contains very informative articles: *Bergakademie Freiberg, Festschrift zu ihrer Zweihundertjahrfeier am 13 November 1965*. Vol. I, *Geschichte der Bergakademie . . . ;* Vol. II, *Geschichte der Lehrstühle, Institute und Abteilungen . . .* (Leipzig: Deutscher Verlag für Grundstoffindustrie, 1965).

[49] There is a full bibliography in the notice on Monnet by Rhoda Rappaport, DSB 9, 478-479. Twenty some volumes of his papers are conserved at the School of Mines.

him notice and attracted the interest of Malesherbes, who set him up in a combination shop and laboratory in the suburb of Vaugirard. Evidently Monnet traveled extensively in Germany, learning mining underground and also learning German.[50] Trudaine took him on in the Bureau of Commerce sometime in the 1760s, and in 1777 Bertin assigned him the task of completing the mineralogical atlas left unfinished by Guettard and Lavoisier. There is every reason to suppose that Monnet was conscientious and accurate in descriptive science and in technology. Unhappily, he had larger ambitions. He long intended to compose a comprehensive work on mineralogy, for which all the substances would be examined chemically in order to range them in a suitable classification. In 1779, he published *Nouveau système de minéralogie*. It is a very large book. An historical preface apologizes for its many imperfections. Monnet tells his reader that he would never have brought it out in so unsatisfactory a state if his hand had not been forced by the "bitter satire" that Macquer had visited upon his views in the guise of criticism in *Dictionnaire de chymie*.[51]

The target of Macquer's irony was the exception Monnet had taken to the program of constructing tables of chemical affinity. This was no mere disagreement among men of science. Monnet says that Macquer has insulted him, has printed indecencies and personal remarks, has set himself up in his dictionary as the lawgiver of chemistry and the judge of chemists, issuing decrees and pronouncing sentence. Behind Macquer loomed the shadow of Monnet's bête-noir, Scheele, pretending to identify in certain minerals a "spathic" acid that was only the vitriol that Scheele had introduced himself.[52] What a shame that French chemistry, too, had fallen under the sway of a superficial coterie: Lavoisier, son of a procurator in the Parlement of Paris and a Jansenist like Guettard; Berthollet, an ignoramus misled by the new chemistry to confuse fulmination with detonation, whence the catastrophe at Essonnes; Darcet, whose only distinction was to be Rouelle's son-in-law; Boulanger, pen-name of the duc de La Rochefoucauld-Liancourt, who published pieces really written by Darcet in a pretense of doing science; Fourcroy, a turncoat. For once upon a time there had been a deep, sound chemistry, started by Stahl and carried

[50] Monnet recalls his early association with Malesherbes in *Démonstration de la fausseté des principes des nouveaux chymistes* (1798), 345-347. He translated several works from the German, notably A. F. Cronsted's treatise (originally in Swedish), *Försök til mineralogie* (Stockholm, 1758), incorporated in *Exposition des mines . . .* (1772), and *Traité de l'exploitation des mines* (1773), originally published by the Council of Mines in Freiberg, by an unidentified author.

[51] Monnet, *Nouveau système de minéralogie* (1779), Avertissement au lecteur. The *Dictionnaire de chymie* was published in numerous editions beginning in 1766, but the first to appear with Macquer's name as author was that of 1778, in which the article "Affinité" gave Monnet offense.

[52] Monnet, *Nouveau système de minéralogie*, 520-524, 566.

on by the two Rouelles and Margraf. There were still a few contemporaries whom Monnet could cite with approval, notably Baumé, Nicolas Leblanc, and Lamarck. Of Lamarck's assault on the new chemistry, Monnet remarked that the author was "as learned in physics as in botany."[53] Like many a self-made scientist of their generation, like Lamarck indeed, Monnet had discerned the identity of a basic principle the transformations of which are responsible for the properties of bodies in general—not fire, not electricity, not aether, but water.[54]

What with the violence of these views, it is not surprising that Monnet should have held bearishly that mining is to be learned in mineshafts, not classrooms, and opposed all suggestions of moving forward with a school of mines. Indeed, he set his face against proceeding further with technical schools of any sort. He admitted that the École des ponts et chaussées gave useful training in the elements of engineering, and that the lectures on chemistry and natural history in the Jardin du roi were informative. A young man needed nothing more before going out into the world to learn by doing. In later years Monnet attributed the mania for schools that infected Paris in the 1780s to the decadence inculcated by the enlightenment (*les lumières*) compounded by the gullibility of ministers ignorant of everything technical.[55] He may have had a point, or part of one. Not only was a school of baking actually started, as well as a drawing school, there were proposals in the Academy of Science in 1784 for a royal swimming school and in 1789 for a glass-making school.[56] In any case, Monnet (at least by his own account) was able to dissuade Bertin from indulging the proponents of a mining school bound to prove frivolous.

When that sensible minister retired in 1780, the mines administration reverted to the Ministry of Finance, and Necker formed a committee consisting of four inspectors attached to the staffs of each of the four intendants of finance who were his immediate subordinates. The mineralogical quadrumvirate consisted of Monnet himself, Guillot-Duhamel (who as a young man had accompanied Jars on the central European expedition of 1756-1759), Jars's brother (also calling himself Gabriel, he who saw *Voyages métallurgiques* through the press), and Pourcher de Bellejeant (a young cousin of Vergennes appointed out of nepotism). In May 1781 a higher inspector, Douet de Laboulaye, of a parliamentary family, was superim-

[53] Monnet, *Démonstration de la fausseté des principes des nouveaux chymistes*, 355-365; 351 (for the Lamarck observation).

[54] Monnet, *Nouveau système de minéralogie*, 72.

[55] Monnet, *Mémoire historique et politique sur les mines de France* (1790), 59-65, 85-87.

[56] Arthur Birembaut, "Les Ecoles gratuites de dessin," and "L'Ecole gratuite de boulangerie," Taton (1964), 441-476, 493-510; for the École de natation, Procès-verbaux de l'Académie des sciences, 13 March 1784; and for the glassmakers, AN, F[12].1486, Caire-Morand to the controller-general.

posed upon these four. He soon came to favor creation of a school, but Necker held off for reasons of economy. Only under the next controller-general, Joly de Fleury, did intrigue and sycophancy carry the day.[57] That is how Monnet perceived the foundation of the School of Mines at any rate, and inasmuch as Sage was the first director, he may again have had a point. It was still not the main point, however, for other and decisive considerations were the influence of Monnet's associates and the growing claims of metallurgy, mineralogy, and extractive industry to rational treatment by science and by government. Haüy, Faujas de Saint-Fond, and Dolomieu lent the support of the new generation of mineralogists from the Jardin du roi. These were also the formative years for Le Creusot and the time when Vandermonde, Berthollet, and Monge were starting their research on steel. The decree authorizing establishment of a School of Mines leading to commissions in a Corps of Mines cleared the Council of State on 19 March 1783. The controller-general who in the next months implemented the legislation was Lefèvre d'Ormesson, the same who in his brief tenure of the Ministry of Finance took title for the crown to Vaucanson's laboratory at the Hôtel de Mortagne and appointed Vandermonde to be curator.

From a scientific point of view, the new School of Mines might be thought even less fortunate in the person of its founder and director, Balthaser Sage, than in the ferocity of Monnet, its foremost enemy and critic.[58] In two respects they were alike—in their incomprehension and resentment of the new chemistry and in their detestation of each other. Monnet never won entry into the Academy. Sage did, in 1770, and in all its long history, there can be no other member whose reputation has suffered from so consistently bad a scholarly and scientific press. His very election was the occasion of a reproach on the part of Turgot over the folly of having preferred him to Darcet.[59] On several occasions, Sage was thought to have falsified his experiments. In 1772 he claimed that an ore of white lead from the Poullaouen mine in Brittany contained marine acid. At another time he reported detecting significant traces of gold in ashes. In neither episode were others able to reproduce his results, and in the latter he was suspected of having planted them. The Academy never quite brought itself to impose formal censure. It did, however, refuse to entertain Sage's explanation of the more serious suspicion.[60] Apparently

[57] Monnet, *Mémoire historique et politique sur les mines de France* (1790), 85-87.

[58] Henry Guerlac, "Sage," DSB *12* 63-69. For a relatively sympathetic account, with documentation, see Dorveaux (1935). In Sage's dossier at the Archives of the Academy of Science there is a manuscript "Notes sur la vie et les travaux de M. G.-B. Sage, extraites de la Notice biographique publiée par lui-même" (1818), and a printed eight-page memoir also by Sage, *Origine de la création de l'École Royale des Mines* (1813).

[59] Turgot to Condorcet, 21 December 1772, Henry (1883), 123.

[60] Procès-verbaux of the Academy of Science, 99, 20 December 1780, fol. 76.

the responsible parties decided that it was the better part of valor to keep the scandal internal. Those matters have often been rehearsed, however, and it will be more to the present point to ask what his merits were? To what did Sage owe the kind of success he did have?

An answer begins with his ability to ingratiate himself about seemingly scientific topics with persons who were not scientists. Louis XV sent a special courier from Versailles to carry word to the Academy of the importance he attached to its electing Sage to membership.[61] A vacancy had been created by the death of the elder Rouelle and promotion to associate rank of Cadet de Gassicourt, the king's bastard son. Sage, too, had floated into the scientific arena from an apothecary's shop on a small tide of mineral water. However incompetent a chemical analyst, he was a keen-eyed collector of mineralogical specimens and precious stones and an artful exhibitor of these objects. He also enjoyed creating the dramatic effects to be obtained in assaying metals, and became a showman in that line. Testimony that he was an extroardinarily effective lecturer and teacher is unanimous. It is not inconsistent with an opinion expressed by Macquer to the effect that Sage never really understood any better than his patrons the difference between the kind of thing that pleased them and serious chemistry.[62] In all probability he was sincere—a quality that combined with enthusiasm often makes for success in teaching.

The chance for a regular forum came in 1778, when the crown created for him an individual chair of mineralogy and "docimastry"—assaying—at the Mint. The government was then installing monetary headquarters in the palace that it still occupies on the left-bank quai a few steps below the Pont-Neuf. The former Hôtel de Conti was renovated for that purpose by the architect J.-D. Antoine. Showcases displayed Sage's collection of gems and minerals in the great hall, and there, in one of the most sumptuous settings of the capital, Sage held forth on the mineral kingdom. Guillot-Duhamel was his assistant, and among his auditors were eminent persons and some who became eminent—for example, Romé de l'Isle (then a man of forty-odd) in crystallography, J.-A. Chaptal in industrial chemistry, and Jean Demeste in surgery.[63]

Sage later called this public course his first "School of Mines," and it was in fact the nucleus of the school created over Monnet's live body in the same quarters five years later when the time was ripe, or ripened. The letters-patent of 1783 established two chairs, one for mineralogy and assaying and one for practical mining. Guillot-Duhamel took the latter and Sage the former. His financial arrangements recall Buffon: Sage valued his mineral collection at 27,400 livres and deeded it to the crown in consid-

[61] Guerlac, "Sage," DSB *12*, 63.

[62] Macquer to Bergman, 28 March 1775, quoted in Guerlac "Sage," DSB *12*, 64.

[63] Birembaut, "L'Enseignement de la minéralogie . . . ," Taton (1964), 388.

eration of a lodging at the Mint and a lifetime annuity of 5,000 livres on top of existing emoluments amounting to 7,000 livres annually.[64] Although inspired by the École des ponts et chaussées, the infant School of Mines followed the example of the military engineering school at Mézières in one cardinal respect. It had a faculty. The teaching was done in the school itself. Besides the two professors, instructors in mathematics, drawing, and foreign languages taught those subjects on the premises. In 1784-1785 pupils also attended the course in physics given by Charles in his demonstration laboratory in the place des Victoires, about ten minutes walk across the Pont Neuf. Thereafter, physics too was taught in the school by Jean-Henri Hassenfratz, a member of the mining service soon to become active politically.[65]

The conduct of Hassenfratz in the Revolution affords a further example of the affinity between mineralogy at this stage and polemical personalities. For reasons inaudible to an Anglo-Saxon ear, Hassenfratz preferred the original Alsatian form of a name (Rabbit-puss) which his father, who ran a bar and café in the present rue Lamartine, had dropped for Lelièvre (Hare). Hassenfratz's early career is an interesting example of what was beginning to open for energetic people with some talent in technical things. Born in 1755, he began life as a woodworker and gave public courses in carpentry, while qualifying himself in surveying. In 1782 he moved into the mining service, fore-runner of the commissioned Corps of Mines. He was admitted to the grade of élève and subsidized on a central European mission of prying and self-education in mining and the metallurgy of steel. Through the latter connection, he came to know Vandermonde, Berthollet, and Monge. In 1785 Hassenfratz was promoted to associate-inspector of mines and led a field trip of students from the new school to Dauphiny. He was learning chemistry all the while, partly by assisting in Lavoisier's laboratory at the Arsenal, and he collaborated with P.-A. Adet on a notation to accompany the new nomenclature. With all this, he found time to teach on the faculty of the School of Mines from 1784 through 1786.[66]

The first class that entered in 1783 consisted of five students who, like Hassenfratz, already had the grade of élève in the existing service; one of them, André Besson, was fifty-eight years old. They were graduated and commissioned in six months. A span of three years was to be the length of studies for the normal class consisting of beginners. The exact number admitted is uncertain, but it appears to have averaged eight a year for the

[64] *Ibid.*, 392; cf. *Archives parlementaires 13*, 523.

[65] Sage published the ideas on which his course was based, *Analyse chimique et concordance des trois règnes de la nature*, 3 vols. (1786).

[66] On Hassenfratz, see Birembaut, DSB 6 (1972), 164-165, with extensive bibliography; Laurent (1924); and his dossier, Archives of the Academy of Science.

next five years. About half of them benefited from partial scholarships. The audience was larger than that would suggest, however, for students from the École des ponts et chaussées attended, as did others auditing courses in the capital. The future geologist, Alexandre Brongniart, for example, followed the lectures of both Sage and Guillot-Duhamel. After 1786, financial stringency reduced the teaching to the sole efforts of Sage. The government stopped supporting students, and the school nearly expired in its cradle. The demand for mineralogy was there, however, if we may anticipate the immediate flourishing of the *Journal des mines* started in 1792 and the rapid success of the school after its reorganization in 1795. In normal circumstances, development of the School and Corps of Mines might probably have proceeded parallel to that of the Ponts et chaussées.

The decree establishing the school provided that graduates should be commissioned in a Corps of Mines, the relation being similar to that between education and practice in the Ponts et chaussées. The corps was still very small just before the Revolution. In 1789 it consisted of 22 technically trained people: two inspectors-general, six inspectors, two associate inspectors, six engineers, and six élèves. With them the government lumped various hold-overs from the old service and two persons who combined notability with knowledge—Faujas de Saint-Fond and Baron Frédéric de Dietrich of Strasbourg, both vested with the title of commissioner.[67]

For purposes of comparison of civil with military engineering, this handful of mining specialists may be put together with the personnel of the Corps des ponts et chaussées. The overall numbers on the two sides then appear to be nearly equal. In 1789, 380 military engineers were on active duty, graduates (except for the fifteen oldest) of the Royal Engineering School at Mézières. Just four years earlier, in 1785, the strength of the Ponts et chaussées was 230.[68] About half as many were in the service of the pays d'états and other agencies. The style of the civil engineers was keyed lower, naturally enough. They wore a drabber uniform, and Perronet never aspired to the same degree of consideration for his charges that in a monarchical state rightfully (he acknowledged) accrued to everything military.[69] Modesty also became them intellectually. Among all the graduates of the École des ponts et chaussées before 1789, Gaspard Riche de Prony alone became notable beyond the realm of mines, roads, and bridges. Even Prony won his way to distinction only in the revolutionary years, although he entered the school in 1776.[70]

[67] Birembaut, "L'Enseignement de la minéralogie . . . ," Taton (1964), 402-405.

[68] Blanchard (1970), 102; above Section 1.

[69] Legendre to Perronet, 17 March 1766, BPC, carton 2636, no. 659 (9).

[70] On the career of Prony, see the notice by R. M. McKeon, DSB *11*, 163-166. His

3. MILITARY ENGINEERING AND MÉZIÈRES

Apart from Prony, no civil engineer attained to the scientific standing of Borda, Dubuat, Coulomb, or Meusnier; to the civic and technological distinction of Lazare Carnot; to the administrative facility of Bureaux de Pusy (son-in-law of Lafayette), Duportail, Letourneur, Dejean, and Millet de Mureau—all ministerial timber; to the political notoriety of Prieur de la Côte-d'Or, with Carnot a member of the Committee of Public Safety; and naturally not to the military eminence of Caffarelli du Falga, Marescot, or Chasseloup-Laubat, Napoleonic officers. Also a graduate of Mézières was the composer of the *Marseillaise*, Rouget de l'Isle. Much involved with the engineers, albeit by a bitter dispute over tactics, was Choderlos de Laclos, artillery officer and author of *Les Liaisons dangereuses* (1782) a fantasy that, like Beaumarchais in *Le Mariage de Figaro*, epitomizes the mentality and morality of an aspect of society. Intimately associated with the academic life of the school were the examiners and teachers: Louis Camus, permanent secretary of the Academy of Architecture; the abbé Antoine Nollet, foremost electrical theorist of the mid-century; the chemist, Jean-François Clouet; two leading authors of mathematical textbooks, the abbé Charles Bossut and Étienne Bézout; and most important of all, Gaspard Monge, whose later prestige reflected something of the grandiose aura of the revolutionary École polytechnique back upon Mézières, where his career began.

In point of contributions to science and to history, then, the comparison between the civil and military engineers is all to the advantage of the latter. In point of contribution to the practice of engineering, however, the situation is ambivalent. Certainly the curriculum at Mézières furnished Monge and his fellow founders of École polytechnique with experience of one direction that the scientific education of engineers would take, namely the high road of mathematics. The careers of graduates of Mézières, on the other hand, involved a tension bordering on conflict between the military and the technical calls on their allegiance. In the direction of the corps, moreover, the military aspects increasingly prevailed in the last two decades of the old regime, so that professionally speaking its later development might be thought regressive. Unlike the Corps des ponts et chaussées, the institutional setting created a disjunction between the organizational and scientific interests of the military engineers. Those who accomplished scientific work did so as an escape from their obligations to the service, or else on leave or release from normal duty, and seldom as a feature of it.

---

earliest public recognition outside the Corps des Ponts et Chaussées appears to have been the invitation in 1788 to contribute on "La science de l'ingénieur" to the *Encyclopédie méthodique*. La Millière to Le Sage, 23 September 1788, BPC, carton 2630, no. 658 (6).

A good deal is known about Mézières.[71] The school has long attracted the attention of historians concerned with military organizations, the social structure of the old regime, and the recruitment of elites, no less than with science, education, engineering, and technology. All those motifs intersected there in a relatively small institution in the old fortified town across the Meuse from Charleville. Hard by the head of the principal bridge at the Port d'Arches, a portion of the building continues in service, having been incorporated into the prefecture of the department of the Ardennes. The Corps of Engineers in the enlightenment traced its inspiration to the master of siegecraft under Louis XIV, Sébastien Le Prestre de Vauban, the first marshal of France to come from the middling ranks of society. Many frontier cities owe their street pattern today to the geometry of the ramparts that Vauban constructed in carrying out a strategy for externalizing and formalizing warfare. Mézières was itself a minor stronghold in his great system. From the War of Devolution in 1667 into the war of the Spanish Succession in 1703, Vauban's was the responsibility for conducting sieges in time of war, and planning them in intervals of peace.[72]

Throughout his career, Vauban sought authority to gather the heterogeneous surveyors, mapmakers, sappers, miners, masons, builders, and architects, whom he and his subordinates directed, into a distinct branch of the army with its own command structure, officers, troops, uniform, and training. In that one objective he failed. The situation of the royal engineers throughout his lifetime, and down to the 1740s, was similar to that of civil engineers before the institution of the Ecole des ponts et chaussées. They served in the army under a general officer, the directeur-général des fortifications des places de terre, which post Vauban occupied with the further title of commissioner. His subordinates were distributed between the three ranks of director (responsible for an entire district), chief engineer (in charge of a single fortress), and engineer (executing specific tasks).[73]

[71] See notably René Taton, "L'Ecole royale du génie de Mézières," in Taton (1964), 559-615, which is based on important archival material. The social history is treated in Blanchard (1970) and Chartier (1973). The experiences of Carnot at Mézières are treated in Reinhard (1950-52), *1*, chs. 1-5 and Gillispie (1971a), ch. 1; of Coulomb in Gillmor (1971), ch. 1; of Prieur in Bouchard (1946), ch. 3; of Meusnier in Darboux (1910). Military history is treated in Augoyat (1860-1864), vol. 2; Dorbeau (1937); and most helpfully in Chalmin (1951), (1954), and (1961). For detail on the over-all history of the Corps Royal du Génie from its origin until the Revolution, see Blanchard (1979), handsomely illustrated and thoroughly documented.

[72] For an essay on Vauban, and an admirable review of the voluminous literature, see Guerlac *DSB 13*, 590-595, and for an example of the esteem in which was held in the tradition of the Corps Royal du Génie, Lazare Canot, *Éloge de Vauban* (1784).

[73] Taton, *op. cit.* note 71, p. 560.

Some among the engineers served on detached service from the infantry and some from the artillery, in both cases with modified commissions. Others were architects or builders directly commissioned from civilian life. A certain proportion were the sons of engineers, admitted without further credentials. In their case technical proficiency was assumed, since morale and tradition also had their claims to recognition. For outsiders, verification of competence was more highly regularized than was recruitment. In his prime Vauban examined candidates himself. In 1702 he delegated the function to Joseph Sauveur, member of the Academy of Science, professor at the Collège de France, and creator of the modern science of acoustics. Among his other writings was a treatise on fortification.[74] In 1720 Sauveur was succeeded in the post of examiner of engineers by a nephew, François Chevallier, altogether less notable a scientist. The qualifications they both tested were elementary but definite: architectural drafting and elevations, mapmaking, arithmetic, geometry, surveying, statics, and hydraulics. Up to forty candidates a year might present themselves. It was for their like that Bélidor composed his classic treatises, *Science des ingénieurs* (1729) and *Architecture hydraulique* (1737-1739). For a century those splendid works were manuals of practical construction, and they still serve scholarship by exhibiting the technical and mathematical level of engineering mechanics in their time, admirable empirically, conventional theoretically.[75]

Into the amorphousness of military engineering came a new secretary of war in 1743, the marquis d'Argenson, appointed when power in the government slipped from the pacific hands of cardinal Fleury. D'Argenson immediately set about ordering the status of the royal engineers. An ordinance of 7 February 1744 laid down regulations. Henceforth, the corps was to consist of three hundred commissioned officers. Candidates had to sustain the examination to be admitted, and only officers already serving in the infantry or cavalry might present themselves. In one respect, Vauban's objectives were still not satisfied. There were to be no troops. Like staff and service officers in a modern army, the engineers never exercised command.

With respect to the more essential matter of a school, however, d'Argenson resolved to go forward. When the aging Chevallier died in 1748, d'Argenson replaced him as examiner with Louis Camus, author of mathematical texts and an experienced teacher of architectural students. The appointment was one element in a comprehensive plan whereby mil-

[74] *Ibid.*, 562, note. The treatise remained unpublished until 1847, when excerpts were printed in P.-M.-T. Choumara, *Mémoires sur les fortifications* (1847), 502-546. On Sauveur's acoustical work, see the notice by Sigalia Dostrovsky, DSB *12*, 127-129.

[75] For a notice on Bélidor, and bibliographical detail of other writings on fortification and artillery practice, see C. C. Gillispie, DSB *1*, 581-582.

itary engineers would be trained in a uniform manner instead of finding the best education they could manage prior to selection by examination. A statute of 10 May 1748 authorized instituting the École royale du génie (Royal Engineering School) at Mézières, and the first cadets were enrolled in 1749 under the command of the chevalier du Chastillon. The academic subjects were much the same as those on which candidates had formerly been tested. Outside examination continued to be decisive. Candidates for admission to the school had to appear before Camus in his quarters in Paris. At first, the commandant at Mézières sought to get the final examinations, which qualified postulants for active duty on completion of the course, into the hands of the military authorities in the school itself. D'Argenson would have none of it, and neither would his successors in the ministry. It was conceded that Camus should travel annually to Mézières to set the examinations there, but no intrusion of military favoritism on his judgment of technical competence was tolerated.[76]

Foundation of the Royal Engineering School almost coincided with Trudaine's conversion in 1747 of the mapping and surveying office in the Ministry of Finance into the École des ponts et chaussées. It is correspondingly important to recognize that in Mézières the Corps of Engineers was furnished with one of the two earliest schools for professional military education, the other having been the artillery school at La Fère. Eighteenth-century usage, in which the artillery and the engineers were called "*les armes savantes*," suggests the obvious reason for their priority in training, if not in martial prestige.[77] Only in those branches—and in the navy, which also had its schools[78]—did an officer need to know something as well as to be somebody. In the infantry and the cavalry, noble lineage was all that mattered. Young officers there learned the art of war by emulation under fire or in garrison while in the service of their regiments.

For architectural reasons, the most familiar of the eighteenth-century military schools is the École militaire in Paris, for which Ange Gabriel designed the masterpiece that completes the vista down the Champ de Mars. That institution was not comparable to Mézières, however, and was in any case of slightly later foundation. Conceived by Madame de Pompadour, and authorized by an edict of 1751, the original École militaire was a charity school at the primary and secondary level intended for scions of impecunious noble and military families. Five hundred boys ranging in age from eight to eighteen were to be taught reading, writing, arithmetic, manners, fencing, horsemanship and so on at the expense of the king. Though there never were that many, and not all were poor, the expense

[76] Taton, *op. cit.* note 71, 569-571.

[77] Chalmin (1951), 165.

[78] R. Hahn, "L'Enseignement scientifique des gardes de la marine au XVIIIe siècle," in Taton (1964), 547-558.

was too great and the results too inconspicuous. In 1776 the pupils were dispersed to be pursued by education in designated colleges kept by the religious orders in the provinces. The main interest for the history of science is that Laplace was briefly and unhappily professor of mathematics there after he came up to Paris in 1768. In 1780 a different École militaire was tried, at a more advanced level, in the same beautiful buildings. It endured until 1788, inculcating discipline and martial skills in cadets who had had preparatory schooling elsewhere, in one of the eleven colleges partially affected to military purposes in 1776. From the new École militaire the most promising candidates might then receive a more specialized training at Mézières or in the artillery.[79] Thus it was that Napoleon Bonaparte came from the college at Brienne to Paris in 1784 and went on to the artillery school at Metz. There he was examined for his commission in 1785 and commended by Laplace, who, having scorned the functions of a teacher in his youth, welcomed the prestige and emoluments of examiner of artillery cadets in his prime.[80]

Training for the artillery was less well integrated, and technically less demanding, than was the engineering education developed in the single institution of Mézières. The school at La Fère, indeed, began giving advanced instruction only after the example of Mézières had begun to be influential. Before that, it was one of five schools of theory and practice in each of the battalion headquarters under which the artillery was grouped in a reorganization of 1720. The others were then at Metz, Strasbourg, Grenoble, and Perpignan, though from time to time one or another was moved elsewhere. The early success of Mézières led d'Argenson to experiment with the notion of uniting the artillery with the Corps of Engineers. Partly his thought was to fortify the technical capacity of artillery officers, and partly to obviate the doctrinal disputes over tactics that kept the two branches at daggers drawn. That forced marriage, formalized on 8 December 1755, lasted a little over two years and was dissolved in 1758. The higher artillery school of La Fère, transferred to Bapaume in 1766, fell victim in 1772 to further discord over strategy raging within the service. It was survived by seven regimental schools, which benefited in their turn from the rising level of technical expectations. A directive of 7 April 1773 on the training to be imparted by mathematics masters in the reorganized schools called for arithmetic and geometry, and put further emphasis on trigonometry and stereotomy together with the elements of statics and fluid mechanics. Cadets were also to learn the metallurgy and practical mechanics required to understand the construction and op-

[79] Chalmin (1954), 132-135; Gambiez (1970).

[80] Duveen and Hahn (1957). Reports of Laplace's examinations are preserved at the Archives de la guerre (Vincennes), X^{d}.249.

eration of artillery pieces. They must find the middle ground between despising mechanical detail and lowering themselves to the level of the workers: "An artillery officer who is well informed about the object of his profession will understand that his ought to be a more elevated outlook. He must not be ignorant of these details, but he should know them in a superior manner, like an architect and not like a mason, and only in order to have them carried into effect by the workers and soldiers whom he employs."[81]

Left to its own devices, Mézières concentrated on the training given the 452 engineering officers who satisfied the examiners in the years before the Revolution.[82] In the record of their background, they offer the social historian a well defined and well documented sample. A graceful analysis of the composition of the sample by Roger Chartier, an exemplar of the quantitative school for whom history is as exact a discipline as ever engineering has been, dispels the remnants of a myth but leaves an accompanying impression largely intact.[83] The myth is that the Royal Corps of Engineers in the reign of Louis XV was an enclave of proto-republican values sheltering bourgeois merit from the norms that governed in the surrounding social system, and that the aristocratic resurgence of the reign of Louis XVI provoked talented technicians into becoming active revolutionaries.[84] The impression remains, even when this political legend is dismissed, that in the 1780s a kind of frustration verging on stultification did afflict military engineers with scientific capability. If so, the genealogical restrictiveness of the decade may have been part of the problem of dysfunction rather than a cause of it. That problem—to anticipate the argument—was that science can and often does serve the military, to their mutual advantage. But it cannot well be military.

Let us recapitulate the facts. Mézières, unlike the École des ponts et chaussées, provided a definite, two-year course of study. The size of the school varied according to estimates of the need for engineers. The target of the ministry was 30 cadets from 1758 through 1762, 50 from 1763 through 1771, 30 from 1772 through 1776, and 20 after 1776. In order to bring the enrollment down, no students were admitted for 1772 and 1788. The entering class usually numbered slightly more than half the

[81] The marquis de Vallière to Peyrand, 7 April 1773, an "Instruction" sent from the ministry to the commandant of the artillery school at Metz, in implementation of the reorganization of the Corps royal d'artillerie ordered by the king on 23 August and 15 December 1772. Archives de la guerre (Vincennes), X [d].248. For an outline of the history of the artillery schools, see Chalmin (1951), 175-181.

[82] The experiences of the entire group in the Revolution have been studied by Blanchard (1973).

[83] Chartier (1973).

[84] The stereotype was already questioned by Chalmin (1951), 173-174, and qualified by Reinhard (1950-52), *1*, 296-297.

total, in order to allow for attrition. Thanks to the researches of Chartier, we know where the engineers came from. Of the entire group of 452 over the 30 year period, 27.1 percent were from village centers of fewer than 2,000 people; 40.9 percent from towns of 2,000 to 10,000; 23.1 percent from small cities of 10,000 to 50,000; and only 8.9 percent from big cities. There were sixteen all told from Paris, eight from Lyons, none from Bordeaux, one from Rouen. Languedoc and Guyenne were the most heavily represented provinces, together with frontier regions in the south, east, and north-east, localities where the presence of the military was always felt. Small-town and country families, then, of the minor nobility and of the upper and middling bourgeoisie, often with a tradition of military service, in relatively unprogressive parts of the country—those were the milieux in which parents found the prospect of hard, technical study and an assured future in the service appropriate for boys who were equal to the demands.

Selection occurred in two stages, social and intellectual. A postulant had first to secure a letter of eligibility signed by the minister of war before being accorded an examination of his competence by Camus, and after 1768 by Bossut. For that purpose he needed to furnish the ministry with certificates exhibiting place and date of birth, the status and circumstances of his family, and evidence of the financial support he could depend upon if accepted in the corps. Candidates who were already officers desiring to transfer their commissions were required to supply a detailed service record. It was certainly true, wrote Camus concerning admission to the competition, that a "gentleman" would be preferred over others, assuming their merits to be roughly equal, and that officers wishing to shift into the engineers would have preference. The most he could say was: "But those who are not gentlemen or who have never served in the army are by no means excluded."[85]

This first stage, then, was the juncture where families of marginal standing would bring to bear what influence they could muster, soliciting the intervention of patrons among the great and standing at the notary's elbow with mildewed parchments showing military and blooded ancestors. For obvious reasons, Lazare Carnot has had more attention than most of the others. His father, Claude, was a lawyer accredited to the Parlement of Dijon, and a considerable personage in the small town of Nolay, on the side of the Côte d'Or where no vines grow and the Morvan stretches stonily westward. The family was bourgeois to the heart, and there was none better in the region. In order to win consideration for Lazare, the second son of nine children who lived to maturity, Claude Carnot turned to the

[85] Camus to Fourcroy de Ramecourt, 16 December 1751, quoted in Chartier (1973), 360, from whose work all these statistics are drawn.

duc d'Aumont, who held the marquisate of Nolay among his many lordships. Carnot père had become his bailiff there in 1756. In Paris, D'Aumont's intendant, or secretary, took charge of the Carnot matter, passing the word to the minister of war and advising the family to make much of three distant cousins who had followed the profession of arms, one of whom the dossier obediently converted into an uncle.[86]

It is not to be supposed that this scale of aristocratic values was imposed on a reluctant student body at Mézières by distant, high-born ministers remote from some comradeship of shared technical capacity. On the contrary, as often happens, and indeed as also happened at the lower stratum of the École des ponts et chaussées, the norms of the larger society were intensified by the intolerance and insecurities of the young. Mézières suffered recurrent incidents of hazing of commoners by their fellow cadets of better lineage, the more virulent (it may be surmised) in that, as nobility went, the perpetrators were themselves pretty small beer. Occasionally an individual student would be singled out for pariah, in one instance because word got about that his father had not only been in trade but had gone bankrupt. Thereupon, his fellow commoners joined in the heaping of scorn. For some, and perhaps for many, personal qualities of camaraderie and charm easily prevailed over the banalities of snobbery. For others, not—in the annals of juvenile barbarism, the pattern is classic, and equally so was the reaction of the authorities called on to punish infractions with which they sympathized. Penalties were soon suspended, and steps taken, not to enforce the rules, but to eliminate the problem. In 1767 the minister resolved that cadets should no longer be given grounds for pleading that the company of people of low birth excused their acts of insubordination. Henceforth, the rosters were to be divided into two categories: one would list those of gentle birth or military extraction on the paternal side, the other those whose background was likely to cause trouble if they were admitted. Into the latter bin went four sons of low-ranking officers, two sons of retainers in princely families, an architect's son, a lawyer's son, and a tradesman's son. None of these was given a letter of examination in 1767, and in future years very few from the high-risk category were allowed to present themselves.[87]

A new regulation of 7 May 1777, promulgated by Saint-Germain, Turgot's appointee to the Ministry of War, formalized the practice of the previous ten years, which had made it virtually impossible for commoners to be admitted to Mézières unless, like Carnot, they could establish or pretend to military connections. On 22 May 1781 a second edict went further. The maréchal de Ségur was then secretary of war. Everyone aspiring to a commission in any branch of the army had thereafter to show four

[86] Reinhard (1950-52), *I*, 18-21.

[87] Chartier (1973), 361-362.

generations of nobility on his father's side, or else to be the son of a chevalier of Saint-Louis—that is to say, an officer or honorably retired officer of field grade. In effect, the legislation of 1781 extended to candidates for Mézières the stipulations of the edict of 1751 establishing the École militaire. It locked a door already nearly shut. Carnot could not then have found a key, and neither could Prieur nor Rouget de l'Isle.

The effect was that gentrification of the Corps of Engineers accelerated in the last decade of the old regime. From 1758 through 1776, approximately 49 percent of the cadets were nobles, 11 percent were previously commissioned officers, 14 percent were the sons of engineers or military administrators, and 18 percent were non-military commoners. (The origin of the remaining eight percent is uncertain.) From 1778 through 1789, 68 percent were noble, 17 percent were officers, 8 percent were sons of engineers or administrators, and fewer than 3 percent were ordinary commoners. This last bourgeois remnant consisted of three students, two of them from the West Indies, out of the 112 accepted in the dozen years before the Revolution.[88]

Whatever the rise in the social threshold for admission to an examination, the postulant still had to pass it in order to gain entrance to Mézières. A pedigree conferred no advantage in the second, intellectual round of the competition. The would-be engineer had to know his Camus, and later his Bossut, for the examination turned on the textbooks written by the examiners. Camus (it will be recalled) had held the post since the creation of the school in 1748, while fulfilling the same function for the artillery schools. He brought out the earliest of many *Cours de mathématiques*, an arithmetic primer, in 1749, adding supplements on geometry and statics in the next few years.[89] The entire set went through four editions by the time of his retirement in 1768, when the treatment was beginning to be deemed too elementary. Bossut, having been professor at Mézières for some fifteen years, started the next generation of textbooks in 1772, also with an arithmetic which he developed into a full *Cours* in 1781. One variant was intended for the École militaire, and a more advanced version for the pupils applying to Mézières. Revisions culminated in a fifth edition, published in 1794-1795.[90] The sequence of topics remained canonical for another two centuries, determining the mathematical order of the secondary education of most persons who com-

[88] *Ibid.*, 372.

[89] The entire series had the generic title *Cours de mathématiques*, and it consisted of three parts (1) *Éléments d'arithmétique* (1749), (2) *Éléments de géométrie théorique et pratique* (1750), (3) *Éléments de méchanique statique*, 2 vols. (1751-52).

[90] *Cours de mathématique, à l'usage des élèves du génie*, 3 vols. (1794-95). For detail of the earlier editions and variants, see Taton, "L'Ecole Royale du Génie de Mézières," in Taton (1964), 584, n. 2.

pleted that process prior to the reforms of the 1960s: arithmetic, algebra, plane and solid geometry, trigonometry, analytic geometry, statics, dynamics, and hydrodynamics.

Comparison of Bossut with Camus gives an index to the process of mathematical sophistication, if the phrase is permissible for the learners' level rather than for higher levels of learning. Though once a pupil of Varignon, Camus gave no evidence in his published writings of ever having employed the calculus. He was of the rough-hewn generation of Hellot in chemistry and Nollet in physics. Bossut, equally the pedagogue and scarcely more creative than Camus, was much smoother in style and more advanced in substance. His course carried students into analytic geometry, where conic sections were the main objects he presented, and also into the mechanics of machines and hydrodynamics. Later editions contained an appendix wherein the author analyzed several basic problems of rational mechanics by means of the differential and integral calculus.[91]

Although the examinations Bossut set for entry to Mézières never reached that far, a youth's appearance before the mathematical abbé was an ordeal about which survivors traded anecdotes. An unkempt, friar-like lackey would open the door and herd the aspirant into the presence. The Saint Peter of military engineering was a burly clergyman, dark of complexion and bushy black of joined eyebrows, garbed in a leather skull cap and an ample brown cloak of clerical cut with a wide fur sash. Bossut would propound problems and require solutions and demonstrations of theorems to be worked on a slate. If responses flowed readily down well-worn channels of the text, the candidate (at least according to student wisdom), could expect passing marks, the degree of merit depending on his facility and self-assurance. Let him risk a different tack, and every step would meet with frowns and objections. All who held steady and won the gamble of thinking for themselves agreed that Bossut was fair, and rewarded successful independence with the highest marks. If you stumbled, however, you would be told, "Why don't you just learn the course, instead of chasing after all these far-fetched proofs?"[92]

Bossut's own account, in a letter to the minister enclosing a rank-ordering of accepted candidates, comes to much the same thing:

> It is clear that the most direct and fairest qualification for being admitted to the School at Mézières is that which is tested in the competition, and clear also that it offers the most reliable criterion by which to compare intellectual merits. But the outcome must not be

[91] The problems are finding the centers of gravity of surfaces and solids, the curvature of a catenary suspended from points in motion, numerical calculations for designing drawbridges, acceleration in general, and finding centers of percussion and oscillation of spheres.

[92] Chalmin (1961), 147.

merely the fruit of dogged study fortified by memory. It must also spring from intelligence and penetration. In accordance with that consideration, Monseigneur, I have made as great an effort to discern the native talents of the students as to satisfy myself of their information at the moment of examination.[93]

Candidates who failed might try a second, third, or even a fourth round. Including repeaters, there were sometimes as many as eighty or ninety candidates competing for fifteen or twenty places annually in the period 1763-1776, and only a third to a fourth of those who received their letter of examination from the ministry ever did manage to satisfy Bossut and pass through the portals of Mézières.[94] In principle, it was possible to prepare privately by studying the published course on one's own. In 1769 a sixteen-year-old Lazare Carnot thought to do so in order to spare his father the expense of further tuition beyond the cost of the Oratorian college in Autun, which he had just completed. With all his intelligence and self-discipline, he nevertheless failed, and failed badly.[95]

That only a very few autodidacts ever succeeded suggests a limitation in the textbooks. Clear and orderly though their presentation was, they were inferior to more recent examples of the genre in one pedagogical respect. The art of incorporating practical exercises in the lessons was still largely undeveloped. In the working of problems, boys needed instruction, supervision, and correction. They needed schooling, and Carnot's father sent his too self-reliant son off to the boarding school kept in Paris by Louis-Siméon de Longpré, where the fees came to 1,600 livres annually. Situated in the Marais, not far from the Pension Blondel, which prepared pupils for the École des ponts et chaussées, the Pension Longpré was the smaller of two that specialized in readying candidates for Mézières. Its competitor was an establishment in the rue du Faubourg Saint-Honoré directed by Claude-Louis Berthaud, and after his death in 1776 by his widow and son-in-law.

In the later 1770s and 1780s, the ministry reduced the enrollment at Mézières to twenty, and each year all but three or four of the thirty-odd candidates came to the examination from one or the other of these cram schools or from the École militaire. As Chartier observes, if the origins of the engineers were provincial, their preparation was Parisian. For once, the cultural attractions of the capital fail to provide the reason. There was no time for such diversions. Pupils were out of bed at 5:30 in the summer and 6:00 in the winter, and the cardinal rule was that they must not be left a moment to themselves, and specially not in the dormitories. No,

[93] Bossut to the Ministry of War, 22 December 1781, quoted in Dorbeau (1937), 336-337, from the Archives de Guerre (Vincennes), X[e].159.

[94] Chartier (1973), 355-356.

[95] Reinhard (1950-1952), *I*, 25.

the advantage of Paris was the presence there of the examiner. Bossut maintained a rapport with Longpré and with the Berthaud directors and would occasionally visit the establishments, the *Cours* made flesh. Similarly Camus in his day had patronized the preparation given for his examination in two clerical colleges at Nanterre and Clamecy.[96]

Let us follow the Class of 1771 from the study where Bossut selected its members into the service of the king (remembering that French usage designates a class by the year of entering an institution). Enrollment was then at its highest, and twenty-two cadets reported to Mézières, the ranking three being Damoiseau, Bénézech de St.-Honoré, and Lazare Carnot. There they received the commission of a second-lieutenant and donned the uniform of the royal engineers—coat of royal blue with facings of black velvet, red doublet and culotte, and gilt buttons. As officers, each had a simple room to himself in the lodgings around the old gubernatorial residence, which housed the commandant and accommodated the workshops, laboratories, and library on the ground floor. (The permanent quarters were completed only in 1780.) The salary supporting the modest style of the student-officers was 720 livres, to which their family usually added a 200-livre supplement. They took their meals at one of four designated inns and could choose their company subject to the requirement that the first- and second-year classes must be represented at every table. An hour and a half were allowed for meals and conviviality, at noon and 8:30 in the evening. No other idleness was tolerated. Entering students soon learned of the punishment visited upon a member of the class of 1766 for neglect of his studies. When the extent of his dereliction was discovered at the time of graduation, he was sentenced to prison. Bossut examined him again in February 1769. After a year of confinement, he proved to be even more ignorant than before so that, finally, he was expelled.[97]

Soon after their arrival, and probably at the opening exercises, the class of 1771 listened to an allocution by Lieutenant-Colonel Antoine Du Vignau, the second-in-command, an engineering veteran who had been with the school since the early days. Youth, he observed, is the time of life for acquiring the knowledge that enables a man to distinguish himself, whatever his field of action, but it is fleeting. The obligation to make the most of one's faculties is specially incumbent upon those fortunately enough placed in society to have an influence on the welfare of others. Common men need think only of their own concerns; the privileged are bound to work harder since more depends on them. Among possible callings, the profession of arms is the noblest since it requires the greatest sacrifice and

[96] On life in these pensions, see Chartier (1973), 365-368; Bouchard (1946), 25-29; Reinhard (1950-1952), *1*, 27-29.

[97] Dorbeau (1937), 332.

the most glorious since the only reward is, precisely, glory. Its attributes are in keeping with "the lively and generous character of the young French nobility," who therefore feel an affinity for the trade of war. It offers possibilities of several sorts. Officers who can never expect to rise above company grade (i.e. captain) can and should qualify themselves in a single skill, for only by dint of deep concentration can they think to distinguish themselves from the run of the mill. Those so placed in the world that they can hope to become general officers should neglect no aspect of the martial arts. It is up to them to master the main branches and have a knowledgeable overview of the others against the day when they will exercise command. And of all the military specialties, the most interesting is siegecraft—"*l'attaque et la défense des places*."[98]

In a school with fifty students, each newcomer soon became known to everyone else, and notably to the commandant, Brigadier Claude Rault de Ramsault de Raulcourt, who also held the office of royal lieutenant, exercising civil and military authority in the town and fortress of Mézières. Ramsault, fifty years of age in 1771, had then completed five years in command of the Royal Engineering School.[99] In that time, he had been able to effect a reorganization of the school he had taken over from Chastillon, the first director, in 1766. Only the operation of the shop, drafting room, and laboratory continued largely unchanged throughout the entire history of Mézières. These installations were headed by three crusty foremen. Though not actually in military service, they were akin in type to indispensable master sergeants who make things work while secretly despising the unhandiness of those set over them. The head draftsman and design-master was called Barré; the master carpenter and builder, Jean-Marie Marion; and the chief stone-cutter, Nicolas Savart (his grandson, Félix, was the nineteenth-century physicist famous in acoustics). Under them served four additional clerks and draftsmen of various skills. From early on, this non-commissioned personnel conducted on the side a trade school for local artisans, as a service to the town and a satisfaction to themselves.[100] One local boy to whom they gave a start, Jean-François Clouet, became a professor in the engineering school in the 1780s and proprietor of the important metallurgical laboratory where Vandermonde, Berthollet, and Monge ran tests on steel.[101] In 1768 Ramsault added a geographer from the parent-service of the Ponts et chaussées to teach topography.

[98] The text was discovered and printed by Reinhard (1950-1952), *1*, 34-36.

[99] For a sketch of his career, see Dorbeau (1137), 330.

[100] Taton, "L'École Royale du Génie de Mézières," in Taton (1964), 572-574, 583-584.

[101] Above, Chapter VI, Section 3; on Clouet, see the notice by Taton in the DSB *3*, 326-327.

When Ramsault became commandant in 1766, he was a seasoned infantry officer and combat engineer who had seen action in Austria, Germany, the Alps, and the Low Countries during the War of the Austrian Succession and the Seven Years War. Camus was then in the last years of his two decades of examining; Bossut was still teaching in the school itself, having been professor of mathematics since 1752; the abbé Nollet alternated with Bézout in giving the course in physics, a very brief affair occupying a few weeks in the spring and early summer; and the young Gaspard Monge, who had been at Mézières for two years, assisted Bossut and also leant a hand with mocking up wooden and plaster models of fortifications and terrain in the *gâche*, as Marion's shop was called. Not surprisingly, Ramsault's first reaction to the academic side of his new responsibility was that practical operations needed far more emphasis, and that mathematical examination by Camus gave an altogether inadequate evaluation of the students. When the latter fell ill in 1768, Ramsault seized the opportunity: "I regard Monsieur Camus as dead, . . ." he wrote the minister, enclosing a plan for revision of the curriculum.[102] The first step would be to require prior certification in elementary mathematics and mechanics and to eliminate instruction in those subjects.

Ramsault took the same occasion to move Bossut into the examiner's seat and out of Mézières. The abbé's presence in the school had never inspired the military authorities with enthusiasm. The first director, the relatively easy-going Chastillon, is said to have complained that he really wanted to make his name in higher mathematics and was lazy about practical engineering.[103] Still, he knew what it was, and by virtue of his very preferences could be counted on to conduct examinations at a higher level than Camus. In thus kicking Bossut upstairs to Paris, Ramsault observed tactfully to the minister that there was no need for so learned a professor: "a good instructor (*répétiteur*) will suffice,"[104] and he recommended Bossut's assistant, the young man who had also been helping the shop stewards. Ramsault was well satisfied with the choice of Monge. When Nollet died less than two years later, in 1770, the natural thing would have been to appoint Bézout, who had been alternating with him, to a professorship of physics. Again, the commandant advised against it on the grounds that there was no point in sending a member of the Academy of Science to Mézières. The "sieur Monge" and the "sieur Savart," who besides demonstrating the techniques of stone-cutting also maintained and prepared Nollet's instruments, could perfectly well give the course in physics. The ministry agreed, assigning the course to Monge with a raise in salary and

[102] Taton, *op. cit.* n. 100, 588-589.
[103] Dorbeau (1937), 324.
[104] Taton, *op. cit.* n. 100, p. 590.

naming Savart assistant. So it happened that in the 1770s the academic staff was reduced to Monge alone.[105]

In the curriculum reformed by Ramsault, the scientific subjects imparted during the two-year sequence at Mézières were dynamics, hydrodynamics, physics, architectural drawing, perspective, the optics of light and shadow, mapmaking, and topography. Those were the topics examined by Bossut in his annual visit to Mézières before graduation. Increasing emphasis, however, was placed on the exercises conducted and evaluated by the commandants themselves: engineering drawing; plans and estimates for the construction of buildings and forts; surveying with traverse and compass; drafting charts and transposing them from one scale to another; the design of a fortification including ground-plan, relief, construction, plan of attack and defense, and expense; a simulated siege, every step specified and followed out on paper; field exercises on irregular terrain; trips to actual engineering installations; visits to industrial establishments near Mézières; reconnaissance of the frontier along the Meuse, with a memoir and recommendations supported by maps and detailed designs.[106] As to mathematics, Ramsault's policy was explicit:

> It would be desirable for engineers to learn algebra, analytic geometry, and the principles of the differential and integral calculus. For there is an infinite number of problems concerning mechanics, strength of materials [la résistance des bois], the thrust of earth against the foundations of retaining walls, the force of water, etc., that cannot be solved without the aid of these sciences. Nevertheless, there will be no public courses on these subjects. Those engineers who wish to learn about them may consult privately with the professor of mathematics, who will help them familiarize themselves.[107]

In passing allusions to Mézières as the first institution of higher education to produce mathematical scientists, historians sometimes leave the impression that it did so on purpose and in large measure. In fact, of course, Ramsault's mission was to train engineers for the army, and the same was true of Chastillon and Villelongue, respectively his predecessor and successor in the post of commandant. The importance Mézières came to hold for the development of mathematical science was owing rather to the potentialities of engineering than of a small military school in the Ardennes. Eventually much of the physics of work and energy did emerge from analysis of problems of construction, power, and machinery by mathematically trained people. Certainly the appointment of Monge at

[105] *Ibid.*, 591-592.

[106] *Ibid.*, 593-595, which prints Ramsault's report to the ministry of 7 March 1772 on the plan of study.

[107] *Ibid.*, 587.

Mézières contributed to the education of several such people there, followed by many more in the later years at École polytechnique. The effect on science was altogether unintended, however. The intention when he was appointed was to favor military training at the expense of mathematics and physics. Courses in the latter subjects were to be restricted to the essential minimum that could be imparted by an unknown young technician with no outside commitments. That was what Ramsault wanted.[108]

What he got in Monge was the founder of modern geometry, a mathematical intelligence the equal of Laplace, one of the great teachers in all the history of science who in later life became a venerated father figure to École polytechnique, and a scientist whose contributions in chemistry, metallurgy, and physics were notable in their own right.[109] Monge's career was on the heroic scale, and it is natural that a certain, harmless mythology should have accrued around the story of his rise to scientific recognition in the 1780s and to political prominence in the Revolution.[110] The embroidery begins in a tendency to exaggerate the poverty and ignorance of his start in life.

Certainly Monge came from more modest circumstances than any other member of the Academy of Science in his generation. In 1737 his father, Jacques, a Savoyard by origin, settled in the thriving little Burgundian city of Beaune at the age of nineteen. He was still a peddler when Gaspard, his oldest, was born on 9 May 1746, but Jacques Monge was even then on the way to succeeding commercially. He became a cloth merchant and member of the mercers' corporation, and was able to see to his children's education, no doubt at considerable sacrifice. All three sons were placed in the admirable Oratorian college in Beaune, one of the best schools in France, which provided primary and secondary education in a twelve-year sequence of classes. Monge graduated in 1726, *puer aureus*,

[108] The motivations are specially explicit in memoirs that Ramsault addressed to the Ministry of War on 24 March and 1 December 1768 concerning curriculum and staff. Archives de guerre, Vincennes, X^{e}.159.

[109] The definitive study of Monge's scientific work is Taton (1951), to be supplemented on personal aspects by Aubry (1954). Taton also contributed the article on Monge in the DSB 9, which takes account of recent scholarship. The bulk of Monge's correspondence remains in the hands of descendants (for whom see Taton [1951], 395). In the middle of the nineteenth century a great-grandson, Baron F.-E. Eschassériaux, prepared a detailed biography which was never published. For that purpose, he copied out many of Monge's letters. His ledgers were deposited in Bibliothèque de l'Institut de France, MSS 2191-2193.

[110] C. Dupin, *Essai historique . . . sur Monge* (1819); Launay (1933); and the éloge by Arago, *Oeuvres de François Arago, notices biographiques* 2 (1853), 426-520. The Arago éloge was reprinted, together with excerpts from Dupin (1819), by Editions Seghers in 1965, under the title *Gaspard Monge, père des polytechniciens*.

golden boy of his year, and gave a public defense in Latin of his thesis on the elements of calculation and of geometry. His younger brothers, Louis and Jean, also went on to useful mathematical careers. So brilliant was Gaspard's success that his teachers urged him on to further education in the more elaborate college in Lyons. There he taught physics briefly and even toyed with the thought of entering the order for the sake of continuing in science, a notion that his father sensibly discouraged.

Instead, fortune stepped in, wearing the uniform of Colonel du Vignau, second-in-command at Mézières. Monge had occupied the leisure of his summer vacation in 1764 by collaborating with a college friend on a topographical and architectural plan of the city of Beaune, a chef-d'oeuvre that may still be inspected there in the public library. Soon after it was completed, du Vignau made a visit, and he was shown Monge's drawing, probably by proud aldermen acting as hosts to a distinguished officer and member of the provincial estates. Impressed by the draftsmanship, du Vignau offered the author a place at Mézières.

A further legendary element in the Monge tradition has it that du Vignau misled the eighteen-year-old innocent into expecting admission as a cadet, and that Monge's high hopes were dashed when he was assigned to the menial work of the shop or gâche. Arago's éloge of Monge retails this story, for which there is not a line of documentation and which is inherently implausible.[111] The same du Vignau who evoked the "lively and generous character of the young French nobility" in a matriculation address could scarcely have led a peddler's son to expect a commission, even if he had had the authority to by-pass the war ministry and the Camus examination. On the other hand, to suppose that du Vignau should have seized the opportunity of hiring a skilled draftsman is as reasonable as it is to imagine that the young Monge, his appetite for a career enlarged by a diet of praise from the Oratorian fathers, should have resented the contrast between his prospects and those of the student-officers whose education he served. Of his frame of mind in these early years, he later said that he often felt like destroying the charts and architectural elevations he had to draw for their lessons, as if he were good for nothing better than draftsmanship and mocking up plaster models.[112]

Monge rescued himself by showing his military superiors that he did, indeed, have something more to offer. He imagined a geometrical construction for solving the problem of defilade, which is to say for determining from a relief map what elevation the ramparts in a proposed fortress must have in order to shelter the defenders from flat trajectory fire or observation from any external terrain feature. Military architects and engineers customarily resolved this sort of problem empirically following rules

[111] Arago, *Gaspard Monge*, 1965 ed., 10-12.

[112] Taton (1951), 11-12.

of thumb that allowed a wide and costly margin of safety. Monge's solution required the architect to substitute a plane of defilade for the horizontal plane relative to which he would establish the elevation of the different parts of the fortress. The plane of defilade was to be constructed tangent to the external vantage point (or points, if there were several) upon a line joining two locations within the fortress so chosen that the angle between the planes of defilade and of the horizon would be the smallest possible and the profile of the construction the lowest possible.[113]

Versions of this episode constitute another set of elaborations, if not quite fabrications, in the biography of Monge. The most exaggerated is the story that, for reasons of military security, his superiors prevented him from publishing his earliest discovery and thus inhibited his mathematical career along with the development of descriptive geometry.[114] Again, there is no contemporary documentation, though an element of confidentiality apparently did surround the methods of fortification taught at Mézières. It is also true that Monge's *Géométrie descriptive* was published only in 1795, in the serial form of lectures given at the ephemeral École normale of the year III, which were supplemented by a more technical series at the infant École polytechnique setting forth the new approach to geometry.[115] It was in fact a renewed approach, or set of practices, some novel and some merely graphical or empirical, elevated to the dignity of mathematics. As Monge's scientific biographer, René Taton, has shown, Monge there gathered up and spun together threads from the largely forgotten geometry of Desargues, from the rules of optical perspective worked out by Dürer, from various methods of cartographic projection, from the science of light and shadow applied in sundials and cadrans, and from techniques for shaping stones in raising arches and vaulted ceilings.[116]

In the definition Monge gave in 1795, "Descriptive geometry has two purposes: first, to impart the methods for representing natural objects which have three dimensions . . . on a sheet of drafting paper which has only two, always provided that the bodies in question are such as are capable of rigorous definition; . . . second, . . . to impart the means for recognizing upon exact description what forms bodies do have and for deducing therefrom all the truths that depend both upon their form and respective positions."[117] The object to be described was represented by

[113] *Ibid.*, 13-14. For Monge's own later account, see Gaspard Monge, *Géométrie descriptive* (3rd ed., 1811), 49-50.

[114] Arago, *Gaspard Monge*, 1965 ed., 20.

[115] For bibliographical detail, see Taton (1951), p. 380, nos. 36, 37.

[116] *Ibid.*, 50-100.

[117] *Journal de l'Ecole polytechnique*, Cahier 1 (1795), 1-2.

projection upon two mutually perpendicular coordinate planes. Unlike classical geometry, proceeding from axioms through deductions to theorems, Monge's exposition moved from conventions through construction of figures to solutions of problems. By one convention, the plane taken as vertical was rotated through a right angle in order to permit figuring the elevation and ground plan (to take the architectural example) on a single sheet. A further convention permitted defining objects bounded by curved surfaces that are engendered by various types of lines in determinate forms of motion. Typical of the problems adduced were the constructions of a plane tangent to a cylindrical surface at a point for which the horizontal projection is given; of a plane simultaneously tangent to three spheres of given size and position; of the intersection of any two surfaces of revolution with axes in the same plane; and—to move into the application with which it all began—determination of the depth to be given to demilunes and communication trenches to provide defilade in a fortress.

The text of the *Géométrie descriptive* separately published as a treatise in 1799 may be read for clues to the motivation and level of Monge's research and teaching in the 1770s and 1780s, but to take it as an index to the content would be risky.[118] The most that can be said is that the mathematics of architectural drawing, engineering, masonry, cartography, and surveying are what he would have taught at Mézières. Success with those subjects might have been expected to win esteem and consideration among students and military superiors, though not among mathematicians even if Monge had then published on them. The treatment was bound to have been elementary, though not like the textbook *Cours* of Camus, Bézout, and Bossut. As to the approach, we may judge from the vein of *Géométrie descriptive*. Those lectures were a summons to a movement in mathematical culture through application of constructive geometry to concrete technical and educational purposes. They belong to the literature of engineering, like the writings of Lazare Carnot and others on the theory of machines. The overtones of social uplift may well have been representative of Monge's thoughts in youth, though there is no written evidence. Such discourse might round out a mathematical reputation, but could never have won the young Monge recognition in the first place. He gained notice among mathematics in the only way possible, through contributions to analysis, in his case analysis of problems involving geometric objects and relations.

[118] The course as published was compiled from Monge's lecture notes and from the notes of auditors, mainly through the good offices of Nicolas Hachette. Hachette had begun his education in the courses for townspeople organized at Mézières. He was a technician and draftsman in the École Royale du génie in the 1780s, when he assisted Monge's successor, C. Ferry, in teaching descriptive geometry. He became Monge's assistant and a wheelhorse at École polytechnique until the end of the Napoleonic period.

The reciprocity of analysis and geometry was the main motif in the series of memoirs on analytic and infinitesimal geometry that Monge addressed to other mathematicians in his most creative years. He began to work seriously and originally in those areas at the time of appointment to his professorship at Mézières, that is to say in 1769 and 1770, when he was in his middle twenties. First he corresponded with Bossut and with a few students and near-contemporaries. Within a short while, he was exchanging letters with Condorcet, d'Alembert, and Vandermonde. In taste and temperament, the last named, with his strong bent for technology, was the most congenial to Monge. The difference in age was ten years, and their association, starting with Monge's approach to the Academy, was the beginning of a lifelong friendship and collaboration on many objects, among them the famous memoir on steel in 1786.

The investigations that Monge put in hand between 1770 and 1776 were eventually written up in some fifteen separate memoirs, which have been admirably analyzed by Taton.[119] Perhaps a précis of one of the more far-reaching, "Memoir on evolutes, radii of curvature, and the inflection of space-curves," will suffice to convey what kind of mathematics this was.[120] As usual, Monge set out from geometric constructions for which he found analytic expressions by extending two-dimensional Cartesian analysis to the three dimensions of physical space. Thus, in his discussion of evolutes (an evolute is the envelope of normals to a curve) he transcended the limitation of plane curves, each of which gives rise to a unique evolute in the plane, and showed that for any curve there is an infinite family of evolutes, all lying on a developable surface (one that may be laid out point for point on a plane). Monge gave the geometric properties of the plane—its relation to the polar axes, the radii of curvature, and the families of normals and tangents to the curve.

The geometric attack distinguished Monge's style from the analytic determination of curvature and surfaces. Once the initial object was accomplished, Monge turned to analysis for the second main topic: given the equations of the curve, to find those of any specified evolute. In solving this problem, and others like it, he applied the most comprehensive of his correlations between analysis and geometry, namely the expressibility by partial differential equations of families of surfaces. Solving partial differential equations requires determination from specified initial conditions of arbitrary functions introduced in integration. He showed that such functions represent surfaces grouped into families by the mode of generation. They may be constructed geometrically from given initial conditions, and it is then possible to specify the particular surface through which a com-

[119] For the bibliographical detail, see Taton (1951), 377-383.
[120] SE *10* (1785), 511-550.

plete solution must pass. In simpler words, an appeal to geometry gives the solution to certain partial differential equations in much the way that in algebra an appeal to reality indicates the choice to be made between the roots of a quadratic expression.

Only one of Monge's early papers, a time-motion study of grading and filling, shows the mark of its origin in an engineering school by its subject (though in fact the analysis is a very abstract exercise in the calculus of extreme values and quite inapplicable to actual earth-moving operations).[121] Nevertheless, the geometric concreteness of all his work, the assumption that mathematical expressions represent physically real spatial quantities and relations, the instinct that mathematics is an instrument for accomplishing mundane purposes—those qualities, together with a certain insensitivity to elegance and economy, bespeak the importance to Monge of the early years at Mézières in forming his style and approach. The innovative researches of his youth were thus expressions of the same mathematical personality as the messiah-like *Géométrie descriptive*, delivered in a maturity bordering on sagehood.

That the spirit of Monge's work should have remained constant is natural enough, but it is surprising to find that by the time these papers were published, his primary attention had already shifted away from doing creative mathematics into his other interests.[122] Normal delays in publication were compounded in his case by disinclination to write up results in finished form, and only between 1776 and 1787 did the researches of the preceding six years appear in print. By then the best days of the Royal Engineering School itself were past. Ramsault died in 1776. His successor, François Rabinel de Villelongue, commanded a school of increasingly aristocratic complexion reduced in numbers to twenty cadets, less than half its fullest strength. Professionalization in the military rather than the scientific sense was the thrust of reform imposed on the Corps of Engineers by Saint-Germain, Turgot's minister of war, and however beneficial to its prestige throughout the army, the emphasis was bound to be uncongenial to Monge.[123]

Moreover, Monge had grown close in sympathy to the patron who appointed him. In 1774, Ramsault took his young professor of mathematics and physics on a three-month holiday to Barèges, a spa in the Pyrenees. Monge met Jean Darcet nearby and collaborated with him on a program of barometric determinations of the altitudes of various peaks.[124] There was always something of the small-town boy discovering the world about

121 "Mémoire sur la théorie des déblais et des remblais," MARS (1781 pt. 2/1784), 666-704.

122 Taton (1951), 23-33, Chapter 7, 310-351.

123 On the decline of Mézières, see Taton (1964), 596-612; Dorbeau (1937), 333-346.

124 Taton (1951), 20.

Monge, and this junket was his introduction to distant travel, for which he developed a strong liking. His life expanded in other ways, despite hints of restiveness and an occasional sense of hostile machinations. A love affair of 1771—"I took a fancy to get married" he sadly wrote a friend—ended three hours short of announcing the engagement.[125] He did not say why. In 1777 Monge found a wife, a young widow of Rocroi, of whom a charming portrait survives from the days of their Napoleonic eminence. Her first husband's smithy drew Monge into management of an iron business, into metallurgy also, and into a greater material prosperity than anyone in his family had yet known.[126]

Mutual attraction between Monge and persons of power and influence was an important factor in his life, beginning with du Vignau and Ramsault. In 1774 a distinguished visitor came to Mézières, the marquis de Castries, marshal of France, together with his son, the comte de Charlus, and the boy's tutor, one Jean-Nicolas Pache, son of the Swiss concierge in the Hôtel de Castries in Paris.[127] The marquis had interested himself in the young Pache, who must have had a certain quality in order to be thus entrusted with companionship to the scion of a noble family. What the quality was is not known, though what Pache became is famous. He was minister of war in the early months of the Republic, from October 1792 to February 1793, when he was driven from the government by a suspicious and insecure Girondist majority. A Jacobin, he was afterward mayor of Paris throughout most of the period of the Terror in 1793 and 1794. His enemies have left the impression of one of the least amiable, least sympathetic of revolutionary figures: dry, abstract, doctrinaire, a sectarian without religion, a Rousseau without emotions, a Robespierre without courage, a scholastic of democracy perfectly representing the common man in two aspects only, his banality and his chance to make banality govern. The picture can only have been one-sided, but somehow Pache's friends never redressed it.[128] The historian of science is left to interpret the intimacy with Monge as one of those unaccountable friendships of a great man, part of his personal life. Personal the matter was, but there are indications of Pache's having been the dominant influence.

However that may be, Castries' visit to Mézières turned Monge's professional life more decisively than any personal encounter before his friendship with Napoleon. Castries was then governor-general of Flanders and Hainaut, and Monge accompanied the party back to Namur. In 1780 Castries became minister of marine affairs, where among vastly more im-

[125] Aubry (1954), 18, quoting Monge to du Marchais, 12 February, 1772.

[126] *Ibid.*, 28-30.

[127] *Ibid.*, 35, 58-59.

[128] For the inimical sense of Pache, see Marie Phlipon Roland, *Mémoires de Madame Roland* (1864), *1*, 3; *2*, 223; C-F.-D. Dumouriez, *Correspondance . . . avec Pache ministre de la guerre, en 1792* (1739); for a sympathetic portrayal, see Pierquin (1900).

portant charges, the chair of hydrography that Turgot had established at the Louvre in 1775 came within his gift. Bossut had held that appointment from the outset. The Academy of Science named Monge a correspondent in 1772. The custom being to affiliate correspondents with a regular member, Monge was naturally assigned to Bossut, his predecessor at Mézières and still his colleague as examiner. In January 1780, the Academy elected Monge to a vacancy in the section of geometry. His sponsors were Bossut, Vandermonde, and Condorcet. Monge had been increasingly in Paris in the late 1770s, revising his mathematical memoirs and seeing them into print, developing his chemical interests in occasional collaboration with Lavoisier, enlarging his acquaintanceship. Regular residence was requisite to membership, however, and Castries obligingly arranged to have him share Bossut's teaching in the chair of hydrography. Bossut was apparently happy at the prospect of some relief.[129]

After election to the Academy, Monge spent half the year in the capital, from November through May, leaving his wife and infant daughters in Mézières. In Paris the friendship with Pache became his closest, the latter having moved from the status of familiar in the Castries household to that of functionary in the naval secretariat. The marquis continued his favor and often allowed the two families the use of his country estate for holidays together. For a few years Monge's teaching at Mézières was supplied in the winter months by his brother, Louis, who taught in Paris at the École militaire and afterward returned to that post. Napoleon Bonaparte was among the pupils of Louis Monge. Inevitably, the division of responsibility and Gaspard Monge's frequent absences proved unacceptable to the commandant at Mézières. The more famous Monge became, Villelongue complained, the less use he was.[130] In 1783 Bézout died, the third of the trinity of textbook authors. He had taken over examination of artillery and naval cadets from Camus in 1768, at the time when Bossut became examiner of engineers. Bossut had hoped to garner back the whole of the Camus legacy. Instead, Castries passed him over for the naval part in favor of Monge, at the special behest of Pache and with the support of Vandermonde in the Academy. Thereupon, Monge settled his family in the capital although he abandoned his salary at Mézières only under du-

[129] On these details, see Taton (1951), 19-23.

[130] *Ibid.*, 27. Perhaps Monge's pedagogical reputation needs to be qualified by what looked to his superiors like neglect of duty at this juncture in his career. In a memoir of 10 October 1784, addressed to the minister of war, De Caux de Blacquetot, directeur de fortifications, attributed the unsatisfactory showing of eight students in physics and chemistry to Monge's dereliction: "Il résulte, Monseigneur, . . . que les élèves n'ont tiré que des secours très insuffisants de M. Monge pour leur instruction . . ." (Archives de guerre, Vincennes, Xe.159).

ress. The type of great teacher who identifies students and protégés with himself is sometimes less considerate of colleagues and of elders than of the younger generation. Bossut never again felt kindly toward Monge. As for the artillery cadets, that part of Bézout's function went to Laplace,[131] to whose conduct of their examinations, in tandem with Monge for the naval training schools, we shall return.

First, however, let us take up the important instances that illustrate the relation between education for military engineering and careers in science. The most famous scientifically of all the graduates of Mézières were Coulomb (entering class of 1760), Carnot (1771), and Meusnier (1774). Borda and Dubuat we may leave out of account for this purpose: Borda (though he and Coulomb were close friends) since he spent only a few months at the school in 1759, when he was already a naval officer and member of the Academy of Science;[132] Dubuat since his sojourn in 1750 was even more fleeting, and his important work in hydrodynamics belongs primarily to engineering.[133] As for the three who did make their mark in exact science, their command of the disciplines imparted at Mézières was clearly the prerequisite to their mature accomplishments. Equally clearly, at a certain juncture, they needed relief from the military duties of the corps for which they were trained. Membership in the Academy required them, like Monge and all their colleagues, to reside regularly in Paris rather than at some scene of fortification. That was as much an intellectual necessity as a formal requirement. To anyone with experience of life in an army, it will be obvious how the alternation beween overwork and idleness, between frustration and inanity, left an unpropitious middle ground for science or for thought of any sort.

In September 1776 Coulomb addressed a memorandum concerning the over-qualification and under-employment of military engineers to the minister of war, the comte de Saint-Germain, whom Turgot had brought into the government in the previous October. Saint-Germain had re-

[131] Duveen and Hahn (1957).

[132] Taton (1964), 586; An anonymous memoir of 5 July 1759 gives a pen-portrait of several fellow students, among them Borda, whom the author felt to be one of the two brightest people in the school: "Mais les qualités de l'esprit ne sont pas sufisantes, il faut encore celles du coeur. Mr de Borda, par exemple, est un homme vraîment savant, appliqué, actif, pénétrant; mais d'une fatuité insuportable, qui méprise tout le genre humain, regardant tous les hommes, excepté quelques-uns de ses confrères comme des animaux bons à etre roulés sous ses piés. Il n'épargne même pas Dieu. Plein d'impiété et de bavardise, il ne fait sans cesse dans nos salles que nous prêcher l'irréligion, tournant en risée tous ceux qui ne tombent pas dans son sens—quelque fois il nous fait trembler par ses discours. Son métier lui fait mal au coeur. Tout ce qui n'est pas sublime géométrie, le révolte: ce sont ses paroles." (Archives de guerre, X^e.159.)

[133] His major work, *Principes d'hydraulique* (1779) had a second edition in 1786 and a third in 1816, and was standard until well into the nineteenth century.

quested suggestions for a reformation of the corps in a manner consonant with the spirit of the Turgot administration. Coulomb was then a captain. The corps of engineers, he reminded his superiors, was both an elite defined by talent and a military organization in which the quality of the elite depended on the functioning of the organization. If promotion was slow and unrelated to ability, ambition would wilt and intellect atrophy. What was the use of running the gantlet of rigorous examinations and acquiring technical knowledge if, during forty years of service, an engineer was called on merely to rehang a gate here, repair a door-frame there, and re-point a crumbling wall in between? "On graduating from the school, a studious young man who would withstand the tedium and monotony of his duties has no choice but to lose himself in some branch of science or literature completely irrelevant to his assignment."[134]

The problem being to find serious professional occupation for 400 military engineering officers in peace time, Coulomb's solution was to put them onto a vast program of public works, which would employ as a labor force the equally idle troops of the combat arms. Technical councils would be created within the corps of engineers in order to plan and pass on these operations both at the district and the national level. Though Coulomb did not say so, these bodies would have resembled the Assemblées de ponts et chaussées, and the recognition was clear that, professionally speaking, the civil engineers might provide a model for their military counterparts. Coulomb's memorandum was as intelligent as it was politically unrealistic. The reform that actually ensued turned engineers into soldiers instead of soldiers into engineers. Coulomb himself applied for the Cross of Saint-Louis and went on inactive duty in 1781, when he was elected to a vacancy in the Academy of Science.

In physics the name of Coulomb has become the most famous to be associated with Mézières, and in science generally it is one of the most familiar among all those on the roster of the Academy. His reputation had not yet matured in the last years of the old regime, and probably not even by the time of his death in 1806. Although his experimental skill was appreciated, it required the investigations of Poisson and Ampère from 1808 through 1825 before physicists could seize on the full importance of Coulomb's extension of the inverse square law to electrostatic and magnetic forces. Coulomb began presenting that research before the Academy in 1785, and published it in a series of seven memoirs from 1787 through 1793.[135]

He was then getting on in years. Born in 1736, Coulomb was forty-

[134] Gillmor (1971), 255-261, where the entire memoir is printed as Appendix C in the standard biography of Coulomb.

[135] *Ibid.*, ch. 6, gives an analysis of the series, with bibliographical detail.

five when he was elected to the Academy, in his fifties when he produced his most important work, and fifty-four when his first son was born in 1790 (he legitimized his marriage to a girl thirty years younger only in 1802). After graduation from Mézières, he had been posted to Martinique in 1764. There he spent his professional youth designing and constructing the fortifications and harbor works of the port now called Fort-de-France, with duty also in Guadeloupe. Those tropical islands were the graveyards of their rulers. Most of Coulomb's associates succumbed to the climate and disease, and he never again enjoyed good health. On his return to France in 1772, he was assigned successively to Cherbourg, Besançon, and Rochefort, and, in the last named, was drawn into the controversy over Montalembert's scheme for a system of quick and easy wooden forts to be dotted about in strategic locations like grounded battleships.

All the while, Coulomb was writing memoirs, even like the typical engineer he described for Saint-Germain. The Academy of Science made him a correspondent in 1774 on the strength of an application of the calculus of extreme values to architectural statics.[136] He further investigated soil mechanics, dredging, windmills, and friction in a series of pieces that he intended to develop into a textbook on engineering mechanics.[137] The approach was as theoretical as the nature of the topics permitted, but his was not a primarily mathematical intelligence. Coulomb could always find and manipulate the expressions he needed to formulate empirically determined relationships, but by nature he thought like a physicist and had, in any case, passed through Mézières before Monge's time. In 1780 he was one of the candidates for the place to which the Academy elected Monge, who was ten years younger. When in 1781 Coulomb was successful in standing for the next vacancy, his strongest claim was having won two of the prizes set annually by the Academy. His entry on "The best method of making magnetic needles" won the contest announced in 1777, and an essay on friction carried the day in the competition of 1781 for a theory of machines applicable to improving the efficiency of naval equipment.[138]

The piece for the 1777 prize started his work in magnetism, and only after Coulomb became a member of the Academy was he free to go forward with electromagnetic researches into pure physics. His election signified, not the advent of a founder of modern physics, but recognition by the Academy of the claims of an engineer, who was thereby able to become a scientist. Clearly, recognition took an engineer rather longer to achieve than it did persons who gained acceptance through the established

[136] Heyman (1972) has translated, edited, and discussed this work for its significance in the history of civil engineering.

[137] Gillmor (1971), p. 38.

[138] "Théorie des machines simples . . . ," SE *10* (1785), 163-332.

disciplines. Even then, Coulomb earned his living in the 1780s as administrator of the water supply for the city of Paris. In that capacity he presided over the commission that approved installation by the brothers Périer of two steam engines with separate condensers of the type invented by James Watt for their pumping station on the Butte de Chaillot.[139]

Lazare Carnot, seventeen years Coulomb's junior, never did win scientific notice until after he became great and famous for political and military reasons in the Revolution. Like Coulomb, Carnot served in a succession of posts, Calais, Le Havre, Cherbourg, Béthune, Arras, and Aire, none of which fully engaged an ambitious young officer's energies or provided opportunity to win distinction and advancement. Also like Coulomb, he occupied his mind in studies of his own, his choice of subject being prompted by academic contests. The two main motifs of Carnot's mature work were, first, application of the principles of mechanics to the analysis of the operation of machinery, and, second, justification of the procedures of the infinitesimal calculus, also in operational terms. Carnot submitted an early essay along the former line to the Academy in Paris in competition for the 1781 prize on friction won by Coulomb. He got an honorable mention, and went on to develop his essay for publication in 1783. This *Essai sur les machines en général* contains statements of all the contributions he ever made to engineering mechanics, notably the principle of continuity in the transmission of power, the definition of the quantity later called work as the measure of the efficacy of machines, and the notion of geometric motion that his son Sadi developed into the reversible processes of thermodynamics. Although his little book contained these nuggets, it attracted no attention whatever. In 1785 Carnot entered a second essay in a contest set by the Prussian Academy in Berlin on the theory of infinitesimals in the calculus. That memoir, which also won an honorable mention, turns out to be an early draft of his most famous treatise, the *Réflexions sur la métaphysique du calcul infinitésimal*, which Carnot brought out in 1797, after having been war leader in the Revolution. He was a member of the ruling Committee of Public Safety at its climax and of the Executive Directory in the aftermath of the Terror. The manuscript of 1785 was completely lost to view until it came to light in the archives of the German Academy of Science in Berlin in 1968. There are few more startling examples of the point that in science, as elsewhere, much depends on entrée, on being known just a little. In the 1780s, Carnot managed to make himself known—just a little—but not by the merit of his scientific writings. He accomplished it by winning another prize, this one set by the provincial Academy in Dijon, for an éloge of Vauban.[140]

[139] Gillmor (1971), 60-65; on the Périer brothers and water supply see Bouchary (1946); Payen (1969).

[140] On the scientific career of Lazare Carnot, see Gillispie (1971a).

Among all the graduates of Mézières none sounded new notes more enthusiastically than Meusnier, and none seems more energetic a harbinger—Jean-Baptiste Meusnier de la Place, to give him his full family name.[141] His was an engineering ardor innocent of those elements of frustration, that sense of unappreciated or unrequited merit in oneself, which sometimes embittered the zeal of innovators of more abstract or introverted a temper. Killed in line of duty in June 1793, a general officer and second-in-command at the defense of Mainz, he has won a kind of immortality as the martyr-paladin of the École polytechnique. In a memorial address there, Monge said of his favorite student that Meusnier's was the most remarkable intelligence he had ever encountered.[142]

On the day Meusnier arrived at Mézières, 1 January 1774, he asked to be given a problem. Monge set him the demonstration of a theorem of Euler specifying the maximum and minimum radii of curvature to certain surfaces. The next morning the new pupil came back with a proof more direct and economical than Euler's own. Monge was just then starting to investigate the integration of Lagrange's second-order partial differential equations. He attributed to Meusnier the root idea of determining which one among a family of surfaces would satisfy a problem (a family being characterized by a common mode of generation). Meusnier exploited the approach for himself in discovering two special cases of a minimum surface, the catenoid and the right helicoid. His derivation of those forms from Lagrange's differential equation of minimal surfaces is contained in the memoir demonstrating a theorem in differential geometry that still goes by his name: "Every element of a surface may be regarded as generated by the rotation of a small arc of a circle around an axis parallel to the plane tangent to the element."[143] Meusnier read the paper before the Academy of Science on 14 February 1776, six weeks after his graduation from Mézières and four months before his twenty-second birthday. At the next election, the Academy named him correspondent. Anyone present at this academic triumph would have predicted a brilliant future for Meusnier in exact science.

He never made another contribution to mathematics. His temperament and milieu led him instead into problems of invention, design, and development, which it is conventional to consider of a lower order. Meusnier felt no such contrast. His taste, indeed, portended the positivist transition in science from contemplation to action. The Academy attached him to Vandermonde for his correspondence. The assignment reinforced the as-

[141] On Meusnier's career, see Darboux (1910), who made much use of biographical notes by Monge *et al.* in *Revue retrospective*, 2nd ser., *4* (1835), 77-99. See also Laissus (1971), and for the works on aerostatics and aeronautics Letonné (1888) and Voyer (1902).

[142] *Revue retrospective*, p. 83; Darboux (1912), iii-v.

[143] "Mémoire sur la courbure des surfaces," SE *10*, pt. 2, 477-510, p. 483.

sociation of both with Monge and favored the tendency among all three to rational technology. Meusnier's sense of the prospect envisioned rather a conquest than an exploitation of nature. Modern instances of the sort of thing to which his spirit of adventure responded might be the exploration of space or the domestication of nuclear fusion, for in the actual event, what appealed most powerfully to his imagination was aeronautics.

In 1779 the Corps of Engineers assigned Meusnier to Cherbourg, his first important post, where Coulomb and Carnot also served. Modernization and militarization of the port was one of the more ambitious engineering projects of the reign of Louis XVI, the goal having been to complete unrealized plans of Vauban on an enlarged scale. There Meusnier made an enemy of no less a personnage than the commandant, François du Perier Dumouriez, the officer who in 1792 saved the Revolution at the Battle of Valmy, led the offensive that conquered the Austrian Netherlands in 1792-1793, and then for political reasons delivered himself and a captive minister of war into Austrian hands, having failed to get his army to go over to the enemy. Others who took military part in the Revolution, notably Carnot and Choderlos de Laclos, conceived their first distrust for Dumouriez at Cherbourg, but Meusnier stood out there as leader in the party of dissident young engineers. Some were of his own corps and others of the Ponts et chaussées. They became skeptical of the design of certain works and suspicious about the propriety of contracts let to construct them. Meusnier carried their complaints over Dumouriez's head to the high command in the person of the duc d'Harcourt, and prevailed. More constructively, a major task was fortification of the Île Pelée, which shelters an important mooring basin. The island had no fresh water, and Meusnier designed a system that would supply the lack from sea water, running the distillation in a partial vacuum to save fuel and harnessing the tides to create the vacuum.[144]

When the Academy named Meusnier a correspondent, he volunteered actually to work and was given charge of the *Recueil des machines approuvées par l'Académie*. The series consisted of six volumes describing inventions that had found favor with the Academy among the much larger number submitted for its approbation. The publication was then forty years behind the times. In 1777, after less than a year, Meusnier put a seventh volume into print, bringing the coverage through 1754. Conscious of its collective guilt in respect to the enormous lag in publication, the Academy adopted a proposal advanced by Vandermonde to the effect that the Ministry of War be requested to allow Meusnier an annual leave of six months in Paris in order to bring the enterprise up to date.[145]

[144] Darboux (1910), vi-ix.

[145] *Ibid.*, ix-x. On the *Machines et inventions approuvées par l'Académie* . . . , 6 vols. (1735), see above, Chapter VI, Section 4.

The démarche succeeded, and Meusnier regularly spent half the year in the company of the scientific community. He was thus in a position to apply his insight and training to analysis of the most exciting of the inventions of the next decade, the flight of balloons, and thereby to inaugurate the discipline of aeronautical engineering. We have already observed the launching of the Montgolfier hot-air balloon in Annonay on 4 June 1783, followed Etienne Montgolfier to Paris, and seen how, in the well-compartmentalized mixture of his motives, he capitalized on the notoriety of the feat to win from the government commercial concessions, a particle of nobility for his father and family, and the appellation "royal" for their paper plant at Vidalon.[146] The take-off of enthusiasm for flight, which Etienne Montgolfier thus skillfully exploited in the summer and autumn of 1783, offers one of the rare instances wherein the scientific community fully shared the pleasure of the general public. The latter motif is fully evident in the charming iconography of aircraft lighter than air, in the new expanse their surface offered the imagination of the decorative artist, in the lift of the spirits experienced even nowadays when a raised glance chances to light on a brightly colored globe drifting all quietly across a summer sky.

The scientific interest prompted by aerostatics (which Meusnier began transforming into aeronautics at the outset) has been less celebrated, although it is agreeable to note that the nearly blind and failing Euler felt it instantly. News of the first balloon in Annonay reached Saint Petersburg in the late summer, and Euler died of a stroke on 7 September. The last calculation found on his slate derived "the laws of vertical motion of a globe rising in calm air in consequence of the upward force owing to its lightness."[147] Johann-Albrecht Euler copied off his father's equations and sent them to the Academy of Science in Paris, which, in paying Euler the tribute of printing his last piece, thus had itself the honor of publishing the first recorded mathematical analysis of the flight of aircraft. The gesture makes a graceful link between one age and another. For the present purpose, however, a full account of the ensuing development would be out of scale, and an outline of the interplay between science and the art of ballooning will have to suffice until another occasion. Here the main thing is to be clear that the matter was more complex than a simple competition between proponents of hot air and of hydrogen.

So soon as the procès-verbal recording the Montgolfier demonstration before the provincial estates of Vivarais reached Paris, the controller-general, Lefèvre d'Ormesson, requested the Academy of Science to create a commission to take cognizance of developments in flight. Its membership

[146] Above, Chapter VI, Section 3.

[147] "Calculs sur les ballons aérostatiques faits par feu M. Léonard Euler . . . ," MARS (1781/84), 264-268.

was named in July—Tillet, Brisson, Cadet, Bossut, Desmarest, Condorcet, Lavoisier, and Leroi, the last being secretary. The president of the Academy, the duc de La Rochefoucauld, was a member ex officio. At the invitation of this group, Etienne Montgolfier came on to Paris from Annonay in mid-July.[148] Meanwhile, experiments of another sort were under way. Let us recapitulate the chronology of the four flights launched before the end of the year 1783.

On 27 August the physics lecturer, J.-A.-C. Charles, together with two shop assistants, the brothers Robert, released a balloon filled with "inflammable air" (hydrogen) from an emplacement where the statue of Joffre now stands in the Champ de Mars. Made of taffeta impregnated with a rubbery gum dissolved in turpentine, it was much smaller than the original Montgolfier hot-air balloon, being twelve feet in diameter and weighing twenty-five pounds as compared to the 35-foot diameter and 450-pound weight of its predecessor in Annonay. The wind was south-south-west and the weather threatening. The little *charlière* rose rapidly to a height of about 1,500 feet, disappeared across the Seine into the rain, reappeared for a moment through a break in the clouds, and fell to earth after forty-five minutes near Ecouen about twenty kilometres north of Paris. Although the astonished peasants nearly destroyed the remnants dragging them to the village of Gonesse, it was possible to determine that the flight had ended when the bag burst from an excess of internal pressure at increasing altitude.[149]

On his arrival a few weeks earlier, Montgolfier had installed himself in the establishment of his great friend and business associate, Jean-Baptiste Réveillon, the manufacturer of wallpaper in the faubourg Saint-Antoine. There he fabricated a balloon of canvas and paper with twice the capacity of the Annonay prototype, that is to say a volume of 45,000 cubic feet and a height of 70 feet. On 12 September the commissioners of the Academy gathered in the courtyard of the Réveillon plant to observe a trial. Montgolfier kindled a fire of straw and woollen shreds beneath the open mouth of the bag draped upon a scaffolding. Retained by cords, the great sack filled and lifted off with a load of 400 pounds, making 1,250 pounds including its own weight. No sooner was it airborne than a gust of wind

[148] "Rapport fait à l'Académie des Sciences sur la machine aérostatique, inventée par MM de Montgolfier," HARS (1783/86), 5-24, and separately printed by the bookseller Moutard (1784), 6.

[149] The fullest eye-witness accounts of the balloon ascensions in Paris in 1783 are in Barthélemy Faujas de Saint-Fond, *Description des expériences de la machine aérostatique de MM de Montgolfier et de celles auxquelles cette découverte a donné lieu* (1783) and *Première suite de la description des expériences aérostatiques* . . . (1784). The 27 August flight is described in the former work, 7-21; cf. Meusnier "Lettre à Monsieur Faujas de Saint-Fond . . ." in the same volume, 49-162.

and sudden downpour tore it loose and destroyed the whole fragile contraption.

The king had set 19 September, just a week later, for a demonstration at Versailles. Working against that deadline, Montgolfier built a second, smaller balloon—some 57 feet high, 41 feet in diameter, and weighing about 700 pounds—and tried it on the 18th. Again a wind sprang up, whipping the globe against the scaffolding and tearing out a seam. There was time only to baste. At Versailles the next day the weather was fine with a light easterly wind. The entire court and the world of fashion assembled on the terrace behind the chateau. The king, queen, and the royal family came down to the launching pad, and Montgolfier explained the design. The balloon filled smoothly over the brazier. At the last moment a basket was suspended from the bag and a sheep, cock, and duck were tethered inside, adding 200 pounds burden. Two astronomers, Le Gentil and Jeaurat, armed with quadrants, prepared to observe the elements of flight from different angles. The balloon reached an altitude of 240 toises—1,440 feet—and traversed a distance of 1,700 toises, say three and a half kilometers, in a flight of ten minutes duration. It would have floated further except that the damaged seam opened a little. Among the observers rushing to the scene were veterinarians who determined that the animals had suffered no ill effects from their passage through the air.[150]

Already at the end of August, François Pilatre de Rozier, a young blade about town on the fringes of science, had approached the academic committee with a request to fly in the first craft to take a man aboard. Pilatre was known as the impresario of the fashionable Musée in the rue de Valois.[151] Immediately after the success at Versailles, Montgolfier set to work to design and construct another "aerostat" on the scale of the one destroyed in the trial of 12 September. It was an affair 70 feet high, and 46 feet in diameter, with a capacity of 60,000 cubic feet. Suspended from the bottom was a circular gallery made of wattles with a three-foot railing and a brazier in the center. The weight without passengers was 1,600 pounds. Pilatre mounted the craft for a captive test on 15 October. In the presence of the committee, he flew the balloon to the limit of the retaining cord, a height of about 100 feet, controlling the altitude by alternately blowing up and damping the fire. After further trials with other would-be aviators, namely the marquis d'Arlandes and one Giroud de Villette, the launching was set for 21 November. Lift-off was in the garden of the Château de la Muette, in the presence of the two-year-old dauphin, his entourage, and another huge and cheerful crowd. With Pilatre and

[150] Faujas, *Description* . . . (1783), 29-48.

[151] Above, Chapter III, Section 1.

d'Arlandes aboard, waving their hats to the people, the balloon rose rapidly to a height of about 1,500 feet, where they caught a westerly wind that took them across the southwestern section of the city for a flight of about seventeen minutes to a soft landing on the road to Fontainebleau.[152]

All this while Charles and the brothers Robert were developing the hydrogen alternative in a new shop at Saint-Cloud. On 1 December, ten days after the Pilatre-d'Arlandes flight, Charles himself and the younger Robert took off from the garden of the Tuileries riding in a gondola suspended from an elongated balloon consisting of a cylindrical part 20 feet in length enclosed between two hemispheres 30 feet in diameter. Their ascent was meteoric. A southeast wind carried them near to the town of Nesles, about 45 kilometers from Paris. There Robert alighted, and Charles took off alone for a second hop, rising into the clouds to an altitude of 9,000 feet and more before throwing over his ballast. When he came down at La Tour-du-Lay, his ears hurt. He never flew again.[153]

So ended the first year of flight, and we shall not follow the repetitions and variations in the provinces, or in Britain, Germany, Italy, and Pennsylvania. Symmetry, however, requires noticing Pilatre's combination of the two lifting agents in an attempt upon the English channel. He designed a compound craft in which a hot air was surmounted by a hydrogen balloon. Having studied the physics of gases, he rejected well-meant warnings and assumed that if any "inflammable air" escaped from the upper envelope, its lightness would cause it to rise faster than any sparks from below. His apparatus was ready at Boulogne-sur-Mer in January 1785. The prevailing winds come from England, and Pilatre was forestalled by J.-F. Blanchard, a Frenchman, and John Jeffries, an American, who crossed from Dover in a simple hydrogen balloon. A good sport, Pilatre persisted. In June, weather seemed propitious. Pilatre took off with a companion on the 16th. They reached an altitude of about 1,700 feet before a sheet of flame engulfed the double balloon, and the first airman fell victim to the first air crash.[154]

Like Etienne de Montgolfier—though unlike his brother Joseph who went up once, early in 1784—Meusnier preferred the design and analysis to the practice of flight. Let us return to the engineering, therefore. Leroi

[152] On the Pilatre-d'Arlandes flight, see Faujas *Première suite* (note 149 above), 11-22, and for d'Arlandes's own account, 23-30.

[153] *Ibid.*, 31-61; cf. Meusnier "Mémoire sur l'équilibre des machines aérostatiques . . ." *Observations sur la physique* . . . 25, pt. 2 (July 1784), 36-69; reprinted in Darboux (1910), 61-91.

[154] Marat's pamphlet on this catastrophe has already been noticed, above Chapter IV, Section 3, n. 197. On the Pilatre debacle, see Smeaton (1955), modified by Birembaut (1958b).

delivered the report of his aerostatical committee to the Academy on 23 December 1783, a little over three weeks after the Charles-Robert exploit. He and his colleagues concluded that it was still too soon to make a choice between hot air and hydrogen. The former being cheaper and easy to produce might prove preferable for civil purposes, and the latter being more efficient and controllable might be the more eligible for scientific purposes. Having been appointed to report on the Montgolfier machine, the commission praised the pioneers and recommended the brothers to the Academy for a special prize.[155] Scientific circles in Paris were clearly more interested in the alternative, however. Indeed, when word first came from Annonay early in the summer, physicists surmised that hydrogen must have been the lifting agent, for the Montgolfiers never made their procedures public prior to Etienne's arrival in Paris in July to demonstrate them. Charles thus began work, not as advocate of a competing method, but as investigator into how the thing was possible at all.

In 1783 Charles, a onetime protégé of Benjamin Franklin, was known for the brilliance of the lectures he had been offering on experimental physics in the two previous years. He consulted with the academic commission from the outset, and also with Meusnier. The latter (it would appear) was drawn in by the intermediary of Barthélemy Faujas de Saint-Fond, the vulcanologist, who had come to know the Montgolfier family during his research on the geology of the Vivarais, even as Desmarest had done both through his geological travels and also through his inspector's responsibility for the paper industry. Desmarest was a member of the balloon commission, while Faujas took it on himself to become the chronicler of these first flights and, at the same time, to be the Montgolfiers' advocate in the capital. He there opened a subscription to finance development of the invention. Charles drew upon the fund to defray the expenses of his first, unmanned balloon.[156] Faujas also invited Meusnier, charged by the Academy with the responsibility for machines and inventions, to consult on the design of the first charlière. Meusnier responded with a proposal to compare its performance with the theoretical motion of a voluminous body governed by the action of two opposing forces, the one varying directly as the density of the atmosphere and the other serving the laws of air resistance.[157] Existing data consisted of Jean-André de Luc's uncertain extrapolation into upper reaches of the atmosphere of Alpine observations

[155] "Rapport fait à l'Académie des sciences sur la machine aérostatique," HARS (1783/86), 26-27.

[156] "Lettre à Monsieur Faujas de Saint-Fond . . ." by an anonymous correspondent, in *Description des expériences* . . . , 196-197.

[157] Meusnier, "Lettre à Monsieur Faujas de Saint-Fond . . ." *ibid.*, 49-50, and in Darboux (1910), 7-8. For analysis and appreciation of Meusnier's work in aeronautical engineering, see also Letonné (1888).

of the variation of air density and temperature with altitude, and also of estimates of the relation between pressure, temperature, and volume in enclosed gases.[158] The law named for Charles governing the latter set of relations was stated in 1787, and certainly this was the research that put him onto it.[159]

Unfortunately, Meusnier had too little time to concert his plan for tracking the Champ de Mars balloon of 27 August, the first to be launched in Paris. The experiment called for observers with astronomical quadrants to take simultaneous reading from several vantage points. Three sets would have sufficed to triangulate successive positions of the balloon, but Meusnier decided on four in order to allow a margin for error. He stationed Le Gentil at the Observatory of Paris, Jeaurat upon the roof of the Garde-Meuble in what is now the place de la Concorde, Prévost on one of the towers of Notre Dame, and d'Agelet, professor of mathematics at the École militaire, upon the dome of that school. Alas, d'Agelet gave the others their instructions verbally and at the last moment. Rain fell. Gusts blew. The crowd grew impatient and demanding, and the balloon was released before the signal could be given to synchronize watches among the four observers. Fortunately, Jeaurat noted the moment when he heard the launching cannon. This good luck permitted his observations to be related to those of d'Agelet, who was right on top of the site. Meusnier retrieved what he could from the near-fiasco by an elaborate calculation establishing the time and location of several positions in the trajectory from two sets of observations made in the rain, haste, and veering winds before the balloon careered off into the clouds.[160] Thereafter, he settled down to work deliberately with Charles in the design of the second hydrogen balloon, the cylinder between hemispheres, that made the spectacular flight of 1 December. There is no record that Meusnier had anything to do with Etienne Montgolfier or took any interest analytically in the flight of the unmanned hot-air balloon at Versailles or its manned successor at La Muette. It was specifically the rupture of the first charlière, resulting from the increase of internal pressure with altitude, that led him into analysis of all the equilibrium considerations involved in ballooning.[161]

On 3 December 1783, two days after the spectacular Charles-Robert flight, Charles and Meusnier appeared before the Academy of Science,

[158] Contained in *Recherches sur les modifications de l'atmosphère*, 2 vols. (Geneva, 1772), expanded and re-issued in a 4-volume edition (Paris, 1784). On De Luc, see the article by Robert P. Beckinsale, DSB *4*, 27-29.

[159] On the career of Charles, see the article by J. B. Gough, DSB *3*, 207-208.

[160] Meusnier's description and calculations constitute the burden of his "Lettre à Monsieur Faujas de Saint-Fond," reprinted in Darboux (1910), 7-56.

[161] *Ibid.*, 55-56.

Charles to tell of his aerial journey and Meusnier to present a memoir developing the principles underlying the design of the elongated craft.[162] The shape was dictated partly by the desirability of reducing the strain of air resistance in lateral displacement and partly for structural convenience in suspending the gondola from the envelope. Decisions about capacity and the trade-off between buoyancy and pressure of the enclosed gas were dictated by calculations derived from the geometry of curved surfaces: the tension being greater in the cylindrical portion, Charles reinforced the taffeta there with ribbing. The memoir offers an especially limpid instance of the carry-over from the new geometry into rational engineering, and its most impressive feature is Meusnier's seizure on the principle by which craft floating submerged in a fluid medium have ever since been stabilized in altitude above the earth or depth beneath the surface of the sea. In the months after the rupture of the Champ de Mars globe, and weeks before anyone had ever actually flown in a balloon, Meusnier on entirely analytic grounds described the principal inconvenience to be encountered by aeronauts aboard the devices imagined up to that time.

Once a simple balloon reached its ceiling, a pilot would be powerless to arrest its descent except by throwing out ballast, whereupon he would bob up a certain way. Nor could he stop his ascent at a given altitude otherwise than by releasing irreplaceable hydrogen, whereupon he would immediately begin descending. Thus must he oscillate, wasting altitude in bounds of increasing amplitude as the proportion of weight to remaining buoyancy increased. The main burden of Meusnier's memoir is to describe the remedy he had imagined and designed a priori for that inconvenience. In order to achieve a steady equilibrium between the weights of the balloon and of the air displaced at any attainable altitude, either the weight or the volume of the machine might be varied. No practical means could be devised to alter the volume of the bag in exactly inverse proportion to the density of the atmosphere. The alternative remained of varying the weight while holding the volume constant. The problem being to eliminate the inequalities set up by jettisoning ballast or releasing gas, "What then could be added to an isolated body, if not the very air in which it swims? That is what no one has thought of, and yet all the difficulties would immediately disappear."[163] Meusnier outlined several pos-

[162] "Mémoire sur l'équilibre des machines aérostatiques . . . avec une addition contenant une application de cette théorie au cas particulier du ballon que MM Robert construisent à Saint-Cloud," *Observations sur la physique* (Rozier's journal), 25, pt. 2 (July 1784), 39-69; reprinted Darboux (1910), 61-91. Meusnier followed this memoir with a brief notice, "Calcul des différentes élévations auxquelles a dû parvenir le globe aérostatique . . . lancé du Jardin des Tuileries . . . d'après la seule consideration du poids . . ." *Journal de Paris*, 29 December 1783, in Darboux (1910), 57-59.

[163] "Mémoire sur l'équilibre des machines aérostatiques . . . ," Darboux (1910), 66. On this design, see Malécot (1910).

sible designs for incorporating in the balloon a compartment, sealed off from the hydrogen, a bladder which might receive or expel atmospheric air to keep the parent craft floating stably: he would have preferred surrounding the hydrogen bag with an envelop of air. The model actually built by the Robert brothers incorporated the air bladder inside at the throat. A bellows served as compressor, and Charles invented a valve for releasing excess air when the whole assemblage needed lightening. Perhaps it will seem no derogation from Meusnier's ingenuity that his principle has proved more valuable in regulating the depth of submarines than in the few blimps that survive the extinction of the dirigible, never a lusty species, even after helium came to replace hydrogen.

That fine memoir was the making of Meusnier scientifically. Clearly a principal factor in the concentration of scientific interest in hydrogen was precisely its strategic importance in the chemical revolution. The experiment in which Lavoisier and Laplace demonstrated the composition of water by synthesis took place on 24 June 1783, almost simultaneously with arrival in Paris of the news from Annonay.[164] This was also the juncture at which Monge was preparing his move to Paris. He too (it will be recalled) had a claim to having established independently that water is a compound of hydrogen and oxygen.[165] The closeness of Meusnier to Monge can only have heightened the young engineer's prospective value in the eyes of Lavoisier, who announced his discovery of the composition of water at the public meeting of the Academy of Science in November.[166] From the point of view of Lavoisier, provision of hydrogen on the scale required for the Charles-Robert balloon was the aspect of aerostatics germane to his own research. Immediately after the meeting of the Academy on 23 December, where Leroi read out his commission's report on the montgolfières with its comparison of hot air to hydrogen, Lavoisier convoked a further meeting of the same group, which might otherwise have been expected to disband. The first session of the continued committee occurred on 27 December in the Hôtel de la Rochefoucauld, the duke presiding and Lavoisier leading the proceedings.

There Lavoisier identified four questions to investigate for the future development of aeronautics: (1) Improving the impermeability and lightening the weight of the fabric; (2) choice and preparation of the lifting gas; (3) perfection of methods for regulating altitude without losing gas or ballast; and (4) studying methods for locomotion and steering. On the third point Meusnier had already charted the way, and since he had quite evidently thought hard about all these matters, the commission would do well to coopt his services.[167] Lavoisier had his way, as he usually did in

[164] Guerlac, "Lavoisier," DSB *8*, 78.

[165] Perrin (1973).

[166] Lavoisier, *Oeuvres*, *2*, 334-359.

[167] Darboux (1910), xiv.

these years. On 28 January the Academy of Science elected Meusnier to membership, preferring his candidacy over that of Charles. Theory thus preceded practice in academic eligibility no less than in the experiment of flight. Thereupon Meusnier could become formally a member of the commission, whose moving spirit he already was. As the time drew near for his return to active duty in the corps of engineers, Lavoisier seconded by La Rochefoucauld requested Meusnier's superiors to extend his leave to a year. "You would be doing the Academy a real service," wrote Lavoisier to Fourcroy de Ramecourt, the chief of engineeers.[168] The military authorities acceded, and allowed Meusnier to remain in Paris to continue aeronautical research, and also to perfect his machine for the distillation of sea-water, in which the naval ministry was specially interested.

The first fruits were the famous experiment in March 1784 in which Lavoisier and Meusnier decomposed water into its elements, thus complementing and confirming Lavoisier's synthesis of the previous year, when he and Laplace had shown that water is the product of the combustion of hydrogen. Working in the laboratory at the Arsenal, Lavoisier and Meusnier extracted eighty-two pints of hydrogen by dripping water through a white-hot gun barrel packed with iron filings. Lavoisier allowed his new associate to present the results to the Academy, for the experiment was designed and prepared by Meusnier, who proceeded to enlarge and modify the apparatus sufficiently to permit collecting hydrogen and oxygen from upwards of forty gallons of water at a single run, a scale approaching the industrial.[169]

Meanwhile, Meusnier was at work on the design of enlarged aircraft and on the problem of locomotion. At a public meeting of the Academy of Science on 13 November 1784, he delivered a précis of the work he had accomplished during his ten months of grace, accompanying it with detailed specifications for construction of the models he had imagined and a splendid atlas of drawings. Although he held but modest hopes for the dirigible, he designed a helical propeller to be worked by a system of cranks. The form, of course, is that which has since served to drive both ships and aircraft, for Meusnier's mechanical instinct was sure. All he lacked was an engine. A speed of one league (three miles) an hour on a still day was the best to which a pilot might aspire, even with a more elongated shape than that of the Charles-Robert craft. Only in selecting landing sites and observation points was locomotion worth attempting, therefore. The effective mode of air travel would always lie in studying to know and take advantage of wind patterns. Not that the limitation seemed severe: Meusnier's specifications call for construction of two types

[168] Lavoisier to Fourcroy de Ramecourt, February 1784, *ibid.*, xvi.
[169] *Ibid.*, 5, 320-339; on this work, see Daumas and Duveen (1959).

of aircraft. One was destined for long voyages, "even overseas and in little known climates" (p. 95). It would carry thirty men and provisions for sixty days and cost more than three million livres. The other, intended for only six men, would serve for experimental cruises on the continent. "Besides the advantage of showing what might be hoped for in aerial navigation," noted Meusnier, "the execution of such a project would secure observations of great importance to science, which is utterly lacking in data on the constitution of the atmosphere."[170]

That project would have cost only 400,000 livres.

Such was Meusnier's style of mind, analytic rather than simply ingenious, his practicality with actual machines graduating by way of rationalism into the higher unrealism of the visionary engineer. He returned to active military duty after his period of scientific leave in Paris, and little more was heard of him before the revolutionary years. Perfecting the harbor works at Cherbourg may well have absorbed his energies. He was not the man, however, to accept the ceiling of a captaincy beyond which a commoner might not aspire in the Corps of Engineers. In 1786, after twelve years as a lieutenant, he had yet to reach even that rank. On 12 July he sent a petition to the maréchal de Ségur requesting transfer to the general staff corps. Only by accident, he observed, had he taken on construction of the Cherbourg roads, and he defined his career as an attempt to perfect his knowledge of all those aspects of the art of war that might have relations with the exact sciences, the subject of his earliest studies. His superiors endorsed his application for transfer: a very talented officer, they said; "it would be a pity if the regime of the corps [of engineers] limited his advancement any longer."[171]

Meusnier withdrew his petition, it seems, when he was promoted captain in May 1787. In the following year he was given the assignment of Maréchal-général des logis with the rank of major. That recognition elicited a further petition from his colleagues and subordinates at Cherbourg, who were fearful lest it entail his separation from the engineers. More important than their work, their pride would suffer: "If we lose him, if we can no longer number him among us, and if we have to add his name to those of so many distinguished men who have already left our ranks, we shall be led to feel that the title of engineer, earned by so much effort, can no longer satisfy our ambition."[172]

Ambition had been whetted by the increasing attractiveness of the learned arms, including the artillery and the navy, and by the growing prestige of the Corps des ponts et chaussées. The number of technically

[170] "Précis des travaux faits à l'Académie des Sciences de Paris pour la perfection des machines aérostatiques," in Darboux (1910), 95-96.

[171] Darboux (1910), xxiii.

[172] *Ibid.*, xxiv.

educated people serving in one or another of those services was in the neighborhood of one thousand in the last decade of the old regime, a figure to be compared to the estimate of three to five thousand fully qualified physicians and to the much smaller population of persons who might reasonably be called practicing scientists.[173] Let us be generous and say that there may have been several hundred in all France at a time when the regular membership of the Academy of Science in Paris was still only forty-eight. Formally, at least, engineering might appear to have overtaken science in vesting itself with certain attributes of a profession: admission was by examination in the course of a prescribed process of education, and not by induction into a privileged body. Nowhere, indeed, does the professional significance of schooling come out more clearly than in the evolution of eighteenth-century engineering. In this respect, as in the shape of careers, the humbler civil engineering seems more inceptive than its military counterpart (the contrast being rather like that between surgery and medicine). For the authorities of the Ponts et chaussées conducted their own examinations while the military engineers relied on external examinations by scientists.

More interesting than that slight difference, however, is the evidence for the growing reliance upon the examining process by high officers of state and the corresponding involvement of important scientists. The first examiners for the military, Camus and Bossut, were men of little moment scientifically. Bézout had greater weight, and we have seen how when he died in September 1783, his responsibilities, respectively, for the navy and the artillery were divided between the two leading mathematicians of the Academy, Monge and Laplace.[174] Both were eager to have the stipends since the duties were less time-consuming than was teaching. In the maneuvering to secure Bézout's succession, both Monge and Laplace undertook (it would appear) to continue examining on the basis of Bézout's textbooks, from the sale of which his widow and children had to live.[175] The arrangement entailed no sacrifice of standards, for Bézout's courses were composed at a higher level of mathematical literacy than Bossut's and were altogether superior in clarity and organization, even as these latter represented a considerable advance upon those of Camus. The matter was more than pedagogical and economic, however. Discharging their duties introduced Monge and Laplace to the interchange of regular communications with persons in high places on questions involving policy and the lives of others.

The duties of Monge were the more demanding, largely because the relatively unsatisfactory state of naval education required far-reaching cor-

[173] Above, Chapter II, Section 2; Chapter III, Section 6, n. 196.

[174] Above, this section.

[175] Duveen and Hahn (1957).

rection. His experience at Mézières, his intimacy with Pache (whom Castries had installed in the ministry), and the minister's confidence in both—those circumstances imposed on Monge a general oversight of technical training for service in the navy and in the colonies.[176] His approach was radical. On 1 January 1786 Castries issued an ordinance abolishing the naval cadet corps—*gardes de la marine*—which for a century and more had failed to provide the navy with well-trained officers.[177] In place of schools badly run by the navy itself in Brest, Rochefort, and Toulon, Castries followed the example of Saint-Germain in the army and looked to certain existing provincial colleges to give future cadets their secondary education, technical as well as general. Candidates for the navy were then to present themselves before Monge for an entrance examination into maritime schools newly established in Vannes in Brittany and in Alès in Languedoc. Cadets to the number of 360 were divided into three classes according to their attainments, and the period of study lasted anywhere between one and three years, including cruises. On its completion, Monge examined the candidates again, certifying those who satisfied him with the technical capacity to become ensigns. Commoners who could not qualify genealogically for the schools might also volunteer for the examinations and be received as officers if they succeeded on their own.

All this took time. Monge passed the months of April and May annually in Vannes and Alès, inspecting the instruction as well as conducting tests. Thereupon, he repaired to Lorient to evaluate candidates for the colonial forces, and thence to Toulon, Brest, and Rochefort, where he set more advanced examinations required of certain categories of officers for promotion to the rank of lieutenant. Finally, in December and January he had to pass on the qualifications of applicants in Paris for commissions in the colonial artillery. Thus would Monge spend a good half the year on the road. In the course of his peregrinations he became a familiar of the entire naval establishment, drawing a stipend of 4,400 livres.[178]

Laplace received approximately the same amount from the war department, 4,000 livres together with travel expenses of 1,500, although his duties in the more stably organized artillery occupied only about a month of his year.[179] His post was no sinecure, however. Students of the École militaire had the privilege of being examined in Paris. Others with appropriate credentials received a letter from the minister of war directing them to report to Metz, where they would be temporarily subject to the orders of the commandant of the artillery school and answer for their tech-

[176] Aubry (1954), 65-67.

[177] R. Hahn, "L'Enseignement scientifique des gardes de la marine au XVIIIe siècle," Taton (1964), 547-558.

[178] Many of Monge's notes on his examinations are in Archives de la Marine, C^8.18.

[179] Laplace to Lagrange, 11 February 1784, quoted in Duveen and Hahn (1957), 424.

nical preparation before Laplace.[180] On his first tour, in July and August of 1784, 87 of the 114 admitted by the ministry actually appeared. The system was to conduct a single examination qualifying "aspirants" for entry to the artillery school and graduates for commissions in the corps. The few aspirants who scored in the range required for lieutenancies might then be exempted from further schooling and commissioned forthwith. On this round, Laplace certified 24 graduating students and eight aspirants for commissions.[181]

In 1785, the ministry authorized 202 men to confront Laplace, of whom 132 actually appeared and 58 passed the examination.[182] Since fifteen of that group had to be content with supernumerary status for lack of assignments in the corps, the minister requested Laplace to raise the standard in the next year. Despite that discouragement, another surge of 140 candidates came forward in 1786. Even though Laplace followed the directive to be "more difficult," he still had to pass 48 for the school and 16 into the corps. The rise in quality he took to be evidence of the increasing excellence of their schooling as well as of the soundness of a proposition to which he had long subscribed, namely "young people will increase their effort to succeed in proportion to the increase in difficulty" of their studies.[183] Laplace filed a report on every candidate, and his covering letter to the ministry also recommended the most successful and conscientious teachers for special recognition and bonuses. In 1785, he mentioned Dom Enard of the Benedictine college at Metz, together with the abbés Plassiard and Thorin of the College of Saint Louis in the same city. In 1786 he singled out Le Brun of the artillery school at Metz. The superiority of that school over the other service institutions, and the excellence of the three preparatory schools there (St. Clément, St. Louis, and the Benedictine college) led Laplace to urge upon the ministry the desirability of concentrating all technical training in that region.[184]

Fiscal stringency closed off all openings in the Artillery in 1787 and 1788, and no action followed upon Laplace's proposals. The backlog of 296 applicants that had accumulated by 1789 called still more urgently for reform, and Laplace took a more assertive lead in proposing measures. He would have had the ministry require all candidates to appear at Metz, since the training at the École militaire in Paris had been transferred nearby to Pont-à-Mousson. He would also separate the admissions from

[180] Laplace's reports and items of correspondence with J.-B. Vacquette de Gribeauval, inspector-general of artillery, and the marquis de Ségur, minister of war, are in the Archives de guerre (Vincennes), X^d.248, X^d.249, and X^d.260.

[181] Gribeauval to Ségur, 30 August 1784, Archives de guerre, X^d.249.

[182] Laplace to Ségur, 12 September 1785, Archives de guerre, X^d.249.

[183] Laplace to Ségur, 15 September 1786, Archives de guerre, X^d.249.

[184] *Ibid.*

the qualifying examination, and would accept for the latter only students who had completed the artillery schools. Their curriculum he would broaden to include physics and chemistry no less than mathematics. In the opinion of Faultrier, the commandant at Metz, who had worked harmoniously with Laplace since the latter's appointment, those proposals made a great deal of sense.[185] They were soon lost from sight, of course, in the vastly greater changes that came with the Revolution.

From a later point of view, the feature to notice is that already on its eve the voices of Laplace and Monge were being raised and even heard in the formation of educational policy. Considering the importance in subsequent French history of the practice of selecting administrative elites by a largely scientific examination of aptitude for specialized schooling, the observation may be an appropriate one with which to conclude a volume concerned with science and polity at the end of the old regime.[186]

[185] Minutes of Gribeauval and Faultrier, 7 July 1789. Laplace to Ministry, 18 August 1789. Archives de guerre, X^d.249.

[186] The role of the École polytechnique in the formation of élites is the main burden of the excellent new history by Shinn (1980).

CONCLUSION

◇◇◇

Assembling the material in this book, and living in the company of the people it treats over a period of some years, has increasingly led me to think that the integration of science into history is to be attempted with better prospects through the medium of events and institutions than through configurations of ideas or culture. Even in the matter of the social sciences, the reality of the return they made on their debt to science is more credible in the design of the reforms Turgot attempted than in the prescriptions proffered by physiocrats and economists. It would not do, however, to leave the impression that the permeation of systematic knowledge began only with his ministry. I have found it necessary to trace to earlier junctures the origin of many of the practices and institutions through which the legacy, augmented and focused by Turgot's innovations, carried over into the fifteen years remaining to the old regime. I think it is broadly correct to say, moreover, that the sectors where science then operated were those in which its public history has transpired ever since in France and elsewhere: public works, both military and civil; education, both as to development of the curriculum and the recruitment of elites; and application to the production of goods in agriculture and industry.

Working out the episodes appears to me to exhibit a pattern in the relations between science and polity, one that I think characteristic of later times and other contexts but nowhere so fully formed in the late eighteenth century as in France. What is it that statesmen have generally wanted of science? They have not wanted admonitions or collaboration, much less interference, in the business of government, which is the exercise of power over persons, nor in the political maneuverings to secure and retain control of governments. From science, all the statesmen and politicians want are instrumentalities, powers but not power: weapons, techniques, information, communications, and so on. As for scientists, what have they wanted of governments? They expressly have not wished to be politicized. They have wanted support, in the obvious form of funds, but also in the shape of institutionalization and in the provision of authority for the legitimation of their community in its existence and in its activities, or in other words for its professional status.

These reflections were first borne in on me in the course of more fragmentary studies of science and polity in the revolutionary and Napoleonic periods,[1] to which I hope to return in order to carry forward themes

[1] Gillispie, 1959 c, 1959 d.

started in the present volume through the next generation of French scientific eminence. During all the changes of those years (as I have remarked elsewhere),[2] the political conduct of the scientific community was in striking contrast with that of other groupings among intellectuals such as writers, artists, philosophers, and social scientists. The scientists, and they alone, pressed into the service of each successive regime, quite without regard to political distinctions between left and right, or to constitutional distinctions between liberty and tyranny, and they received back increasing institutional benefits from each government in turn. Their behavior, I have come to think, was characteristic of the general relation between science and the state, which has been one of partnership rather than one of partisanship, whatever the strife of factions within the political process. So matters stood in Turgot's time and earlier. Science was not the source of a reform movement or of liberalism. Its role was to provide the monarchy with the services and knowledge of experts and in return to draw advantages from the state for the furthering of science. The specification furnishes a scheme within which to follow in actual detail what the services were and what the advantages, or how the transactions have worked in concrete instances.

They have worked, of course, through the agency of institutions, and for science the professional bodies have been the most significant medium in which these transactions between power and knowledge occur. That is why I have found myself putting special emphasis on the emergence of professionalization in the arrangements that scientists were winning from the bureaucrats for the conduct of their affairs, and not only in Part II where that topic is featured. (The division of the book into three parts is for distribution of emphasis rather than assignment of subject matter to exclusive categories.) Probably it would be too procrustean an exercise to take each of the institutions (Academy of Science, Observatory, Jardin des plantes, etc.) and each of the occupations (scientist, physician, apothecary, engineer, etc.) and range them along a scale according to the degree of their approximation to the definition of professional ventured in the second section of the second chapter. Two features of the evolution in that direction do invite a further word, however. The first is schooling and the second service.

When writing down the three attributes of professionalism—learning, livelihood, and legitimation—in the passages just mentioned, I noted the requirement that the learning in question be acquired through a formal process of education. I confess that I was not then prepared to have that observation as fully confirmed as it has been by the centrality of educational developments to all of the institutions and occupations that come

[2] Gillispie, 1968.

into view, whether it be the possibility that the courses at the Collège de France really did signal the combination of teaching science with its advancement, whether it be the aspirations of surgeons and apothecaries to conduct their own colleges, or whether it be the background in engineering for the *grandes écoles* of modern times, in their relation to science and to technocracy. As to the motif of service, it is equally clear in every instance that enhancement of public welfare was the justification of the claim for privilege of a professionalism informed by new knowledge. The point comes out most explicitly in the bid of the Royal Society of Medicine to take over that old profession in the name of science.

In the course of a discussion of the occupations of men of science, a colleague asked how they felt about these civic duties? How did they see themselves in relation to knowledge, industry, and government? And I think that what was said of Berthollet[3] in this respect is largely true of all, namely that he would not have divided his life into these three compartments. The duties of a man of science entailed placing his knowledge at the disposal of the authorities, and he was eager to advance his interests in that service in the normal fulfillment of ambition. Perhaps the impatience of academicians with craftsmen and inventors was as much that of the bureaucracy as of pure science. Scientific reputation was one thing, of course, and technological innovation or control another, but those two aspects of a career were complementary and not conflicting. Indeed, if we were to estimate how the people in this book distributed their time, for many the most considerable portion must clearly have been committed to their official duties.

On another occasion, I once wrote of those duties that they represented the application, not of science, but of scientists to industry.[4] Now I should like to modify that phrasing. It is too categorical to say that the development of theory had nothing to do with the increasing sophistication of the chemistry of dyes in the hands successively of Hellot, Macquer, and Berthollet, and I think that the distinction ventured here between an encyclopedic or descriptive and a bureaucratic or positive application of science to industry gets closer to what was happening. Those three at the Gobelins, Macquer again and Darcet at Sèvres, Jars at Le Creusot, Mignot de Montigny in his steering of Holker, Desmarest in the Montgolfier mill at Vidalon, not to mention Lavoisier at the Arsenal and Vicq d'Azyr quarantining the cattle plague—theirs was the role of impresario with respect to technical initiatives. The managerial initiative, on the other hand, the choice of enterprises, was bureaucratic and governmental. Even here, moreover, the motif of schooling reappears, though in the quite different

[3] Chapter 6, Section 2.

[4] Gillispie, 1957 b, p. 404.

sense of a would-be education of industry. For the instinct of government officials was always to industrialize by means of establishing a model factory, a pilot plant—the Gobelins, Sèvres, Aubenas, Saint-Sever, Le Creusot, Vidalon-le-haut—whether state-owned or state-encouraged, and then to create incentives to bring producers around to emulating the enlightened leader.

BIBLIOGRAPHY

◇◇

The entries in the bibliography give the publisher in the case of titles that have appeared since 1940. For works published in French, the place of publication is Paris unless otherwise indicated. For works in English, "Cambridge" refers to England unless Massachusetts is specified.

Abrahams, H. J. "A Summary of Lavoisier's Proposals for Training in Science and Medicine," *Bulletin of the History of Medicine* 32 (1958), 389-407.

Ackerknecht, Erwin H. *Medicine at the Paris Hospital, 1794-1848* (Baltimore: Johns Hopkins Press, 1967).

Aguillon, Louis. "L'Ecole des Mines de Paris, notice historique," *Annales des mines*, 8th ser., *15* (1889), 433-686.

Alengry, Frank. *Condorcet, guide de la Révolution française* (1904).

Allard, Michel. *Henri-Louis Duhamel du Monceau, et le Ministère de la marine* (Montreal: Leméac, 1970).

Amadou, Robert, ed. *F.-A. Mesmer, le magnétisme animal* (Payot, 1971).

Amiable, Louis. *Le Franc-maçon Jérôme Lalande* (1889).

———. *Une Loge maçonnique avant 1789* (1897).

Andoyer, H. *L'Oeuvre scientifique de Laplace* (1922).

Ardashev, Pavel N. *Les Intendants de province sous Louis XVI* (1909).

Armitage, Angus. "The Pilgrimage of Pingré: An Astronomer-Monk of Eighteenth-Century France," *Annals of Science* 9 (1953), 47-63.

———. "Chappe d'Auteroche: A Pathfinder for Astronomy," *Annals of Science 10* (1954), 227-293.

Aron, Jean-Paul. "Les Circonstances et le plan de la nature chez Lamarck," *Revue générale des sciences pures et appliquées* 64 (1957), 243-250.

Aubry, P. V. *Monge, le savant ami de Napoléon Bonaparte: 1746-1818* (Gauthiers-Villars, 1954).

Aucoc, Léon, *L'Institut de France: Lois, statuts et règlements* (1889).

Audry, J. *Le Mésmerisme et le somnambulisme à Lyon avant la Révolution* (Lyons, 1922).

Augoyat, Antoine-Marie. *Aperçu historique sur les fortifications, les ingénieurs, et sur le corps du génie en France*, 3 vols. (1860-64).

Bacquié, Franc. *Les Inspecteurs des manufactures sous l'ancien régime, 1669-1792*, comprising *11e série des mémoires et documents pour servir à l'histoire du commerce et de l'industrie en France*, ed. Julien Hayem (1927).

Baker, Keith M. "Scientism, Elitism, and Liberalism: The Case of Condorcet," *Studies on Voltaire and the Eighteenth Century* 55 (1967a), 129-165.

———. "Les Débuts de Condorcet au secrétariat de l'Académie Royale des Sciences, 1773-1776," *Revue d'histoire des sciences* 20 (1967b), 229-280.

———. *Condorcet, From Natural Philosophy to Social Mathematics* (Chicago: University of Chicago Press, 1975).

Baker, Keith M. "French Political Thought at the Accession of Louis XVI," *Journal of Modern History* 50 (1978), 279-301.

Baker, Keith M., and Smeaton, W. A. "The Origins and the Authorship of the Educational Proposals published in 1793 by the Bureau de Consultation des Arts et Métiers and generally ascribed to Lavoisier," *Annals of Science* 21 (1965), 33-46.

Ball, Walter William Rouse. *An Essay on Newton's Principia* (London and New York, 1893).

Balland, Antoine. *La Chimie alimentaire dans l'oeuvre de Parmentier* (1902).

Ballot, Charles. *L'Introduction du machinisme dans l'industrie française* (1923). Reprinted (Slatkine) Geneva, 1978.

Barritault, Georges. *L'Anatomie en France au XVIIIe siècle* (1940).

Barrucand, Dominique. *Histoire de l'hypnose en France* (Presses universitaires de France, 1967).

Bechler, Zev. "Newton's Search for a Mechanistic Model of Colour Dispersion: A Suggested Interpretation," *Archive for the History of the Exact Sciences* 11 (1973), 3-37.

———. "Newton's Law of Forces Which Are Inversely as the Mass," *Centaurus* 18 (1973-74), 184-222.

———. "A Less Agreeable Matter: The Disagreeable Case of Newton and Achromatic Refraction," *British Journal for the History of Science* 8 (1975a), 103-126.

———. "Newton's 1672 Optical Controversies: A Study in the Grammar of Scientific Dissent," *Interaction Between Science and Philosophy*, ed. Y. Elkana (Humanities Press, 1975b), 115-142.

Ben-David, Joseph. "Scientific Growth: A Sociological View," *Minerva* 2 (1964), 455-476.

———. "The Scientific Role: The Conditions of Its Establishment," *Minerva* 4 (1965), 15-54.

———. "The Profession of Science and Its Powers," *Minerva* 10 (1972), 362-383.

Bergasse, Louis. *Un Defenseur des principes traditionnels sous la Révolution, Nicolas Bergasse* (1910).

Berman, Morris. *Social Change and Scientific Organization: The Royal Institution, 1799-1844* (Ithaca, N.Y.: Cornell University Press, 1978).

Berthaut, Henri M.-A. *La Carte de France, 1750-1898*, 2 vols. (1898).

Berthelot, M. "Notice sur les origines et sur l'histoire de la Société Philomatique" in *Mémoires . . . a l'occasion du centenaire de sa fondation* (1888), i-xvii.

Bertrand, Alexandre. *Du Magnétisme animal en France, et des jugements qu'en ont portés les sociétés savantes, avec le texte des divers rapports faits en 1784 par les commissaires de l'Académie des Sciences, de la Faculté et de la Société Royale de Médecine* . . . (1826).

Bigourdan, G. *Le Système métrique des poids et mésures* (1901).

———. "L'Astronomie à Béziers, l'Observatoire. La querelle Cassini-Lalande," *Comptes rendus du Congrès des Sociétés savantes en 1926*. Sciences, 26-42. (1927).

———. "La Jeunesse de P.-S. Laplace," *La Science moderne* 8 (1931a), 377-384.

———. "L'Observatoire de Paris sous les directions de F. Tisserand, M. Loevy et M.-B. Baillaud," *La Science moderne 8* (1931b), 121-130, 187-192.

Biollay, Léon. *Etudes économiques sur le XVIIIe siècle. Le Pacte de famine, l'administration du commerce* (1885).

Birembaut, Arthur. "Précisions sur la biographie du mathématicien Vandermonde et de sa famille," *Actes de la 72e session de l'Association Française pour l'Avancement des Sciences* (1953a), 530-533.

———. "Les Préoccupations des minéralogistes français au 18e siècle," *Actes de la 72e session de l'Association Française pour l'Avancement des Sciences* (1953b), 534-538.

———. "L'Académie Royale des Sciences en 1780 vue par l'astronome suédois Lexell (1740-1784)," *Revue d'histoire des sciences 10* (1957), 148-166.

———. "La Contribution de Réaumur à la thermométrie," *Revue d'histoire des sciences 11* (1958a), 302-329.

———. "W. A. Smeaton, 'Jean-François Pilâtre [*sic*] de Rozier,' " *Archives internationales d'histoire des sciences 11*(1958b), 100-101.

———. "Une Lettre inédite de Marat à Roume," *Annales historiques de la Révolution française 39* (1967), 398-399.

Black, Duncan. *The Theory of Committees and Elections* (Cambridge: Cambridge University Press, 1958).

Blaessinger, Edmond. *Quelques grandes figures de la chirurgie et de la médecine militaires* (J. B. Baillière et Fils, 1946).

———. *Quelques grandes figures de la pharmacie militaire* (J. B. Baillière et Fils, 1948).

———. *Quelques grandes figures de la chirurgie, de la médecine et de la pharmacie militaires* (Blanchard, 1952).

Blanchard, A. "Les Ci-devant ingénieurs du roi." *Revue internationale d'histoire militaire 30* (1970), 97-108.

———. *Les Ingénieurs du "Roy" de Louis XIV à Louis XVI: Etude de corps des fortifications* (Montpellier, 1979).

Bloch, Marc. "La Lutte pour l'individualisme agraire dans la France du XVIIIe siècle," *Annales d'histoire économique et sociale* (1930), 329-383, 511-556.

———. *Les Caractères originaux de l'histoire rurale française*, Oslo 1931. Instituttet for sammenlignende Kulturforskning, Serie B, XIX. A second edition, incorporating revisions by the author and commentary in the light of later research was published by Robert Dauvergne, 2 vols. (A. Colin, 1955-56).

———. *Les Caractères originaux de l'histoire rurale française*, 2 vols. (A. Colin, 1960, new edition).

Boehm, Klaus. *The British Patent System: Administration* (Cambridge: Cambridge University Press, 1967).

Boissier, Gaston. *L'Académie Française sous l'ancien regime* (1909).

Bolton, H. C. "A Catalogue of Scientific and Technical Periodicals, 1665-1882," *Smithsonian Miscellaneous Collections 29* (Washington, D.C., 1885).

Bondois, Paul M. "Le Privilège exclusif au 18e siècle," *Revue d'histoire économique et sociale 21* (1933), 140-189.

Bonnassieux, Pierre, and Lelong, Eugène, eds. *Conseil de Commerce et Bureau de Commerce, 1700-1791: Inventaire analytique des procès-verbaux* (1900).

Bonno, Gabriel, ed. "Notes et documents. Une Lettre inédite de Marat," *Révolution française* 85 (1932), 350-353.

Borda, Jean-Charles de. *Description et usage du cercle de réflexion* (1816).

Bordes, Maurice. "Les Intendants de Louis XV," *Revue historique* 223 (1960), 45-62.

Bordier, Henri, and Brièle, Léon. *Les Archives hospitalières de Paris*, 2 vols. (1877).

Borgnis, J.-A. *Traité complet de méchanique appliquée aux arts contenant toutes les espèces de machines*, 8 vols. (1819-20).

Bottée, J.J.H., and Riffault, J.R.D.A. *Traité de l'art de fabriquer la poudre à canon . . . précédé d'un exposé historique sur l'établissement du service des poudres et salpêtres en France*. 2 vols., 1st vol. text; 2nd plates (*Recueil de planches relatif à l'art de fabriquer la poudre à canon*.) (1811).

Bouchard, Georges. *Guyton-Morveau* (1938).

———. *Prieur de la Côté-d'Or* (R. Clavreuil, 1946).

Bouchard, Marcel. *L'Académie de Dijon et le premier discours de Rousseau* (Les Belles Lettres, 1950).

Bouchary, Jean. *L'Eau à Paris à la fin du XVIIIe siècle* (Rivière, 1946).

Bougeart, Alfred. *Marat, l'ami du peuple*, 2 vols. (1865).

Bouillier, Francisque. *L'Institut et les académies de province* (1879).

Bouis, R. "Un Echo d'une réclamation de Lavoisier en 1793." *Annales historiques de la Révolution française* 27 (1955), 168-169.

Bouissounouse, Janine. *Condorcet, le philosophe dans la Révolution* (Hachette, 1962).

Bourde, André J. *The Influence of England on the French Agronomes* (Cambridge: Cambridge University Press, 1953).

———. *Agronomie et agronomes en France au XVIIIe siècle*, 3 vols. (S.E.V.P.E.N., 1967).

Bourdier, Franck. "Le Cabinet d'Histoire Naturelle du Muséum, 1635-1935," *Sciences*, no. 18 (1962), 35-50.

Bourgin, Hubert and Georges. *L'Industrie sidérurgique en France au début de la Révolution. Collection de documents inédits sur l'histoire économique de la Révolution française* (1920).

Bouvet, Maurice. *Histoire de la pharmacie en France des origines à nos jours* (1936a).

———. "L'inspection des hopitaux militaires avant la Révolution," *Revue du service de santé militaire* 104 (1936b), 299-354.

Bowers, Claude G. *Pierre Vergniaud, Voice of the French Revolution* (New York: Macmillan, 1950).

Boyer, Carl B. *The Rainbow from Myth to Mathematics* (New York: Yoseloff, 1959).

Brandenburg, David J. "Agriculture in the *Encyclopédie:* An Essay in French Intellectural History," *Agricultural History* 24 (1950), 96-108.

———. "French Agriculture: Technology and Enlightened Reform, 1750-1789." Thesis, Columbia University, 1954. University Microfilms, Ann Arbor, Michigan, Doctoral Dissertation Series, Publication No. 8616.

Bréhier, E. *Chrysippe et l'ancien stoicisme* (Presses universitaires de France, 1951).

Brièle, Léon, ed. *Collection de documents pour servir à l'histoire des hôpitaux de Paris*, 4 vols. (1881-1887).

Brongniart, Alexandre. *Traité des arts ceramiques* (1844).

Brown, Harcourt. *Scientific Organizations in Seventeenth-Century France (1620-1680)* (Baltimore, 1934; reprinted New York: Russel and Russel, 1967).

Brown, Lloyd A. *The Story of Maps* (Boston: Little, 1949).

Brucker, Gene A. *Jean-Sylvain Bailly, Revolutionary Mayor of Paris* (Urbana, Illinois: University of Illinois Press, 1950).

Brunel, Lucien. *Les Philosophes et l'Académie Française au dix-huitième siècle* (1884).

Buffon, a commemorative volume with contributions by Lèon Bertin, Franck Bourdier, Ed. Dechambre, Yves François, E. Genet-Varcin, Georges Heilbrun, Roger Heim, Jean Pelseneer, Jean Piveteau (Publications françaises, 1952).

Buranelli, Vincent. *The Wizard from Vienna, Franz Anton Mesmer* (New York: Coward, McCann and Geoghegan, 1975).

Burke, J. G. *Origins of the Science of Crystals* (Berkeley: University of California Press, 1966).

Burkhardt, Richard W., Jr. "Lamarck, Evolution and the Politics of Science," *Journal of the History of Biology 3* (1970), 275-298.

———. "The Inspiration of Lamarck's Belief in Evolution," *Journal of the History of Biology 5* (1972), 413-438.

Burr, Charles W. "Jean Paul Marat, Physician, Revolutionist, Paranoiac," *Annals of Medical History* 2 (1919), 248-261.

Cabanès, Auguste. *Marat inconnu, l'homme privé, le médecin, le savant* (1891; 4th ed., 1911). The so-called 4th edition is really a second, the other two having been re-printings. It does not so much revise as it does complement the first edition.

Cahen, Léon. *Condorcet et la Révolution française* (1904).

Cardwell, D.S.L. *The Organisation of Science in England: A Retrospect* (Melbourne, 1957; revised edition, London: Heinemann, 1972).

Cassini, J. D. *Mémoires pour servir à l'histoire des sciences et à celle de l'Observatoire Royal de Paris* (1810).

Catterall, R.C.H. "The Credibility of Marat," *American Historical Review 16* (1910), 24-35.

Caulle, Joseph. "Delambre: Sa participation à la détermination du mètre," *Recueil des publications de la Société Havraise d'etudes diverses 103* (1936), 141-157.

Cavanaugh, Gerald J. "Turgot: The Rejection of Enlightened Despotism," *French Historical Studies* 6 (1969), 31-58.

Cazé, Michel. *Le Collège de Pharmacie de Paris, 1777-1796*. Thesis, Université de Strasbourg, 1942. Fontenay-aux-Roses (Seine), 1943.

Challinor, John. "The Early Progress of British Geology," *Annals of Science 10* (1954), 107-148.

Chalmin, Pierre. "La Formation des officiers des armes savantes sous l'ancien régime," *Actes du 76ème congrès des Sociétés Savantes* (Rennes, 1951), 165-182.

———. "Les Écoles militaires françaises jusqu'en 1914," *Revue historique de l'armé 10* (1954), 129-166.

———. "L'Ecole du Génie de Mézières 1784-1794," *Revue historique de l'armée 17* (1961), 141-154.

Champeix, Robert. *Savants méconnus* (Dunod, 1966).

Chapin, Seymour L. "The Academy of Sciences during the Eighteenth Century: An Astronomical Appraisal," *French Historical Studies* 5 (1968), 371-404.

Charliat, P.-J. "L'Académie Royale de Marine et la révolution nautique au XVIIIe siècle," *Thalès 1* (1934), 71-82.

Chartier, Roger. "Un recrutement scolaire au XVIIIe siècle: L'Ecole Royale du Génie de Mézières," *Revue d'histoire moderne et contemporaine* 20 (1973), 353-375.

Chevalier, Auguste. *La Vie et l'oeuvre de René Desfontaines, fondateur de l'herbier du Muséum* (1939).

Chevalier, Jean. "La Mission de Gabriel Jars dans les mines et usines brittaniques en 1764," *Transactions of the Newcomen Society* no. 26 (1947-48 and 1948-49), 57-68.

Chèvremont, F. *Marat, index du bibliophile et de l'amateur* (1876).

———. *Jean-Paul Marat, esprit politique accompagné de sa vie scientifique, politique et privée*, 2 vols. (1880).

Chinard, Gilbert. *The Correspondence of Jefferson and Dupont de Nemours* (Baltimore, 1931).

Choffel, Jean. *Saint-Gobain: Du Miroir à l'atome* (Plon, 1960).

Clos, Dominique. "Lamarck botaniste, sa contribution à la méthode dite naturelle et à la troisième édition de *La Flore Française*," *Mémoires de l'Académie des Sciences, Inscriptions et Belles-lettres de Toulouse*, 9th ser. *8* (1896), 202-225.

Cocheris, Hippolyte. *Table méthodique et analytique des articles parus dans le Journal des savants, . . . précédée d'une notice sur ce journal depuis sa fondation jusqu'à nos jours* (1860).

Cochin, Augustin. *Les Manufactures des glaces de Saint-Gobain de 1665 à 1865* (1865).

Cole, Arthur H., and Watts, George B. *The Handicrafts of France, as Recorded in the Descriptions des arts et metiers, 1761-1788* (Boston, Mass., The Kress Library of Business and Economics, publication no. 8, Harvard Business School, 1952).

Conant, J. B., ed. *Harvard Case Histories in Experimental Science, 1*, case 2, *The Overthrow of the Phlogiston Theory: The Chemical Revolution of 1775-1789* (Cambridge, Mass.: Harvard University Press, 1957).

Condorcet, A.-N.-C. de. *Oeuvres*, 12 vols., ed. Mme Condorcet-O'Connor and F. Arago (1847-1894).

Contant, J.-P. *L'Enseignement de la chimie au Jardin Royal des Plantes de Paris* (Cahors: A. Coueslant, 1952).

Corlieu, A. *L'Ancienne Faculté de Médecine de l'Université de Paris* (1877).

Corvisier, André. "Hiérarchie militaire et hiérarchie sociale à la veille de la Révolution," *Revue internationale d'histoire moderne contemporaine* 30 (1970), 77-92.

Coury, Charles. *L'Enseignement de la médecine en France des origines à nos jours* (L'Expansion scientifique française, 1968).

———. *L'Hôtel-Dieu de Paris, treize siècles de soins, d'enseignement, et de recherche* (L'Expansion scientifique française, 1969).

Cowan, C. F. "The Daubentons and Buffon's Birds," *Journal of the Society for the Bibliography of Natural History* 5 (1968-1971), 37-40.

Crosland, Maurice. *Historical Studies in the Language of Chemistry* (Cambridge, Mass.: Harvard University Press 1962).

———. *The Society of Arcueil* (Cambridge, Mass.: Harvard University Press, 1967).

———. *Science in France in the Revolutionary Era* (Cambridge, Mass., and London: Society for the History of Technology, 1969). An edition of Thomas Bugge's journal of 1798-1799.

———. *The Emergence of Science in Western Europe* (London: Macmillan, 1975).

Cuvier, Georges. *Rapport historique sur les progrès des sciences naturelles depuis 1789, et sur leur état actuel* (1810).

———. *Recueil des éloges*, 3 vols. (Strasbourg, 1819-1827).

Cuzacq, René. *Un Savant chalossais: Le Chimiste Jean Darcet et sa famille* (Mont de Marsan: Jean-Lacoste, 1955).

Dakin, Douglas. *Turgot and the Ancien Régime in France* (London, 1939).

Darboux, Gaston. "Notice historique sur le Général Meusnier," *Mémoires de l'Académie des Sciences de l'Institut de France*, 2nd ser., *51* (1910), i-xxxviii.

———, ed. "Mémoires et travaux de Meusnier relatifs à l'aérostation," *ibid.*, 3-128. Darboux published these papers separately in the same year, together with a photographic reproduction of the Atlas containing Meusnier's engineering drawings of the two proposed aircraft that he designed.

Darnton, Robert. "Marat n'a pas été un voleur," *Annales historiques de la Révolution française 38* (1966), 447-450.

———. *Mesmerism and the End of the Enlightenment in France* (Cambridge, Mass.: Harvard University Press, 1968).

———. "The High Enlightenment and the Low-Life of Literature in Pre-Revolutionary France," *Past and Present 51* (1971), 81-115.

———. *The Business of Enlightenment, A Publishing History of the Encyclopédie, 1775-1800* (Cambridge, Mass.: Harvard University Press, 1979).

Dartein, F. de. "Notice sur le régime de l'ancienne Ecole des Ponts et Chaussées et sur sa transformation à partir de la révolution," *Annales des Ponts et Chaussées*, 8th ser., 22 (2e trimestre, 1906), pp. 5-123.

Dauben, Joseph W. "Marat: His Science and the French Revolution," *Archives internationales d'histoire des sciences* 22 (1969), 235-261.

Daudin, Henri. *De Linné à Jussieu, méthodes de la classification et l'idée de série en botanique et en zoologie, 1740-1790* (1926a).

———. *Cuvier et Lamarck: Les Classes zoologiques et l'idée de série animale (1790-1830)*, 2 vols. (1926b).

Daumas, Maurice. "Le Corps des Ingénieurs brevetés en instruments scientifiques (1787)," *Archives internationales d'histoire des sciences* 5 (1952), 86-96.

———. *Les Instruments scientifiques aux XVIIe et XVIIIe siècles* (Presses universitaires de France, 1953).

———. *Lavoisier, théoricien et expérimentateur* (Presses universitaires de France, 1955).

Daumas, Maurice, and Duveen, Denis I. "Lavoisier's Relatively Unknown Large-Scale Decomposition and Synthesis of Water, February 27 and 28, 1785," *Chymia* 5 (1959), 113-129.

De Beer, Gavin. "The Volcanoes of Auvergne," *Annals of Science 18* (1962), 49-61.

Delambre, J.-B.-J. *Rapport historique sur le progrès des sciences mathématiques depuis 1789, et sur leur état actuel* (1810).

———. *Histoire de l'astronomie au 18 ^e^ siècle* (1827), which forms vol. 6 of his

Histoire de l'astronomie (1817-1827). This volume was published posthumously by a colleague, Mathieu. It had been Delambre's intention to follow it with a vol. 7, on the history of geodesy, leading up to his *Base.* The materials he had assembled were published long after by Bigourdan, see Delambre (1912).

———. "Notice sur la vie . . . de Lagrange" from *Oeuvres de Lagrange* (1867), 50, xxxviii.

———. *Grandeur et figure de la terre*, ed. G. Bigourdan (1912).

Delambre, J.-B.-J., and Méchain, P.-F.-A. *Base du système métrique décimal, ou mesure de l'arc du méridien compris entre les parallèles de Dunkerque et Barcelone*, 3 vols. (1806-1810).

Delaunay, Paul. *Le Monde médical Parisien au dix-huitième siècle* (1906).

Deleuze, J.P.F. *Histoire critique du magnétisme animal* (1813).

———. *Histoire et description du Muséum Royal d'Histoire Naturelle*, 2 vols. (1823).

Delorme, Suzanne. "Une Famille de grands commis de l'Etat: Les Trudaine," *Revue d'histoire des sciences* 3 (1950), 101-109.

Demuth, Norman, *French Opera: Its Development to the Revolution* (Sussex: Artemis Press, 1963).

Desaive, J. P., Goubert, J. P., Ladurie, E. Le Roy, Meyer, J., Muller, O., Peter, J. P. *Médecins, climat et épidémies à la fin du XVIIIe siècle* (Mouton, 1972).

Desboves, Adolphe. *Delambre et Ampère* (Amiens, 1881).

Devic, J.-F.-S. *Histoire de la vie et des travaux scientifiques et littéraires de J. D. Cassini IV* (Clermont, 1851).

Dorbeau, General, "L'Ecole de Mézières," *Revue du génie militaire* 76 (1937), 311-346, 443-469.

Dorveaux, Paul. "Inventaire des archives de la compagnie des Marchands Apothicaires de Paris et du Collège de Pharmacie de Paris dressé en 1786," *Extrait de la revue des bibliothèques* (1893).

———. "Apothicaires membres de l'Académie Royale des Sciences, XI: Balthazar-Georges Sage," *Revue d'histoire de la pharmacie* 23 (1935), 152-166, 216-232, and also printed separately.

Doyon, André, and Liaigre, Lucien. *Jacques Vaucanson, mécanicien de génie* (Presses universitaires de France, 1966).

Drapeyron, Ludovic. "Enquête à instituer sur l'exécution de la grand carte topographique de France de Cassini de Thury," *Revue de géographie* 38 (1896a), 1-16.

———. "La Vie et les travaux géographiques de Cassini de Thury," *Revue de géographie* 39 (1896b), 241-251.

Dreux, Th. *Le Cadastre et l'impôt foncier* (1933).

Dronne, Michel. *Bertin et l'élevage française au XVIIIe siècle.* Thèse pour le doctorat vétérinaire, Ecole nationale vétérinaire d'Alfort, no. 49 (Alfort, 1965).

Dubois, Frederick. "Recherches historiques sur les dernières années de Louis et de Vicq d'Azyr," *Gazette médicale de Paris* (1866), 6 October, 641-650; 13 October, 655-667; 20 October, 669-680. BN f° T^{33}.25.

Dufresne, A.-J.-L.-M. *Notes sur la vie et les oeuvres de Vicq d'Azyr (1748-1794).* Thèse medicale no. 65 (Bordeaux, 1906).

Dujarric de la Rivière, R., *Lavoisier économiste* (Masson, 1949).

———. *E.-I. DuPont de Nemours, élève de Lavoisier* (Librairie des Champs-Elysées, 1954).

Dulieu, Louis. "Le Mouvement scientifique montpéllierain au XVIIIe siècle," *Revue d'histoire des sciences 11* (1958a), 227-249.

———. "La Contribution montpéllieraine à l'Académie des Sciences," *Revue d'histoire des sciences 11* (1958b), 255-262.

Dupuy, Paul, ed. *Centenaire de l'Ecole Normale Supérieure* (1895).

Dussaud, René. *La Nouvelle Académie des Inscriptions et Belles-Lettres* (P. Geuthner, 1946).

Duval, Henri. "Marat et l'abbé Bertholon (1785)," *Revue historique de la Révolution française 3* (1912), 461-462.

Duveen, Denis I. "Augustin François Silvestre and the Société Philomathique," *Annals of Science 10* (1954), 339-341.

———. "Lavoisier Writes to Fourcroy from Prison," *Notes and Records of the Royal Society 13* (1958), 59-60.

———. *Supplement to a Bibliography of the Works of Antoine-Laurent Lavoisier* (London: Dawson, 1965).

Duveen, Denis I., and Klickstein, Herbert S. *A Bibliography of the Works of Antoine-Laurent Lavoisier* (London: Dawson, 1954a).

———. "Le Journal polytype des sciences et des arts," *The Papers of the Bibliographical Society of America 48* (1954b), 402-410.

———. "Antoine Laurent Lavoisier's Contributions to Medicine and Public Health," *Bulletin of the History of Medicine 29* (1955), 164-179.

Duveen, Denis, and Roger Hahn. "Laplace's Succession to Bezout's Post of *examinateur des élèves de l'artillerie*," *Isis 48* (1957), 416-427.

———. "A Note on Some Lavoisiereana in the *Journal of Paris*," *Isis 51* (1960), 64-66.

Ellenberger, Henri F. "Mesmer and Puységur: From Magnetism to Hypnotism," *Psychoanalytic Review 52* (1965), 281-291.

———. *Discovery of the Unconscious. The History and Evolution of Dynamic Psychiatry* (New York: Basic Books 1969).

Ewan, Joseph. "Fougeroux de Bondaroy (1732-1789) and His Projected Revision of Duhamel du Monceau's *Traité* (1755) on Trees and Shrubs." *Proceedings of the American Philosophical Society 103* (1959), 807-818.

Faddegon, J.-M., and Boizard de Guise. "L'Astéréomètre de Jeaurat," *Société Astronomique de France 50* (1936), 553-559.

Fages, Urbain. "Les Débuts de l'industrie cotonnière en France: Jean Holker," extrait de *Nouvelle Revue de Paris 4* (date unknown), 47-79.

Falls, William Franklin. *Buffon et l'agrandissement du Jardin du Roi.* (Philadelphia, 1933). Also published in *Archives du Muséum d'Histoire Naturelle*, 6th ser., *10* (1933), 131-200. Citations are to Philadelphia edition.

Faure, Edgar. *La Disgrâce de Turgot* (Gallimard, 1961).

Favre, Adrien. *Les Origines du système métrique* (1931).

Fayet, Joseph. *La Révolution française et la science* (Rivière, 1960).

Fenaille, Maurice. *Etat-général de la manufacture des Gobelins depuis son origine, jusqu'à nos jours, 1600-1900*, 4 vols. in 80 (1903). A final volume on the 19th century was prepared in two parts by Fernand Calmettes (1912).

Festy, Octave. *L'Agriculture pendant la Révolution française: Les Conditions de production et de récolte des céréales* (Gallimard, 1947).

———. *L'Agriculture pendant la Révolution française: L'Utilisation des jachères* (Revière, 1950).

Filion, Maurice. *Maurepas, ministre de Louis XV (1715-1949)* (Montreal, Les Editions Leméac, 1967).

Finot, A. *Les Facultés de médecine de province avant la Révolution* (A. Legrand, 1958).

Fling, Fred Morrow. "Une Pièce fabriquée: Le Troisième volume des *Mémoires de Bailly*," *La Révolution française* 43 (1902), 466-474.

———. "The Mémoires de Bailly," *University studies published by the University of Nebraska* 3 No. 4 (1903), 331-353.

Flourens, P. *Eloges historiques*, 2nd ser. (1857).

Foncin, P. *Essai sur le ministère de Turgot* (1877).

Fordham, Herbert George. *A Study in the History of Cartography* (Cambridge, 1929).

Forster, Robert. "Obstacles to Agricultural Growth in Eighteenth-Century France," *American Historical Review* 75 (1970), 1600-1615.

Fosseyeux, M. *L'Hôtel-Dieu de Paris au XVIIe et au XVIIIe siècle* (1912).

Foucault, Michel. *Naissance de la clinique, une archéologie du regard médical* (Presses universitaires de France, 1963).

Fréchet, Maurice. "Biographie du mathématicien alsacien Arbogast," *Thalès* 4 (1937-1939), 43-55.

Frémy, Elphège. *Histoire de la Manufacture royale des glaces de France au XVIIe et au XVIIIe siècles* (1909).

Freycinet, Louis de. *Essai sur la vie, les opinions et les ouvrages de Barthélemy Faujas de Saint-Fond* (Valence, 1820).

Gallois, L. "L'Académie des Sciences et les origines de la carte de Cassini," part one, *Annales de géographie 18* (1909), 193-204. Part two, *Annales de géographie 18* (1909), 289-307.

Gambiez, F. "L'Ecole Militaire," *Revue internationale d'histoire militaire* 30 (1970), 23-39.

Garat, Dominique-Joseph. *Mémoires historiques sur la vie de M. Suard, sur ses écrits, et sur XVIIIe siècle*, 2 vols. (1821).

———. *Mémoires de Garat* (1862).

Garnier, Edouard. *The Soft Porcelain of Sèvres, with an Historical Introduction* (London, 1892).

Gauja, Pierre. "Les Origines de l'Académie des Sciences de Paris," *Académie des Sciences, Institut de France, troisième centenaire, 1666-1966* (1967), 1-51.

Geikie, Archibald. *The Founders of Geology* (London, 1905).

Gelfand, Toby. "The Training of Surgeons in 18th-century Paris and Its Influence on Medical Education." Thesis, The Johns Hopkins University, Baltimore (1973a).

———. "A Confrontation over Clinical Instruction at the Hôtel-Dieu of Paris during the French Revolution," *Journal of the History of Medicine and Allied Sciences* 28 (1973b), 268-282.

———. "The Hospice of the Paris College of Surgery (1774-1793), 'A Unique and Invaluable Institution,' " *Bulletin of the History of Medicine* 47 (1973c), 375-393.

Germain, A. "Histoire de l'Université de Montpellier," *Cartulaire de l'Université de Montpellier 1* (Montpellier, 1890), 1-176.

Gillispie, C. C. "The Work of Elie Halévy, A Critical Appreciation," *Journal of Modern History 22* (1950), 232-250.

———. "The Formation of Lamarck's Evolutionary Theory," *Archives internationales d'histoire des sciences 9* (1956a), 323-338.

———. "Notice biographique de Lavoisier par Mme Lavoisier," *Revue d'histoire des sciences 9* (1956b), 52-61.

———. "The Discovery of the Leblanc Process," *Isis 48* (1957a), 152-170.

———. "The Natural History of Industry," *Isis 48* (1957b), 398-407.

———. *A Diderot Pictorial Encyclopedia of Trades and Industry: Manufacturing and Technical Arts in Plates from l'Encyclopédie*, 2 vols. (New York: Dover, 1959a).

———. "Lamarck and Darwin in the History of Science" in Bentley Glass, ed., *The Forerunners of Darwin* (Baltimore: Johns Hopkins Press, 1959b).

———. "Science in the French Revolution," *Behavioral Science 4* (1959c), 67-73.

———. "The *Encyclopédie* and the Jacobin Philosophy of Science," Chapter IX in *Critical Problems in the History of Science*, Marshall Clagett, ed. University of Wisconsin Press (Madison, 1959d), pp. 255-290.

———. *The Edge of Objectivity, An Essay in the History of Scientific Ideas* (Princeton: Princeton University Press, 1960).

———. "Intellectual Factors in the Background of Analysis by Probabilities," in A. C. Crombie, ed., *Scientific Change* (London: Heinemann, 1963), 431-453.

———. "Remarks on Social Selection as a Factor in the Progressivism of Science," *American Scientists,* 56 (December, 1968), pp. 439-450.

———. *Lazare Carnot, Savant* (Princeton: Princeton University Press, 1971a).

———. "Probability and Politics: Laplace, Condorcet, and Turgot," *Proceedings of the American Philosophical Society 116* (1972), 1-20.

———. "The Liberating Influence of Science in History," *Aspects of American Liberty, Philosophical, Historical and Political*, Memoirs of the American Philosophical Society *118* (1977), 37-46.

———. "Mémoires inédits ou anonymes de Laplace sur la théorie des erreurs, les polynomes de Legendre, et la philosophie de probabilité," *Revue d'histoire des sciences 33* (1979), 223-279.

Gillmor, C. Stewart. *Charles-Augustin Coulomb* (Princeton: Princeton University Press, 1971).

Gingerich, Owen, "Messier and His Catalogue—I," *Sky and Telescope 12* (1953), 255-258, 288-291.

Goodwin, A. "Calonne, the Assembly of Notables and the Origins of the Révolte nobiliaire," *English Historical Review 61* (1946), 202-234, 327-377.

Gottschalk, Louis, "Du Marat inédit," *Annales historiques de la Révolution française 3* (1926), 209-216.

———. *Jean Paul Marat: A Study in Radicalism* (New York, 1972; 2nd ed., Chicago, 1967).

Goubert, Jean-Pierre. *Malades et médecins en Bretagne, 1770-1790* (Rennes: C. Klincksieck, 1974).

Gough, J. B. "Lavoisier's Early Career in Science: An Examination of Some New Evidence," *British Journal for the History of Science 4* (1968), 52-57.

Granger, G.-G. *La Mathématique sociale du Marquis de Condorcet* (Presses universitaires de France, 1956).
Green, F. C. *Rousseau and the Idea of Progress*, The Zaharoff Lecture for 1950 (Oxford: Clarendon Press, 1950).
Greenbaum, Louis S. "The Commercial Treaty of Humanity: La Tournée des hôpitaux anglais par Jacques Tenon en 1787," *Revue d'histoire des sciences 24* (1971), 317-350.
———. "The Humanitarianism of Antoine Laurent Lavoisier," *Studies on Voltaire and the Eighteenth Century 88* (1972), 651-675.
———. "Jean-Sylvain Bailly, the Baron de Breteuil, and the 'Four New Hospitals' of Paris," *Clio Medica 8* (1973), 261-284.
———. "Tempest in the Academy: Jean-Baptiste Le Roy, the Paris Academy of Sciences and the Project of a new Hôtel-Dieu," *Archives internationales d'histoire des sciences 24* (1974), 122-140.
Grimaux, Edouard. *Lavoisier, 1743-1794* (1888).
Grosclaude, Pierre. *Malesherbes, témoin et interprète de son temps* (Fischbacher, 1961).
Gruder, Vivian R. *The Royal Provincial Intendants, A Governing Elite in Eighteenth-Century France* (Ithaca, N.Y.: Cornell University Press, 1968).
Guédès, Michel. "La Méthode taxonomique d'Adanson," *Revue d'histoire des sciences 20* (1967), 361-386.
Gueneau, Louis. "L'Usage industriel de la houille au XVIIIe siècle," *VIIIe série des mémoires et documents pour servir à l'histoire du commerce et de l'industrie en France*, ed. Julien Hayem (1924), 320-334.
Guerlac, Henry. "Lavoisier and His Biographers," *Isis 45* (1954), 51-62.
———. "A Note on Lavoisier's Scientific Education," *Isis 47* (1956), 211-216.
———. "Joseph Priestley's First Papers on Gases and Their Reception in France," *Journal of the History of Medicine and Allied Sciences 12* (1957), 1-12.
———. "A Lost Memoir of Lavoisier," *Isis 50* (1959a), 125-129.
———. "Some French Antecedents of the Chemical Revolution," *Chymia 5* (1959b), 73-112.
———. "A Curious Lavoisier Episode," *Chymia 7* (1961a), 103-108.
———. *Lavoisier, the Crucial Year: The Background and Origin of His First Experiments on Combustion in 1772* (Ithaca, N.Y.: Cornell University Press, 1961b).
———. "Chemistry as a Branch of Physics: Laplace's Collaboration with Lavoisier," *Historical Studies in the Physical Sciences 7* (1976), 193-276.
Guilbaud, G.-T. "Les Théories de l'intérêt général," *Economique appliquée 5* (1952), 501-584.
Guillaume, James. *Etudes révolutionnaires*, 2 vols. (1908-1909).
Guillois, Antoine. *La Marquise de Condorcet* (1897).
Guitard, Eugene-Humbert. *Index des travaux d'histoire de la pharmacie de 1913 à 1963* (Société d'histoire de la pharmacie, 1963).
Günther, L. "Jean-Paul Marat, der 'ami du peuple,' als Criminalist," *Der Gerichtssaal 61* (1902), 161-252, 321-388.
Hahn, Roger. "Quelques nouveaux documents sur Jean-Sylvain Bailly," *Revue d'histoire des sciences 8* (1955), 338-353.
———. "The Chair of Hydrodynamics in Paris, 1775-1791," *Acts of . . . IX*

International Congress of the History of Science (Ithaca, N.Y., 1962), 751-754.
———. "L'Hydrodynamique au XVIIIe siècle, aspects scientifiques et sociologiques," Conference at Palais de la Découverte (1964).
———. "Laplace's First Formulation of Scientific Determinism in 1773," *Actes du XIe congrès international d'histoire des sciences* (Warsaw, 1967), 167-171.
———. *The Anatomy of a Scientific Institution: The Paris Academy of Sciences, 1666-1803* (Berkeley: University of California Press, 1971).
———. "Sur les débuts de la carrière scientifique de Lacepède," *Revue d'histoire des sciences* 27 (1974), 347-353.
———. "L'Autobiographie de Lacepède retrouvée," *Dix-huitième siècle*, no. 7 (1975), 49-85.

Halévy, Élie. *The Growth of Philosophical Radicalism*, trans. M. Morris (New York: Faber, 1949).

Hall, A. R. "Newton in France," *History of Science* 13 (1975), 233-250.

Hamy, E.-T. *Les Derniers jours du jardin du roi et la fondation du Muséum d'Histoire Naturelle*. Extrait du volume commémoratif du Centenaire de la fondation du Muséum d'Histoire Naturelle (1893).
———. *Les Débuts de Lamarck, matériaux inédits pour servir à sa biographie* (1907).
———. *Les Débuts de Lamarck, suivis de recherches sur Adanson, Jussieu, Pallas, etc.* (1909).

Hanks, Lesley. *Buffon avant l' "Histoire naturelle"* (Presses universitaires de France, 1966).

Harnack, Adolf. *Geschichte der königlich preussischen Akademie der Wissenschaften zu Berlin*, 3 vols. (Berlin, 1900).

Hartemann, B., and Ducousset, R. *B.S.N. contre Saint-Gobain* (Editions et publications premières, 1969).

Havard, Henry, and Vachon, Marius. *Les Manufactures nationales: Les Gobelins, La Savonnerie, Sèvres, Beauvais* (1889).

Haug, Geneviève Levallet. *Claude-Nicolas Ledoux, 1736-1806* (Paris and Strasbourg, 1934).

Hautecoeur, Louis. "Pourquoi les académies furent-elles supprimées en 1793?" *Revue des deux mondes* (December 15, 1959), 593-604.

Hecht, Jacqueline. "La Vie de François Quesnay," *François Quesnay et la physiocratie 1* (1958), 211-286.

Heim, Roger et al. *Michel Adanson, 1727-1806* (1963).

Hellman, C. Doris. "Legendre and the French Reform of Weights and Measures," *Osiris 1* (1936), 314-340.

Henry, Charles, ed. *Correspondance inédite de Condorcet et de Turgot, 1770-1779* (1883).

Héron de Villefosse, René. *Histoire des grandes routes de France* (Librairie académique Perrin, 1975).

Heyman, Jacques. *Coulomb's Memoir on Statics: An Essay in the History of Civil Engineering* (Cambridge: Cambridge University Press, 1972).

Hippeau, C. *L'Instruction publique en France pendant la Révolution, débats législatifs* (1883).

Hodge, M.J.S. "Lamarck's Science of Living Bodies," *British Journal for the History of Science* 5 (1971), 323-352.

Holmes, Frederick. "From Elective Affinities to Chemical Equilibria: Berthollet's Law of Mass Action," *Chymia* 8 (1962), 105-145.

Hooykaas, R. "Les Débuts de la théorie cristallographique de R.-J. Haüy, d'après les documents originaux," *Revue d'histoire des sciences* 8 (1955), 317-337.

Hours, Henri, *La Lutte contre les épizootics et l'Ecole Vétérinaire de Lyon a XVIIIe siècle* (Presses universitaires de France, 1957).

Huard, P. "Les Papiers de Jacques Tenon, 1724-1816," *Hôpitaux et aide sociale à Paris* 7 (1966), 627-633.

Humbert-Bazile. *Buffon, sa famille, ses collaborateurs et ses familiers*, ed. Henri Nadault de Buffon (1863).

Hunter, Alfred. *J.B.A. Suard* (1925).

Imberdis, Franck. *Le Réseau routier de l'Auvergne au XVIIIe siècle* (Presses universitaires de France, 1967).

Isoré, Jacques. "De l'Existence des brevets d'invention en droit français avant 1791," *Revue historique du droit français et étranger*, 4th ser., *16* (1937), 94-130.

Jansen, Albert. *Rousseau als Botaniker* (Berlin, 1885).

Johnson, Terence J. *Professions and Power* (London: Macmillan, 1972).

Jolly, Pierre, *Dupont de Nemours, Soldat de la liberté* (Presses universitaires de France, 1956).

Jommard, M. *Notice sur la vie et les ouvrages de Claude-Louis Berthollet* (Annecy, 1844).

Juergens, Ralph E. "Minds in Chaos: A Recital of the Velikovsky Story," *American Behavioral Scientist* 7 (1963), 4-17.

Juskiewenski, G. S. *Jean-Paul Marat* (Bordeaux, 1933).

Justin, Émile. *Les Sociétés Royales d'Agriculture au XVIIIe (1757-1793)* (Saint-Lo, 1935).

Kahane, Ernest. *Parmentier, ou la dignité de la pomme de terre* (A. Blanchard, 1978).

Kaplan, Steven L. *Bread, Politics and Political Economy in the Reign of Louis XV*, 2 vols. (The Hague: Martinus Nijhoff, 1976).

Kapoor, Satish C. "Berthollet, Proust and Proportions," *Chymia 10* (1965), 53-110.

Kennelly, Arthur E. *Vestiges of Parametric Weights and Measures* (New York, 1928).

Kersaint, Georges. "Notes pour l'éloge de Fourcroy," *Revue d'histoire de la pharmacie* (1957), 157.

———. "Antoine François de Fourcroy (1755-1809), sa vie et son oeuvre," constituting vol. 2, ser. D (Sciences physico-chimiques), *Mémoires du Muséum National d'Histoire Naturelle* (1966).

Kington, J. A. "A Late Eighteenth-Century Source of Meteorological Data," *Weather* 25 (1970), 169-175.

Koestler, Arthur. *The Case of the Midwife Toad* (New York: Random House, 1971).

Kronick, David A. *A History of Scientific and Technical Periodicals . . . 1665-1790* (New York: Scarecrow Press, 1962).

Kunz, George Frederick. "The Life and Work of Haüy," *American Mineralogist 3* (1918), 61-89.

Lacroix, Alfred. *Déodat Dolomieu, membre de l'Institut National (1750-1801): Sa vie aventureuse, sa captivité, ses oeuvres, sa correspondance*, 2 vols. (1921).

———. *Figures des savants*, 4 vols. (1932-1938).

Lagrange, Joseph Louis, *Oeuvres de Lagrange*, ed. J. A. Serret (14 volumes, 1867-1892).

Laignel-Lavastine, Maxime, ed. *Histoire générale de la médecine, de la pharmacie, de l'art dentaire, et de l'art vétérinaire*, 3 vols. (1949).

Laissus, Joseph. "Antoine-Laurent de Jussieu, 'l'aimable professeur,' " *Comptes-rendus du 89e Congrès des Sociétés Savantes* (Lyons, 1964), 27-39.

———. "A propos de Jean-Dominique Cassini," *Comptes-rendus du 90e Congrès des Sociétés Savantes 3* (Nice, 1965), 9-16.

———. "La Succession de Le Monnier au Jardin du Roi: Antoine-Laurent de Jussieu et René Louiche-Desfontaines," *Comptes-rendus du 91e Congrès national des Sociétés Savantes* (Rennes, 1966), Section des sciences *1*, Histoire des sciences, 137-152.

———. "Le Général Meusnier de la Place, membre de l'Académie Royale des Sciences," *Comptes-rendus du 93ème Congrès national des Sociétés Savantes* (Tours, 1968), Section des sciences 2 (1971), 75-101.

Laissus, Yves. "Lettres inédites de René Desfontaines à Louis-Guillaume Le Monnier," *Comptes-rendus du 91e Congrès national des Sociétés savantes* (Rennes, 1966), Section des sciences *1*, Histoire des sciences, 153-169.

Lamontagne, Roland. "Chronologie de la carrière de La Galissonière," *Revue d'histoire des sciences 14* (1961), 255-256.

———. "Rapport sur le *Traité des arbres et arbustes* . . . de Duhamel du Monceau," *Revue d'histoire des sciences 16* (1963), 221-225.

Landrieu, Marcel. *Lamarck, le fondateur du transformisme, sa vie, son oeuvre* (1909), constituting *21* of *Mémoires de la Société zoologique de France*.

Laplace, Pierre Simon. *Oeuvres complètes*, 14 vols., ed. l'Académie des Sciences (1878-1912).

Laugier, A., and Duruy, V. *Les Pandectes pharmaceutiques* (1837).

Launay, Louis de. *Un Grand Français: Monge, fondateur de l'Ecole Polytechnique* (1933).

———. *Une Grande Famille des savants: Les Brongniart* (1940).

Laurent, Gustave. "Une Mémoire historique du chimiste Hassenfratz," *Annales historiques de la Révolution française 1* (1924), 163-164.

Lavergne, Léonce de. *Les Économistes français au XVIIIe siècle* (1870).

Lavoisier, A.-L. *Oeuvres*, ed. J.-B. Dumas, *1-4* (1864-1868) and Edouard Grimaux, 5-6 (1892-1894).

Lavoisier, A.-L. *Oeuvres de Lavoisier—Correspondance*, ed. René Fric, 3 fascicules (1855-64); fasc. IV to appear.

Lebesgue, Henri. "L'Oeuvre mathématique de Vandermonde," *Thalès 4* (1937-1939), 28-42.

Leclainche, E. *Histoire de la médecine vétérinaire* (Toulouse, 1936).

Lefranc, Abel. *Histoire du Collège de France depuis ses origines jusqu'à la fin du premier empire* (1893).

Le Guin, Charles A. "Roland de La Platière, A Public Servant in the Eighteenth

Century," *Transactions of the American Philosophical Society*, new ser., vol. 56, Part 6 (Philadelphia, 1966).

Lemay, Pierre. "Berthollet et l'emploi du chlore pour le blanchiment des toiles," *Revue d'histoire de la pharmacie* 3 (1932), 79-86.

Lemay, Pierre, and Oesper, Ralph. "Claude-Louis Berthollet," *Journal of Chemical Education* 23 (1946), 158-165, 230-236.

Lemonnier, Henry, ed. *Procès-verbaux de l'académie Royale d'Architecture*, 10 vols. (1911-1929).

———. *L'Art français au temps de Louis XIV (1661-1690)* (1911).

———. *L'Art français au temps de Richelieu et de Mazarin* (1913).

Lenglen, M. *Lavoisier agronome* (1936).

Le Paute Dagelet. "Notice bibliographique sur Le Paute Dagelet," *Bulletin de la Société de Géographie*, 7th ser. 9 (1888), 293-302.

Leroy, Jean François. "Adanson dans l'histoire de la pensée scientifique," *Revue d'histoire des sciences* 20 (1967), 349-360.

Letonné, F. "Le Général Meusnier et ses idées sur la navigation aérienne," *Revue du génie militaire* 2 (1888), 247-258.

Lhomer, Jean. *François de Neufchateau, 1750-1828* (1913).

Lintilhac, Eugène. *Vergniaud* (1920).

Lohne, J. A. "Newton's 'Proof' of the Sine Law and His Mathematical Principles of Colors," *Archive for History of Exact Sciences* 8 (1961), 389-405.

———. "Isaac Newton: The Rise of a Scientist, 1661-1671," *Notes and Records of the Royal Society* 20 (1965), 125-139.

———. "Experimentum Crucis," *Notes and Records of the Royal Society* 23 (1968), 169-199.

Lopez, Claude-Anne. "Saltpetre, Tin and Gunpowder: Addenda to the Correspondence of Lavoisier and Franklin," *Annals of Science* 16 (1960), 83-94.

———. *Mon cher papa: Franklin and the Ladies of Paris* (New Haven and London: Yale University Press, 1966).

Lovejoy, Arthur O. *The Great Chain of Being* (Cambridge, Mass., 1936).

Lubimenko, I. I. "Un Académicien russe à Paris (d'après ses lettres inédites, 1780-1781)," *Revue d'histoire moderne* (1935), 415-447.

———, ed. *Uchenaia Korrespondentsiia Akademii Nauk, XVIII veka,* 2 of Akademiia Nauk Soiuza Sovetskikh Sotsialisticheskikh Respublik, *Trudy Arkhiva*, ed. D. S. Rozhdestvensky (Moscow and Leningrad, 1937).

Lublinski, V. S. "Voltaire et la guerre des farines," *Annales historiques de la Révolution française* 31 (1959), 127-145.

Lüthy, Herbert. *La Banque protestante en France de la Révocation de l'edit de Nantes à la Révolution*, 2 vols. Vol. 1, *Dispersion et regroupement (1685-1730)* (S.E.V.P.E.N., 1959). Vol. 2, *De la Banque aux finances (1730-1794)* (S.E.V.P.E.N., 1961).

Lyons, Sir Henry George. *The Royal Society, 1660-1940, A History of Its Administration under Its Charters* (Cambridge: Cambridge University Press, 1944).

MacAuliffe, Léon. *La Révolution et les hôpitaux (années 1789, 1790, 1791)* (1901).

Maindron, Ernest. *Les Fondations de prix à l'Académie des Sciences: Les Lauréats de l'Académie 1714-1880* (1881).

———. *L'Académie des Sciences* (1888).

Malécot, H. L. *Ballons à ballonets, dits sphériques modernes . . . Théorie de Meusnier, 1783-1909* (1910).

Marguet, F. *Histoire de la longitude à la mer au XVIIIe siècle en France* (1917).

Martin, Germain. *La Grande industrie en France sous le règne de Louis XV* (1900).

Martin-Allanic, Jean-Etienne. *Bougainville, navigateur et les découvertes de son temps* (Presses universitaires de France, 1964).

Marx, C. M. *Geschichte der Kristallkunde* (Karlsruhe, 1825).

Mascart, Jean. *La Vie et les travaux du chevalier Jean-Charles de Borda, 1733-1799* (Paris and Lyon, 1919).

Massin, Jean. *Marat* (Le Club français du livre, 1970).

Masson, Frédéric. *L'Académie Française, 1629-1793* (1912).

Matthews, George T. *The Royal General Farms in Eighteenth-Century France* (New York: Columbia University Press, 1958).

Maurice, Frédéric. "Mémoire sur les travaux et les ecrits de M. Legendre," *Bibliothèque universelle, sciences et arts* 52 (1833), 45-82.

Maury, Alfred. *Les Académies d'autrefois: L'Ancienne Académie des Inscriptions et des Belles-lettres* (1864a).

———. *L'Ancienne Académie des Sciences* (1864b).

Mauskopf, Seymour H. "Crystals and Compounds: Molecular Structure and Composition in Nineteenth-Century French Science," *Transactions of the American Philosophical Society* (new ser., vol. 66, part 3, Philadelphia, 1976).

———. "Thomson before Dalton," *Annals of Science* 25 (1969a) 229-242.

———. "The Atomic Structural Theories of Ampère and Gaudin: Molecular Speculation and Avogadro's Hypothesis," *Isis* 60 (1969b), 61-73.

———. "Minerals, Molecules and Species," *Archives internationales d'histoire des sciences* 23 (1970a), 185-206.

———. "Haüy's Model of Chemical Equivalence," *Ambix* 17 (1970b), 182-191.

Mazon, A. *Histoire de Soulavie (naturaliste, diplomate, historien)*, 2 vols. (1893).

McClellan, James E. "The International Organization of Science and Learned Societies in the Eighteenth Century." Thesis, Princeton University, 1975.

———. "Rozier's Journal Revisited: A Case Study in the History of the Scientific Press," *Annals of Science* 36 (1979a).

———. "The Scientific Press in Transition: Rozier's Journal and the Scientific Societies in the 1770's," *Annals of Science* 36 (1979b), 425-449.

McCloy, Shelby T. *Government Assistance in 18th-Century France* (Durham, N.C.: Duke University Press, 1946).

———. *French Inventions of the 18th Century* (Lexington, Kentucky: University of Kentucky Press, 1952).

McDonald, Eric. "The Collaboration of Bucquet and Lavoisier," *Ambix* 13 (1966), 74-83.

McKie, Douglas. *Antoine Lavoisier, The Father of Modern Chemistry* (London, 1935).

———. "The Scientific Periodical from 1665 to 1798," *The Philosophical Magazine, 7th ser.*, 39 (1948), 122-132.

———. *Antoine Lavoisier, Scientist, Economist, Social Reformer* (London: H. Schuman, 1952).

McKie, Douglas. "The *Observations* of the abbé François Rozier," *Annals of Science 13* (1957), 73-89.

Meek, Ronald L., ed. and trans. *Turgot on Progress, Sociology and Economics* (Cambridge: Cambridge University Press, 1973).

Merton, Robert K. "Science, Technology and Society in Seventeenth-Century England," *Osiris 4* (1938), 360-632. Reprinted, New York, 1970.

———. "Priorities in Scientific Discovery: A Chapter in the Sociology of Science," *American Sociological Review 22* (December 1957), 635-659.

———. "Singletons and Multiples in Scientific Discovery: A Chapter in the Sociology of Science," *Proceedings of the American Philosophical Society 105* (1961), 470-486.

———. *On the Shoulders of Giants; A Shandean Postscript* (New York: Free Press, 1965).

Mesnard, Paul. *Histoire de l'Académie Française depuis sa fondation jusqu'en 1830* (1857).

Metzger, Hélène. *La Genèse de la science des cristaux* (1918).

Middleton, W. E. Knowles. *The History of the Barometer* (Baltimore: Johns Hopkins Press, 1964).

———. *A History of the Thermometer and Its Use in Meteorology* (Baltimore: Johns Hopkins Press, 1966).

———. *The Experimenters: A Study of the Accademia del cimento* (Baltimore: Johns Hopkins Press, 1971).

Monod-Cassidy, Hélène. "Un Astronome-philosophe, Jérôme de Lalande," *Studies on Voltaire and the Eighteenth Century* 56 (1967), 907-930.

Morellet, l'abbé de. *Mémoires de l'abbé Morellet*, 2 vols. (1821-1823).

———. *Lettres de l'abbé Morellet à Lord Shelburne* (1898).

Morgan, Betty T. *Histoire du Journal des sçavans depuis 1665 jusqu'en 1701* (1929).

Morineau, M. "Y-a-t-il eu une révolution agricole en France au XVIIIe siècle?" *Revue historique* 239 (1968), 229-326.

Mornet, Daniel. *Les Sciences de la nature en France au XVIIIe siècle* (1911).

———. *Les Origines intellectuelles de la Révolution française*. (5th edition, A. Colin, 1954).

Mosser, Françoise. *Les Intendants des finances au XVIIIe siècle* (Genève: Droè, 1978).

Multhauf, Robert P. "The French Crash Program for Saltpeter Production, 1776-1794," *Technology and Culture 12* (1971), 163-181.

Musson, A. E., ed. *Science, Technology, and Economic Growth in the Eighteeenth Century* (London: Methuen, 1972).

Musson, A. E., and Robinson, E. *Science and Technology in the Industrial Revolution* (Manchester: University Press, 1969).

Mutel, D.-P. *Vie d'Antoine-Augustin Parmentier, pour faire suite à la galérie des hommes illustres du département de la Somme* (1819).

Nadault de Buffon, Henri. *Correspondance inédite de Buffon*, 2 vols. (1860).

Neave, E.W.J. "Chemistry in Rozier's Journal," *Annals of Science* 6 (1950), 416-421; 7 (1951), 101-106, 144-148, 284-299, 393-400; 8 (1952), 28-45.

North, John D. "Venus, by Jupiter," review of *Velikovsky Reconsidered* by the editors of *Pensée* (London, 1976), in *Times Literary Supplement* no. 3, 876 (25 June 1976), 770-771.

Oestreich, Gerhard. "Justus Lipsius als Theoretiker des neuzeitlichen Machstaates," *Historische Zeitschrift 181* (1956), 31-78.

Olschki, Leonardo. *Geschichte der neusprachlichen wissenschaftlichen Literatur. Galilei und seine Zeit* (Halle [Saale], 1927).

Orcel, J. "Daubenton (1716-1799) organisateur de Cabinet d'Histoire Naturelle et créateur de l'enseignement de la minéralogie au Jardin du Roi, puis au Muséum," *Comptes-rendus du Congrès des Sociétés Savantes . . . tenu à Dijon en 1959* (1960), Section des sciences, 41-62.

Ornstein, Martha. *The Role of Scientific Societies in the Seventeenth Century* (New York, 1913; 2nd ed., Chicago, 1928).

Palmer, Robert R. *Twelve Who Ruled* (Princeton: Princeton University Press, 1941).

———. *The Age of the Democratic Revolution*, 2 vols. (Princeton: Princeton University Press, 1959-1964).

———, ed. and trans. *The School of the French Revolution: A Documentary History of the College of Louis-le-Grand and its Director, Jean-François Champagne, 1762-1814* (Princeton: Princeton University Press, 1975).

Pariset, Etienne. "Éloge de Berthollet," *Mémoires de l'Académie Royale de Médecine* (1826), 157-187.

Parker, Harold T. "French Administrators and French Scientists during the Old Regime and the Early Years of the Revolution," Richard Herr and Harold T. Parker, eds., *Ideas in History* (Durham, N.C., 1965), 85-109.

Partington, J. R. "Lavoisier's Memoir on the Composition of Nitric Acid," *Annals of Science 9* (1953), 96-98.

———. "Berthollet and the Antiphlogistic Theory," *Chymia 5* (1959), 130-137.

Passy, Louis. *Histoire de la Société Nationale d'Agriculture de France, Tome premier, 1761-1763* (1912). No further volumes appeared.

Pattie, Frank A. "Mesmer's Medical Dissertation and Its Debt to Mead's *De Imperio Solis ac Lunae*," *Journal of the History of Medicine and Allied Sciences 11* (1956), 275-287.

Payan, Jacques. *Capital et machine au XVIIIe siècle. Les Frères Périer et l'introduction en France de la machine à vapeur de Watt* (Paris and La Haye: Mouton, 1969).

Payan, Régis. *L'Evolution d'un monopole: L'Industrie des poudres avant la loi du 13 Fructidor An V* (1934).

Payenneville, J., ed. "Deux lettres inédites de Marat," *Revue historique de la Révolution française 14* (1919), 198-202.

Pearson, Karl. "Laplacé," *Biometrika 21* (1929), 202-216.

Perrin, C. E. "Lavoisier, Monge, and the Synthesis of Water, a Case of Pure Coincidence?" *British Journal for the History of Science* 6 (1973), 424-428.

Perroud, Claude, ed. *Lettres de Madame Roland*, 2 vols. Collection de documents inédits sur l'histoire de France, . . . (1900-1902).

Petot, Jean. *Histoire de l'administration des Ponts et Chaussées, 1599-1815* (Revière, 1958).

Philippe, A. *Histoire des apothicaires* (1853).

Phipson, S. L. *Jean Paul Marat, His Career in England and France before the Revolution* (London, 1924).

Pierquin, Louis. *Mémoires sur Pache* (Charleville; 1900).

Pigeire, Jean. *La Vie et l'oeuvre de Chaptal, 1756-1832* (1932).

Pigeonneau, Henri, and de Foville, Alfred. *Procès-verbaux de l'administration de l'agriculture au contrôle général des finances, 1785-1787* (1882).

Pillas, Albert, and Balland, Antoine. *Le Chimiste Dizé, sa vie, ses travaux, 1764-1852* (1906).

Piveteau, Jean, ed. *Oeuvres philosophiques de Buffon* (Presses universitaires de France, 1954).

Planchon, G. "Le Jardin des apothicaires de Paris," *Journal de pharmacie et de chimie*, ser. 5, *28* (1893); ser. 5, 29 (1894); ser. 5, 30 (1894); ser. 6, *1* (1895).

———. "L'Enseignement de l'histoire naturelle au Jardin des apothicaires," *Journal de pharmacie et chimie*, ser. 6, *3* (1896).

———. "L'Enseignement des sciences physico-chimiques au Jardin des apothicaires," *ibid.*, ser. 6, *5* (1897).

———. "L'Enseignement de la pharmacie au Jardin des apothicaires," *ibid.*, ser. 6, 7 (1898).

Plantefol, Lucien. "Du Hamel du Monceau," *Dix-huitième siècle* no. 1 (1969), 123-137.

Pottinger, David T. *The French Book Trade in the Ancien Régime, 1500-1791* (Cambridge, Mass.: Harvard University Press, 1958).

Poynter, F.N.L., ed. *The Evolution of Hospitals in Britain* (London: Pittman Medical Club, 1964).

Prevet, François. *Contribution à l'étude de l'évolution historique des techniques d'organisation sociale appliquées à la pharmacie* (Recueil Sirey, 1940).

Procès-verbaux du comité d'instruction publique, ed. J. Guillaume (1894).

Procès-verbaux des comités d'agriculture et de commerce publiés et annotés par Gerbaux et Schmidt (1906-1910).

Proteau, Pierre. *Etude sur Morellet* (Laval, 1910).

Pueyo, Guy. "Du Hamel du Monceau, précurseur des études climatiques et microclimatiques," *Actes du XIIe Congrès international de l'histoire des sciences* (Paris, 1968), 63-68.

Purver, Margery. *The Royal Society: Concept and Creation* (London: Routledge and Kegan Paul, 1967).

Railliet, A., and Moulé, L. *Histoire de l'Ecole d'Alfort* (1908).

Rappaport, Rhoda. "G. F. Rouelle: An Eighteenth-Century Chemist and Teacher," *Chymia* 6 (1960), 68-101.

———. "Government Patronage of Science in Eighteenth-Century France," *History of Science 8* (1969), 119-136.

———. "The Geological Atlas of Guettard, Lavoisier, and Monnet: Conflicting Views of the Nature of Geology," in *Toward a History of Geology,* ed. Cecil J. Schneer (Cambridge, Mass., 1969).

Raspail, Julien. "Les Papiers de Lalande," *La Révolution française* 74 (1921), 236-254.

Raugel, Félix. *Recherches sur quelques maîtres de l'ancienne facture d'orgues française* (1919).

Razy, C. *Etude analytique des petits modèles de métiers du musée historique des tissus de Lyon* (Lyon, 1913).

Reinhard, Marcel. *Le grand Carnot*, 2 vols. (Hachette, 1950-52).

———. "La Population de la France et sa mesure, de l'ancien régime au consulat," *Contributions à l'histoire démographique de la Révolution française*, 2nd ser.; in Commission d'histoire économique et sociale de la Révolution française, *Mémoires et documents 18* (1965), 259-274.

Rémond, André. *John Holker, manufacturier et grand fonctionnaire en France au XVIIIe siècle, 1719-1786* (Revière, 1946).

Richmond, Phyllis Allen. "The Hôtel-Dieu of Paris on the Eve of the Revolution," *Journal of the History of Medicine and Allied Sciences 16* (1961), 335-353.

Robida, Michel. *Ces Bourgeois de Paris; trois siècles de chronique familiale, de 1675 à nos jours* (Julliard, 1955).

Roche, Daniel. "Milieux académiques provinciaux et société des lumières; trois académies provinciales au 18e *siècle: Bordeaux, Dijon, Châlons-sur-Marne*," *Livre et société dans la France du XVIIIe siècle*, ed. François Furet (Paris and The Hague, 1965).

———. *Le Siècle des lumières en province: Académies et académiciens provinciaux, 1680-1789*, 2 vols. (Mouton, 1978).

Roger, Jacques, ed. *Buffon: Les Epoques de la nature, édition critique avec le manuscrit, introduction et des notes* (1962): *Mémoires du Muséum d'Histoire Naturelle* (New ser. C, *10*).

———. *Les Sciences de la vie dans la pensée française du XVIIIe siècle* (A. Colin, 1963).

Rosenband, Leonard N. "Work and Management in the Montgolfier Paper Mill," Thesis, Princeton University, 1980.

Rostaing, Léon. *La Famille de Montgolfier, ses alliances, ses descendants* (Lyons, 1910).

Rouff, Marcel. *Les Mines de charbon en France au XVIIIe siècle* (1922).

Roule, Louis. "La vie et l'oeuvre de Lacépède," *Mémoires de la Société Zoologique de France 27* (1917), 139-237.

———. *Daubenton et l'exploitation de la nature* (1925).

———. *Lacépède et la sociologie humanitaire selon la nature* (1932). 6 in Roule's series, L'Histoire de la nature vivante d'après l'oeuvre des grands naturalistes français.

Rouquette. "Jérôme Dizé," *Revue d'histoire de la pharmacie 17* (1965), 411-418.

Rousse, Emile. *La Roche-Guyon, châtelains, château et bourg* (1892).

Rozbroj, Hugo, *J. P. Marat, ein Naturforscher und Revolutionär, sein Zusammentreffen in der Geisteswelt mit Goethe, Lamarck, Rousseau, u.a.* (Berlin, 1937).

Rudé, George. "La Taxation populaire de mai 1775 à Paris et dans la région parisienne," *Annales historiques de la Révolution française 28* (1956), 139-179.

Sadoun-Goupil, Michelle. "Science pure et science appliquée dans l'oeuvre de Claude-Louis Berthollet," *Revue d'histoire des sciences 27* (1974), 127-145.

———. *Le Chimiste Claude-Louis Berthollet, 1784-1822, sa vie, son oeuvre* (Vrin, 1977).

Saint-Lambert, Jean-François. *Oeuvres philosophiques de Saint-Lambert* (1801).

Saricks, Ambrose. *Pierre Samuel Du Pont de Nemours* (Lawrence, Kansas: University of Kansas Press, 1965).

Sarton, George. "Montucla (1725-1799), His Life and Works," *Osiris 1* (1936), 519-567.

———. "Vindication of Father Hell," *Isis 35* (1944), 98-99.

Saunders, Richard M. "The Provincial Academies of France during the 17th and 18th Centuries," Thesis, Cornell University, 1931.

Schapiro, J. Salwyn. *Condorcet and the Rise of Liberalism* (New York, 1934).

Scheler, Lucien. *Lavoisier et la Révolution française*. Vol. 2: *Le Journal de Fougeroux de Bondaroy* (1960).

———. "Antoine Laurent Lavoisier et Michel Adanson, rédacteurs de programme des prix à l'Académie des Sciences," *Revue d'histoire des sciences 14* (1961a), 257-284.

Scheler, Lucien, and Smeaton, W. A. "An Account of Lavoisier's Reconciliation with the Church a Short Time before His Death," *Annals of Science 14* (1961b), 148-153.

Schelle, Gustave. *Vincent de Gournay* (1897).

———. *Le Docteur Quesnay* (1907).

Schelle, G., and Grimaux, E. *Lavoisier—statistique agricole et projets de réformes* (1894).

Schenk, Jerome M. "The History of Electrotherapy and Its Correlation with Mesmer's Animal Magnetism," *American Journal of Psychiatry 116* (1959), 463-464.

Schiller, Joseph. "Physiologie et classification dans l'oeuvre de Lamarck," *Histoire et biologie 2* (1969), 35-57.

Schofield, Robert. *A Scientific Autobiography of Joseph Priestley, 1733-1804, Selected Scientific Correspondence*, edited with commentary by Robert E. Schofield (Cambridge, Mass.: MIT Press, 1966).

Schumpeter, Joseph. *Capitalism, Socialism, and Democracy* (New York: Harper, 1942).

Sedillot, L. A. "Les Professeurs de mathématiques et de physique générale au Collège de France," *Bullettino di bibliografia e di storia delle scienze matematiche e fisiche*, ed. B. Boncompagni, *2* (Rome, 1869), 343-368, 387-448, 461-510; *3* (1870), 107-170.

Serbos, Gaston. "L'Ecole Royale des Ponts et Chaussées." In Taton (1964), 354-363.

Shapiro, Alan E. "The Evolving Structure of Newton's Theory of White Light and Color," *Isis 71* (1980), 211-235.

Shils, Edward. "The Profession of Science," *Advancement of Science 24* (1968), 469-480.

Shinn, Terry. *L'École polytechnique, 1794-1914*. (Presse de la Fondation nationale des sciences politiques, 1980).

Silvestre, A. F. "Notices biographiques sur MM Journu-Auber, Cotte, Allaire, Desmarest, et Tenon," *Mémoires d'agriculture, d'économie rurale et domestique, publiés par la Société Royale et Centrale d'Agriculture* (1816), 80-123.

———. "Notice biographique sur M. André Thouïn" (1825), originally published in *Mémoires de la Société Royale d'Agriculture.*

Silvestre de Sacy, Jacques. *Le Comte d'Angiviller* (Plon, 1953).

Simon, G.-A. "Les Origines de Laplace: sa généalogie,—ses études," *Biometrika*

21 (1929), 217-230. Published with Pearson (1929), which article was based largely on information supplied by the Abbé Simon.

———. "Laplace, ses origines familiales et ses premiers débuts," *Normannia* (Revue . . . d'histoire de Normandie) 9e année (January-March 1936), 477-496.

Smeaton, W. A. "The Early Years of the Lycée and the Lycée des Arts: I, The Lycée of the Rue de Valois," *Annals of Science 11* (1955a), 257-267.

———. "The Early Years of the Lycée and the Lycée des Arts: II, The Lycée des Arts," *Annals of Science 11* (1955b), 309-319.

———. "Jean-François Pilâtre de Rozier, The First Astronaut," *Annals of Science 11* (1955c), 349-355.

———. "Lavoisier's Membership of the Société Royale de Médecine," *Annals of Science 12* (1956a), 228-244.

———. "Lavoisier's Membership of the Société Royale d'Agriculture and the Comité d'Agriculture," *Annals of Science 12* (1956b), 267-277.

———. "The First and Last Balloon Ascents of Pilâtre de Rozier," *Archives internationales d'histoire des sciences 11* (1958), 263-269.

———. *Fourcroy, Chemist and Revolutionary, 1755-1809* (Cambridge: Heffer, 1962).

———. "New Light on Lavoisier: The Research of the Last Ten Years," *History of Science* 2 (1963), 51-69.

Smith, Edwin Burrows. "Jean-Sylvain Bailly, Astronomer, Mystic, Revolutionary, 1736-1793," *Transactions of the American Philosophical Society* (new ser., vol. 44, part 4, Philadelphia, 1954).

Smith, John Graham. *The Origins and Development of the Heavy Chemical Industry in France* Oxford: Oxford University Press, 1979).

Spillman, Ramsay. "Félix Vicq d'Azyr and Benjamin Franklin," *Journal of Nervous and Mental Disease* 94 (1941), 428-444.

Sprat, Thomas. *History of the Royal Society, for the Improving of Natural Knowledge* (1st ed., London, 1667). Edited with critical apparatus by Jackson I. Cope and Harold Whitmore Jones (St. Louis, 1958, 1966).

Steinheil, G., ed. *Commentaires de la Faculté de Médecine de Paris, 1777 à 1786*, 2 vols. (1903). The documents are supplemented by a volume of notes, containing a detailed history by Henri Varnier.

Stigler, Stephen M. "Napoleonic Statistics: The Work of Laplace," *Studies in the History of Probability and Statistics*, XXXIV, *Biometrika* 62 (1975), 503-517.

Storrs, F. C. "Lavoisier's Technical Reports, 1768-1794," *Annals of Science* 22 (1966), 251-275, and 24 (1968), 179-197.

Taton, René. *L'Oeuvre scientifique de Gaspard Monge* (Presses universitaires de France, 1951).

———. "Laplace et Sylvestre-François Lacroix," *Revue d'histoire des sciences* 6 (1953), 350-360.

———, ed. *Enseignement et diffusion des sciences en France au XVIIIe siècle* (1964), Ecole pratique des hautes études, Histoire de la pensée No. 11.

Taylor, Kenneth L. "Nicolas Desmarest and Geology in the Eighteenth Century," in Cecil J. Schneer, ed., *Toward a History of Geology* (Cambridge, Mass., 1969), 339-356.

Thackray, Arnold. "Natural Knowledge in Cultural Context: The Manchester Model," *American Historical Review* 79 (1974), 672-709.

Théodoridès, Jean. "Le Comte de Lacépède (1756-1825) naturaliste, musicien, et homme politique," *Comptes-rendus du 96e Congrès national des Sociétés Savantes*. (Toulouse, 1971), Section des Sciences (Paris, 1974), *1*, Histoire des Sciences, 47-62.

———. "The History of Rabies," *Korean Journal of Infectious Diseases* 6 (1974a), 27-35. An abstract is published in *Proceedings of the XXIII Congress of the History of Medicine* (London, 1972), 1252-1257.

———. "Quelques travaux concernant la rage publiès ou réalisés dans l'Empire Austro-Hongrois aux 18e et 19e siècles," *Wien und die Weltmedizin*, ed. Erna Lesky. Symposium der internationalen Akademie für Geschichte der Medizin der Universitat Wien, 17-19 September 1973 (Vienna, 1974b).

Thibaudeau, A.-C. *Mémoires sur la convention et le directoire*, 2 vols. (1824).

Thiébaut de Berneaud, A. *Eloge historique d'André Thoüin* (1825).

Tischner, Rudolf. *Franz Anton Mesmer, Leben, Werk und Wirkungen* (Munich, 1928).

Tisserand, Roger. *Au temps de l'Encyclopédie: l'Académie de Dijon* (1936).

Todériciu, Doru. "Chimie appliquée et technologie chimique au milieu du XVIIIe siècle: Oeuvre et vie de Jean Hellot, 1685-1766." Thèse de doctorat du 3e cycle, VI Section (1977), Ecole pratique des hautes études, Th 35 II.

Todhunter, Isaac. *A History of the Mathematical Theory of Probability from the Time of Pascal to that of Laplace* (London, 1865).

Toraude, L. G. *Etude fantaisiste sur les almanachs du Collège de Pharmacie Paris, 1780-1810* (1904).

———, ed. *J. F. Demachy, Histoire et contes, précédés d'une étude historique, anecdotique, et critique sur sa vie et ses oeuvres* (1907).

Torlais, Jean. "Chronologie de la vie et des oeuvres de René-Antoine Ferchault de Réaumur," *Revue d'histoire des sciences 11* (1958), 1-12.

Tournier, Clément. *Le Mesmérisme à Toulouse, suivi de lettres inédites sur le XVIIIe siècle, d'après les archives de l'Hôtel du Bourg* (Toulouse, 1911).

Les Travaux de l'Académie des Inscriptions et Belles-lettres, Histoire et inventaire des publications (Klincksieck, 1947).

Truesdell, C. A. "A Program toward Rediscovering the Rational Mechanics of the Age of Reason," *Archive for History of Exact Sciences 1* (1960), 3-36.

Turgot, A.-R.-J. *Oeuvres de M. Turgot, ministre d'état, précédées et accompagnées de Mémoires et de Notes sur sa vie, son administration et ses ouvrages*, 9 vols. (1808-1811), ed. Pierre-Samuel Dupont de Nemours, who reprinted in the first volume with minor revisions his *Mémoires sur la vie . . . de Turgot* (Philadelphia, 1782).

———. *Oeuvres de Turgot et documents le concernant*, ed. G. Schelle, 5 vols. (1913-1923). Cited as Turgot *Oeuvres*.

Vachon, Max; Rousseau, Georges; Laissus, Yves. "Liste complète des manuscrits de Lamarck conservés à la Bibliothèque centrale du Muséum National d'Histoire Naturelle," *Bulletin du Muséum National d'Histoire Naturelle*, 2nd ser., 40 (1969), 1093-1102.

Vallery-Radot, René. "La Vie ardente de P. I. Poissonier," *Bulletin de la Société Française d'Histoire de la Médecine 32* (1938), 44-54.

Valmy-Baysse, Jean. *Naissance et vie de la Comédie-française* (Floury, 1945).

Van den Berg, Hendrik K. "Etienne de Lacépède, een grondlegger van de moderne dierentuin," *Zoo* (May 1961).

Vellay, Charles, ed. *La Correspondance de Marat recueillie et annotée* (1908).

———, ed. "Supplément à la correspondance de Marat," *Revue historique de la Révolution française 1* (1910), 81-95, 219-235. These pieces were separately published under the same title (Le Puy, 1910).

———, ed. *Les Pamphlets de Marat, avec une introduction et des notes* (1911).

———. "Marat et l'abbé Bertholon," *Revue historique de la Révolution française 3* (1912a), 294-297.

———, ed. "Lettres inédites de Marat à Benjamin Franklin, 1779-1783," *Revue historique de la Révolution française 3* (1912b), 353-361. The originals are in the library of the American Philosophical Society and the Library of Congress.

Verdière, L. de. *Biographie de Vergniaud* (1866).

Vergnaud, Marguerite. "Science et progrès d'après Lavoisier," *Cahiers internationaux de sociologie 15* (1953), 174-186.

———. "Un Savant pendant la Révolution," *Cahiers internationaux de sociologie 17* (1954), 123-139.

Véri, Joseph Alphonse de. *Journal de l'abbé de Véri*, 2 vols. (1928-1930).

Vernière, Paul, ed. *Oeuvres philosophiques de Diderot* (Garnier, 1956).

Vicq d'Azyr, Félix. *Oeuvres*, ed. Jacques L. Moreau, 6 vols. (1805).

Vignery, John Robert. "Jacobin Educational Theories and Policies in the French National Convention." Thesis, University of Wisconsin, 1960. Dissertation Abstracts (December, 1960), 1547.

Vignon, Eugène. *Etudes historiques sur l'administration des voies publiques en France*, 3 vols. (1862).

Villenave, G. T. *Eloge historique de M. le comte de Lacépède* (1826).

Vinchon, Jean. *Mesmer et son secret*, ed. R. de Saussure, Collection "Rhadamanthe" (Privat: Toulouse, 1971). First published, 1936.

Vitet, L. *L'Académie Royale de Peinture et de Sculpture* (2nd ed., 1880).

Vovelle, Michel, ed. *Textes choisis de Marat* (Editions sociales, 1963).

Voyer, J. S. "Histoire de l'aérostation: (1) Les Lois de Meusnier," *Revue du génie militaire 23* (1902), 421-430; (2) "La Ballonet de Meusnier," *ibid.* 23, 521-532; (3) "Le Général Meusnier et les ballons dirigeables," *ibid.* 24 (1902), 135-156.

Walmsley, D. M. *Anton Mesmer* (London: Hale, 1967).

Walter, Gérard. *Marat* (1933; 2nd ed. A. Michel, 1960).

Watson, James D. *The Double Helix: A Personal Account of the Discovery of the Structure of DNA* (New York: New American Library, 1968).

Watts, George B. "The *Supplément* and the *Table analytique et raisonnée* of the *Encyclopédie*," *French Review 28* (1954), 3-19.

———. "The *Encyclopédie Méthodique*," *Publications of the Modern Language Association, 73*, IV, Pt. 2 (1953), 348-366.

Watts, George B. "Thomas Jefferson, the *Encyclopédie*, and the *Encyclopédie méthodique*," *French Review* 38 (1965), 318-325.

Wehnelt, B. "Matthieu Tillet," *Beiträge zur Geschichte der Phytopathologie* 3 (1937), 45-146.

Westfall, R. S. "The Development of Newton's Theory of Color," *Isis* 53 (1962), 339-358.

Weulersse, George. *Le Mouvement physiocratique en France de 1756 à 1770*, 2 vols. (1910).

———. *La Physiocratie sous les ministères de Turgot et de Necker* (Presses universitaires de France, 1950).

———. *La Physiocratie à la fin du règne de Louis XV, 1770-1774* (Presses universitaires de France, 1959).

Wilson, Arthur M. *Diderot*, 2 vols. (New York: Oxford University Press, 1957-1972).

Wohl, Robert. "Buffon and His Project for a New Science," *Isis* 51 (1960), 186-199.

Wolf, C. *Histoire de l'Observatoire de Paris, de sa fondation à 1793* (1902).

Woolf, Harry. *The Transits of Venus, A Study of 18th-Century Science* (Princeton: Princeton University Press, 1959).

INDEX

◇◇

Note: Footnotes containing extensive bibliographical information are indicated here by an italicized page and note number.

LIBRARY OF CONGRESS CATALOGING IN PUBLICATION DATA

Gillispie, Charles Coulston.
Science and polity in France at the end of the old regime.

Bibliography: p.
Includes index.
1. Science—France—History. 2. Science and state—France. I. Title.
Q127.F8G53 509.44 80-7521
ISBN 0-691-08233-2